Developments in Sedimentary Provenance Studies

Geological Society Special Publications
Series Editor J. BROOKS

GEOLOGICAL SOCIETY SPECIAL PUBLICATION NO 57

Developments in Sedimentary Provenance Studies

EDITED BY

A. C. MORTON
British Geological Survey
Keyworth
Nottingham, UK

S. P. TODD
BP Exploration Ltd
London, UK

P. D. W. HAUGHTON
Department of Geology and Applied Geology
University of Glasgow, Glasgow, UK

1991

Published by

The Geological Society

London

THE GEOLOGICAL SOCIETY

The Geological Society of London was founded in 1807 for the purposes of 'investigating the mineral structures of the earth'. It received its Royal Charter in 1825. The Society promotes all aspects of geological science by means of meetings, special lectures and courses, discussions, specialist groups, publications and library services.

It is expected that candidates for Fellowship will be graduates in geology or another earth science, or have equivalent qualifications or experience. All Fellows are entitled to receive for their subscription one of the Society's three journals: *The Quarterly Journal of Engineering Geology*, *Journal of the Geological Society* or *Marine and Petroleum Geology*. On payment of an additional sum on the annual subscription, members may obtain copies of another journal.

Membership of the specialist groups is open to all Fellows without additional charge. Enquiries concerning Fellowship of the Society and membership of the specialist groups should be directed to the Executive Secretary, The Geological Society, Burlington House, Piccadilly, London W1V 0JU.

Published by the Geological Society from:
The Geological Society Publishing House
Unit 7
Brassmill Enterprise Centre
Brassmill Lane
Bath
Avon BA1 3JN
UK
(*Orders*: Tel. 0225 445046)

First published 1991

Distributor USA:
AAPG Bookstore
PO Box 979
Tulsa
Oklahoma 74101-0979
USA
Tel. (918) 584-2555

Distributor Australia:
Australian Mineral Foundation
63 Conyngham Street
Glenside
South Australia 5065
Tel. (08) 379-0444

Printed in Great Britain at the Alden Press, Oxford

British Library Cataloguing in Publication Data
Developments in sedimentary provenance studies.
 1. Sedimentary rocks. Origins
 I. Morton, A. C. *1954*- II. Todd, S. P. *1964*- III. Haughton, P. D. W. *1959*- IV. Series 552.5

ISBN 0–903317–56–7

Contents

2 plates in pocket at back

Sedimentary provenance studies

P. D. W. HAUGHTON[1], S. P. TODD[2] & A. C. MORTON[3]

[1] *Department of Geology and Applied Geology, University of Glasgow, Glasgow G12 8QQ, UK*

[2] *BP Exploration, Britannic House, Moor Lane, London, EC2Y 9BU, UK*

[3] *British Geological Survey, Keyworth, Nottingham NG12 5GG, UK*

The study of sedimentary provenance interfaces several of the mainstream geological disciplines (mineralogy, geochemistry, geochronology, sedimentology, igneous and metamorphic petrology). Its remit includes the location and nature of sediment source areas, the pathways by which sediment is transferred from source to basin of deposition, and the factors that influence the composition of sedimentary rocks (e.g. relief, climate, tectonic setting). Materials subject to study are as diverse as recent muds in the Mississipi River basin (Potter *et al.* 1975), Archaean shales (McLennan *et al.* 1983), and soils on the Moon (Basu *et al.* 1988).

A range of increasingly sophisticated techniques is now available to workers concerned with sediment provenance. Provenance data can play a critical role in assessing palaeogeographic reconstructions, in constraining lateral displacements in orogens, in characterizing crust which is no longer exposed, in testing tectonic models for uplift at fault block or orogen scale, in mapping depositional systems, in sub-surface correlation and in predicting reservoir quality. On a global scale, the provenance of fine-grained sediments have been used to monitor crustal evolution.

We introduce below some of the novel techniques which are currently being used in provenance work, and some of the areas in which provenance studies are making, and promise to make, an important contribution to our understanding of earth processes. Many of the techniques and applications are covered by papers collected in this volume. These papers represent a selection of those contributed to a joint British Sedimentological Research Group/Petroleum Group meeting on 'Developments in Sedimentary Provenance Studies' convened by A. C. Morton and S. P. Todd at the Geological Society in June, 1989. In an area as diverse as sediment provenance, it is not surprising that the coverage of papers is incomplete, and we hence include some of the developments and applications which do not appear in this volume, but which also represent frontier areas in the study of sediment provenance. We start this short review by looking briefly at the framework within which provenance studies are undertaken.

A requisite framework for provenance studies

The validity and scope of any provenance study, and the strategy used, are determined by a number of attributes of the targeted sediment/ sedimentary rock (e.g. grain-size, degree of weathering, availability of dispersal data, extent of diagenetic overprint etc.). For most applications, the location of the source is critical, and ancillary data constraining this are necessary. These data should limit both the direction in which the source lay with respect to the basin of deposition, and some estimate of the distance of transport. In addition, it is important to have some constraint on the degree to which the composition of sediment has been biased away from the original source material(s) by weathering/ erosion, abrasion, hydraulic segregation, diagenesis and/or sediment recycling.

Palaeoflow direction may be obtained by judicious use of directional structures, where outcrop is available. In the subsurface, palaeoflow directions may be determined by reconstruction of regional facies patterns using drill core and seismic configuration, together with dipmeter records. A degree of circumspection is necessary when determining palaeoflow as there are instances where this may not be straightforward, even with outcrop data. Low stage fluvial cross-strata may diverge systematically from palaeo-slope by up to 90° (Bluck 1976), whilst turbidity currents are often deflected to flow axially with no hint of from where, or from which, flank of the basin margin the sediment was originally shed. Palaeoflow data collected in areas of structural complexity have the additional problem of uncertainties inherent in the structural correction procedure. This is exacerbated in zones of steep or overturned dip, and where rotations about a vertical axis have taken place. As the latter rotations are more commonplace than was previously thought (Kissel & Lau 1989), a

From Morton, A. C., Todd, S. P. & Haughton, P. D. W. (eds), 1991, *Developments in Sedimentary Provenance Studies.* Geological Society Special Publication No. 57, pp. 1–11.

marriage of palaeomagnetic and palaeoflow data may preface future provenance studies in areas of suspected rotation.

In considering the distance to source, the scale of the dispersal system from which the sediment was deposited must be addressed. Some types of depositional environment will be more amenable to this sort of analysis than others. For example, it is unlikely that angular fanglomerates containing labile clasts could have travelled any great distance (>10 km), but it is difficult to limit the transport scale of dropstones in a tillite. Theoretically, it should be possible to place some numerical limits on scale for alluvial systems, using a palaeohydrological approach. However, in practice, the plan-form and cross-sectional geometry of alluvial channels are often difficult to define precisely (Bridge 1985), and the validity of predictive palaeohydraulic equations for gravels remains uncertain (Reid & Frostick 1987). High sinuosity rivers are more conducive to palaeohydraulic estimation of discharge than are those of low sinuosity, either by an empirical approach (see Ethridge & Schumm 1978 for a critical review) or by employing mathematical models of channel bend flow (Bridge 1978; Bridge & Diemer 1983). Discharge may be related to drainage basin size and stream length (Leopold *et al.* 1964) thus providing some control on sediment transport distance.

Whilst some estimate of transport distance is included in many types of provenance study, others invert the problem and attempt to use the provenance data to limit the transport distance or scale of a dispersal system (e.g. Cliff *et al.* this volume). To use provenance data in this way, an upslope source must retain a similar distribution of lithologies to that present during sedimentation. However, as uplift and erosion expose progressively deeper levels of the source, a different suite of lithologies may be brought to the surface, obscuring source correlations. Linking detritus to source in these instances must consider the configuration of the 'lost' cover, for it may be that source mismatches simply reflect different structural levels of the same source block. This point is developed in the paper by Graham *et al.* (this volume).

Another potential pitfall is the fact that the immediate provenance of most detritus is either a pre-existing sediment or a soil profile, and not bedrock. Climatic factors can exert an important influence on mineralogical and geochemical transformations during soil formation (Singer 1980; Curtis 1990), and on the composition of sediment passed through river drainage basins (Franzinelli & Potter 1983). These modifications can obscure the ultimate source of the sediment, but the sediment composition may still be a useful palaeoclimatic indicator. Velbel & Saad (this volume) explore the climatic control on sediment composition by comparing detritus shed from the same source under different climatic conditions (an arid Triassic and a humid Holocene setting).

Sediment recycling can (1) bias the composition towards mature grains which are less amenable to source discrimination; (2) produce complex mixtures involving different sources; and (3) mask the relationship between scale of the dispersal system and the ultimate source of the detritus. Textural and/or petrographic evidence may identify a source in pre-existing sediment e.g detrital 'cement' grains (Zuffa 1987); broken and re-rounded clasts (Tanner 1976) and textural inversion (Haughton 1989). Zuffa (1987, this volume) provides a useful inventory of features which can be used to distinguish multi- from first-cycle sand grains. Geochronological data can be important in limiting the time available for such recycling.

Subsurface studies have reinforced the significance of diagenesis in modifying detrital grain assemblages (Morton 1984; Milliken 1988, Humphreys *et al.* this volume; Valloni *et al.* this volume). Many of these modifications are now predictable, and mineral assemblages and grain surface textures can be used to monitor diagenetic effects, and to ensure that these are minimized in provenance work.

Lastly, it is worth highlighting an obvious but commonly encountered problem in matching detritus to a prospective source block. Measured attributes for detrital grains (particularly those relating to the newer single grain studies) may provide no new insight unless a comparable dataset exists for the potential source block. Modern stream sediment samples can provide a rapid means of characterising the types of sediment grains expected from a particular basement block (Haughton & Farrow 1989).

Development of new techniques

Over the last decade, classical petrography (of clasts and sand grains) has given way to an increasingly sophisticated range of geochemical and isotopic techniques. Basic petrography has remained an important tool, particularly in mixed clastic/carbonate or calcarenitic provenance work, and when coupled with the newer techniques in clastic systems (Nelson & DePaolo 1988; McCann this volume; Floyd *et al.* this volume). A quantification of sandstone petrography took place in the 1970s and heralded the diversification of provenance techniques. A key factor in this quantification was a drive towards using sandstone composition in tectonic discrimination, thus linking sediment provenance to major plate setting (Dickinson & Suczek 1979; Ingersoll 1983). A similar motive drove subsequent attempts to use the bulk geochemistry of sedimentary rocks to look at provenance (e.g. Bhatia 1983). However, tectonic discrimination studies now look set to be eclipsed by techniques which exploit the characteristics of individual grains, rather than bulk populations (see below). This trend towards extracting information from single grains is perhaps the single most important development in recent years, and has been made possible by parallel developments in chemical and isotopic microanalysis.

Provenance of conglomerates and breccias

Conglomerates are particularly useful in looking at sediment provenance in that they provide intact samples of proximal source areas. Clasts thus provide evidence for mineral assemblages which are otherwise lost (or ambiguous) in disaggregated sands. Cuthbert (this volume) shows how pressure–temperature estimates using these clast mineral assemblages can be used to characterize the provenance of metamorphic detritus, and to reconstruct the P–T–t path for the source block. The geochronology of co-existing minerals can also provide a cooling history for the source uplift and this can be important in assessing whether or not basins were syn-orogenic.

Igneous clasts yield much useful information. Their geochemistry can be used to explore the relationship between various clast types (were they all derived from the same intrusion or volcanic centre?) and to identify the setting of the magmatism (rift, arc, etc.). Leitch & Cawood (1987) and Heinz & Loesche (1988) describe volcanogenic conglomerates which were derived from cryptic volcanic arcs. Igneous detritus can also be used to characterize the deeper structure of the source area, as revealed through inherited zircon grains and isotopic compositions. Geochronological data for clasts can provide a time frame for the magmatism and this is critical to any attempt to infer tectonic setting from petrology of the detritus. A flaw in many conventional petrographic studies is uncertainty as to whether or not the igneous detritus was derived from *coeval* volcanoes or intrusions. Floyd *et al.* (this volume) describe an example of foreland sediments with arc petrographic and geochemical signatures produced by recycling of much older arc terranes. The timing of magmatism in the

source can also be important in the search for displaced source areas (see Haughton *et al.* 1990). New high precision methods of U-Pb dating (e.g. Rogers *et al.* 1989) offer considerable scope for reliable age determinations on clasts. However, Rb-Sr mineral ages for single granite clasts have proved useful (and have been confirmed by U-Pb dating). The use of composite Rb-Sr whole-rock isochrons for groups of clasts is discouraged, as the clasts may not have been co-magmatic, and even if so, may still have had heterogeneous initial Sr isotopic compositions.

Much information may still be derived from the analysis of clast types in the field (see Cuthbert this volume; Graham *et al.* this volume; Garden this volume). Direct evidence for the age of the source can be supplied by fossiliferous clasts (Haughton 1988; Todd 1989). Any analysis of vertical and/or lateral compositional variation should include some assessment of how grain size controls composition. In reconstructing a source area from clasts, attention should be given to how the various clast types might relate to one another. For example, if granite clasts are found in association with sandstone clasts, did the former intrude the latter? Hornfels clasts might have some bearing on this, as might xenoliths preserved within granite boulders. Todd (1989) interpreted clasts of spotted slate and a two-mica, garnetiferous granite as having a source in the aureole and roof of a pluton.

Sandstone provenance

Many of the more recent developments concern sandstone provenance. The high level of interest in this area is mirrored by the predominance of papers dealing with sandstones in this volume. Since sandstones almost invariably comprise mixtures of source materials, sandstone provenance is often best tackled using a range of techniques rather than relying on any one method, an approach emphasized by Humphreys *et al.* (this volume).

Zuffa (this volume) describes recent progress in the petrographic analysis of sandstones (specifically turbiditic arenites) and emphasises that provenance work must go beyond the framework compositions of the QFL approach. Important temporal (coeval versus non-cocval, first cycle versus recycled) and spatial (intrabasinal versus extrabasinal) factors must also be addressed. The value of this approach is evident in studies such as that by Thornburg & Kulm (1987) who show how QFL data do not adequately discriminate modern sand samples from the Chile Trench, demonstrating significant hydraulic sorting of scoriaceous volcanic grains, and the role of onshore forearc basins in supressing the supply of volcanic detritus to the trench. Arribas & Arribas (this volume) show that the QFL approach correctly identifies the tectonic setting of the northern Tajo Basin (Spain), but only if calcareous rock fragments are included in the total lithics. Sandstone compositions in the Larsen Basin, Antarctica (Pirrie, this volume) evolve from undissected, transitional and dissected arc provenance to a recycled orogen setting, suggesting unroofing of an arc, but tectonic setting of the northern Tajo Basin (Spain), but only if calcareous rock fragments are included in the total lithics. Sandstone compositions in the Larsen Basin, Antarctica (Pirrie, this volume) evolve from undissected, transitional and dissected arc provenance to a recycled orogen setting, suggesting unroofing of an arc, but tectonic setting of the northern Tajo Basin (Spain), but only if calcareous rock fragments are included in the total lithics. Sandstone compositions in the Larsen Basin, Antarctica (Pirrie, this volume) evolve from undissected, transitional and dissected arc provenance to a recycled orogen set- Tortosa *et al.* (this volume) re-examine the use of quartz grain types in provenance analysis, and show that the distinction of plutonic and high grade metamorphic source rocks using the Basu *et al.* (1975) method requires caution.

Conventional heavy mineral analysis has been revitalized by studies of compositional variation within a single mineral species, thus circumventing the detrimental effect of intrastratal solution, often the dominant control on subsurface heavy mineral distribution. Amphibole, pyroxene, epidote, staurolite, monazite, zircon, garnet, spinel, chloritoid, mica and tourmaline are all amenable to this sort of analysis and the development of the approach is discussed by Morton (this volume). Basu & Molinaroli (this volume) explore the use of the opaque heavy mineral phases often disregarded in provenance work. They show that detrital Fe-Ti minerals can retain a provenance record, although no individual character is diagnostic and diagenetic alteration can occur in some instances.

In addition to looking at sediment distribution, specific mineral compositions can have important petrogenetic implications for source areas. Detrital pyroxene and amphibole compositions can be used as petrogenetic tracers in volcaniclastic sequences (Cawood 1983 and this volume; Morris 1988; Styles *et al.* 1989), whilst certain white mica compositions and the presence of glaucophane can identify erosion of high pressure metamorphic rocks (Sanders & Morris 1978). The isotopic composition of detrital mineral grains (using stable and/or radiogenic isotope ratios) can provide additional petrogenetic constraints on source rocks (Heller *et al.* 1985). Geochemical analysis of sandstones has largely concentrated on tectonic discrimination, following a suggestion by Crook (1974) that

active and trailing margin sandstones might be distinguished on the basis of their SiO_2 contents and K_2O/Na_2O ratios. The advantage here is that geochemistry might allow the tectonic setting of metasediments to be identified despite the loss of original petrographic detail (assuming isochemical metamorphism). More complex multivariate techniques were accordingly developed using both major (Bhatia 1983, Roser & Korsch 1988) and trace element concentrations (Bhatia 1985; Bhatia & Crook 1986). As with all discriminant techniques, the methods are only as good as the data base used to erect them, and problems have been encountered distinguishing sediments from different plate settings e.g. Van der Kamp & Leake (1985). Again, the problem of recycling rears its head. Reworking of older arcs can produce a spurious arc chemistry. McCann (this volume) illustrates the problem of recycling by demonstrating how the provenance of Ordovician–Silurian sediments of the Welsh basin fails to reflect accurately the palaeotectonic setting of the area. An alternative way of utilizing sediment chemistry is the use of specific provenance tracers e.g. the high Cr and MgO contents of sediments with a significant ultramafic source. This approach has been used to trace the original distribution of Caledonian ophiolites with some success (Hiscott 1984; Wrafter & Graham 1989). The geochemistry of modern sands and soils (e.g. Cullers *et al.* 1988) can be used to evaluate provenance signatures in different tectonic and climatic settings.

The geochronology of single sand-sized grains promises to revolutionize the study of sandstone provenance. Three main techniques are currently available: fission track dating of detrital grains; U-Pb dating of U-bearing minor phases e.g. zircon, monazite, titanite; and argon laser probe dating of detrital micas and amphiboles.

Hurford & Carter (this volume) summarize several applications of fission track dating to provenance work. The main limitation of this technique is that the subsequent thermal history of the sediment (following deposition) may reset fission track ages by partial or complete annealing of tracks. Apatite, with its low annealing temperature for tracks ($< 100°C$) is particularly susceptible to this resetting, but detrital zircon (with a closure temperature of 200–250°C) is more likely to preserve original crystallization ages, and consequently has the greater potential for provenance work.

U-Pb dating of single grains has been made possible by the development of low blank, micro-chemical separation procedures (Krogh 1973). When combined with abrasion techniques which minimise the discordance of analysed grains (Krogh 1982), precise ages may be determined without the interpretative problems posed by the common high discordance of multi-grain or unabraded single grain data. In addition, detrital monazite grains (see paper by Cliff *et al.* this volume) are particularly useful in that they are generally concordant. Cliff *et al.* show how single grain data for zircons and monazites can help to resolve some of the problems posed by multi-grain populations from the same suite of Carboniferous sandstones. Single grains can also be analysed by ion microprobe and it is possible to derive complex multistage histories from single grains and populations of grains in this way (Compston & Pidgeon 1986).

The $^{40}Ar/^{39}Ar$ laser probe also allows dating of a single, or portions of a single detrital grain. A laser is used to ablate a small pit (40–100 μm in diameter) in a grain which has been previously irradiated, and the extracted sample passed to a gas source mass-spectrometer. Kelley & Bluck (1989) present laser probe ages for detrital muscovites and volcanic rock fragments from Lower Palaeozoic greywackes in southern Scotland, identifying a source for the muscovites in an uplifting basement to an arc complex flanking the basin, but a remnant source for at least some of volcanic detritus previously thought to be contemporaneous with deposition.

Mudrock provenance

In spite of being the most abundant sedimentary rock, logistical problems mean that mudrocks have been relatively poorly studied in terms of provenance. In principle, it is possible to study the constituent clay minerals by X-ray diffraction analysis and make some inference of provenance. However, the approach is in practice fraught with difficulties, not least the rapid diagenetic alteration of primary clay mineral species on burial (Humphreys *et al.* this volume). Although the primary mineralogy of mudrocks is easily modified, a useful axiom is to assume that the primary rock chemistry remains unaltered, allowing major, trace and isotopic analyses of mudrocks to discriminate provenance (e.g. Humphreys *et al.* this volume; McCann this volume). Blatt (1985) also recommends that greater attention be paid to the non-clay mineralogy of mudrocks.

Developments in REE and trace element geochemistry, and isotopic techniques, have been particularly important in fine-grained rocks, and have also been widely used in sandstone provenance work. The uniformity of REE patterns in post-Archaean shales has been used to

estimate the composition of the upper crust, and to contrast this with an Archaean upper crustal composition revealed by distinct REE patterns for sediments in that era (Taylor & McClennan 1985). Various ratios of the REE, Th, Sc and Co in pelites exhibit secular changes across the Archaean-Proterozoic boundary and these have been related to a worldwide change in upper crustal composition at this time (Condie & Wronkiewicz 1990).

The uniform REE patterns, implying little fractionation of REE in sedimentary environments, underpins a second development, the use of Sm-Nd model ages to constrain the provenance of mudrocks and sandstones. Significant fractionations of Sm from Nd are thought to occur during addition of material to the crust from the mantle, but not during subsequent reprocessing of this material in the crust. Measured Sm/Nd ratios can thus be used to back correct the Nd isotopic composition (which is controlled by the time integrated decay of ^{147}Sm to ^{143}Nd) until it overlaps with the composition of one of a series of modelled mantle reservoirs. The age of the overlap is a crustal residence age and it is obviously dependent on the type of mantle model used.

From a provenance perspective, several aspects of the procedure are pertinent. First, the residence age returned by a sediment is a *weighted* average of the different source contributions. Petrographic or other constraints on the mixture of various components present may aid interpretation (Nelson & DePaolo 1988; Evans *et al.* this volume; Floyd *et al.* this volume). Secondly, fractionation of REEs, by hydraulic segregation and concentration of heavy minerals or pumaceous lithic fragments may occur in some sandstones (Frost & Winston 1987; McLennan *et al.* 1989) and may bias residence ages. No significant fractionation according to grain-size occurs in other sequences (e.g. Mearns *et al.* 1989). Thirdly, the behaviour of the REEs during diagenesis needs to be further investigated. Milodowski & Zalasiewicz (this volume) identify REE mobilization and fractionation during diagenesis of a mud-dominated turbidite–hemipelagic sequence. Awwiler & Mack (1989) have recently described an apparent diagenetic control on Sm-Nd model ages from the Wilcox Formation in Texas, where sandstones buried at depths of less than 3000 m have model ages of 1.4–1.5 Ga, and those below this level have model ages of 1.5–2.0 Ga. The depth at which this change takes place coincides with the onset of major diagenetic effects, especially dissolution. A similar pattern was noted in the associated shales, but with model ages changing over depths of 1500–3000 m, coinciding with the illitization of smectite over this interval; (4) crustal residence ages can be relatively insensitive to mixing of young mafic material with older recycled continental detritus (Haughton 1988). Nd data for sediments from the South Island, New Zealand (Frost & Coombs 1989) illustrate the potential for sediment contributions from contemporary mantle additions in active continental margin settings; and (5) the sites at which the REE reside in sedimentary rocks are poorly known. Model ages only constrain the origin of REEs themselves. If REEs are dominantly held in minor phases, the provenance of the bulk of the sediment remains unconstrained.

Problem-solving using provenance data

The development of these new techniques has widened the scope of provenance studies, allowing more precise source reconstruction/correlation, and opening up new areas in which provenance data can make a contribution. Some of the more important applications are considered briefly below.

Palaeogeography and palaeogeology

Provenance data can be usefully incorporated in palaeogeographic reconstructions (Allen this volume), particularly in areas of complex tectonics. Heller *et al.* (1985) ruled out the conventional source for turbidites of the Tyee Formation, Oregon from the upslope Klamath terrane on the basis isotopic data from detrital minerals. Instead, a provenance from the inboard Idaho batholith was preferred, with the basin originally located much farther east, closer to Idaho. Its present position was achieved during a subsequent history of tectonic rotation. In a second example, Bluck (1983) and Dempster & Bluck (1989) show an incongruity between the provenance of Ordovician sediments now in fault contact with the Dalradian in central Scotland, and the Ordovician history of the Dalradian metamorphic rocks. Despite evidence for rapid uplift of the Dalradian at this time, no detritus of Dalradian provenance reached the flanking Ordovician sediments. What metamorphic detritus there is was shown to have a provenance in a much older Proterozoic belt. Consequently, the Ordovician palaeogeography of central Scotland must include a separation of the Dalradian and the Ordovician sediments.

Strike-slip deformation: timing and scale of displacements

Provenance data can play a key role in unravelling the history of strike-slip deformation. Provenance mismatches across basin margins can identify structures on which lateral displacements have taken place (Crowell 1982). This is conditional on being able to demonstrate that the mismatch is not merely a product of erosion level (Graham *et al.* this volume). In favourable instances, the distribution of clast types in marginal fanglomerates may be used together with lithosome structure to infer the sense of strike-slip displacement (Ballance 1980; Todd 1989). Displaced source areas can also constrain the scale of displacement (Ross *et al.* 1973), but as large strike-slip faults often trend parallel to regional strike, the resolution may be poor. If the provenance can be tied to subtle lateral differences in the timing of uplift or the nature of the magmatism along the length of a narrow fault-parallel terrane, resolution may be improved. This also applies to using sediment provenance to demonstrate juxtaposition of terranes by establishing transfer of sediment from source to flanking basin across the trace of a major fault (Haughton *et al.* 1990).

Nature of uplift: regional and fault scale

Although unroofing sequences have been at the forefront of provenance studies since the last century, surprisingly few records of the uplift of regional metamorphic belts have been identified. Modern ideas on the evolution of orogenic belts indicate why this is so. Tectonic erosion by late orogenic extension may excise large crustal sections and can reduce the yield of sediment produced during uplift. Strike-slip displacements are also important in many orogenic belts and these can mean that an unroofing record will not be a simple vertical clast stratigraphy. Instead, different stages in the unroofing process may be preserved at separate strike locations. Orogenic sediment tends to be recycled rapidly through temporary basins which are continually being reworked as deformation propagates towards the foreland, so tending to mask simple unroofing sequences. Lastly, sediment is often transferred through basins close to mountain belts to more distant locations.

The paper by Cuthbert (this volume) illustrates the fragmentary nature of the clastic record of uplift which may be typical of synorogenic basins. The Middle Devonian Hornelen basin of Norway records no obvious unroofing history, despite a thick fill which was coeval with uplift. Instead, the basin reveals erosion of lithologies which were at a relatively high level in the orogen, with subsequent displacement on the detachment which generated the basin juxtaposing the sediments against deeper level, high pressure metamorphic rocks whose exhumation to the surface left no clastic record. Miller & John (1988) have also used provenance to trace the history of a basin underlain by a low-angle detachment. In this instance, clast assemblages in a Tertiary basin in SE California reveal the progressive unroofing of hanging wall-rocks to reveal the detachment zone itself, with the youngest sediments derived from both the hanging wall and the footwall.

A useful provenance record can also be obtained from basins in front of, or riding on, thrust sheets. Graham *et al.* (1986) describe an inverted clast sequence in the proximal Laramide foreland basin, and show how sedimentation can be controlled by variable resistance of lithologies progressively exposed in the encroaching thrust sheet. Times at which more resistant lithologies dominate will favour deposition of conglomerates in the basin, whereas erosion of mudstones can suppress conglomerate deposition. Evans & Mange-Rajetzky (this volume) integrate facies, palaeocurrent, heavy mineral and structural evidence to evaluate the provenance of sediments in the Barreme thrust top basin, Hauge-Provence (France), and a record of Alpine metamorphic and structural events is deduced on the basis of these data.

Igneous evolution deduced from provenance record

A detrital record may be all that remains of some crustal blocks, and of the higher crustal levels of others. Provenance data may thus be the only means of redressing this bias, and can be critical in looking at the evolution of ancient destructive plate margins where often only a partial record of the associated arc magmatism is preserved in situ. A more complete picture of the magmatism may be derived by combining data from igneous detritus with that preserved in what (if any) remains of the arc basement. This has been the case in Scotland, where the Lower Palaeozoic magmatic record south of the Highland Boundary Fault is largely a detrital one. Longman *et al.* (1979) established the presence of an Ordovician magmatic arc in central Scotland on the basis of large boulders preserved in a fore-arc basin. Similarly, Leitch & Willis (1982) interpreted Devonian conglomerates with a

variety of plutonic and volcanic clasts as having a source in the upper levels of a lost volcanic arc marginal to the New England Fold Belt, SE Australia. Cawood (this volume) presents data on mineral grains and volcanic glass from the Tonga arc, showing a uniform, low-K tholeiitic source supplied sediment from the Oligocene to Recent. Nichols *et al.* (this volume) describe unusual sandstone compositions from eastern Indonesia with minimal continental input. Both volcanic arc and ophiolitic source terrains dominate the provenance.

Tectonic setting

Misgivings about tectonic discrimination based solely on petrographic or geochemical data have already been voiced above and elsewhere (e.g. Girty *et al.* 1988; Mack 1984; Van de Kamp & Leake 1985; Zuffa 1987) and will not be reiterated. The advent of more precise dating techniques for mineral and lithic grains should allow more meaningful tectonic inferences to be made in that it is now possible to identify recycled signatures and to assign precise ages to grains with geotectonic significance.

Crustal evolution

Fine-grained sediments can sample large continental areas and their provenance has been used to track the evolution of the upper crust through time. Whilst trace-element variations tell us something of the changing composition of the upper crust (see above), isotopic data can constrain the pattern of crustal growth and the importance of sediment recycling. Sm-Nd isotopic data for shales have been used to examine the relationship between time, crustal growth and periods of orogeny (Andre *et al.* 1986; Michard *et al.* 1985; Miller *et al.* 1986). These studies suggest that while orogenies are not always sites of substantial new crustal additions, growth has been episodic with *c.* 90% of the crust existing by the end of the Proterozoic. Veizer & Jansen (1985) use the excess of Sm-Nd residence age over stratigraphic age to predict that recycling is *c.* 90% cannibalistic for the post-Archaean sedimentary mass. Sr isotopic compositions for sediments pose a problem in that $^{87}Sr/^{86}Sr$ ratios for younger sediments are unusually low, given their high Rb/Sr ratios, and the evidence for recycling of older crust. This can be explained by either buffering by the return of Sr to the mantle (Goldstein 1988) or by a secular increase in Rb/Sr of the upper crustal source of clastic sediments (McDermott & Hawkesworth 1990).

Another aspect of provenance and crustal evolution relates to the earliest preservation of crust. Detrital zircons incorporated in the Jacks Hill Metasedimentary Belt, Western Australia have ages close to 4.2 Ga, older than any so far measured from in-situ crust (Compston & Pidgeon 1986). These imply that parts of the crust existed since this time and were preserved from recirculation through the mantle. The provenance of Archaean sediments can thus provide an important window on the earliest evolution of the Earth's crust.

Sediment recycling

The extent of Phanerozoic recycling means that the distribution of characteristic detrital grains (for instance, grains of zircon of known age) must be sought first in the oldest sediments in which they might be expected to occur, and then in successively younger formations into which the grains may have been recycled. Only then can the provenance of grains in the younger sediments be interpreted. Although single grain studies are still in their infancy, this approach promises to tell us much about the complex prehistory of sediment grains. In the meantime, derived fossils place important constraints on the age of precursor sedimentary sequences. Batten (this volume) shows how reworked plant microfossils can be used to infer derivation from deposits of more than one age or source area, and to constrain the thermal history of the source(s).

Analysis of depositional systems

Sedimentary provenance can be used at a variety of scales to analyse ancient depositional systems. Provenance data can distinguish different alluvial systems in the same basin. Hirst & Nichols (1986) demonstrate a petrographic distinction between two fluvial distributory systems and marginal alluvial fans in the Ebro Basin, Spain. Separate basins may have shared the same antecedent rivers and again this possibility can be explored. Did, for example, large early Devonian rivers in northern Britain supply sediment to coastal alluvial plains in southern Britain (see Haughton & Farrow 1989)? Another problem which can be tackled is the nature of axial drainage in fluvial basins. Were axial rivers fed by transverse rivers or is the axial system antecedent and hence the basin architecture open to hydrological imbalance effects (cf. Blair & Bilodeau 1988)?

Reservoir models can also benefit from a detailed understanding of provenance. Apart from the obvious implications for primary porosity and subsequent diagenesis, sediment compositions can be used to assess sandbody connectivity. Hurst & Morton (1988) use detrital garnet compositions to recognize Ness Formation fluvial sandstones downcutting into the Etive Formation shoreline complex in the Oseberg Field of the northern North Sea, with implications for reservoir simulation. Provenance can also be useful in correlation and Mearns *et al.* (1989) show how Sm-Nd provenance ages display similar vertical patterns in different wells and may be used to correlate barren strata.

Climatic implications

Climate can play an important role in determining the composition of sedimentary rocks and it may be possible to make palaeoclimatic inferences on the basis of provenance data (Velbel & Sand this volume). Climate is particularly important in considering the origin of first-cycle quartz-arenites (Johnsson *et al.* 1988). These are produced where there is intense chemical weathering (generally under tropical weathering conditions) and in environments where such weathering can operate on sediments over an extended period of time. Evidence from coeval palaeosols can be important in assessing the connection between contemporary weathering and resulting sediment composition (Russell & Allison 1985). In fine-grained marine sequences, mineralogical (kaolinite/smectite ratios) and chemical (Th/K ratios) parameters have been used to monitor climate change (e.g. Wignall & Ruffell 1990).

Concluding statement

The London Sedimentary Provenance meeting was timely in that it drew together workers developing and applying techniques which promise to shift the emphasis of provenance work away from tectonic discrimination. Improved petrographic, geochemical and isotopic methods mean that it is now possible to extract a lot more information from sedimentary rocks, and in particular from single grains and clasts. Several contributors stressed the benefit of a multidisciplinary approach, and the value of applying several techniques as part of the same study. We should now be able to achieve a better understanding of how the grain components which comprise a sedimentary rock were assembled, to reconstruct source areas with greater confidence, and to use provenance data more effectively to test tectonic models.

References

ANDRE, L., DEUTSCH, S. & HERTOGEN, J. 1986. Trace-element and Nd isotopes in shales as indexes of provenance and crustal growth: the early Paleozoic from the Brabant Massif (Belgium). *Chemical Geology*, **57**, 101–115.

AWWILER, D. N. & MACK, L. E. 1989. Diagenetic resetting of Sm-Nd isotope systematics in Wilcox Group sandstones and shales, San Marcos Arch, south-central Texas. *Transactions of the Gulf Coast Association of Geological Societies*, **39**, 321–330.

BALLANCE, P. F. 1980. Models of sediment distribution in non-marine and shallow marine environments in oblique slip fault zones. *In*: BALLANCE, P. F. & READING, H. G. (eds) *Sedimentation in oblique-slip mobile zones.* Special Publication of the International Association of Sedimentologists, **4**, 229–236.

BASU, A., MCKAY, D. S. & GERKE, T. 1988. Petrology and provenance of Apollo 15 drive tube 15007/8. *Proceedings of the Eighteenth Lunar and Planetary Science Conference*, Cambridge University Press, 283–298.

——, YOUNG, S. W., SUTTNER, L. J., JAMES, W. C. & MACK, G. H. 1975. Re-evaluation of the use of undulatory extinctions and polycrystallinity in detrital quartz for provenance interpretation. *Journal of Sedimentary Petrology*, **45**, 873–882.

BHATIA, M. R. 1983. Plate tectonics and geochemical composition of sandstones. *Journal of Geology*, **91**, 611–627.

—— 1985. Rare earth element geochemistry of Australian Paleozoic greywackes and mudrocks: provenance and tectonic control. *Sedimentary Geology*, **45**, 97–113.

—— & CROOK, K. A. W. 1986. Trace element characteristics of greywackes and tectonic setting discrimination of sedimentary basins. *Contributions to Mineralogy and Petrology*, **92**, 181–193.

BLAIR, T. C. & BILODEAU, W. L. 1988. Development of tectonic cyclothems in rift, pull-apart, and foreland basins: Sedimentary response to episodic tectonism. *Geology*, **16**, 517–520.

BLATT, H. 1985. Provenance studies and mudrocks. *Journal of Sedimentary Petrology*, **55**, 69–75.

BLUCK, B. J. 1976. Sedimentation in some Scottish rivers of low sinuosity. *Transactions of the Royal Society of Edinburgh: Earth Sciences*, **70**, 29–46.

—— 1983. Role of the Midland Valley of Scotland in the Caledonian Orogeny. *Transactions of the Royal Society of Edinburgh: Earth Sciences*, **74**, 119–136.

BRIDGE, J. S. 1978. Paleohydraulic interpretation using mathematical methods of contemporary flow and sedimentation in meandering channels. *In*: MIALL, A. D. (ed.) *Fluvial Sedimentology.* Canadian Society of Petroleum Geologists, Memoir, **5**, 401–416.

—— 1985. Paleochannel patterns inferred from alluvial deposits: a critical evaluation. *Journal of Sedimentary Petrology*, **55**, 579–589.

—— & DIEMER, J. A. 1983. Quantitative interpretation of an evolving river system. *Sedimentology*, **30**, 599–623.

CAWOOD, P. A. 1983. Modal compositions and detrital pyroxene geochemistry of lithic sandstones from the New England Fold Belt (east Australia): a Paleozoic forearc terrane. *Geological Society of America Bulletin*, **94**, 1199–1214.

COMPSTON, W. & PIDGEON, R. T. 1986. Jack Hills, evidence of more very old detrital zircons in Western Australia. *Nature*, **321**, 766–769.

CONDIE, K. C. & WRONKIEWICZ, D. J. 1990. The Cr/Th ratio in Precambrian pelites from the Kaapvaal Craton as an index of craton evolution. *Earth and Planetary Science Letters*, **97**, 256–267.

CROOK, K. A. W. 1974. Lithogenesis and geotectonics: the significanse of compositional variations in flysch arenites (graywackes). *In*: DOTT, R. H. & SHAVER, R. H. (eds) *Modern and ancient geosynclinal sedimentation.* Society of Economic Paleontologists amf Mineralogist Special Publication, **19**, 304–310.

CROMWELL, J. C. 1982. Tectonics of Ridge basin, Southern California, *In*: CROWELL, J. C. & LINK, M. H. (eds) *Geologic history of Ridge basin, Southern California.* Society of Economic Paleontologists and Mineralogists, Pacific Section, Guidebook, **25–41**.

CULLERS, R. L., BASU, A. & SUTTNER, L. J. 1988. Geochemical signature of provenance in sand-size material in soils and stream sediments near the Tobacco Root Batholith, Montana, U. S. A. *Chemical Geology*, **70**, 335–348.

CURTIS, C. D. 1990. Aspects of climatic influence on the clay mineralogy and geochemistry of soils, palaeosols and clastic sedimentary rocks. *Journal of the Geological Society*, London, **147**, 351–357.

DEMPSTER, T. J. & BLUCK, B. J. 1989. The age and origin of boulders in the Highland Border Complex: constraints on terrane movements. *Journal of the Geological Society*, London, **146**, 377–379.

DICKINSON, W. R. & SUCZEK, C. A. 1979. Plate tectonics and sandstone compositions. *American Association of Petroleum Geologists Bulletin*, **63**, 2164–2182.

ETHRIDGE, F. G. & SCHUMM, S. A. 1978. Reconstructing paleochannel morphologic and flow characteristics. *In*: MIALL, A. D. (ed.) *Fluvial Sedimentology.* Canadian Society of Petroleum Geologists, Memoir, **5**, 703–721.

FRANZINELLI, E. & POTTER, P. E. 1983. Petrology, chemistry, and texture of modern river sands, Amazon river system. *Journal of Geology*, **91**, 23–39.

FROST, C. D. & COOMBS, D. S. 1989. Nd isotope

character of New Zealand sediments: implications for terrane concepts and crustal evolution. *American Journal of Science*, **289**, 744–770.

FROST, C. D. & WINSTON, D. 1987. Nd isotope systematics of coarse- and fine-grained sediments: examples from the Middle Proterozoic Belt-Purcell Supergroup. *Journal of Geology*, 95, 309–329.

GIRTY, G. H., MOSSMAN, B. J. & PINCUS, S. D. 1988. Petrology of Holocene sand, Penninsula Ranges, California and Baja Norte, Mexico: implications for provenance discrimination. *Journal of Sedimentary Petrology*, **58**, 881–887.

GOLDSTEIN, S. L. 1988. Decoupled evolution of Nd and Sr isotopes in the continental crust and mantle. *Nature*, **336**, 733–738.

GRAHAM, S. A., TOLSON, R. B., DECELLES, P. G., INGERSOLL, R. V., BARGAR, E., CALDWELL, M., CAVAZZA, W., EDWARDS, D. P., FOLLO, M. F., HANDSCHY, J. F., LEMKE, L., MOXTON, I., RICE, R., SMITH, G. A. & WHITE, J. 1986. Provenance modelling as a technique for analysing source terrane evolution and controls on foreland sedimentation. *In*: ALLEN, P. A. & HOMEWOOD, P. (eds) *Foreland Basins*. Special Publications of the International Association of Sedimentologists, **8**, 425–436.

HAUGHTON, P. D. W. 1988. A cryptic Caledonian flysch terrane. *Journal of the Geological Society, London*, **145**, 685–703.

—— 1989. Structure of some Lower Old Red Sandstone conglomerates, Kincardineshire, Scotland: deposition from late orogenic antecedent streams. *Journal of the Geological Society, London*, **146**, 509–525.

—— & FARROW, C. M. 1989. Composition of Lower Old Red Sandstone detrital garnets from the Midland Valley of Scotland and the Anglo-Welsh Basin. *Geological Magazine*, **126**, 373–396.

——, ROGERS G. & HALLIDAY, A. B. 1990. Provenance of Lower Old Red Sandstone conglomerates, SE Kincardineshire: evidence for the timing of Caledonian terrane accretion in central Scotland. *Journal of the Geological Society, London*, **147**, 105–120.

HEINZ, W. & LOESCHKE, J. 1988. Volcanic clasts in Silurian conglomerates of the Midland Valley (Hagshaw Hills Inlier) Scotland, and their meaning for Caledonian plate tectonics. *Geologische Rundschau*, **75**, 333–340.

HELLER, P. L., PETERMAN, Z. E., O'NEIL, J. R. & SHAFIQULLAH, M. 1985. Isotopic provenance of sandstones from the Eocene Tyee Formation, Oregon Coast Range. *Geological Society of America Bulletin*, **96**, 770–780.

HIRST, J. P. P. & NICHOLS, G. J. 1986. Thrust tectonic controls on Miocene alluvial distribution patterns, southern Pyrenees. *In*: ALLEN, P. A. & HOMEWOOD, P. (eds) *Foreland Basins*. Special Publications of the International Association of Sedimentologists, **8**, 247–258.

HISCOTT, R. N. 1984. Ophiolitic source rocks for Taconic-age flysch: trace element evidence. *Geological Society of America Bulletin*, **95**, 1261–1267.

HURST, A. R. & MORTON, A. C. 1988. An application of heavy mineral analysis to lithostratigraphy and reservoir modelling in the Oseberg Field, Northern North Sea. *Marine and Petroleum Geology*, **5**, 157–169.

INGERSOLL, R. V. 1983. Petrofacies and provenance of late Mesozoic forearc basin, northern and central California. *American Association of Petroleum Geologists Bulletin*, **67**, 1125–1142.

JOHNSSON, M. J., STALLARD, R. F. MEADE, R. H. 1988. First cycle quartz arenites in the Orinoco River Basin, Venezuela and Colombia. *Journal of Geology*, **96**, 263–277.

KELLEY, S. & BLUCK, B. J. 1989. Detrital mineral ages from the Southern Uplands using ^{40}Ar-^{39}Ar laser probe. *Journal of the Geological Society*, London, **146**, 401–403.

KISSEL, C. & LAJ, C. (eds) 1989. *Paleomagnetic rotations and continental tectonics*. Kluwer, Dordrecht.

KROGH, T. E. 1973. A low-contamination method for hydrothermal decomposition of zircon and extraction of U and Pb isotopic age determinations. *Geochimica et Cosmochimica Acta*, **46**, 631–635.

—— 1982. Improved accuracy of U-Pb zircon dating by the creation of more concordant systems using an air abrasion technique. *Geochimica et Cosmochimica Acta*, **46**, 637–649.

LEITCH, E. C. & CAWOOD, P. A. 1987. Provenance determination of volcaniclastic rocks: the nature and tectonic significance of a Cambrian conglomerate from the New England Fold Belt, eastern Australia. *Journal of Sedimentary Petrology*, **57**, 630–638.

—— & WILLIS, S. G. A. 1982. Nature and significance of plutonic clasts in Devonian conglomerates of the New England Fold Belt. *Journal of the Geological Society of Australia*, **29**, 83–89.

LEOPOLD, L. B., WOLMAN, M. G. & MILLER, J. P. 1964. *Fluvial processes in geomorphology*. Freeman & Co., San Francisco.

LONGMAN, C. D., BLUCK, B. J. & VAN BREEMEN, O. 1979. Ordovician conglomerates and the evolution of the Midland Valley. *Nature*, **280**, 578–581.

MCDERMOTT, F. & HAWKESWORTH, C. 1990. The evolution of Sr isotopes in the upper continental crust. *Nature*, **344**, 850–853.

MCLENNAN, S. M., MCCULLOCH, M. T., TAYLOR, S. R. & MAYNARD, J. B. 1989. Effects of sedimentary sorting on neodymium isotopes in deep-sea turbidites. *Nature*, **337**, 547–549.

——, TAYLOR, S. R. & ERIKSSON, K. A. 1983. Geochemistry of Archean shales from the Pilbara Supergroup, Western Australia. *Geochimica et Cosmochimica Acta*, **47**, 1211–1222.

MACK, G. H. 1984. Exceptions to the relationship between plate tectonics and sandstone composition. *Journal of Sedimentary Petrology*, **54**, 212–220.

MEARNS, E. W., KNARUD, R., RA STAD, N., STANLEY, K. O. & STOCKBRIDGE, C. P. 1989. Samarium-neodymium isotope stratigraphy of the Lunde and Statfjord Formations of Snorre Oil Field, northern North Sea. *Journal of the Geological*

Society, London, **146**, 217–228.

MICHARD, A., GURRIET, P., SOUDANT, M. & ALBAREDE, F. 1985. Nd isotopes in French Phanerozoic shales: external vs. internal aspects of crustal evolution. *Geochimica et Cosmochimica Acta*, **49**, 601–610.

MILLER, M. G. & JOHN, B. E. 1988. Detached strata in a Terhàry low-angle normal fault terrane, south eastern California a sedimentary record of unrooting, breaching, and continued slip. *Geology*, **16**, 645–648.

MILLER, R. G., O'NIONS, R. K., HAMILTON, P. J. & WELIN, E. 1986. Crustal residence ages of clastic sediments, orogeny and continental evolution. *Chemical Geology*, **57**, 87–99.

MILLIKEN, K. L. 1988. Loss of provenance information through subsurface diagenesis in Plio-Pleistocene sandstones, Northern Gulf of Mexico. *Journal of Sedimentary Petrology*, **58**, 992–1002.

MORRIS, P. A. 1988. Volcanic arc reconstruction using discriminant function analysis of detrital clinopyroxene and amphibole from the New England Fold Belt, Eastern Australia. *Journal of Geology*, **96**, 299–311.

MORTON, A. C. 1984. Stability of detrital heavy minerals in Tertiary sandstones from the North Sea basin. *Clay Minerals*, **19**, 287–308.

NELSON, B. K. & DEPAOLO, D. J. 1988. Comparison of isotopic and petrographic provenance indicators in sediments from Tertiary continental basins of New Mexico. *Journal of Sedimentary Petrology*, **58**, 348–357.

POTTER, P. E., HELING, D., SHRIMP, N. F. & VAN WIE, W. 1975. Clay mineralogy of modern alluvial muds of the Mississippi River Basin. *Bulletin des centres de Recherches exploration-production Elf Aquitaine*, **9**, 353–389.

ROGERS, G., DEMPSTER, T. J., BLUCK, B. J. & TANNER, P. W. G. 1989. A high precision U-Pb age for the Ben Vurich granite: implications for the evolution of the Scottish Dalradian Supergroup. *Journal of the Geological Society*, London, **146**, 789–798.

ROSER, B. P. & KORSCH, R. J. 1988. Provenance signatures of sandstone-mudstone suites determined using discriminant function analysis of major element data. *Chemical Geology*, **67**, 119–139.

ROSS, D. C., WENTWORTH, C. M. & MCKEE, E. H. 1973. Cretaceous mafic conglomerate near Gualala offset 350 miles by San Andreas fault from oceanic crustal source near Eagle Rest Peak, California. *U.S. Geological Survey Journal of Research*, **1**, 45–52.

RUSSELL, M. J. & ALLISON, I. 1985. Agalmatolite and the maturity of sandstones of the Appin and Argyll groups and Eriboll Sandstone. *Scottish Journal of Geology*, **21**, 113–122.

SANDERS, I. S. & MORRIS, J. H. 1978. Evidence for Caledonian subduction from greywacke detritus in the Longford-Down inlier. *Journal of Earth Sciences*, Royal Dublin Society, **1**, 53–62.

SINGER, A. 1980. The palaeoclimatic interpretation of clay minerals in soils and weathering profiles. *Earth Science Reviews*, **15**, 303–326.

STYLES, M. T., STONE, P. & FLOYD, J. D. 1989. Arc detritus in the Southern Uplands: mineralogical characterization of a 'missing' terrane. *Journal of the Geological Society, London*, **146**, 397–400.

TANNER, W. F. 1976. Tectonically significant pebble types: sheared, pocked and second-cycle examples. *Sedimentary Geology*, **15**, 69–83.

TAYLOR, S. R. and MCCLENNAN, S. M. 1985. *The Continental Crust: its composition and evolution.* Blackwell.

THORNBURG, T. M. & KULM, L. D. 1987. Sedimentation in the Chile Trench: petrofacies and provenance. *Journal of Sedimentary Petrology*, **57**, 55–74.

TODD, S. P. 1989. Role of the Dingle Bay Lineament in the evolution of the Old Red Sandstone of southwest Ireland. *In*: ARTHURTON, R. S., GUTTERIDGE, P. & NOLAN, S. C. (eds) *The role of tectonics in Devonian and Carboniferous sedimentation in the British Isles.* Yorkshire Geological Society Occasional Publication, **6**, 35–54.

VAN DE KAMP, P. C. & LEAKE, B. E. 1985. Petrography and geochemistry of feldspathic and mafic sediments of the northeastern Pacific margin. *Transactions of the Royal Society of Edinburgh: Earth Sciences*, **76**, 411–449.

VEIZER, J. & JANSEN, S. L. 1985. Basement and sediment recycling—2: Time dimension to global tectonics. *Journal of Geology*, **93**, 625–643.

WIGNALL, P. B. & RUFFELL, A. H. 1990. The influence of a sudden climatic change on marine deposition in the Kimmeridgian of northwest Europe. *Journal of the Geological Society, London*, **147**, 365–371.

WRAFTER, J. P. & GRAHAM, J. R. 1989. Ophiolitic detritus in the Ordovician sediments of South Mayo, Ireland. *Journal of the Geological Society, London*, **146**, 213–215.

ZUFFA, G. G. 1987. Unravelling hinterland and offshore palaeogeography from deep-water arenites. *In*: LEGGETT, J. K. & ZUFFA, G. G. (eds) *Marine clastic sedimentology.* Graham & Trotman, 39–61.

Provenance research: Torridonian and Wealden

P. ALLEN

Postgraduate Research Institute for Sedimentology, University, Reading, RG6 2AB, UK

Abstract: Torridonian regional flow-directions are indicated unequivocally by sediment geometries and palaeocurrent structures. In the Wealden more reliance has to be placed on textural and compositional gradients. Flow directions in both guide the search for source-massif 'stumps': uncovered, covered or inferred. 'Putting back' the detritus on to the stumps leads to broad palaeogeographies; but erosional histories, recycling, basin budgets and changes in basinal configuration are major uncertainties. Diagenetic interference with the evidence is minimal. Overlapping opposing flows are consistent with extensional tectonic regimes.

Torridonian waste provides evidence on the timing of mantle contributions to the Laurentian crust. Refreshingly new Hercynian petrographies in the Wealden may bear on a similar problem in European Gondwana.

Single-grain dating (perhaps to include quartz) and rapid and ultra-precise geochemical and isotopic analysis (including whole-rock) make the prospects for provenance research exciting. Tourmaline, zircon and glauconite have considerable potential for dating and justify fuller investigation of their isotopic and mineral chemistries.

Major objectives of provenance research are the location of detrital sources (immediate and ultimate), determination of their palaeotopographies, palaeoweathering and palaeogeologies and the bearing of these on lithospheric evolution. Problems of source identification are simplified in nonmarine rocks such as the Torridonian and Wealden by the coarser clastics still bordering their source-stumps (plate-tectonics permitting) and the ease with which diagenetic damage can be recognized and allowed for. Recycling and loss of fine clastics, however, pose severe difficulties, especially in very coarse relics like the Torridonian.

Ideally, source directions in sedimentary basins are indicated by regional flow and facies patterns. These point towards the 'stumps' (real or inferred), the contemporary geology and relief of which are to be reconstructed. This is accomplished by estimating the amount of post-stump erosion, matching clastic and geochemical 'markers' against putative source rocks, and 'putting the detritus back'.

Work on the late Proterozoic Torridonian of NW Scotland and early Cretaceous Wealden of southern England serves to illustrate some of the provenance problems clastic sedimentologists are heirs to.

Torridonian Supergroup

Flow directions

Fluvial flow-paths, local and regional, are frequently unequivocal, being well shown by 3-D exposures of body-geometries and large palaeocurrent structures. Lateral changes of lithofacies (e.g. red breccia → red sandstone → red or grey mudstone and shale) are useful in both the Torridon and Stoer Groups. In the former they are seen as small fans coming off the source-stumps in all directions, like the screes forming around them today. Some, at least, of the grey illitic shales and mudstones into which the Torridonian fans pass are, boron-wise, lake deposits (Stewart & Parker 1979), but the possibility of brackish or marine incursions (Allen *et al.* 1960) has not been eliminated.

Regional flow was broadly towards the ESE and WNW (Stewart 1969), at right angles to the Caledonian grain and indicating opposing source-massifs. Stewart (1988) has suggested a rift-basin scenario, bounded on the west and east by normal faults which subsequently became the Outer Hebridean and Moine thrusts. In this model reversals of flow, well documented in the Stoer, are attributed to tilting of the rift valley floor by faulting. Any S–N or N–S flows would be along-rift and therefore unlikely to indicate the directions of ultimate sources.

In the south, the sub-Torridon Group stumps form a buried mountain landscape. Some are visibly over 0.5 km high, and possibly more than 1 km originally (see e.g. Peach *et al.* 1907, p. 311 and fig. 11; but cf. Stewart 1972, p. 113 and fig. 4). Northwards, the land surface is more subdued, even pedimented (Williams 1969a). Relief on the sub-Stoer surface is poorly known owing to the restricted outcrop, and a similar change cannot be demonstrated. Perhaps related to

From Morton, A. C., Todd, S. P. & Haughton, P. D. W. (eds), 1991,
Developments in Sedimentary Provenance Studies.
Geological Society Special Publication No. 57, pp. 13–21.

pedimentation, both sequences broadly fine upwards as though the basin widened while the fault scarps retreated and the source mountains were lowered (Williams 1969a).

Torridon Group

Burial of the local source-topography beneath Applecross Formation sands of two 100 km sized fan-piles (Williams 1969a) raises fascinating problems of provenance. The early geologists recognized that a substantial portion (up to more than 50% locally) of the post-Diabaig Formation detritus came from unmetamorphosed rocks and could not be accounted for from local Lewisian sources (Williams 1969b). Williams's fans prograded ESE across the Minch, so their apexes must have backed across the Lewisian of the Outer Hebrides into a terrain of supracrustal rocks. The latter were either folded or thrust deeply into the basement (Stewart 1988) or formed an incomplete cover on the Lewisian, because outcrops of both are needed to explain the binary origin of the Torridon detritus. Therefore more Laurentian source-mountains lay beyond the Outer Hebrides. Present knowledge of plate-tectonics suggests that the stumps persist in southern Greenland and Labrador.

Some of the Torridon clasts show similar characteristics to supracrustal rocks in situ there (e.g. Gardar volcanic rocks), or inferred to be or have been there arguing from dated matches in North America (e.g. the cherts with ?algae and ?fungi). Moorbath et al. (1967) suggested that many, perhaps most, are 1.8–1.5 Ga old. Two isotopic Sm-Nd crustal residence ages (1.99, 1.87 Ga) from shales (O'Nions et al. 1983) are not inconsistent with this but not necessarily supportive. A broadly similar early Proterozoic ('Laxfordian') clustering is shown by the 9 tourmaliniferous pebbles so far dated (Table 1). Four of the mean values lie in the 1.7–1.5 Ga range (raw dates). They came, like the two late Proterozoic (?Grenville) dates, from the southern flank of Williams's southern fan (1969a, fig. 17). The two Archaean ('Scourian') dates are from the northern fan. This N–S variation, if regionally true, must relate to higher level rocks further west, though paralleled by major events in the local Lewisian. (Recently, Cliff & Rex (1989) have recognized a 'Grenville' event in the Lewisian of the Outer Hebrides as far north as Carloway in the Isle of Lewis.)

Can it be more than coincidence that the reassembled fragments of Laurentia show the same N–S age change (Allen et al. 1974)? The Labrador stump in Canada and the pre-Ketili-

Table 1. *Tourmaline dates (total degassing ^{40}Ar-^{39}Ar) from Torridon Group (Applecross Formation) pebbles*

'Scourian' (Ma)		'Laxfordian' (Ma)	'Grenvillian' (Ma)
2684 ± 29 } a 2857 ± 19 }	1938 ± 24 } c 1961 ± 12 }	1382 ± 10 } d 1896 ± 10 }	1052 ± 5 } h 1031 ± 8 }
2509 ± 13 } b 2923 ± 10 }		1705 ± 10 e	881 ± 20 } i 801 ± 23 }
		1648 ± 13 } f 1611 ± 12 }	
		1571 ± 7 } g 1565 ± 5 }	

Locations of pebbles

N fan of Williams 1969a, fig. 17:

    a, N Sutherland (NC 183608)   (=S19121 of Allen et al. 1974, table 1)
    b, N Sutherland (NC 183608)   (=S19118 of Allen et al. 1974, table 1)
    c, N Sutherland (NC 183608)   (=S19117 of Allen et al. 1974, table 1)
    d, N Sutherland (NC 183608)   (=S19119 of Allen et al. 1974, table 1)

S fan of Williams 1969a, fig. 17:

    e, Beinn Alligin, Wester Ross (NG877610)
    f, Beinn Alligin, Wester Ross (NG873614)
    g, Beinn Alligin, Wester Ross (NG877610)
    h, Beinn Alligin, Wester Ross (NG873614)
    i, Applecross Valley, Wester Ross   (=S6179 of Allen et al. 1974, table 1)

Analysts: J. A. Miller and F. J. Fitch (1974 dates recalculated 1989 using latest constants).

dean and Ketilidean stumps in Greenland apparently have no protective covers sufficiently old to allow useful guesses about the sourcelands' geology during Applecross times.

Stoer Group

The earliest Torridonian fan-pile, again associated with probable lake deposits of shales, mudstones and evaporites (Stewart & Parker 1979), unconformably underlies the Torridon Group and overlies the Lewisian. Both unconformities are buried landscapes and the Stoer lithofacies closely resemble those of the Torridon. However, the Stoer Group is less completely preserved, and the source-stumps of its west-north-westerly flows must be hidden by the Moine thrust-mass to the east. Possibly significant in the tectonic and topographic setting, is evidence of local volcanism (Lawson 1965; Stewart 1982). Little is known about the clasts, fully half of which in the trough cross-bedded sandstones appear to be of Scourian age (c. 2600 Ma) and lithology (Stewart 1982) and therefore local. Exotic smoothed pebbles and cobbles of unmetamorphosed sandstones and quartzites of unknown origin occur in the sandstones deposited from both the WNW and ESE flows. It is not known if any closely resemble Torridon Group clasts derived from the west. Presumably of eastern provenance, they might indicate repeated cannibalization during the to-and-fro of flow reversals. Detailed investigation of the composition of the sediments in relation to current directions and their Torridon successors could be rewarding.

Two isotopic Sm-Nd crustal residence ages (2.00, 2.33 Ga) have been obtained from the siltstones (O'Nions et al. 1983). Future research might well show that the Stoers relate to the Proterozoic movements of Baltica (see e.g. Anderton 1982) rather than Laurentia.

Future research

The previous picture of Torridonian origins is based on agonizingly few clasts and whole-rock samples, and is by no means compensated by the variety of sophisticated techniques employed (petrographic, mineralogical, geochemical, isotopic, palaeontological). It must be simplistic and may be wide of the mark. Many of the data are based on outmoded methods. Application of new techniques is needed for a massive assault producing large numbers of more precise single-grain/crystal dates, isotopic ratios, geochemical analyses (especially whole-rock), and chert floras. At present only whole-rock and 'average'

ages of the exotic lithoclasts and crystal-grains are available. A detailed 3-D survey of them has not begun. Bearing in mind the vast quantities of sandstone, single-grain dating (laser ^{40}Ar-^{39}Ar, U-Pb, EMS) should revolutionize the work and lend extra precision to it. Much effort will be needed to search for matches with possible source-stumps through petrographic, mineralogical, geochemical, palaeontological and age characteristics. There will be surprises. We know little more about the sand, silt and clay grades than was disclosed by the pioneers Mackie (1899, 1923, 1928), Teall (in Peach et al. 1907) and Gilligan (1920).

Tourmaline should continue to play a useful role in dating. On the basis of comparing like with like, it was originally chosen because, while usually crystallizing at supracrustal depths, it is highly stable in sedimentary environments, occurs widely in the Torridonian as grains and in pebbles, and could in principle reveal its recycling histories through dated outgrowths and overgrowths. In view of the debate on its reliability (see e.g. Henry & Guidotti 1985), particularly a suspected propensity for excess argon, comparisons were made between the ages of in situ tourmalines and their geological contexts over a wide time range (c. 3.8–0.3 Ga). For the Archaean sourcelands of the Torridonian D. Bridgwater and A. D. Gibbs supplied tourmalines, background information and personal judgements and J. A. Miller and F. J. Fitch dated the crystals 'blind' under code names (Table 2). Again, like most Torridonian datasets (excepting of course Williams's palaeocurrents), these are far too few. But surely worth following up? Similar comparisons at the young (Hercynian) end of the time-scale also suggest that tourmaline may pass the 'proof-of-the-pudding' test (Allen 1972, 1975, 1981). Zircon is another obvious target, as Drewery et al. (1987) have shown in their study of Precambrian grains from the Carboniferous.

Much provenance research will concern stable isotopes and their ratios, which can survive weathering and throw light on the evolutionary history and recycling in the Torridonian lithosphere. Initial ^{87}Sr/^{86}Sr data and isotopic Sm–Nd crustal residence times also help to constrain the dating of the detritus.

Wealden Series

Source directions

Wealden flow-patterns are more difficult to determine because of the scarcity of conglomerates, predominance of mud, and poor exposures

Table 2. *Tourmaline dates (total degassing ^{40}Ar-^{39}Ar) from N Atlantic Archaean*

GGU sample number and locality	Rock	Petrographic history of rock, and references	Tourmaline dates (Ma)
117931 N of 'lake 678 m' (ex Imarssuaq) Usukasia, West Greenland. (Isua supracrustal belt)	'Acid volcanic sediment'	Crystallization (zircon and ?tourmaline): 3.8 Ga Metamorphism (amphibolite facies): 3.7, 2.8 Ga Shearing (greenschist facies): 1.8 Ga (Bridgwater et al. 1973; Nutman et al. 1984; Baadsgaard et al. 1986)	3418 ± 27 3143 ± 17
JE69577 Buksefiord area, S. of Ameralik, West Greenland	Granite facies pegmatite	Crystallization (intrusion): 2.9–2.8 Ga Metamorphism (granulite facies): 2.82 Ga Metamorphism (retrograde): 2.8–2.5 Ga (Bridgwater et al. 1973 Schiøtte et al. 1989a)	2763 ± 19 2758 ± 19
171420 Buksefiord area, West Greenland	Margin of Qorqut pegmatite swarm	Crystallization (Qorqut granite): 2.58 Ga (Pankhurst et al. 1973)	2566 ± 37 (excess argon problem handled)
KC-DB 74–16 W side Shuldam Island, Saglek Bay, Labrador	Garnet-pelite	Crystallization (zircon): c. 3.4 Ga Metamorphism (granulite facies): 2.77–2.71 Ga Intrusion (granite): 2.75–2.5 Ga Shearing (recrystallization): 1.8 Ga (Shiøtte et al. 1989b, c)	2424 ± 22 2148 ± 23

Samples and other information supplied by D. Bridgwater and A. D. Gibbs
Analysts: J. A. Miller and F. J. Fitch (1974 dates recalculated 1989 using latest constants)

(usually small, 2-D and few). Much reliance has to be placed on textural and compositional gradients. These suggest four major source-directions, three leading back towards massifs in the southwest, west and northeast (Armorica, Cornubia and Londinia) and one towards the northern (Boreal) sea. Evidence for the oft-postulated Welsh and Pennine massifs has not been found.

Clastics and detrital clays from the massifs are interpreted as the distal deposits of small fans (sandy braidplains → muddy meanderplains → lake, lagoon and bay clays). The Wessex sub-basin (not equivalent to the 'Wessex Basin' of P. E. Kent) was dominated almost throughout (Valanginian to earliest Aptian) by detritus from Cornubia, with minor amounts from Armorica and, late on, the ?Boreal sea. The Weald sub-basin received most of its sediment from Londinia (late Berriasian to earliest Aptian), with early contributions (in some Hastings Group sands) from Armorica and later ones (Weald Clay sands) from Cornubia and the northern sea.

Armorican sources

Detritus from Brittany–Normandy is recognized in the Weald by compositional gradients of high-grade metamorphic clasts (e.g. Allen 1949, 1959). These are streamlined SW/NE in the eastern part of the sub-basin and S/N in the centre. None of the sand grains (staurolite, kyanite, sillimanite, monazite, ceylonite, purple zircon, etc.) has been dated, but 'average' ages of the micas, tourmalines and K-feldspars hint at a Cadomian–early Hercynian–late Hercynian profile, held by some workers to be typical of Brittany (Allen 1981, p. 393, fig. 10). The possibility of matching staurolites through their trace-elements was explored many years ago (Allen 1959), but the quartz-spectograph of the day was not up to the task.

The patches of Armorican sand in Sussex merge into finer sands poorer in the above Armorican minerals but richer in garnet and apatite. The latter were first interpreted as derived from the Kentish sector of the Londinia–Brabant–Ardennes massif. Accepting the statistical dangers of patchy distributions, trend analysis of their garnet frequencies seemed supportive (Allen & Krumbein 1962). But the garnets showed no enhanced Mn contents like those in the Wealden facies on the Dutch–German side of the watershed (Allen 1967a), and it now seems likely that they are Armorican. This is strengthened by the associated 'cored' apatites, which closely resemble those in the Brittany granites (Groves 1930). Probably the two groups of minerals were kept largely apart by particle size controls on their frequencies at source, the fine garnet–apatite facies being deposited as fringes round the coarser staurolite–kyanite sands. Modern dating techniques (U-Pb, laser ^{40}Ar-^{39}Ar, and possibly fission track) could settle the matter. Non-mixing of the detritus with that of Cornubian origin traversing Wessex may be explained either by a more southerly route round or over the Portsdown–South Downs intra-basinal swell (Allen 1975), or by recycling from older Mesozoic rocks on the latter.

'Putting back' the detritus on to the Armorican stump and reconstructing its contemporary geology is hopeless because there is no post-Wealden cover old or extensive enough. One can only guess at a Precambrian core of coarser staurolite–kyanite schists and finer grained garnet schists, with granites and marginal and/or intrabasinal Mesozoic rocks supplying Jurassic volcanogenic debris (Allen 1981) and glauconite.

Cornubian sources

Dated pebbles and sand, with flow patterns showing derivation from the W or SW, abound in Wessex from the Valanginian (basal Wealden) upwards (Wessex Formation), and in the Weald thin sands are common from the Hauterivian upwards (Weald Clay). Compositional gradients arise from mixing with small quantities of Armorican and Boreal materials in the Wessex sub-basin and Londinian and Boreal in the Weald. Dates are palaeontological (e.g. radiolarian chert of Carboniferous Culm facies, Oakley 1947) and radiometric (tourmalines, micas, K-feldspars), the latter being average values when from the sands. The radiometric dates (Allen 1975, pp. 430–431) appear to identify early and late Hercynian (>300, $\leqslant 300$ Ma) and some older crystalline rocks. This may bear on the question of a Start Point–Cotentin source ridge connecting Cornubia with Armorica. Sparse current directions from about SW in Dorset provide some support. Associated glauconite was presumably eroded from Jurassic rocks exposed on intrabasinal and/or marginal uplifts (e.g. South Dorset High).

Some of the Cornubian detritus in the Wealden seems to have been recycled through the New Red Sandstone. Hence the varied palaeogeology of the Cornubian stump, as recorded by the clasts, belongs partly to an earlier time. During the Wealden it was partially blanketed by outcrops of Permo-Triassic red beds (including unweathered screes), relieved here and there by relict volcanic and thermal activity (Allen 1972) and perhaps some marginal Jurassic. Absence of a protective cover older than Tertiary (though the Chalk may have been removed quite recently) makes it difficult to test this scenario by 'putting back' the detritus. Any hope of tracing progressive changes in the outcrop pattern during Wealden times is slim.

Nevertheless, some changes in denudation rates on the stump can be detected from evidence in the sediments. Periodically, erosion was briefly enhanced (presumably due to upfaulting, with or without higher rainfall) when violent flushes of coarse sediment were shot across the Wealden plains. The gravelly Coarse Quartz Grit is a good example. This member traverses the whole of Dorset, and its distal fringe, if not halted by the intrabasinal high mentioned, might well lie in the Weald. Perhaps it is one of the sands of Topley's Bed 5 (Hughes & McDougall 1990; Allen 1990) in the Weald Clay. It could be a useful time plane. On the other hand the sands could have been brought in from the Boreal sea, though this seems unlikely. How far east the Cornubian sands went is an interesting question. Certainly their materials are recognizable in the southeasternmost Weald Clay 250 km away from source (Allen 1990). Whole-rock geochemical work (e.g. on Sn, Wo) would probably extend the range. Traces of Armorican detritus in the sands suggest recycling from the Cornubian NRS cover, or even from the Midlands NRS (cf. that in the Quaternary 'northern drift' on the Chilterns).

Like the sands from Armorica, several of the Cornubian aprons reached almost to the feet of Londinia. Cornubia itself may have been surrounded by them because facies like the Coarse Quartz Grit are known in the Western Approaches and Celtic Sea basins (Allen 1981, pp. 400–401). Whether the events to which they testify were local and uncoordinated or regional and contemporaneous is a problem for the future.

No isotopic Sm-Nd crustal residence ages suggesting mantle contributions to the crust during the Hercynian orogeny have been published (Miller & O'Nions 1984).

Londinian sources

The buried stump of the London massif, which dominated sedimentation in the Weald sub-basin, is clearly signposted by flow patterns and some (but all too few) large sedimentary structures. Distal deposits of small fans emerging from the faulted margin are suggested in the lower Wealden (Hastings Sands Group). Pebbles and sand from Lower Palaeozoic, Devonian (ORS), Carboniferous and Upper Jurassic rocks are recognizable with certainty through their contained fossils or radiometric dates (Allen 1960, 1961, 1967b, 1975, 1981). A little Triassic material may be present. Most of the clasts appear to have come direct from rocks of these ages outcropping on the massif (albeit sometimes leached in the acid soils). There are, however, severe problems of recycling through Old Red Sandstone conglomerates (some of the jaspers from which might be ultimately Torridonian!) and Upper Jurassic pebble beds; and the ages and deeper-basin affinities of the glauconites (Portlandian or approximately contemporaneous?) combined with the absence of cherty Portlandian in SE England are puzzling (Allen et al. 1964; Allen 1981, 1990). The clays seem to be mostly detrital and formed by weathering of clays in the same Mesozoic and Palaeozoic formations (Sladen 1987). Smectite-rich facies in the upper Wealden (Weald Clay Group) could be volcanogenic, a possibility enhanced perhaps by euhedral biotites in the predominantly Cornubian sands on the north crop (Allen 1990). Recently, S. P. Kelley (then at the Scottish Universities Research and Reactor Centre, East Kilbride) obtained a laser probe ^{40}Ar-^{39}Ar date of 179 ± 30 Ma from the central portion of a single flake of an optically extremely dispersed rimmed biotite (dim. $200 \times 200\,\mu m$) from the Weald Clay Grayswood Sand (BGS Bed 7i) at Grayswood (SU924349). This is almost certainly a minimum age. The biotite is associated with Cornubian/Hercynian detrital materials (Allen 1975, 1981).

Putting the detritus back on to the buried stump and reconstructing its Wealden geology has great possibilities. First, the N–S and NE–SW streamlines of sand and gravel with different compositions must line up, however crudely, with the outcrops from which they came. Second, the stump has an oldish cover (Albian), leaving little more than 6–17 Ma for post-Wealden erosion. Third, sufficient is known about the present subcrops beneath the cover to attempt subcrop maps (Sellwood et al. 1986, fig. 2 and refs). Fourth, some borehole cores are available for petrographic and geochemical matching. Fifth, the time is ripe for quantitative work on intra-Wealden erosion of the stump and the sub-basin's sedimentary profit-and-loss account. A largely intuitive Wealden geological map of Londinia was (rashly) published many years ago (Allen 1967b, fig. 3).

Boreal oceanic sources

Several semi-marine beds, some containing sand and pebbles, occur in the Weald Clay (Hauterivian–Barremian) of the N to NW Weald (Allen 1975, fig. 3, inset A; 1990, fig. 8.). The pebbles include quartz, Jurassic phosphorite and Lower Carboniferous chert. Marker grains are acid plagioclase, microcline, epidote, bluish amphibole, garnet, kyanite > staurolite, monazite and ?contemporary glauconite. Frequency ratios of the same species as contributed by the massifs lie in separate fields from the latter (e.g. Allen 1981, fig. 12). Neither the pebbles nor the sands, nor their marine contemporaries in eastern England, nor possible Jurassic source-rocks around the western end of Londinia (Allen 1967b) have been investigated in a modern mode (as deployed, for example, by Morton (1985a, b)). The sands are presumably Caledonian in ultimate origin, but there are signs of subtle differences which only single-grain dating, microprobe analysis and determination of stable isotope ratios will illuminate. The outlook is promising because the grains are large and easy to clean and separate. Whether any of the Cornubian detritus in the other Weald sands was washed in (from the W or N?) by the Boreal sea is an open question.

In Wessex, at least some of the clay in the Vectis Formation is also of marine origin, but whether from the Boreal sea or an arm of Tethys to the SE or SW is uncertain.

Overview

The Weald sub-basin is an arena where brief interventions from three other sources periodically polluted, and locally interrupted, the accumulation of Londinian detritus (Allen 1990). If the episodes were generated by vertical movements of the crust then these were not coordinated, and the relative heights of the three massifs must have changed considerably from time to time. Most dramatic was the refreshingly new Cornubian petrography which spread across both sub-basins and into the Western Approaches and Celtic Sea. This effectively ended the predominance of ultimately Caledonian detritus (involving repetitive recycling) which had prevailed over Britain since Silurian times.

Much remains to be done to test and firm up the picture and extend it deeper into the lithosphere. Modern techniques will be needed: laser ^{40}Ar-^{39}Ar and U-Pb for single-grain and overgrowth dating, and exploitation of neglected minerals such as tourmaline, zircon and glauconite; microprobe analysis for 'tagging' marker clasts; whole-rock geochemical analysis; Sm-Nd isotopic ratios for testing the apparent absence of Hercynian mantle contributions to the crust between Laurentia and European Gondwanaland. All should be used in a continuing context of basin-and-massif budget analysis and well-tried field and laboratory techniques, and directed strongly at possible source-rocks. Curiously, the latter seem often underused in provenance research.

Reading University PRIS contribution No. 038.

References

ALLEN, P. 1949. Wealden petrology: the Top Ashdown Pebble Bed and the Top Ashdown Sandstone. *Quarterly Journal of the Geological Society of London*, **104**, 257–321.
—— 1959. The Wealden environment: Anglo-Paris basin. *Philosophical Transactions of the Royal Society of London*, **B242**, 283–346.
—— 1960. Strand-line pebbles in the mid-Hastings Beds and the geology of the London uplands. General features. Jurassic pebbles. *Proceedings of the Geologists' Association*, **71**, 156–168.
—— 1961. Strand-line pebbles in the mid-Hastings Beds and the geology of the London uplands. Carboniferous pebbles. *Proceedings of the Geologists' Association*, **72**, 271–285
—— 1967a. Origin of the Hastings facies in northwestern Europe. *Proceedings of the Geologists' Association*, **78**, 27–105.

—— 1967b. Strand-line pebbles in the mid-Hastings Beds and the geology of the London uplands. Old Red Sandstone, New Red Sandstone and other pebbles. Conclusion. *Proceedings of the Geologists' Association*, **78**, 241–276.
—— 1972. Wealden detrital tourmaline: implications for northwestern Europe. *Journal of the Geological Society, London*, **128**, 273–288.
—— 1975. Wealden of the Weald: a new model. *Proceedings of the Geologists' Association*, **86**, 389–437.
—— 1981. Pursuit of Wealden models. *Journal of the Geological Society, London*, **138**, 375–400.
—— 1990. Wealden research–ways ahead. *Proceedings of the Geologists' Association*, **100**, 529–564.
—— & KRUMBEIN, W. C. 1962. Secondary trend components in the Top Ashdown Pebble Bed: a case history. *Journal of Geology*, **70**, 507–538.

——, ALLEN, J. R. L., GOLDRING, R. & MAYCOCK, I. D. 1960. Festoon bedding and "mud-with-lenticles" lithology. *Geological Magazine*, **97**, 261–262.

——, DODSON, M. H. & REX, D. C. 1964. Potassium-argon dates and the origin of Wealden glauconites. *Nature*, **202**, 505–586.

——, SUTTON, J. & WATSON, J. V. 1974. Torridonian tourmaline-quartz pebbles and the Precambrian crust northwest of Britain. *Journal of the Geological Society, London*, **130**, 85–91.

ANDERTON, R. 1982. Dalradian deposition and the late Precambrian-Cambrian history of the N Atlantic region: a review of the early history of the Iapetus Ocean. *Journal of the Geological Society, London*, **139**, 421–431.

BAADSGAARD, H., NUTMAN, A. P. & BRIDGWATER, D. 1986. Geochronology and isotopic variation of the early Archaean Amitsoq genisses of the Isukasia area, southern West Greenland. *Geochimca et Cosmochimica Acta*, **50**, 2173–2183.

BRIDGWATER, D., WATSON, J. & WINDLEY, B. F. 1973. The Archaean craton of the North Atlantic region. *Philosophical Transactions of the Royal Society of London*, **A273**, 493–512.

CLIFF, R. A. & REX, D. C. 1989. Evidence for a 'Grenville' event in the Lewisian of the Outer Hebrides. *Journal of the Geological Society, London*, **146**, 921–924.

DREWERY, S., CLIFF, R. A. & LEEDER, M. R. 1987. Provenance of Carboniferous sandstones from U-Pb dating of detrital zircons. *Nature*, **325**, 50–53.

GILLIGAN, A. 1920. The petrography of the Millstone Grit of Yorkshire. *Quarterly Journal of the Geological Society of London*, **75**, 251–294.

GROVES, A. W. 1930. The heavy mineral suites and correlation of the granites of nothern Brittany, the Channel Islands, and the Cotentin. *Geological Magazine*, **67**, 218–240.

HENRY, D. J. & GUIDOTTI, C. V. 1985. Tourmaline as a petrogenetic indicator mineral: an example from the staurolite-grade metapelites of NW Maine. *American Mineralogist*, **70**, 1–15.

HUGHES, N. F. & McDOUGALL, A. B. 1990. New Wealden correlation for the Wessex basin. *Proceedings of the Geologists' Association*, **101**, 85–90.

LAWSON, D. E. 1965. Lithofacies and correlation within the Lower Torridonian. *Nature*, **207**, 706–708.

MACKIE, W. 1899. The felspars present in sedimentary rocks as indicators of the conditions of contemporaneous climates. *Transactions of the Edinburgh Geological Society*, **7**, 443–468.

—— 1923. The source of the purple zircons in the sedimentary rocks of Scotland. *Transactions of the Edinburgh Geological Society*, **11**, 200–213.

—— 1928. The heavy minerals in the Torridon sandstones and metamorphic rocks of Scotland and their bearing on the ages of these rocks. *Transactions of the Edinburgh Geological Society*, **12**, 181–182.

MILLER, R. G. & O'NIONS, R. K. 1984. The provenance and residence ages of British sediments in relation to palaeogeographical reconstructions. *Earth and Planetary Science Letters*, **68**, 459–470.

MOORBATH, S., STEWART, A. D., LAWSON, D. E. & WILLIAMS, G. E. 1967. Geochronological studies on the Torridonian sediments of north-west Scotland. *Scottish Journal of Geology*, **3**, 389–412.

MORTON, A. C. 1985a. Heavy minerals in provenance studies. *In*: ZUFFA, G. G. (ed.) *Provenance of Arenites*, D. Reidel, Dordrecht. 249–277.

—— 1985b. A new approach to provenance studies: electron microprobe analysis of detrital garnets from Middle Jurassic sandstones of the North Sea. *Sedimentology*, **32**, 553–566.

NUTMAN, A. P., ALLAART, J. H., BRIDGWATER, D., DIMROTH, E. & ROSING, M. 1984. Stratigraphic and geochemical evidence for the depositional environment of the Early Archaean Isua supracrustal belt, southern West Greenland. *Precambrian Research*, **25**, 365–396.

OAKLEY, K. P. 1947. A note on Palaeozoic radiolarian chert pebbles found in the Wealden Series of Dorset. *Proceedings of the Geologists' Association*, **58**, 255–258.

O'NIONS, R. K., HAMILTON, P. J. & HOOKER, P. J. 1983. A Nd isotope investigation of sediments related to crustal development in the British Isles. *Earth and Planetary Science Letters*, **63**, 229–240.

PANKHURST, R. J., MOORBATH, S., REX, D. & TURNER, G. 1973. Mineral age patterns in ca. 3700 my old rocks from West Greenland. *Earth and Planetary Science Letters*, **20**, 157–170.

PEACH, B. N., HORNE, J., GUNN, W., CLOUGH, C. T., HINXMAN, L. W. & TEALL, J. J. H. 1907. The geological structure of the north-west highlands of Scotland. *Memoir of the Geological Survey of the United Kingdom*.

SCHIØTTE, L., COMPSTON, W. & BRIDGWATER, D. 1989a. U-Pb single zircon age for the Tinissaq gneiss of southern West Greenland: a controversy resolved. *Chemical Geology (Isotope Geosciences section)*, **79**, 21–30.

——, —— & —— 1989b. Ion probe U-Th-Pb zircon dating of polymetamorphic orthogneisses from northern Labrador, Canada. *Canadian Journal of Earth Sciences*, **26**, 1533–1566.

——, —— & —— 1989c. U-Th-Pb ages of single zircons in Archaean supracrustals from Nain Province, Labrador, Canada. *Canadian Journal of Earth Sciences*, **26**, 2636–2644.

SELLWOOD, B. W., SCOTT, J. & LUNN, G. 1986. Mesozoic basin evolution in Southern England. *Proceedings of the Geologists' Association*, **97**, 259–289.

SLADEN, C. P. 1987. Aspects of the clay mineralogy of the Wealden and upper Purbeck rocks. *In*: LAKE, R. D. & SHEPHARD-THORN, E. R. 1987. *Geology of the Country around Hastings and Dungeness*. Memoir of the British Geological Survey.

STEWART, A. D. 1969. Torridonian rocks of Scotland reviewed. *In*: KAY, M. (ed.) *North Atlantic—Geology and Continental Drift*. American Association of Petroleum Geologists Memoir, **12**, 595–608.

—— 1972. Pre-Cambrian landscapes in Northeast

Scotland. *Geological Journal*, **8**, 111–124.

—— 1982. Late Proterozoic rifting in NW Scotland: the genesis of the 'Torridonian'. *Journal of the Geological Society, London*, **139**, 413–420.

—— 1988. The Sleat and Torridon Groups. *In*: WINCHESTER, J. A. (ed.) *Later Proterozoic stratigraphy of the northern Atlantic region*. Blackie, Glasgow, 104–112.

—— & PARKER, A. 1979. Palaeosalinity and environmental interpretations of red beds from the late Precambrian ('Torridonian') of Scotland. *Sedimentary Geology*, **22**, 229–241.

WILLIAMS, G. E. 1969*a*. Characteristics and origin of a Precambrian pediment. *Journal of Geology*, **77**, 183–207.

—— 1969*b*. Petrography and origin of pebbles from Torridonian strata (Late Precambrian), northwest Scotland. *In*: KAY, M. (ed.) *North Atlantic— Geology and Continental Drift*. American Association of Petroleum Geologists Memoir, **12**, 609–629.

On the use of turbidite arenites in provenance studies: critical remarks

GIAN GASPARE ZUFFA

Dipartimento di Scienze Geologiche, Università di Bologna, Via Zamboni, 67, 40127
Bologna, Italy

Abstract: When we determine framework petrography of deep-sea arenites for provenance
studies we should concentrate not only on what particles make up sedimentary beds but also
when, where and how they form. Taking such information into account, grains can be
classified according to their temporal, spatial and compositional characteristics. This will
allow a better understanding of palaeogeography and palaeotectonics of the basin/source
system. Two examples from the literature that support this approach are reported.

Knowledge of arenite petrology has advanced
over the last fifteen years as a result of research
and publications that have strongly emphasized
both diagenesis and petrofacies relations in various geological settings. A better understanding
of diagenetic modifications has been of great importance in the exploration and exploitation of
hydrocarbons and is essential in determining the
original composition of the arenite framework
for palaeogeographic and palaeotectonic reconstructions (Helmold 1985; McBride 1985). Petrofacies of both modern and ancient arenites have
received a great deal of attention and, as a result,
we can relate the major provenance types to the
main plate-tectonic settings (Dickinson 1985;
Valloni 1985).

Unfortunately, little interest has been devoted
to the petrology of arenites in terms of detailed
basin analysis; this is because we have traditionally considered sedimentology and sedimentary
petrology as quite separate research areas (Zuffa
1985). Field sedimentologists of both modern
and ancient systems often deal with transport/
depositional processes, and resulting sedimentary facies, perhaps without considering the fact
that sedimentary grains are not glassy spheres.
In contrast, sedimentary petrographers often
deal with sediments without closely relating their
results to the architecture of the sedimentary
body and its various facies. In order better to
integrate the above approaches, sedimentary
petrologists should concentrate not only on
establishing what kind of particles make up
sedimentary beds, but they should also focus on
when, where and how grains form, because all
these are linked with the development of facies
and, ultimately, with source/basin evolution.

General framework

The following questions are critical in the basin
analysis of deep-sea arenites:

> what kind of grains, compositionally and
> texturally, are produced;
> how grains are generated, e.g. deep weath-
> ering of low relief areas versus intense ero-
> sion of high-relief areas; biogenic produc-
> tion versus pure chemical precipitation;
> where grains come from, i.e. extrabasinal
> versus intrabasinal sources;
> when grains are generated, i.e. age of grains:
> (i) generated 'ex novo' by chemical–bio-
> chemical processes or (ii) by breaking up of
> lithified rock units (grains coeval or non-
> coeval with respect to a considered deposi-
> tional sequence).

All these questions are essential and their
relevance can be illustrated by the following
circumstances.

(a) Grain size, shape and density control the
hydraulic differentiation occurring during transport, and, in turn, influence the resulting sedimentary structure and facies. For example,
turbidite beds may locally lack the Bouma
c-division due to the scarcity of detritus of the
proper grain size (i.e. fine-grained sand/silt: this is
particularly true for carbonate turbidites), or,
grading may be absent simply because of very
good sorting (oolitic or some quartz arenitic
turbidites).

(b) The balance between intrabasinal generation of new (coeval) grains in shallow-water
areas and terrigenous input of old (non-coeval)

grains can be strongly affected by eustatic/tectonic variations of the sea-level; as a consequence, a turbidite body may display sequences with superimposed distinct petrofacies recording the history of the basin/source system (e.g. the Hecho Formation, Eocene, Spain: Fontana *et al.* 1987).

(c) The recognition of sequences characterized either by compositionally hybrid beds or alternating beds of pure siliciclastic detritus and of calciclastic coeval grains may, respectively, indicate: (i) the existence of marginal shallow-water 'parking areas' where extrabasinal and intrabasinal components can mix before resedimentation (e.g. the Bismantova Formation, Miocene, northern Apennines: Zuffa 1969), and (ii) distinct terrigenous and intrabasinal sources playing an independent role in supplying the deep-sea basin, such as in the Miocene Marnoso–Arenacea Formation (Gandolfi *et al.* 1983; Ricci Lucchi 1985).

(d) The coexistence in the same arenite sequence of neovolcanic and palaeovolcanic grains, if detected, is decisive for the correct understanding of the geotectonic setting in which sediments were deposited (e.g. the Quaternary volcaniclastic turbidite sands of the Nankai Trough: De Rosa *et al.* 1986).

Thus, a knowledge of origin, timing, and composition of grains can be critical in unravelling source/basin palaeogeography and tracing its evolution through time.

Standard modal analyses and classifications used in provenance studies mainly focus on the siliciclastic framework of arenites and have proved a powerful tool in unravelling the tectonic setting in which turbidite sequences were deposited. However we should consider a scale problem: (i) there is no doubt that it is crucial to know whether a turbidite terrane, accreted to the west America continental margin or to the British Caledonides, was deposited in a back-arc or a fore-arc setting; but, in contrast, (ii) petrographical analyses are unnecessary to find out the geological setting of, for example, the Apennine Marnoso–Arenacea or the Swiss Alpine Molasse. In the latter case, since we already know the geological setting in which the formations were deposited, we need to reconstruct the configuration in time of the source/basin system. This entails trying to solve problems such as:

(a) which sources (intrabasinal or extrabasinal) fed the basin and how grains were generated;
(b) how sea-level changes have influenced production and dispersal of grains;

(c) whether sea-level changes were eustatically or tectonically induced;
(d) how these have affected the architecture of the sedimentary body.

However, these issues cannot be dealt with only on the basis of routine petrology which commonly ends up with QFL diagrams. We need to separate the information carried by the various types of grains about hinterland sources versus basin sources; this involves the detection of the age of particles versus age of the deposit. Critical grains for this purpose are not the common quartz and feldspar but carbonate and volcanic grains since the latter can be intra- or extrabasinal, coeval or non-coeval with the deposit.

Taking into account the relevant issues listed above, sedimentary grains forming deep-sea turbidite bodies have been classified, for source/basin reconstructions, according to their temporal, spatial, and compositional characteristics (Zuffa 1987). A slightly modified version of this classification scheme is reported in Fig. 1.

Applying these distinctions, naturally occurring deep-sea arenites can be described according to their grain characteristics and grouped in the following main types.

(1) Pure extrabasinal. Grains derived from weathering and erosion of non-coeval hinterland rock units (siliciclastic and calciclastic turbidites); examples of these deposits can be found in laterally-confined elongated depositional bodies of trench, foreland settings and of some strike-slip basins.

(2) Pure intrabasinal. Mostly coeval carbonate grains derived from margin-linked ramps and intrabasinal isolated highs deposited as confined and non-confined turbidite bodies supplied by generally complex intra-oceanic linear sources (e.g. carbonate turbidites of the Bahamas area: Mullins 1983; Late Cretaceous carbonate Helminthoid Flysch of northern Apennine: Scholle 1971).

(3) Mixture of extrabasinal (mainly siliciclastic) with intrabasinal (mainly carbonate) grains, the latter produced in shelf areas by biogenic activity: 'foramol-type' associations (Lees & Buller 1972); examples may be found in turbidite bodies chiefly of passive continental margins, in 'piggy-back' basins (Zuffa 1969) or in key beds of some foreland sequences (Gandolfi *et al.* 1983). This main type also includes mixtures of intrabasinal and extrabasinal volcanic resedimented grain deposits adjacent to active arc settings which can derive by erosion of ancient (non-coeval) or penecontemporaneous (coeval)

CRITERIA	TEMPORAL	SPATIAL	COMPOSITIONAL	
TYPES OF GRAINS (classification mainly applies to deep-marine arenitic sequences)	COEVAL	EXTRABASINAL	CARBONATE	ooids bioclasts caliches travertine
			NONCARBONATE	VOLCANICS (V_{2a} - V_{2b} - V_3) vegetation particles
		INTRABASINAL	CARBONATE	OOIDS BIOCLASTS INTRACLASTS PELOIDS
			NONCARBONATE	glauconite iron-oxides gypsum phosphates rip-up clasts volcanics (V_4)
	NONCOEVAL	EXTRABASINAL	CARBONATE	LIMESTONES DOLOSTONES
			NONCARBONATE	SILICICLASTS (including V_1)
		INTRABASINAL	CARBONATE	limestones dolostones
			NONCARBONATE	siliciclasts (including V_1)

Fig. 1. Classification of sand particles for palaeogeographic reconstructions of source area and depositional basin (modified after Zuffa 1987). Column 2: terms refer to the time interval of deposition of the sedimentary sequence under consideration. Column 5: the terms in capital letters indicate major arenite components; those in small letters indicate minor components. V_1: derived from erosion of old volcanic suites; V_{2a}, V_{2b}: derived from active volcanism located in the source area; V_3: derived from active volcanoes not located in the main source area; V_4: derived from active volcanoes located within the depositional basin. Definitions, assumptions and limitations concerning this scheme are discussed in Zuffa (1987).

volcanic products. The tremendous complexity as regards types of particles, fragmentation processes, dispersal mechanisms (direct dispersion, surge and flow processes, subaqueous or subaerial volcanic activity, multiple resedimentation), frequent changes of source area locations, complex basin morphologies, and mixing of volcanic particles with penecontemporaneous carbonate grains, make the composition and geometry of these little-studied systems very difficult to tackle (e.g. Busby-Spera 1988).

This subdivision is not meant to be a formal classification but only a basis for raising questions as to possible relationships between geometry and facies of turbidite bodies versus time, place of production, and composition of sandy particles affected by resedimentation processes. Indeed, deep-sea arenite bodies are seldom entirely expressed by only one of the described main types; instead, what we commonly observe is a more or less complicated alternation (both as single beds and bed packages), reflecting distinct source areas and sediment entry-points which evolve through time in relation to tectonic/eustatic sea-level variations affecting the basin/source system.

When arenite grains are entirely extrabasinal or intrabasinal (i.e. non-coeval or coeval respectively) classification is relatively easy. Serious problems arise when the arenite framework is a mixture of intrabasinal and extrabasinal grains (mostly coeval and non-coeval carbonate and volcanic clasts).

Some criteria and methods for distinguishing between coeval and non-coeval carbonate sand particles and the various types of volcanic grains are discussed in Zuffa (1987). However, we still need to improve techniques to determine the age of grains in order to infer their origin. Palaeontological methods combined with the observation of textural characteristics of grains allow age determination of carbonate grains in most cases. On the contrary, age determination of volcanic particles is generally more difficult as it implies the adoption of absolute dating methods. However, field observations combined with compositional, textural and alteration characteristics of grains (see Zuffa 1987, fig. 5) are in some cases decisive in detecting the coexistence of coeval grains originated by penecontemporaneous activity, with palaeovolcanic grains derived by erosion of ancient volcanic terranes incorporated in the orogenic belts (see second illustrative example presented in this paper).

Illustrative examples

Two examples from the literature have been chosen to support aspects of the above approach.

The Marnoso–Arenacea Formation (Miocene, northern Apennines): Gandolfi et al. (1983).

The Marnoso–Arenacea Formation is a thick (up to 3500 m) foreland turbidite sequence deposited during the Miocene Epoch on a northwest-trending, laterally confined, almost flat, elongated basin (60 × 350 km) located on the thinned continental crust of the Adriatic continental margin.

By taking into account the framework grain characteristics three main types of turbidite beds can be distinguished with some simplification.

(1) Pure terrigenous beds made up of siliciclastic and calciclastic grains (litharenites).
(2) Hybrid beds made up of terrigenous (non-coeval) and intrabasinal (coeval) non-carbonate and carbonate grains (hybrid arenites; Zuffa 1980); one of these strata is well known as the 'Contessa key bed'.
(3) Pure intrabasinal beds made up of coeval carbonate grains (calcarenites); these beds are known as the 'Colombine key beds'.

Integration of petrofacies with palaeocurrents and sedimentary facies analysis allows us to outline a complex palaeogeographic scenario of the evolving source/basin system (Fig. 2).

The Alpine orogenic neoformed relief, located on the northern end of the basin, was the main extrabasinal (non-coeval) source giving rise to the pure terrigenous beds (non-carbonate and carbonate extrabasinal). The Apenninic forming relief, which constituted the southwest basin margin moving toward the northeast, was a 'parking area' where hybrid sands formed in piggy-back basins and were later remobilized into the deep-sea Marnoso–Arenacea basin. A carbonate platform, locking the southeastern end of the basin, delivered coeval carbonate grains, thus giving rise to pure intrabasinal carbonate key beds (carbonate intrabasinal). The stacking of these turbidite beds derived from different extra- and intrabasinal sources, both transversal and axial to the basin, occurred because of the absence of topographic gradient in the basin.

The reconstruction of the palaeogeography outlined above was facilitated by the recognition of temporal, spatial and compositional grain signatures. It is important to emphasize that a standard petrography of these sediments by using the classical QFL diagram (which includes only the extrabasinal siliciclastic framework grains) would only have allowed the authors to conclude that the Marnoso–Arenacea Formation is of 'recycled orogenic provenance' (Fig. 2; plot on the lower left).

Quaternary sands from the Nankai Trough (southwestern Japan): De Rosa et al. (1986).

Quaternary turbidite sands from sites 582 and 583 (DSDP Leg 87) show 70–80% of volcanic component which comprises both palaeovolcanic and neovolcanic grains. The palaeovolcanic grains, despite the fact that they are of acidic composition, are altered, whereas the neovolcanic grains consist of both basic and acidic unaltered types. Composition, texture, degree and type of alteration allow a relatively easy distinction between the two different types. The remaining minor terrigenous component is made up of sedimentary, metamorphic and plutonic grains.

Intrabasinal grains such as carbonate-spar aggregates, nodules of blue phosphate (vivianite), pyrite and intraclasts (made up of volcanic ash, foraminiferal planktonic tests, radiolarian remains and carbonate peloids), are present in significant amounts in some samples from site 583.

Determination of temporal ('palaeo' versus 'neo'), spatial (extra- versus intrabasinal) and compositional grain characteristics support the following interpretation. Detritus is derived mainly from the Tokai drainage basin where both Cretaceous–Early Tertiary volcanic forma-

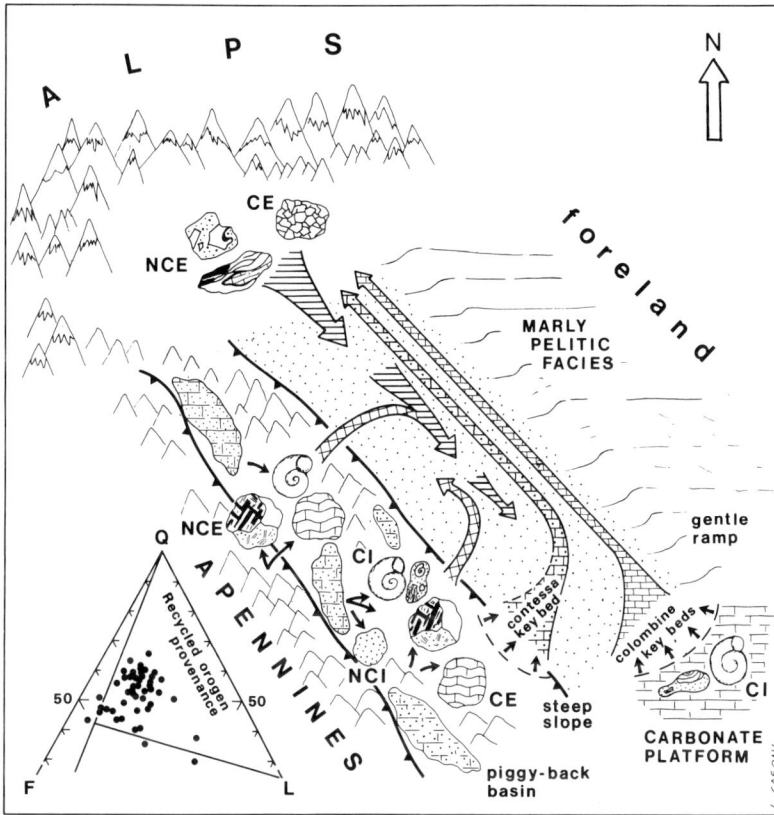

Fig. 2. Palaeogeographic sketch of the source/basin systems and sand dispersal pattern in the Marnoso–Arenacea basin (Miocene, northern Apennine): after Gandolfi *et al.* (1983) and Ricci Lucchi & Ori (1985), modified (not to scale). The triangle in the lower left is a QFL plot of 47 samples (fields after Dickinson 1985). Grain types: (NCE) non-carbonate extrabasinal; (CE) carbonate extrabasinal; (CI) carbonate intrabasinal; (NCI) non-carbonate intrabasinal.

tions and Quaternary volcanic rocks from the Izu–Bonin and Honsu arcs are present. Sediments accumulated in a proximal marine 'parking area' where early diagenetic processes were active in producing intrabasinal grains; they were later emplaced, via the Suruga Trough, by turbidity currents to the Nankai Trough more than 500 km to the southwest. This interpretation is also supported by the fact that most of the sediments, derived from the adjacent Nankai Trough-hinterland are ponded and trapped in the fore-arc region (Fig. 3).

In conclusion, (a) recognition of palaeovolcanic and neovolcanic particles allows the detection of the presence in the source area of old volcanic terranes together with products of active volcanism, (b) recognition of coeval intrabasinal grains, probably originated in relatively shallow waters, allows an understanding of some of the basin processes. The mere determination of the quartz–feldspar–lithic component would tells us simply that we are dealing with sands of 'magmatic arc' provenance (see QFL plot in Fig. 3).

The approach used for the Quaternary Nankai Trough sands would prove very useful in the case of an ancient terrane where the palaeotectonic setting is unknown. However, successful results depend, to a large extent, on the degree to which diagenetic processes have obscured the original features of palaeo- and neovolcanic grains.

The author is grateful to T. H. Nielsen, G. Shanmugam and W. Cavazza for critically reviewing this contribution.

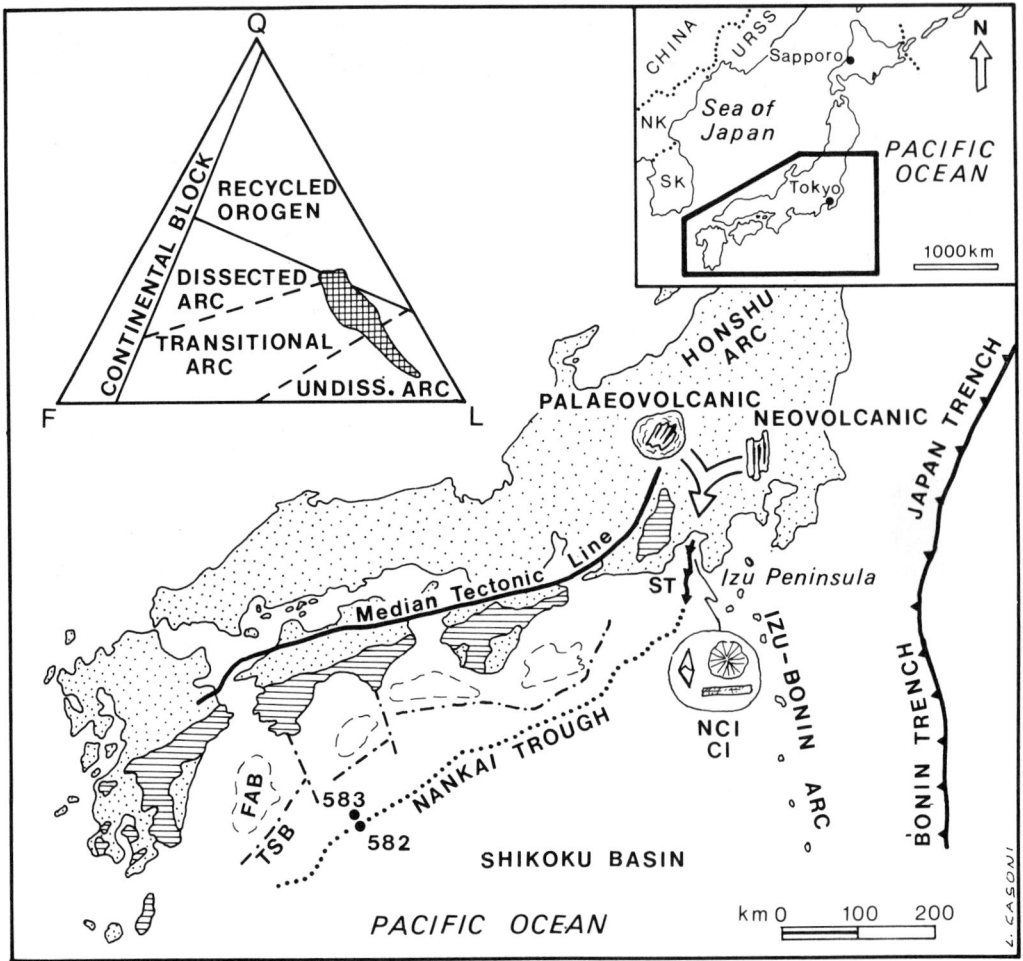

Fig. 3. Simplified regional setting of the Nankai Trough (southwest Japan) and location of DSDP sites 582 and 583. The dashed pattern onland is the Shimanto terrane; all other terrains are dotted. FAB: fore-arc basin; TSB: trench–slope break; ST: Suruga Trough; NCI: non-carbonate intrabasinal grains; CI: carbonate intrabasinal grains. Provenance categories in the triangular plot after Dickinson (1985). From De Rosa *et al.* (1986), modified.

References

BUSBY-SPERA, C. J. 1988. Evolution of a Middle Jurassic back-arc basin, Cedros Island, Baja California: Evidence from a marine volcaniclastic apron. *Geological Society of America Bulletin*, **100**, 218–233.

DE ROSA, R., ZUFFA, G. G., TAIRA, A. & LEGGETT, J. K. 1986. Petrography of trench sands from the Nankai Trough, southwest Japan: implications for long-distance turbidite transportation. *Geological Magazine*, **123**, 477–486.

DICKINSON, W. R. 1985. Interpreting provenance relations from detrital modes of sandstones. *In*: ZUFFA, G. G. (ed.) *Provenance of Arenites.*

NATO-ASI Series 148, D. Reidel, Dordrecht, 333–361.

FONTANA, D., GARZANTI, E., VALLONI, R. & ZUFFA, G. G. 1987. Evoluzione composizionale di un sistema di avanfossa: le torbiditi di Hecho (Eocene, Pirenei). *Riassunti Convegno della Società Geologica Italiana, Naxos-Pergusa, 22–25 Aprile 1987*, 54–55.

GANDOLFI, G., PAGANELLI, L. & ZUFFA, G. G. 1983. Petrology and dispersal pattern in the Marnoso-Arenacea Formation (Miocene, northern Apennines). *Journal of Sedimentary Petrology*, **53**, 493–507.

HELMOLD, K. P. 1985. Provenance of feldspathic sandstones—The effect of diagenesis on provenance interpretations: a review. *In*: ZUFFA, G. G. (ed.) *Provenance of Arenites*. NATO-ASI Series 148, D. Reidel, Dordrecht, 139–163.

LEES, A. & BULLER, A. T. 1972. Modern temperate-water and warm-water shelf carbonate sediments contrasted. *Marine Geology*, **13**, M67–M73.

McBRIDE, E. F. 1985. Diagenetic processes that affect provenance determinations in sandstones. *In*: ZUFFA, G. G. (ed.) *Provenance of Arenites*, NATO-ASI Series 148, D. Reidel, Dordrecht, 95–113.

MULLINS, H. T. 1983. Modern carbonate slopes and basins of the Bahamas. *Society of Economic Palaeontologists and Mineralogists Short Course No.* **12**, 4-1/4-138.

RICCI LUCCHI, F. 1985. Influence of transport processes and basin geometry on sand composition. *In*: ZUFFA, G. G. (ed.) *Provenance of Arenites*. NATO-ASI Series 148, 19–45.

—— & ORI, G. G. 1985. Field excursion D: Synorogenic deposits of a migrating basin system in the NW Adriatic Foreland: Examples from Emilia-Romagna region, northern Apennines. *International Symposium on Foreland Basins, Fribourg, Switzerland, 2–4 September 1985*, 137–176.

SCHOLLE, P. A. 1971. Sedimentology of fine-grained deep-water carbonate turbidites, Monte Antola Flysch (Upper Cretaceous), Northern Apennines, Italy. *Geological Society of America Bulletin*, **82**, 629–658.

VALLONI, R. 1985. Reading provenance from modern marine sands. *In*: ZUFFA, G. G. (ed.) *Provenance of Arenites*. NATO-ASI Series 148, D. Reidel, Dordrecht, 309–332.

ZUFFA, G. G. 1969. Arenarie e calcari arenacei miocenici di Vetto-Carpineti (Formazione di Bismantova, Appennino settentrionale). *Mineralogica et Petrographica Acta*, **15**, 191–212.

—— 1980. Hybrid arenites: their composition and classification. *Journal of Sedimentary Petrology*, **50**, 21–29.

—— 1985. Optical analyses of arenites: influence of methodology on compositional results. *In*: ZUFFA, G. G. (ed.) *Provenace of Arenites*. NATO-ASI Series 148, D. Reidel, Dordrecht, 165–189.

—— 1987. Unravelling hinterland and offshore palaeogeography from deep-water arenites. *In*: LEGGETT, J. K. & ZUFFA, G. G. (eds.), *Marine clastic sedimentology*. Graham & Trotman, London, 39–61.

Geochemical studies of detrital heavy minerals and their application to provenance research

ANDREW C. MORTON

British Geological Survey, Keyworth, Nottingham NG12 5GG, UK

Abstract: Although heavy mineral analysis is a sensitive and well-proven technique for determining the provenance of clastic sediments, the interpretation of the data is considerably enhanced by determining the composition of individual detrital grains. Many heavy mineral species, including pyroxene, amphibole, epidote, staurolite, garnet, tourmaline, monazite, chloritoid and spinel, show significant variations in composition that are related to the conditions under which their parent rocks were formed. Thus, as well as giving greater confidence in identification, geochemical analysis of detrital minerals adds precision to the evaluation of the relative contributions of potential source lithologies. Furthermore, by concentrating on stable minerals, geochemical studies avoid, or at least minimize, the problems caused by diagenetic and hydraulic processes.

An understanding of sand provenance is critical to the evaluation of clastic depositional systems, placing important constraints on transport, dispersal and depositional patterns that must be considered in the generation of sedimentological models, on both regional and local scales. In addition, variations in provenance within a depositional system provide a basis for correlation of siliciclastic sequences. This is of particular importance in non-marine or marginal marine sequences where biostratigraphic control may be poor, and when correlating marine and non-marine sequences.

Heavy mineral analysis has a long history of application to sand provenance evaluation. A large number of detrital mineral species with specific gravity greater than 2.80 have been recorded from sandstones and many of these are exceptionally source-diagnostic. However, identification of provenance from heavy mineral suites is often problematic, leading in many cases to erroneous conclusions. This stems from a variety of factors that alter the mineral assemblage from that present in the source rock to that identified under the microscope, namely:

(i) the changes made to source rock mineralogy by weathering at outcrop prior to incorporation into the transport system;
(ii) the changes caused by mechanical processes (abrasion) during transportation prior to final deposition;
(iii) the hydraulic behaviour of the depositional medium, reflected by bulk sediment granulometric parameters such as grain size and sorting;
(iv) the effects of diagenesis, caused by the

post-depositional percolation of pore fluids through the sediment;
(v) changes in mineralogy of the sediment by weathering at outcrop prior to sampling;
(vi) the variations in laboratory procedures, including chemical pretreatment, separation methods and mineral identification.

The effects of these processes and procedures have been widely documented previously and need not be repeated here: for further discussion, see Luepke (1984), Morton (1985a) and Mange & Maurer (1990). The purpose of this paper is to demonstrate how single-grain geochemical analysis of heavy minerals can not only aid identification of provenance but also add a considerable degree of sophistication to that determination. Single-grain analyses are now readily obtainable by electron microprobe, and most heavy mineral species show sufficient diversity in chemical composition for microprobe analysis to be used advantageously in their study.

In the first instance, microprobe analysis is useful in confirming optical identification. Despite the widespread availability of excellent identification manuals (e.g. Milner 1962; Parfenoff *et al.* 1970; Mange & Maurer 1990), every analyst, however experienced, realises that a degree of subjectivity is involved with many identifications. Microprobe analysis is a rapid and effective method of confirming optical identification, particularly of rare or problematic grains. A recent example of the use of the microprobe to confirm an optical identification is the lawsonite found by Mange-Rajetsky & Oberhänsli (1982) in Miocene Molasse sediments of Savoy, France. In an ongoing study of Palaeo-

From Morton, A. C., Todd, S. P. & Haughton, P. D. W. (eds), 1991,
Developments in Sedimentary Provenance Studies.
Geological Society Special Publication No. 57, pp. 31–45.

cene sandstone provenance of the northern North Sea, microprobe analysis of a greenish isotropic grain, tentatively identified optically as a spinel, proved it to be the extremely rare vanadium garnet goldmanite. In theory, such an identification should be extremely source-diagnostic: however, the only documented occurrences of this mineral are in the eastern USSR, China, Japan and New Mexico (Deer *et al.* 1982). It is unlikely that the grain is so long-travelled! The microprobe has also been used effectively in the identification of composite grains consisting of microcrystalline or crypto-crystalline aggregates of material unidentifiable by optical methods. In the peri-Alpine Molasse, such grains have been shown to consist of pumpellyite (Mange-Rajetsky & Oberhänsli 1986), an identification that is obviously of prime importance in provenance reconstruction.

However, the real value of the microprobe is to provide information on the geochemistry of mineral grains that is unobtainable by any other means. This aids provenance studies in two significant ways. First, it provides more sophisticated data that can be used to test provenance hypotheses. Secondly, and possibly more importantly, it provides a way to compare sand-stones with different transportation, deposition and diagenetic histories. Variation in the effects of these processes are minimized by comparing geochemical data, because the range of density and mechanical and chemical stability shown by a single mineral species is far less than that displayed by the entire heavy mineral assemblage.

A considerable number of heavy mineral species are potentially useful in provenance studies. This paper reviews the application of microprobe studies to a number of mineral species found as detrital components of sandstones, including pyroxene, amphibole, epidote, staurolite, garnet, spinel, chloritoid, monazite, tourmaline and zircon, using both new and previously published data.

Pyroxene

A large number of minerals belong to the pyroxene group. In most cases, these have distinctive optical properties: for example, the orthopyroxenes (enstatite–ferrosilite) and the sodic pyroxenes (aegirine, acmite) are readily distinguishable optically from the more common

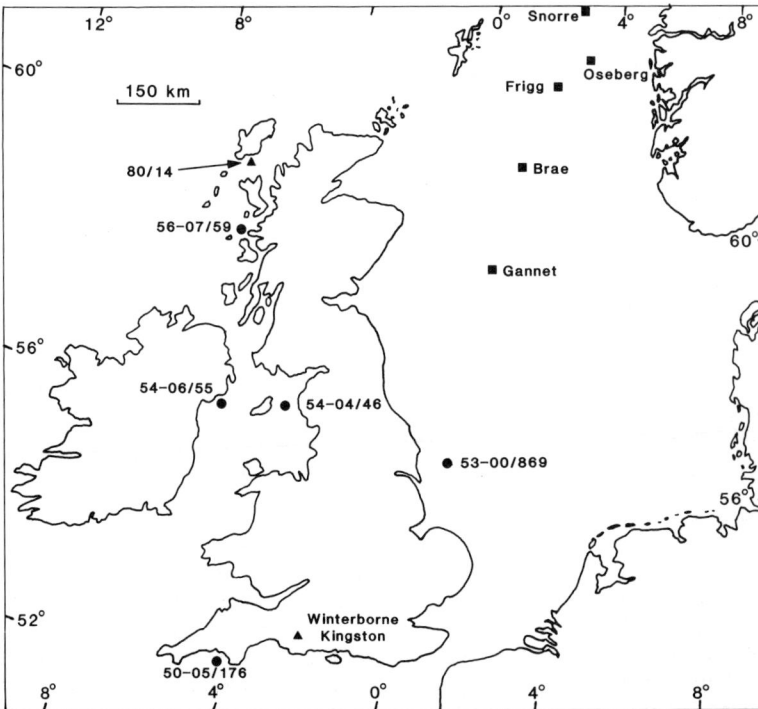

Fig. 1. Location of sample points discussed in this paper. Triangles, BGS Boreholes; circles, BGS seabed sampling stations; squares, North Sea oilfields.

clinopyroxenes. However, optical differentiation of clinopyroxene types is less easy, and microprobe analyses begin to play a significant role here. For example, clinopyroxenes in the Oligocene sediments of BGS Borehole 80/14, in the Minch (Fig. 1), are readily distinguishable from seabed sediment pyroxenes from the Sea of the Hebrides, Irish Sea and North Sea (Fig. 2), consisting of diopside–endiopside compared with salite–augite. The clinopyroxenes in 80/14 are associated with orthopyroxene, garnet, amphibole and epidote, an assemblage believed to represent detritus from Lewisian granulite and amphibolite facies rocks of the Outer Hebrides.

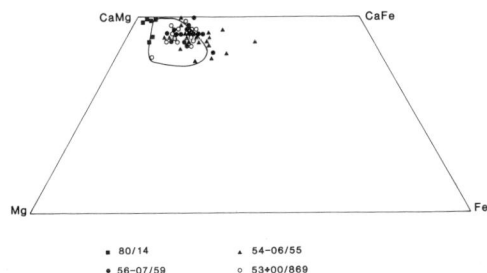

Fig. 2. Composition of pyroxenes from BGS Borehole 80/14 and three seabed sediment sites on the UK continental shelf, compared with the range of pyroxenes from the Portpatrick Formation (enclosed area) of the Southern Uplands of Scotland (Styles pers. comm. 1989).

Despite such examples, the great majority of detrital pyroxenes have an igneous parentage and fall, by and large, into the area in Fig. 2 occupied by the seabed pyroxenes. Determination and discrimination of the provenance of these pyroxenes is possibly by reference to Ca, Ti, Cr, Na and Al contents following the scheme of Letterrier et al. (1982). This was developed to assist the discrimination of tectonic setting of altered basaltic rocks. Figure 2 shows that the pyroxenes from the Ordovician Portpatrick Formation of the Southern Uplands (Styles et al. 1989) are comparable to the seabed pyroxenes described above. The two fields are not readily distinguishable, and apart from indicating a broadly igneous parentage, the pyroxene compositions tell little of the provenance of the two sedimentary units. By recourse to the discriminant plots of Leterrier et al. (1982), however, the two groups of pyroxene can be distinguished and their provenance determined. The Leterrier scheme is a three-stage process. In the first instance (Fig. 3a), tholeiitic and alkalic provenances are distinguished using Ti and (Ca + Na)

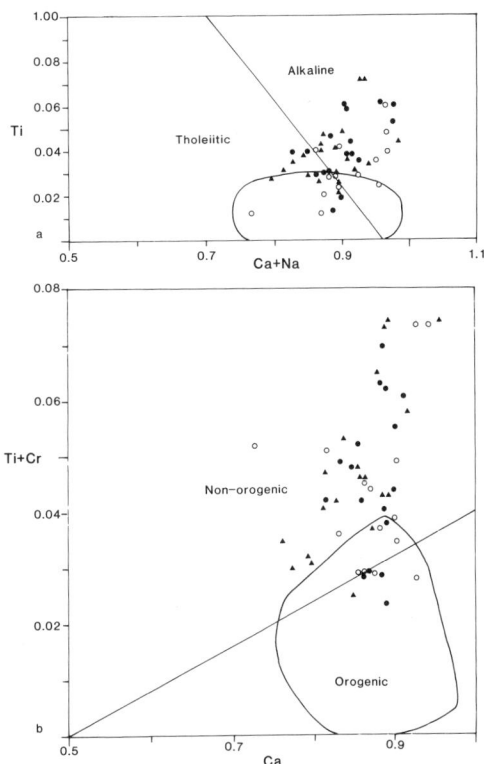

Fig. 3. Discrimination of provenance of UK continental shelf pyroxenes (same symbols as Fig. 2) from Portpatrick Formation pyroxenes (enclosed area) using the method of Leterrier et al. (1982). Data for the Portpatrick Formation are from Styles et al. (1989). (**a**) discriminates tholeiitic and alkaline sources, and (**b**) discriminates pyroxenes of orogenic and non-orogenic settings.

contents. In this plot, both sets of pyroxenes straddle the boundary but occupy largely different areas, with seabed pyroxenes having higher Ti than those of the Portpatrick Formation (Styles et al. 1989). This is followed by a (Ti + Cr)–Ca crossplot, which distinguishes pyroxenes from non-orogenic settings from those of orogenic source (Styles et al. 1989) whereas those from the seabed are typically non-orogenic. The third stage of the Leterrier scheme, not shown here, is a distinction of orogenic pyroxenes into calcalkaline and tholeiitic suites, which shows that the Portpatrick pyroxenes have calcalkaline affinities (Styles et al. 1989). Evidently, Portpatrick Formation pyroxenes were derived from a calcalkaline volcanic arc terrane, whereas the seabed pyroxenes are from a non-orogenic source that includes both tholeiitic and alkalic basalts. The geographic location of the samples (Fig. 1) indicates that this source is most likely to be the British Tertiary Volcanic Province. There is no evidence supporting an Ordovician source for the seabed pyroxenes. Neither is there any evidence to suggest that the North Sea pyroxenes were derived from a different area from those of the Sea of the Hebrides and Irish Sea.

Another approach to pyroxene provenance determination is to use the abundances of the trace elements Ti, Na and Mn (Nisbet & Pearce 1977). As with the Leterrier scheme, the method was devised to allow tectonic setting of altered basalts to be assessed, but it has been successfully applied in provenance studies. Cawood (1983) showed that individual formations in a Palaeozoic clastic succession from eastern Australia can be readily distinguished (Fig. 4) and that volcanic arc sources dominated the sequence throughout.

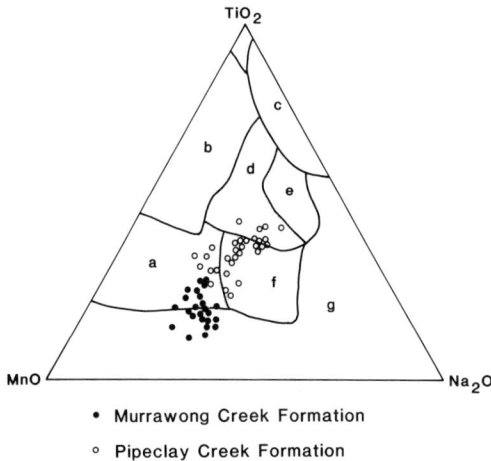

• Murrawong Creek Formation

○ Pipeclay Creek Formation

Fig. 4. Discrimination of provenance of detrital pyroxenes from two formations belonging to a Palaeozoic clastic sequence in eastern Australia, using the plot described by Nisbet & Pearce (1977). Ocean floor basalts plot in fields b and d. Volcanic arc basalts plot in fields a, d, e and f. Within-plate alkalic basalts plot in fields c, d, f and g. Within-plate tholeiites plot in fields d and e.

The use of pyroxene as a provenance indicator is severely limited by its stability. It is rapidly corroded and dissolved at comparatively low porefluid temperatures and thus rarely appears in heavy mineral assemblages in porous sandstones at even modest burial depth (Morton 1984a). Indeed, many pyroxene grains in the seabed sediment samples described above are etched, presumably through dissolution at ambient temperatures at the sea floor. However, in sandstones with very low porosity, pyroxene is frequently preserved, as in the two cases of Palaeozoic clastic successions cited above.

Amphibole

Amphibole geochemistry is exceptionally complex, with the group showing a very large range in composition (see Leake 1978 for discussion). To an extent, the chemical variations are mirrored by changes in optical properties; for example, alkali amphiboles such as glaucophane and riebeckite are readily distinguishable optically from calcic amphiboles such as edenite and pargasite. Thus, in studies of detrital sediments, all green calcic amphiboles are generally termed 'hornblende' and blue sodic amphiboles 'glaucophane'. Application of microprobe techniques adds precision to such identifications. Furthermore, microprobe analysis of detrital amphiboles has considerable potential in provenance studies, because the wide range in possible compositions reflects a wide variety of parageneses.

Both calcic and alkali detrital amphiboles have been studied by microprobe methods. Morton (1984b) differentiated two calcic amphibole suites in early Palaeogene sediments from the SW Rockall Plateau area, one comprising actinolite, actinolitic hornblende and magnesio-hornblende, the other edenite and pargasite (Fig. 5). The former is associated with epidote group minerals including piemontite, and is believed to have a southern Greenland provenance, whereas the latter is associated with clinopyroxene, garnet and apatite and was derived from Precambrian basement of the Rockall Plateau. The composition of alkali amphiboles from Oligo-Miocene Molasse sediments of Savoy, France, has been studied by Mange-Rajetzky & Oberhänsli (1982), who showed that they comprise glaucophane, ferroglaucophane and crossite. A general progression toward less aluminous compositions occurs with time from the Chattian to the Burdigalian (Fig. 6), reflecting changing metamorphic conditions in the Western Alps source area. Amphibole geochemistry has also been used to determine the provenance of archaeological material. For example, the sand found in Iron Age pottery of probable Roman origin from SE England contains two amphibole populations, one with low Na and Ti, the other with high Na and Ti (Fig. 7). This indicates that both volcanic and plutonic/metamorphic rocks were present in the source area, and consideration of the most likely sources of Roman pottery led Freestone & Middleton (1987) to conclude that the pottery originated in the Massif Central of France.

As with pyroxene, the limiting factor in the use of amphibole is its instability even in relatively low temperature porefluids (Morton 1984a), causing it to be lost from porous sand-

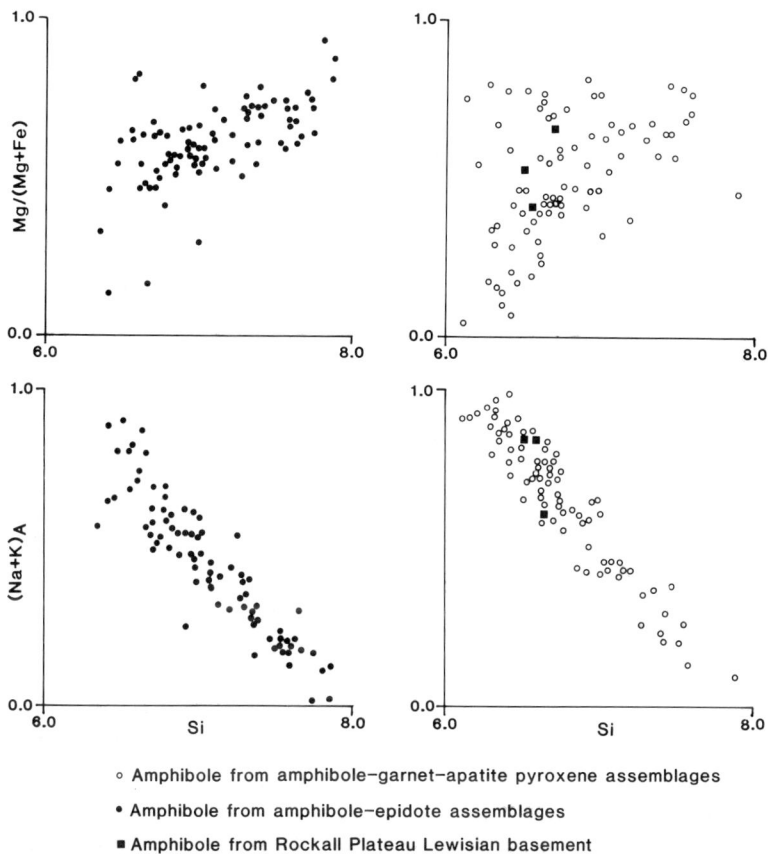

○ Amphibole from amphibole–garnet–apatite pyroxene assemblages

• Amphibole from amphibole–epidote assemblages

■ Amphibole from Rockall Plateau Lewisian basement

Fig. 5. Discrimination of two detrital calcic amphibole populations from DSDP Leg 81 sediments from SW of the Rockall Plateau, NE Atlantic. Composition of Rockall Bank amphiboles are shown for comparison. Data from Morton (1984b). The amphiboles from amphibole–epidote assemblages (believed to have a SE Greenland provenance) are markedly more Si- and Mg-rich and lower in alkalis than those from amphibole–garnet–apatite ± pyroxene assemblages (believed to have a Rockall Bank provenance).

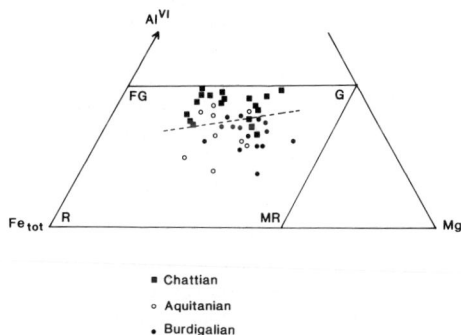

■ Chattian
○ Aquitanian
• Burdigalian

Fig. 6. Variation in composition of detrital sodic amphiboles from Molasse sediments of Savoy, France, from data of Mange-Rajetzky & Oberhänsli (1982). R, riebeckite; MR, magnesio-riebeckite; G, glaucophane; FG, ferroglaucophane.

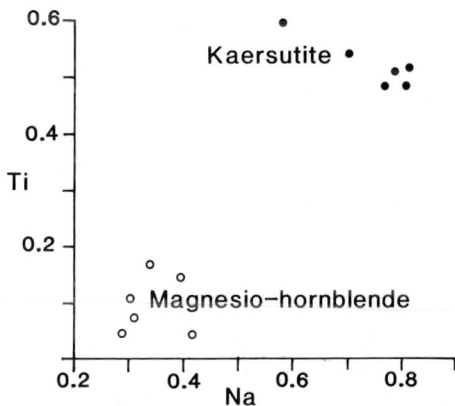

Fig. 7. Variation in Ti and Na contents of amphiboles in Iron Age pottery from SE England (from Freestone & Middleton 1987).

stones by intrastratal solution at shallow burial depths. However, in ancient low porosity sandstones, amphibole may be preserved, as in the Lower Palaeozoic sediments of the Southern Uplands (Styles *et al.* 1989); its study in such cases would undoubtedly add increased sophistication to provenance models.

Epidote

The epidote group minerals (zoisite, clinozoisite, epidote, allanite and piemontite) show a wide compositional range, but to a large extent these are manifested by optical variations that readily permit categorisation of grains. Possibly for this reason, little attention has been paid to the composition of detrital epidotes. Epidote stability is also a problem. Although it is more stable than amphibole, epidote nevertheless lies low in the relative order of mineral stability (Morton 1984*a*). Variations in epidote compositions in Bengal Fan sediments have been described by Yokoyama *et al.* (in press). These sediments were largely derived from the Himalayas, and variations in geochemistry are relatively limited. However, one zone containing abundant zoisite occurs in early Miocene sediments, and another zone has a more limited range of $Fe/(Fe + Al)$ than epidotes in other zones (Fig. 8). These variations relate both to heterogeneities in the Himalayan source area and to subsidiary input from the Indian subcontinent.

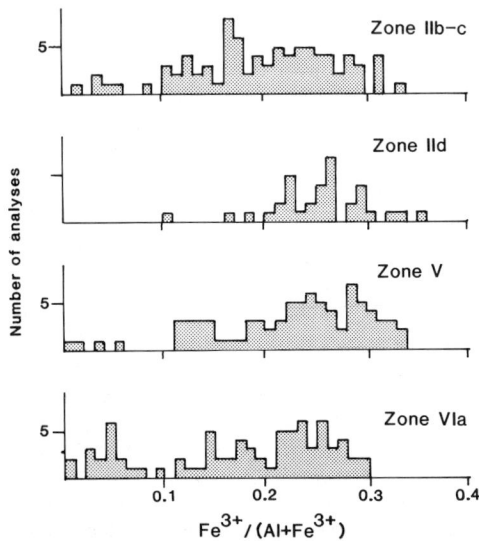

Fig. 8. Composition of detrital epidotes in Neogene sediments from ODP Leg 116 sites on the Bengal Fan. Data, including zonation, from Yokoyama *et al.* (in press).

Staurolite

Staurolite shows a relatively limited amount of compositional variation, but Kepezhinskas & Koryluk (1973) suggest that 'ferruginosity' $(Fe^{2+} + Mn + Fe^{3+}/Fe^{2+} + Mn + Fe^{3+} + Mg)$ is controlled by metamorphic grade. Investigations of detrital staurolites in North Sea sediments (Fig. 9) demonstrate that there are regional and stratigraphic variations in composition related to provenance. Middle Jurassic Tarbert Formation (Brent Group) staurolite from the Oseberg Field (Fig. 1) has a normal distribution of ferruginosity, with a mode between 89 and 91. This contrasts with Lower Eocene sandstones, which have bimodal distributions. The patterns in the Gannet and Brae areas, central North Sea (Fig. 1) are similar, with modal ferruginosity values between 89 and 90 and 91 to 93. The Eocene of the Frigg area (northern Viking Graben) also has staurolites grouping at 91–92, but has a more significant grouping between 87 and 88. The more northern sand body also differs from the area to the south in containing detrital chloritoid. These variations indicate that at least two distinct sources supplied sands to the North Sea Basin during the early Eocene, and that the staurolites in the Middle Jurassic of the northern North Sea had a different source again.

Substitution of Co and Zn also occurs in staurolite, and recognition of detrital cobalt- or zinc-rich staurolites would be particularly source-diagnostic. For example, zinc-rich staurolites with *c.* 2.5% ZnO occur in Dalradian metasediments of NE Scotland (Ashworth 1975). Similar Zn-rich staurolites have been identified in the Lower Eocene of the adjacent North Sea Basin, although they are considerably less abundant than Zn-poor varieties. This indicates that one of the main sources of the Lower Eocene sands was the Grampian area.

Staurolite is relatively stable during the early stages of diagenesis, but suffers dissolution in high temperature porefluids. In porous North Sea Palaeocene sandstones, total dissolution of staurolite occurs at about 2.4 km burial depth (Morton 1984*a*).

Garnet

Compositional variations shown by detrital garnet have proved particularly useful in the study of North Sea sandstones, derived essentially from metamorphic basement sources, either directly or through successive recycling episodes. Many different detrital garnet suites have been

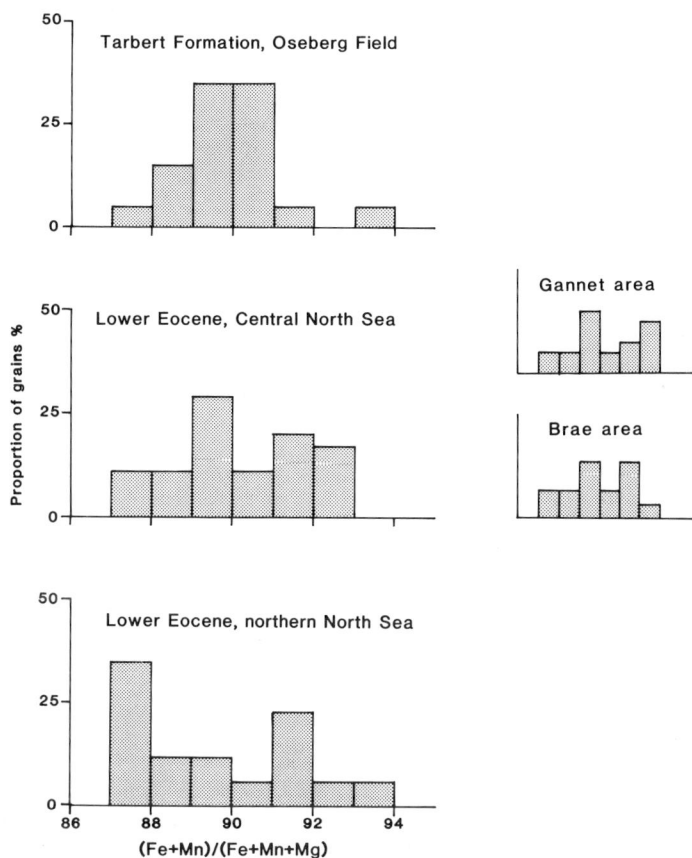

Fig. 9. Comparison of staurolite compositions from North Sea sandstones.

identified in North Sea sandstones (Fig. 10), allowing increased sophistication in studies of provenance and dispersal of important hydrocarbon reservoir sequences such as the Middle Jurassic Brent Group (Morton 1985b, 1987a; Hurst & Morton 1988; Morton et al. 1989) and the Palaeocene Forties Formation (Morton 1987b). Garnet geochemistry has been used to evaluate provenance elsewhere, notably of Tertiary sediments offshore New Zealand (Smale & Morton 1988) and in the Bengal Fan (Yokoyama et al. in press), Jurassic sandstones offshore mid-Norway (Gjelberg et al. 1987), Triassic sandstones of southern Germany (Borg 1986) and Devonian sandstones of the Midland Valley of Scotland (Haughton & Farrow 1989). Garnets of different parageneses occupy different compositional fields (Wright 1938), as shown in Fig. 11, and so identification of source type is possible, although mixing from different sources, either because of source area complexity or because of recycling and diagenesis, may intro-

duce considerable complications. For example, the assemblage in Fig. 10a is likely to have been sourced by amphibolite-facies rocks (Fig. 11), but the assemblage in Fig. 10e probably represents a mixture of source types.

Garnet is relatively stable in diagenesis, although North Sea sandstones buried to greater than 3–3.5 km are generally garnet-free due to dissolution by high-temperature porefluids. Because of the chemical variations shown by the group, different garnets have different stabilities, with low-Ca garnets being more stable than high-Ca garnets (Morton 1987b). Thus sandstones with similar garnet provenance could now show substantial variations in garnet geochemistry as a result of regional variations in diagenetic history. Therefore, diagenetic history must be considered when assessing provenance using garnet geochemical methods. Relative garnet stability may be the reason for the somewhat enigmatic recurrence of assemblages dominated by low-Ca garnets in North Sea sandstones (Fig.

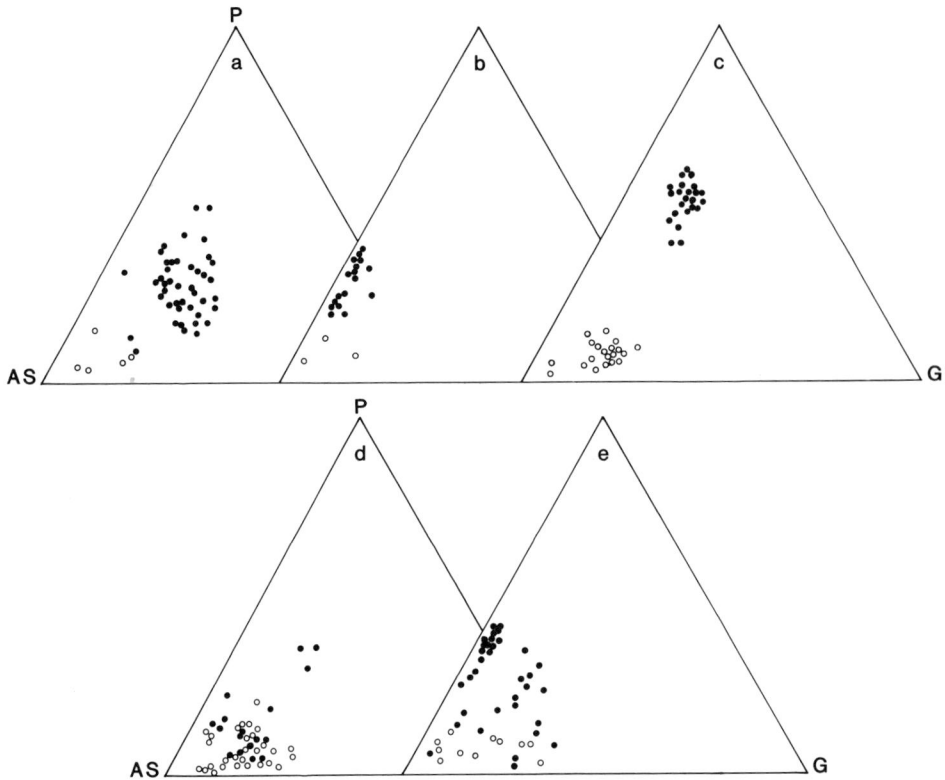

Fig. 10. Illustration of the variety of compositions shown by detrital garnets of North Sea sediments.
(**a**) Oseberg Formation (Middle Jurassic), Oseberg Field (from Hurst & Morton 1988). (**b**) Broom Formation
(Middle Jurassic), Murchison Field (from Morton 1985*b*). (**c**) Etive Formation (Middle Jurassic), Murchison
Field (from Morton 1985*b*). (**d**) Ness Formation (Middle Jurassic), Oseberg Field (from Hurst & Morton
1988). (**e**) Forties formation (Palaeocene), Forties Field (from Morton 1987*b*). AS, almandine + spessartine;
P, pyrope; g, grossular. Open circles have spessartine > 5%, closed circles have spessartine < 5%.

9*b*), as seen, for example, in Triassic and Upper
Jurassic sandstones of the central North Sea
(Morton 1987*a*), Broom Formation sandstones
of the northern North Sea (Morton 1985*b*) and
Palaeocene sandstones of the Viking Graben
(Morton & Hallsworth, unpublished data). Garnets such as these have few counterparts in the
compilation of Wright (1938), as shown in Fig.
10. Deer *et al.* (1982) identify some garnets of
this type (table 55, analyses 23, 24, 27, 28 and
30), found in granulites, gneisses and migmatites, and it is likely that such rocks (probably
from the Lewisian basement) provided these
garnets in the North Sea area. However, the
dominance of such low-Ca garnets in North Sea
sediments probably reflects the relative instability of high-Ca garnet: thus, assemblages dominated by low-Ca garnets may indicate recycling
of older sandstones that have been subjected to a
phase of diagenesis which has removed the less
stable garnets.

Spinel

Spinel group minerals have a considerable range
in optical properties, to the extent of including
opaque varieties, reflecting a wide range in composition. This account deals only with the translucent heavy minerals, although geochemical
studies of the opaques also have considerable
provenance potential (Darby & Tsang 1987;
Basu & Molinari this volume). Microprobe
analysis of the translucent spinels is of value in
identifying spinel types; for example, grains of
blue-green spinel in North Sea Jurassic sandstones, identified optically as ceylonite or pleo-
naste (an Mg-Fe spinel) have been shown to the
the zinc spinel gahnite. However, most attention
has been paid to the geochemistry of detrital
chrome spinels. The geotectonic significance of
chrome spinels in sediments was first discussed
by Zimmerle (1984) and, as Press (1986) discussed, their geochemical characteristics are use-

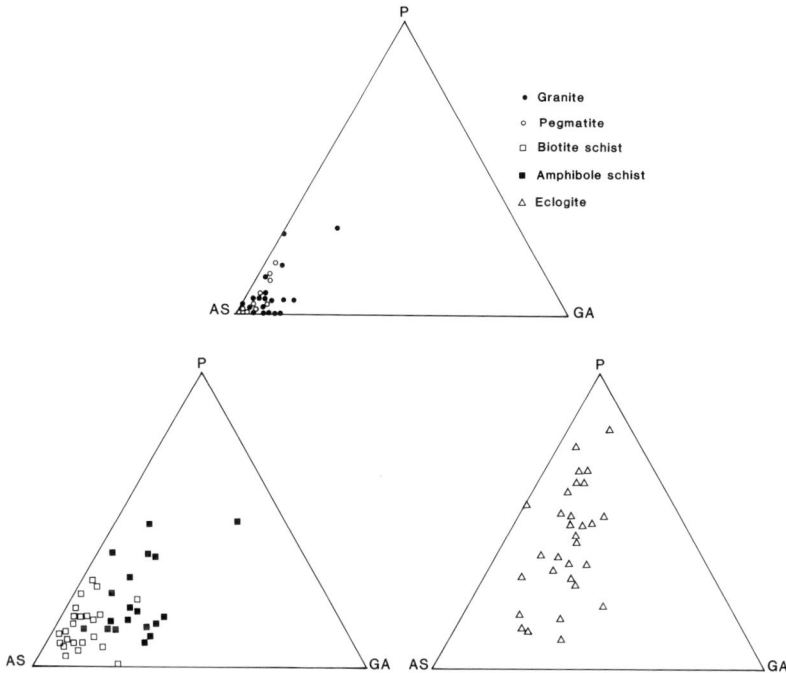

Fig. 11. Variation in composition of garnets from different source lithologies, from Wright (1938). AS, almandine + spessartine; P, pyrope; GA, grossular + andradite.

ful in diagnosing the tectonic setting of their mafic or ultramafic source. To exemplify this, Press (1986) examined detrital chrome spinels from Middle Devonian sediments of the Rhenish Massif (Fig. 12), and considered that they were derived from an alpinotype ophiolite complex, rather than a mid-ocean ridge or a stratiform intrusion.

Although little detailed data is available on stability of detrital spinels, studies of North Sea Jurassic reservoir rocks have shown that both the chromite and gahnite varieties are present to great depth (at least 4 km) without displaying any evidence of corrosion. However, corroded gahnite spinels have been observed in zones subjected to flushing by low-pH groundwaters (Morton & Humphreys 1983).

Chloritoid

Chloritoid displays major variations in the abundance of Mg, Mn and Fe, and can therefore be used to test provenance models. The Lower Jurassic Bridport Sands of southern England are unusual in containing a small but significant proportion of chloritoid (Boswell 1924; Davies 1969; Morton 1982). The provenance of the Bridport Sands is enigmatic. The nearest outcrop of chloritoid-bearing rocks is in the Ile de Groix (Brittany), and thus this area has been frequently regarded as the source (Boswell 1924; Davies 1969) although this diachronous sand unit becomes younger from north to south (Knox *et al.* 1982). Chloritoids from the Bridport Sands of the Winterborne Kingston Borehole (Fig. 1) compare favourably with those of the Ile de Groix (Fig. 13), with a similar compositional range and clustering at similar MnO values. Another alternative, the London–Brabant Massif, is readily ruled out (Fig. 13), as chloritoids of the Ottre area are distinctly different, with very high MnO values (Fransolet 1978). For comparison, Fig. 13 also shows chloritoid anlayses from Lower Eocene sandstones of the Frigg area (northern North Sea), derived from the Shetland region. These have distinctly higher MnO values compared to Bridport Sands chloritoid.

As with the spinel group, little detail is available on stability of detrital chloritoid, although North Sea Jurassic sandstones buried in excess of 4 km contain chloritoid grains that lack any evidence of corrosion. In contrast, corroded chloritoid has been observed in zones subjected to flushing by low pH groundwaters (Morton & Humphreys, 1983).

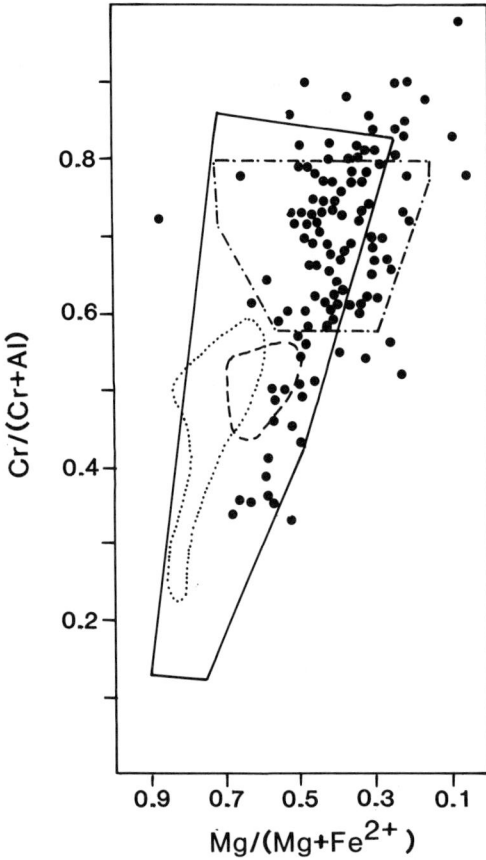

Fig. 12. Composition of detrital chromite from
Middle Devonian sediments of the Rhenish Massif,
from Press (1986). The data are shown compared
with spinels from alpinotype ophiolite complexes
(area enclosed by solid line), those from stratiform
intrusions (area enclosed by dashed-dotted line),
those from mid-ocean ridge basalts from the mid-
Atlantic Ridge (area enclosed by dotted line) and
those from mid-ocean ridge basalts from the
FAMOUS area (area enclosed by dashed line).

Monazite

Monazite contains the light rare earth elements
(REE) La, Ce, Pr, Nd, Sm and Gd in sufficient
abundance for detection by energy-dispersive
microprobe analysis. Eu is close to detection
limit and the heavier REEs can only be detected
using the wavelength-dispersive technique. The
degree of light REE enrichment, as exemplified
by the ratio La/Nd, is provenance-sensitive
(Fleischer & Altschuler 1969). La/Nd values of
monazites from granites and granite pegmatites
are low (means of 1.23 and 1.06 respectively),
whereas those from alkalic igneous rocks and

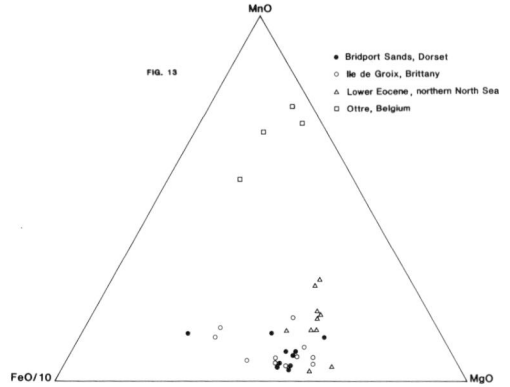

Fig. 13. Composition of detrital chloritoids from the
Bridport Sands of the Winterborne Kingston Borehole, Dorset, compared with chloritoids from the Ile
de Groix, Brittany (data of Makanjuola & Howie
1972). Also shown are chloritoids from the Ottre
area of Belgium and Lower Eocene sediments of the
Frigg area, northern North Sea.

carbonatites are high, averaging 3.11. The
Lower Jurassic Statfjord Formation of the
Snorre Field (northern North Sea, Fig. 1) contains some sandstone units rich in monazite.
Microprobe studies of these show them to be
only moderately light REE enriched (Fig. 14),
with a mean La/Nd ratio of 1.09. Comparison
with the data of Fleischer & Altschuler (1969)
suggests that these monazites were derived from
granite pegmatites or granites.

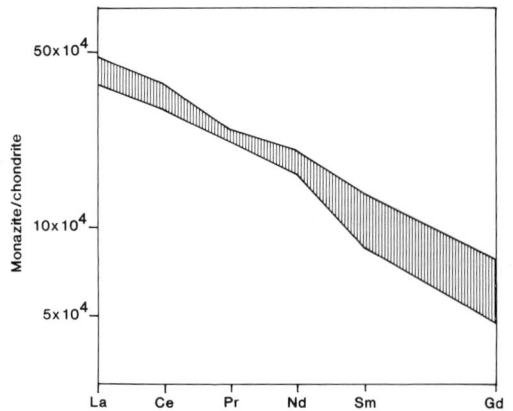

Fig. 14. Chondrite-normalized rare earth element
plot showing range in composition of detrital monazites from Statfjord Formation sandstones of the
Snorre Field. The pattern is characteristic of a
granitic or pegmatitic source, rather than an alkalic
source.

Monazite stability in sediments is not known in detail, but there is evidence for its dissolution in Jurassic sandstones of the North Sea and mid-Norwegian shelf (Milodowski & Hurst in press). In these sandstones secondary REE phosphates are also present as rhabdophane, and elsewhere secondary monazites have been observed, as in the Lower Palaeozoic of central Wales (Milodowski & Zalasiewicz this volume). There is, therefore, considerable evidence to suggest that monazite can suffer dissolution, with the REE mobilized and subsequently precipitated, in some cases as authigenic monazite.

Tourmaline

Tourmaline chemistry is relatively complex, so much so that its basic formula was uncertain until fairly recently. Although there are a large number of end-member compositions (see Deer *et al.* 1986), most of the variations can be described in terms of the abundance of Fe and Mg. Figure 15 shows the fields of the most common tourmaline types (dravite, uvite, schorl and elbaite) on a binary plot of FeO vs. MgO, using the analyses of Deer *et al.* (1986). Dravite and uvite are typically metamorphic or metasomatic in origin, whereas schorl and elbaite have granitic or pegmatitic parageneses.

Tourmaline varieties have been used in the interpretation of provenance for a considerable length of time, but this has relied heavily on their optical properties (e.g. Krynine 1946), and little attention has been paid to geochemical characteristics. The potential value of this approach is illustrated in Fig. 15, which shows the compositions of tourmalines from sea bed sediments

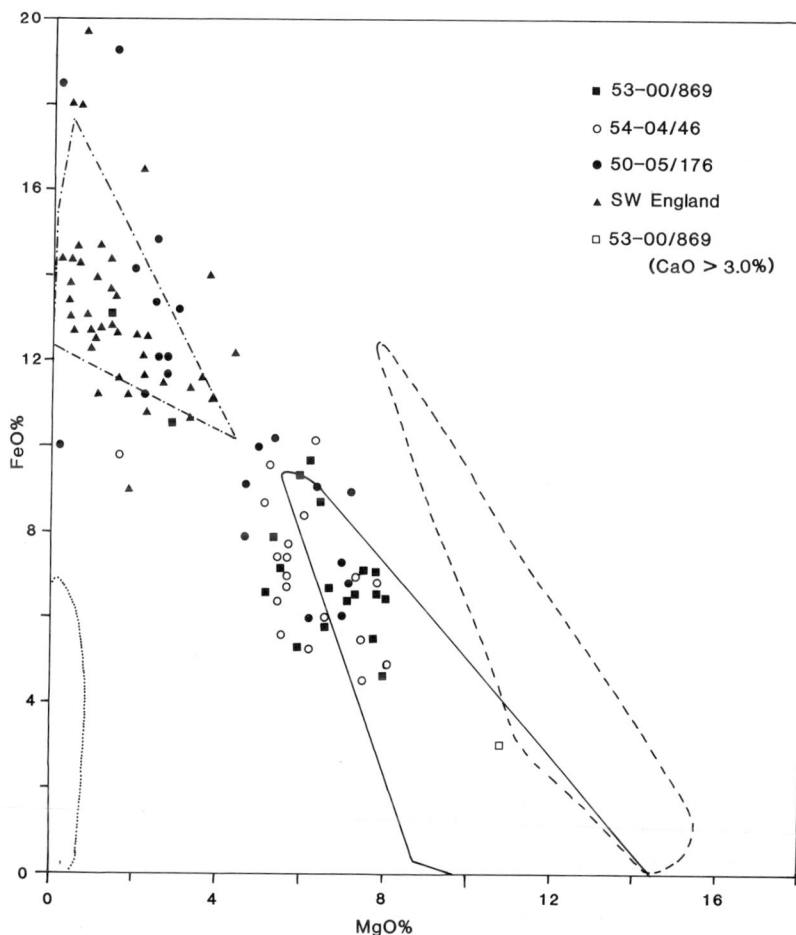

Fig. 15. Variation of FeO and MgO contents of detrital tourmalines from seabed sediments on the UK continental shelf, compared with Cornubian tourmalines (Power 1968) and ranges of dravite (solid line), elbaite (dotted line), schorl (dotted-dashed line) and uvite (dashed line), from analyses of Deer *et al.* (1986).

from three sample stations on the UK continental shelf (Fig. 1), 50-05/176 (offshore Cornwall), 54-04/46 (eastern Irish Sea) and 53-00/869 (southern North Sea). The sample from offshore Cornwall has a large schorl population, with a lesser number of dravites; the compositions of the schorl matches the compositional range of Cornubian tourmalines analysed by Power (1968). As might be predicted, therefore, many of the tourmalines in this sample have a local Cornubian provenance. In contrast, the other two samples are largely composed of dravite, and both have similar compositional ranges. These tourmalines are probably of metamorphic or metasomatic origin, ultimately from basement rocks of the Scottish landmass although the degree of recycling is uncertain.

The potential value of tourmaline compositional data in provenance studies is high, not least because of its great stability; it is considered to be one of the so-called 'ultrastable' heavy minerals. Tourmaline geochemistry could be particularly powerful when used in combination with optically-based varietal studies.

Zircon

Zircon, like tourmaline, is another mineral whose optical characteristics are used to identify provenance. Unlike tourmaline, however, zircon shows only limited chemical variation. Zircon is the Zr end-member of the zircon-hafnon solid solution series, but in general Hf substitution is minor: mean Zr/Hf in zircon is 40, with the normal Hf content between 0.6% and 3.0% by weight (Ahrens & Erlank 1969). There is evidence that Zr/Hf in zircons varies in different igneous rock types (Butler & Thompson 1965). Accurate determination of such trace amounts of Hf cannot be acheived using energy-dispersive analytical systems, and thus wavelength-dispersive techniques must be employed. Owen (1987) tested the hypothesis that the upper Jackford Sandstone of Arkansas and the Parkwood Formation of Alabama (both mid-Carboniferous) were derived from the same source, by comparing the Hf contents of detrital zircon populations (Fig. 13). Mean Hf is 1.37% in the Parkwood and 1.34% in the upper Jackford (Owen 1987), supporting the same-source hypothesis. Although the upper Jackford zircon population appears to have a bimodal Hf distribution, unlike that of the Parkwood (Fig. 16), statistical tests indicate that the difference between the populations is insignificant (Owen 1987). Other trace elements such as Ta, Th, U, W and Sr are also present in zircon; determination of such

elements in detrital zircons from Oligocene sediments from the Lough Neagh Group of Northern Ireland has greatly aided the understanding of the provenance of these sediments (Shukla 1988). The potential of this approach is particularly significant as zircon is a ubiquitous and extremely stable mineral in sandstones.

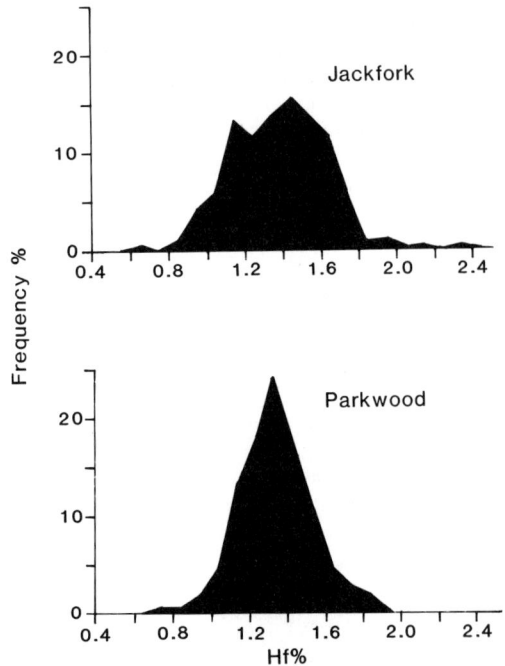

Fig. 16. Variations in hafnium content of detrital zircons from mid-Carboniferous sandstones in Arkansas and Alabama, USA, from data of Owen (1987).

Conclusions

This paper has shown that the microprobe analysis of detrital heavy minerals adds considerable sophistication to provenance determinations. The microprobe provides, firstly, confirmation of optical identifications, and secondly, greater accuracy in identifying source lithologies. In some cases, identification of the tectonic setting of the provenance, with consequent implications for the nature of the depositional basin, is possible.

A large number of detrital minerals can be used in this way; examples to date include pyroxene, amphibole, epidote, staurolite, garnet, spinel, chloritoid, monazite, tourmaline and zircon. Other minerals with potential include apatite, rutile, titanite and olivine. Variations of REE and Y content in detrital apatite and titanite are worthy of examination; as Fleischer & Altschuler (1969) discuss, both of these minerals show variations both in total REE content and in relative REE abundance that are petrogenetically significant. Furthermore, apatite is a stable mineral except in low pH groundwaters, where it suffers rapid dissolution (Morton 1984a, 1986). The value of geochemical studies of titanite is likely to be more limited as titanite is relatively unstable and suffers dissolution at relatively low porefluid temperatures (Morton 1984a). In rutile, Nb and Ta may substitute for Ti. Thus, investigations of the abundance of these trace elements in detrital rutile may well prove useful for provenance analysis, especially as rutile is one of the 'ultrastable' minerals. Variations in Fe/Mg ratio in olivine are in many cases related to paragenesis (Deer et al. 1984), and are thus source-diagnostic. However, olivine is even less stable in sediments than pyroxenes; consequently, the value of olivine in provenance studies is likely to be strictly limited.

The geochemical approach described herein clearly plays a very important role in the determination of the ultimate provenance of the analysed grain. It cannot, however, directly indicate the degree to which that grain has been recycled. For example, the ZnO-rich staurolites described earlier from the Lower Eocene of the North Sea almost certainly had their ultimate provenance in the Grampian Highlands, but it is not immediately apparent whether they have been reworked from earlier sediments such as the Permo-Trias. Recycling provides a possible explanation for the presence of southerly-derived chloritoid in the southward-younging Bridport Sands of southern England: the chloritoid may have been transported northward from the Ile de Groix in a previous phase of sedimentation (perhaps the Permo-Trias). These sediments could then have been recycled to generate the Bridport Sands. Many heavy mineral species, notably those that have high chemical stabilities such as zircon, rutile and tourmaline, are most commonly polycyclic, and at first sight this appears to be an important limiting factor in the application of the geochemical technique. However, in many cases it is possible to identify the sandstones that were potentially involved in the recycling process. By analysing and comparing the compositional ranges of one or more mineral species from the sediment and its potential sedimentary precursors, this technique clearly provides a very powerful tool for assessing the recycling paths of sandstones, for distinguishing sand bodies and formations, for evaluating the degree of mixing from different sources and for identifying fresh sediment influx.

With the recent acquisition of considerable amounts of geochemical data on detrital mineral grains, it is rapidly becoming apparent that suitable data sets are not available from the ultimate source regions, i.e. the basement terranes. This is exemplified by the discovery of detrital goldmanite in the North Sea Palaeocene: this must have been derived from sources in NW Europe, although it has not yet been recorded in basement rocks of the area. It is vital, for accurate provenance reconstruction, that during evaluation of any particular sedimentary unit, comparative data are acquired from the potential basement regions. This point has already been made in connection with possible sedimentary precursors to the interval in question. Despite the importance of this comparative process, it should always be recognised that direct comparisons are not possible, simply because the process of sedimentation has irrevocably altered the geology of the source.

In conclusion, therefore, a large number of detrital heavy mineral species have sufficiently wide geochemical variations to repay microprobe study. Consequently, it is recommended that the microprobe plays a major role in future heavy mineral studies. The advantages not only include the confirmation of optical identifications and the identification of rare minerals, but also the determination of the compositional range of a wide variety of more common heavy minerals. By comparing these data with those from basement rocks and possible sedimentary precursors, the technique not only offers an accurate and objective method of determining the nature of the ultimate source of the sediment, but also allows us to assess the recycling paths of sandstones, to distinguish sand bodies and formations, to evaluate the degree of mixing from different sources and to identify influx of fresh sediment. The method also helps us to overcome two of the most serious difficulties with heavy mineral analysis, the comparison of suites from sandstones that were deposited under different hydraulic regimes and the comparison of suites from sandstones that have suffered variable degrees of diagenetic alteration.

I am grateful to many people for their help and encouragement in the process that has led up to the publication of this paper, both during the research and in the preparation of the manuscript. However, I am especially indebted to C. Hallsworth and M. Styles for their help with microprobe analysis, and to M. Mange, P. Cawood and S. Todd for their thoughtful comments on the text. The paper is published with the approval of the Director, British Geological Survey (NERC).

References

AHRENS, L. H. & ERLANK, A. J. 1969. *In*: WEDEPOHL, K. H. (ed.) *Handbook of Geochemistry*. Springer-Verlag, New York, Volume II/5, sections B–O.

ASHWORTH, J. R. 1975. Staurolite at anomalously high grade. *Contributions to Mineralogy and Petrology* **53**, 281–291.

BASU, A. & MOLINAROLI, E. 1990. Provenance of detrital Fe–Ti oxides. This volume.

BORG, G. 1986. Facetted garnets formed by etching. Examples from sandstones of Late Triassic age, South Germany. *Sedimentology*, **33**, 141–146.

BOSWELL, P. G. H. 1924. The petrography of the sands of the Upper Lias and Lower Inferior Oolite in the west of England. *Geological Magazine*, **31**, 246–264.

BUTLER, J. R. & THOMPSON, A. J. 1965. Zirconium: hafnium ratio in some igneous rocks. *Geochimica et Cosmochimica Acta*, **29**, 167–175.

CAWOOD, P. A. 1983. Modal composition and detrital clinopyroxene geochemistry of lithic sandstones from the New England Fold Belt (east Australia): a Paleozoic forearc terrane. *Geological Society of America Bulletin*, **94**, 1199–1214.

DARBY, D. A. & TSANG, Y. W. 1987. Variation in ilmenite element composition within and among drainage basins: implications for provenance. *Journal of Sedimentary Petrology*, **57**, 831–838.

DAVIES, D. K. 1969. Shelf sedimentation: an example from the Jurassic of Britain. *Journal of Sedimentary Petrology*, **37**, 1344–1370.

DEER, W. A., HOWIE, R. A. & ZUSSMAN, J. 1982. *Rock-forming minerals, volume 1A: Orthosilicates*. Longman, London.

——, —— & —— 1986. *Rock-forming minerals, volume 1B: Disilicates and ring silicates*. Longman, London.

FLEISCHER, M. & ALTSCHULER, Z. S. 1969. The relationship of the rare-earth composition of minerals to geological environment. *Geochimica et Cosmochimica Acta*, **33**, 725–732.

FRANSOLET, A. M. 1978. Donnees nouvelles sur ottrelite d'Ottre, Belgique. *Bulletin Mineralogique*, **101**, 548–557.

FREESTONE, I. C. & MIDDLETON, A. P. 1985. Mineralogical applications of the analytical SEM in archaeology. *Mineralogical Magazine*, **51**, 21–31.

GJELBERG, J., DREYER, T., HØIES, A., TJELLAND, T. and LILLENG, T. 1987. Late Triassic to Mid Jurassic sandbody development on the Barents and mid-Norwegian shelf. *In*: BROOKS, J. & GLENNIE, K. W. (eds) *Petroleum Geology of Northwest Europe*. Graham & Trotman, London, 1105–1129.

HAUGHTON, P. D. W. & FARROW, C. M. 1989. Compositional variations in Lower Old Red Sandstone garnets from the Midland Valley of Scotland and the Anglo-Welsh Basin. *Geological Magazine*, **126**, 373–396.

HURST, A. R. & MORTON, A. C. 1988. An application of heavy-mineral analysis to lithostratigraphy and reservoir modelling in the Oseberg Field, northern North Sea. *Marine and Petroleum Geology*, **5**,

157–169.

KEPEZHINSKAS, K. B. & KORYLUK, V. N. 1973. Range of variation in the composition of staurolite from typical metapelites with pressure and temperature of metamorphism. *Doklady, Academy of Sciences of the USSR, Earth Science Section*, **212**, 121–125.

KNOX, R. W. O'B., MORTON, A. C. & LOTT, G. K. 1982. Petrology of the Bridport Sands in the Winterborne Kingston Borehole, Dorset. *Report of the Institute of Geological Sciences*, **81/3**, 107–121.

KRYNINE, P. D. 1946. The tourmaline group in sediments. *Journal of Geology*, **54**, 65–87.

LEAKE, B. E. 1978. Nomenclature of amphiboles. *American Mineralogist*, **63**, 1023–1052.

LETERRIER, J., MAURY, R. C., THONON, P., GIRARD, D. & MARCHAL, M. 1982. Clinopyroxene composition as a method of identification of the magmatic affinities of paleo-volcanic series. *Earth and Planetary Science Letters*, **59**, 139–154.

LUEPKE, G. (ed.) 1984. *Stability of heavy minerals in sandstones*. Benchmark Papers in Geology Series 81. Van Nostrand Reinhold, New York.

MAKANJUOLA, A. A. & HOWIE, R. A. 1972. The mineralogy of the glaucophane schists and associated rocks from the Ile de Groix, Brittany, France. *Contributions to Mineralogy and Petrology*, **35**, 83–118.

MANGE, M. A. & MAURER, H. F. 1990. *Heavy minerals in colour*. Unwin Hyman, London.

MANGE-RAJETZKY, M. A. & OBERHANSLI, R. 1982. Detrital lawsonite and blue sodic amphibole in the Molasse of Savoy, France and their significance in assessing Alpine evolution. *Schweizerische Mineralogische und Petrographische Mitteilungen*, **62**, 415–436.

—— & —— 1986. Detrital pumpellyite in the peri-Alpine Molasse. *Journal of Sedimentary Petrology* **56**, 112–122.

MILNER, H. B. 1962. *Sedimentary Petrography*, 4th edition. Allen & Unwin, London.

MILODOWSKI, A. E. & HURST, A. in press. The authigenesis of phosphate minerals in some Norwegian hydrocarbon reservoirs: evidence for the mobility and redistribution of rare earth elements (REE) and Th during sandstone diagenesis. *Proceedings of the Sixth International Symposium on Water–Rock Interaction*, Malvern, UK, 3–12 August 1989.

—— & ZALASIEWICZ, J. 1990. Diagenesis and redistribution of rare earth elements in turbidite/hemipelagite mudrocks from central Wales: implications for provenance studies. This volume.

MORTON, A. C. 1982. Heavy minerals from the sandstones of the Winterborne Kingston Borehole, Dorset. *Report of the Institute of Geological Sciences*, **81/3**, 143–148.

—— 1984*a*. Stability of detrital heavy minerals in Tertiary sandstones from the North Sea Basin. *Clay Minerals*, **19**, 287–308.

—— 1984*b*. Heavy minerals from Paleogene sediments, Deep Sea Drilling Project Leg 81: their bearing on stratigraphy, sediment provenance and the evolution of the North Atlantic. *In*: ROBERTS, D. G., SCHNITKER, D., *ET AL*. *Initial Reports of the Deep Sea Drilling Project*, **81**, 653–661.

—— 1985a. Heavy minerals in provenance studies. *In*: ZUFFA, G. G. (ed.) *Provenance of Arenites*. Reidel, Dordrecht, 249–277.

—— 1985b. A new approach to provenance studies: electron microprobe analysis of detrital garnets from Middle Jurassic sandstones of the northern North Sea. *Sedimentology*, **32**, 553–566.

—— 1986. Dissolution of apatite in North Sea Jurassic sandstones: implications for the generation of secondary porosity. *Clay Minerals*, **21**, 711–733.

—— 1987a. Detrital garnets as provenance and correlation indicators in North Sea reservoir sandstones. *In*: BROOKS, J. & GLENNIE, K. W. (eds) *Petroleum Geology of Northwest Europe*. Graham & Trotman, London, 991–995.

—— 1987b. Influences of provenance and diagenesis on detrital garnet suites in the Forties sandstone, Paleocene, central North Sea. *Journal of Sedimentary Petrology*, **57**, 1027–1032.

—— & HUMPHREYS B. 1983. The petrology of the Middle Jurassic sandstones from the Murchison Field, North Sea. *Journal of Petroleum Geology*, **5**, 245–260.

——, STIBERG, J. P., HURST, A. & QVALE, H. 1989. Lithostratigraphic correlation using heavy minerals: the Brent Group, Oseberg Field, North Sea. *In*: COLLINSON, J. (ed.) *Correlation in Hydrocarbon Exploration*. Graham & Trotman, London, 217–230.

NISBET, E. G. & PEARCE, J. A. 1977. Clinopyroxene compositions in mafic lavas from different tectonic settings. *Contributions to Mineralogy and Petrology*, **63**, 149–160.

OWEN, M. R. 1987. Hafnium content of detrital zircons, a new tool for provenance study. *Journal of Sedimentary Petrology*, **57**, 824–830.

PARFENOFF, A., POMEROL, C. & TOURENQ, J. 1970. *Les Mineraux en Grains*. Masson et Cie, Paris.

POWER, G. M. 1968. Chemical variations in tourmalines from south-west England. *Mineralogical Magazine*, **36**, 1078–1089.

PRESS, S. 1986. Detrital spinels from alpinotype source rocks in Middle Devonian sediments of the Rhenish massif. *Geologische Rundschau*, **75**, 333–340.

SHUKLA, B. 1988. Provenance of Oligocene Lough Neagh Group, Northern Ireland, United Kingdom (abstract). *In*: JAMES, D. P. & LECKIE, D. A. (eds) *Sequences, stratigraphy, sedimentology: surface and subsurface*. Memoir of the Canadian Society of Petroleum Geologists, **15**, 583–584.

SMALE, D. & MORTON, A. C. 1988. Heavy mineral suites of core samples from the McKee Formation (Eocene-Lower Oligocene), Taranaki: implications for provenance and diagenesis. *New Zealand Journal of Geology and Geophysics*, **30**, 299–306.

STYLES, M. T., STONE, P. & FLOYD, J. D. 1989. Arc detritus in the Southern Uplands: mineralogical characterization of a 'missing' terrane. *Journal of the Geological Society*, London, **146**, 397–400.

WRIGHT, W. I. 1938. The composition and occurrence of garnets. *American Mineralogist*, **23**, 436–449.

YOKOYAMA, K., AMANO, K., TAIRA, A. & SAITO, Y. in press. Petrological study of silts in the Bengal Fan. *In*: COCHRAN, J. R., STOW, D. A. V., AUROUX, C. A. *ET AL*. *Proceedings of the Ocean Drilling Programme* 116B.

ZIMMERLE, W. 1984. The geotectonic significance of detrital brown spinel in sediments. *Mitteilungen Geologische-Palaontologische Institut Universitat Hamburg*, **56**, 337–360.

Quartz grain types in Holocene deposits from the Spanish Central System: some problems in provenance analysis

AMPARO TORTOSA, MARTA PALOMARES & JOSÉ ARRIBAS

Dpto. Petrologia y Geoquímica, Universidad Complutense de Madrid, 28040 Madrid, Spain

Abstract: Holocene sands of the Spanish Central System were exclusively derived from plutonic, middle–upper grade and low-grade metamorphic rocks. Modal composition of studied sands is mostly controlled by grain size and source area lithology. Thus, sands derived from slates and schists plot near the QR edge on the QFR diagram for all grain size fractions. Sands derived from granitic or gneissic rocks have a wide dispersion on the QFR diagram, from the R pole to the QF edge, depending on sand grain size.

Percentages of quartz types in granitic-derived sands are Qnu_{42}, Qu_{40}, $Qp2–3_{14}$ and $Qp>3_4$. Sands of gneissic origin have Qnu_{51}, Qu_{15}, $Qp2–3_{23}$ and $Qp>3_{11}$. Sands derived from slates and schists have Qnu_{20}, Qu_{12}, $Qp2–3_5$ and $Qp>3_{63}$. Quartz types easily discriminate sands of low-grade origin, but distinction of sands derived from plutonic rocks from those derived from middle–upper grade metamorphic rocks is difficult because of the highly variable Qu content of plutonic rocks related to strain history and crystallization conditions. Thus, quartz types must be used with caution in source discrimination if plutonic rocks are present in the source area.

The main purpose of provenance analysis on sandstones is to recognize the source area lithology. A lot of studies are based on specific analysis of some detrital minerals, as heavy minerals (Morton 1985), feldspars (Helmold 1985) and quartz grains (Basu *et al.* 1975). Detrital quartz is the most common constituent of the sandy deposits, because of its high mechanical and chemical stability. Thus, quartz grains have been frequently studied in provenance analysis (Krynine 1946; Blatt & Christie 1963; Conolly 1965; Folk 1965; Blatt 1967*a, b*). Basu *et al.* (1975) first established a method of distinguishing source areas by the analysis of quartz grain types. Their statistical method is now commonly used for characterization of source area lithology. However, the data used to generate the method come only from first-cycle Holocene sands in first order streams in local geographical areas (E and W of the United States of America). Because of the obvious geographical limitations, their results must be compared with data from similar deposits in other geographical areas. Only then will it be possible to extrapolate to older deposits. The aim of this paper is to test the Basu *et al.* (1975) method on modern sands from the Spanish Central System.

The Central System consists of large exposures of granites of Hercynian age intruding low- to high-rank metamorphic rocks. According to Aparicio (1975), and Casillas & Peinado (1988), the granite massifs include biotite leuco-granites (La Pedriza, Peña del Hombre and El Quintanar), adamellites (La Enebrosilla), pegmo-aplitic leucogranites (Los Rosados), porphyritic adamellites (El Canto del Fraile) and foliated granodiorites (El Jornillo and El Sobaquillo) (Fig. 1). The emplacement of the granitoids in the Central System was linked to a late-Hercynian regional extensional tectonics (Casquet *et al.* 1988). El Jornillo and El Sobaquillo massifs are the oldest ones and were strongly affected by deformation during and after emplacement. The other granitic massifs considered here are younger but were still emplaced during the final stages of the extensional event. Post-Hercynian deformation is brittle and consists only of discontinuous fractures and faults late Permian and younger in age (Vegas *et al.* 1986). Therefore we consider that with the exception of El Jornillo and El Sobaquillo massifs, the granites dealt with here did not undergo post-emplacement deformation to a large extent. In addition to sands derived from these granites, we collected sands derived from the slates and schists of the Morequero and Prádena de Atienza massifs, which formed at temperatures below 450°C (Lopez Ruiz *et al.* 1975) and sands derived from leuco-gneiss and augen-gneiss of both low-rank (Prado Redondo) and middle–upper rank metamorphic origins (Cerro de San Pedro, La Mujer Muerta and El Purgatorio), according to Aparicio & Galan (1980), Navidad & Peinado (1981), and Villaseca (1984) (Fig. 1).

From Morton, A. C., Todd, S. P. & Haughton, P. D. W. (eds), 1991,
Developments in Sedimentary Provenance Studies.
Geological Society Special Publication No. 57, pp. 47–54.

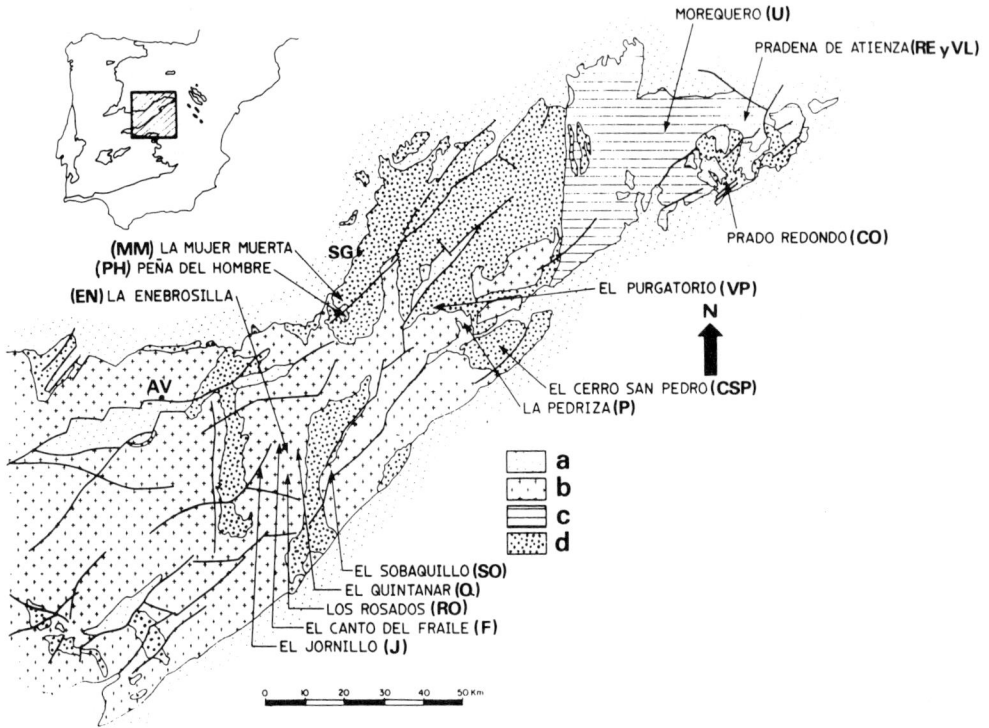

Fig. 1. Geological setting of sampled massifs in the spanish Central System. SG., Segovia; AV, Avila; (a) sedimentary rocks, (b) plutonic rocks, (c) slate, schistose and quartzite rocks, (d) gneissic rocks (after Capote *et al.* 1981).

The sands that were sampled are of Holocene age, and come from first-order streams derived exclusively from known granitic or metamorphic rocks. Thus, a close relation between the quartz types of the source lithology and their erosion products may be established. In this respect our sampling plan is identical to that of Basu *et al.* (1975) and our results may be compared directly to theirs.

Methodology

Forty-one samples were collected from first-order streams, draining eight granitic massifs (22 samples), four gneissic massifs (11 samples) and three slate and schist massifs (8 samples) (Table 1).

The sand-sized fraction (2 mm to 0.062 mm) was removed by dry sieving into five 'phi' fractions. All samples were cast into blocks with epoxy resin and thin sectioned for microscopic study. Thin sections were stained with Na cobaltinitrite solution to aid K-feldspar identification. Modal analysis of thin sections used the point-counting method of Chayes (1956). Four hundred points were counted per thin section. Each specimen was counted for Q (Qnu, Qu, Qp2–3

and Qp > 3, Basu *et al.* 1975), F (plagioclase and K-feldspar) and R (rock fragments).

Modal composition

Rock fragments of the sands derived from the granitic and gneissic sources are principally composed of quartz–feldspar aggregates, dominantly with between four and eight crystals. Labile rock fragments (Dickinson 1970) are more frequent in sands derived from the slates and schists. These also contain idiomorphic–subidiomorphic K-feldspars and plagioclases, both twinned and untwinned, and angular–subangular quartz. Recrystallized quartz grains occur in sands originated from deformed granitic rocks (El Jornillo and El Sobaquillo), and in some sands from gneissic rocks. These consist of large crystals with smooth crystal–crystal boundaries and interfacial angles of 120° at triple junctions. Young (1976) interpreted such recrystallization as characteristic of granitic and high-rank metamorphic rocks.

Table 1. *Samples, massifs and their lithology*

Samples (No.)	Lithology	Massifs
P (4)	Biotite leucogranite of coarse to very coarse grain size	La Pedriza
PH (5)	Very homogeneous granite of coarse grain size	Peña del Hombre
EN (4)	Adamellite of medium to coarse grain size	La Enebrosilla
Q (3)	Biotite leucogranite of medium to fine grain size	El Quintanar
RO (2)	Pegmatitic leucogranite	Los Rosados
F (2)	Porphyritic adamellite	El Canto del Fraile
J (1)	Adamellite	El Jornillo
SO (1)	Foliated granodiorite	El Sobaquillo
CSP (3)	Augen-gneiss of coarse grain size	Cerro de San Pedro
MM (4)	Leucogneiss of medium to fine grain size	La Mujer Muerta
VP (2)	Ocellar gneiss of coarse grain size	El Purgatorio
CO (2)	Augen-gneiss of coarse grain size	Prado Redondo
U (2)	Slate–schist and quarzite beds	Morequero
RE (3)	Homogeneous slates	Prádena de Atienza
VL (3)	Slates and schists	Prádena de Atienza

The composition of the sandy deposits is shown in Table 2 and on a QFR diagram (Fig. 2). Their composition varies systematically with respect to grain size and source-area lithology.

Sands derived from slates and schists

All fractions of sands derived from slates and schists group near the QR edge, clearly separated from the other analysed sands. Quartz-grain proportions increase at the expense of rock fragments with decreasing grain size. Feldspar grains occur only in the 0.125–0.062 mm fraction, with a percentage <10%. Variation of feldspar content is very low, but amounts of both quartz grains and rock fragment are very variable, controlled by the grain size of the source rock. In very fine-grained source rocks, the rock fragment percentage is high even in the fine-grained fractions.

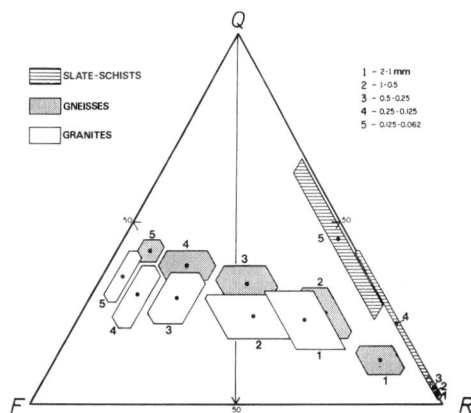

Fig. 2. Mean and standard deviation of QFR compositions of Holocene sands from the Central System.

Table 2. *Quartz, feldspar and rock-fragment percentages of size fractions of Holocene sands in Central System*

	2–1 mm	1–0.5	0.5–0.25	0.25–0.125	0.125–0.062
Granites					
Q	23 ± 8	25 ± 6	29 ± 7	31 ± 9	35 ± 7
F	22 ± 6	33 ± 9	50 ± 9	59 ± 9	60 ± 7
R	55 ± 15	42 ± 14	21 ± 5	10 ± 3	5 ± 2
Gneisses					
Q	12 ± 4	25 ± 7	32 ± 5	37 ± 4	41 ± 3
F	9 ± 5	16 ± 4	32 ± 7	44 ± 7	50 ± 4
R	79 ± 7	59 ± 9	36 ± 8	19 ± 7	9 ± 3
Slate–schists					
Q	3 ± 4	3 ± 4	4 ± 3	22 ± 20	45 ± 24
F	0	0	0	0.5	3 ± 2
R	97 ± 4	97 ± 4	96 ± 3	78 ± 20	52 ± 20

Sands derived from granites and gneisses

Modal compositions of sands derived from granitic and gneissic sources are similar, varying from close to the R pole to near the QF margin. This trend is produced by decreasing rock-fragment contents in finer grain size fractions. All fractions of sands from the gneissic source are displaced toward the R pole compared with sands of granitic origin (Fig. 2).

Quartz grain percentages are less dependent on grain size in granite-sourced sands because of its high mechanical and chemical stability. However, these percentages increase with decreasing grain size in sands derived from metamorphic sources because of the high quartz content in coarse rock fragments.

In medium-grained fractions (0.5–0.25 mm), the mean composition of sands from granitic sources is $Q_{30}F_{50}R_{20}$, of sands from gneissic origin is $Q_{30}F_{30}R_{40}$, and is $Q_5F_0R_{95}$ in sands derived from slates and schists. Although the choice of a certain sand-size is useful for distinction of different source lithologies, that choice risks the loss of information about their overall composition. Further, Basu (1976) has suggested that variations in sand-fractions composition ('tie-lines') are useful to determine the degree of weathering in different climatic regions.

Quartz types: results and discussion

The content of the four quartz types varies with grain size and source lithology (Table 3). Qp content decreases with decreasing grain size (Fig. 3). This trend is produced by grains breaking along crystalline boundaries. Thus, $Qp > 3$ break down to Qp2–3 and, finally, to monocrystalline quartz types in the fine fractions. The Qp2–3 type increases from the 2–1 mm fraction to the 1–0.5 mm fraction, but decreases in the finer sand fractions. The Qu content tends to decrease in the fine sand fractions in samples of granitic and gneissic origin (Fig. 3). Basu (1976) has related this to their low mechanical and chemical stability compared with Qnu. However, we consider that the difficulty of observation of this character in fine sand fractions is a factor that may also be involved.

The influence of source lithology on quartz types is considered for the medium sand fraction (Fig. 4). Granitic derived sands have Qnu_{42}, Qu_{40}, $Qp2–3_{14}$ and $Qp > 3_4$. Sands of gneissic origin have Qnu_{51}, Qu_{15}, $Qp2–3_{23}$ and $Qp > 3_{11}$. Sands derived from slates and schists have Qnu_{20}, Qu_{12}, $Qp2–3_5$ and $Qp > 3_{63}$. Monocrystalline quartz types are predominant in sands from granitic and gneissic sources, whereas polycrystalline quartz types constitute the main population of quartz grains in sands derived from slates and schists.

In contrast, the Qp grains consist of few crystals (five at most) in sands of granitic and gneissic origin, while polycrystalline quartz grains in sands derived from low-rank metamorphic source are dominated by those with more than five fine–very fine crystals. Qnu contents are higher than Qu content in all three sand types, being particularly high in sands derived from gneissic rocks.

The Qp contents of sands from plutonic and gneissic sources are very similar to those proposed by Basu et al. (1975). However, sands

Table 3. *Detrital quartz types percentages of size fractions of Holocene sands from granites, gneisses and slate–schists in Central System.*

	2–1 mm	1–0.5	0.5–0.25	0.25–0.125	0.125–0.062
Granites					
Qnu	15.7	32.8	41.5	47.7	62.9
Qu	25.6	34.2	39.6	42.6	33.4
Qp2–3	23.6	19.5	13.7	7.5	3
Qp > 3	35.3	13.7	5	2.1	0.4
Gneisses					
Qnu	25.2	26.6	50.1	68.3	84.7
Qu	7.3	8.5	15.3	16	9.9
Qp2–3	12.8	26	22.3	12.7	4.9
Qp > 3	54.6	38.9	12.2	2.9	0.4
Slate–schists					
Qnu	0.8	14.4	19.4	35.8	61.8
Qu	3.3	12	12.6	9.2	18.6
Qp2–3	0	0	5.2	13.8	10.9
Qp > 3	95.9	73.4	62.8	41.1	8

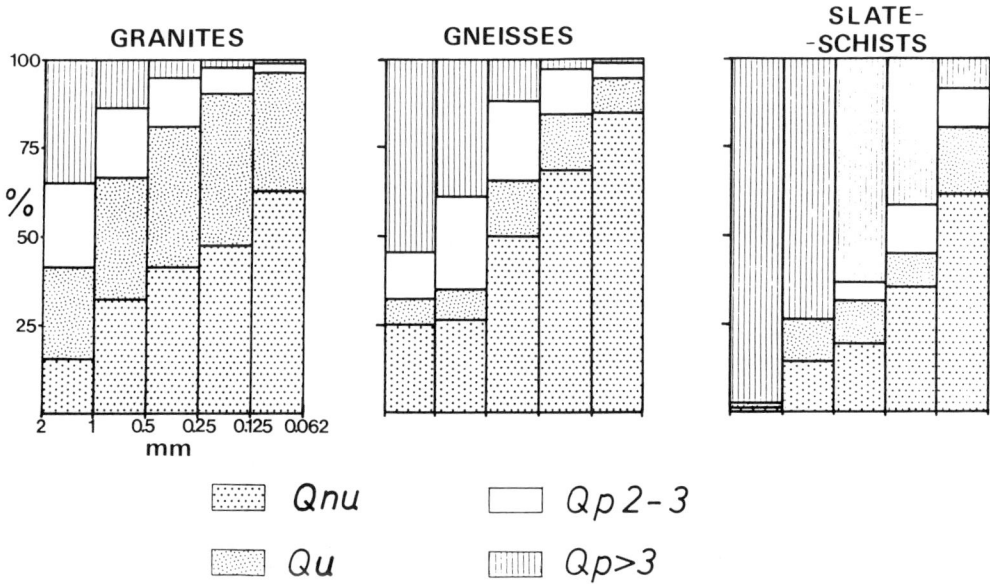

Fig. 3. Relative percentage of the four quartz-grain types in Holocene sands from the Central System versus grain size.

from low-rank metamorphic rocks contain a higher amount of polycrystalline grains with over three crystals than those observed by Basu et al. In spite of these differences, the criterion established by Basu et al., that Qp content is useful to differentiate sands derived from low-rank metamorphic rocks and sands from plutonic and high-rank metamorphic origin, remains valid.

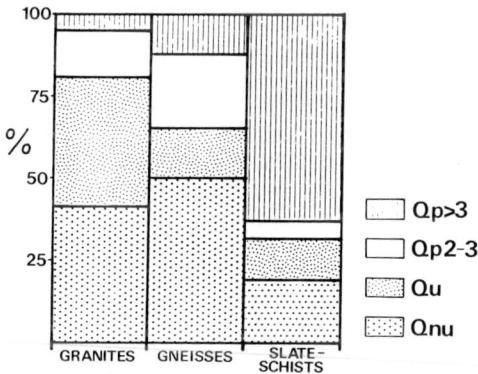

Fig. 4. Mean percentages of medium-grained detrital quartz populations of Holocene sands from Central System versus source lithology.

Sands from gneissic sources have similar contents of monocrystalline quartz types to those observed in the earlier work on sands of middle and high-rank metamorphic parentage (Basu et al. 1975). The overall content of monocrystalline quartz types is similar in sands of granitic origin, but the relative percentages of each varies considerably from that reported by Basu et al. (1975). For example, in this study the Qu percentage is 40% in sands from granitic source, compared with 4% in the study of Basu et al. Therefore, the use of Qu to distinguish between plutonic areas and high-rank metamorphic areas is not applicable to the Spanish Central System. We consider that the differences in Qu contents in sands of granitic origin are related to a number of factors, including strain history (El Jornillo and El Sobaquillo massifs), magma crystallization conditions, particularly high viscosity (adamellites, leucogranites without pegmatites) (Arzi 1978), emplacement of the pluton, uplift and decompression.

Diamond-shaped diagram

The relative percentages of the four quartz types have been plotted in the diamond-shaped provenance-discrimination diagram of Basu et al. (1975). The main content of each type and its variations are represented, producing a dispersal area for each massif (Table 4).

Samples derived from slates and schists plot in the low-rank metamorphic field, in the lower triangle with high contents of Qp (Fig. 5). These sands were derived from massifs with formation

Table 4. *Detrital quartz types percentages in medium-grained sands from granites, gneisses and slate–schists in Central System. Data were recalculated for plotting on Fig. 5*

Samples	Qnu	Qu	Qp2–3	Qp > 3
Granites				
P	82.2 ± 0.9	5.3 ± 0.9	—	6.4 ± 1.5
PH	71.7 ± 5.4	12.5 ± 1.3	16 ± 6.1	—
EN	43.1 ± 8.9	42.3 ± 3.7	14.5 ± 5.2	—
EN	10.1 ± 8.1	80 ± 4	—	9.8 ± 4.2
Q	3.7 ± 4	79 ± 0.8	17.3 ± 4.5	—
RO	46	39.7	14.3	—
RO	0	94.6	—	5.4
F	18.1	75.2	6.6	—
F	20.7	76.1	—	3.2
SO	9.1	63.6	—	27.3
J	56.5	23.2	—	20.2
Gneisses				
CSP	92.5 ± 2.4	1.1 ± 1.1	—	6.4 ± 1.3
MM	52.8 ± 6.9	11.8 ± 1	35.4 ± 5.9	—
MM	54.9 ± 12	28 ± 10.4	—	17 ± 1.6
PV	67 ± 4.9	17.2 ± 4.7	—	15.8 ± 0.2
CO	17.2 ± 3.1	49.6 ± 4	—	33.1 ± 7.2
Slate–schists				
VL	23.7 ± 14.9	9.8 ± 1.1	—	66.5 ± 15.7
RE	17.3 ± 15.4	22.6 ± 17.6	—	60.1 ± 28.3
U	21.6 ± 4.3	3 ± 3	—	74.7 ± 8

temperatures lower than 450°C, comparable with formation temperature data of low-rank metamorphic lithologies in the work of Basu (1985).

Sands from gneissic sources plot variably on this diagram according to metamorphic rank rather than lithology. For example, samples derived from La Mujer Muerta (MM) and El Purgatorio (PV) massifs (high-rank metamorphic) plot in the middle and upper-rank metamorphic field, whereas sands derived from Prado Redondo (CO) (low-rank metamorphic massif) plot in the low-rank metamorphic field. The latter has lower polycrystalline quartz contents than sands from slates and schists. An exception is the sand derived from Cerro de San Pedro (CSP) (high-rank metamorphic) which plots in the plutonic field because of its large amount of Qnu (Fig. 5).

Sands of granitic origin have remarkable variations due to different Qu contents of different massifs. These sands plot in all three fields of the diagram, although they always plot in the central area with low amounts of polycrystalline quartz. For example, the samples from La Pedriza (P) and Peña del Hombre (PH) massifs plot within the plutonic field, whereas sands from Los Rosados (RO), El Jornillo (J) and La Enebrosilla (EN) massifs plot in the middle and upper-rank metamorphic field. Finally, sands from El Sobaquillo (SO), El Canto del Fraile (F)

and El Quintanar (Q) massifs plot in the low-rank metamorphic field (Fig. 5).

In short, the diamond-shaped diagram of Basu *et al.* (1975) acceptably discriminates sands derived from slates and schists from sands of other origins, but sands of plutonic origin show wide dispersion.

Summary and conclusions

The modal composition of sands derived from slates and schists provenance is dominated by rock fragments, although its percentage decreases with decreasing grain size. Rock fragment diminution with decreasing grain size is exclusively bound to increasing feldspar content in granitic-derived sands, while quartz contents remain practically constant in all five size fractions. Rock fragment diminution is bound to increases of both feldspar and quartz grains in sands of gneissic origin. The variation of the light fraction components is a function of original lithology, but the trends of these constituents in terms of grain size are similar for all detrital sediments analysed.

Following the method of Basu *et al.* (1975), we have calculated the content of the four quartz types in first cycle detrital deposits derived from different lithologies in the Spanish Central Sys-

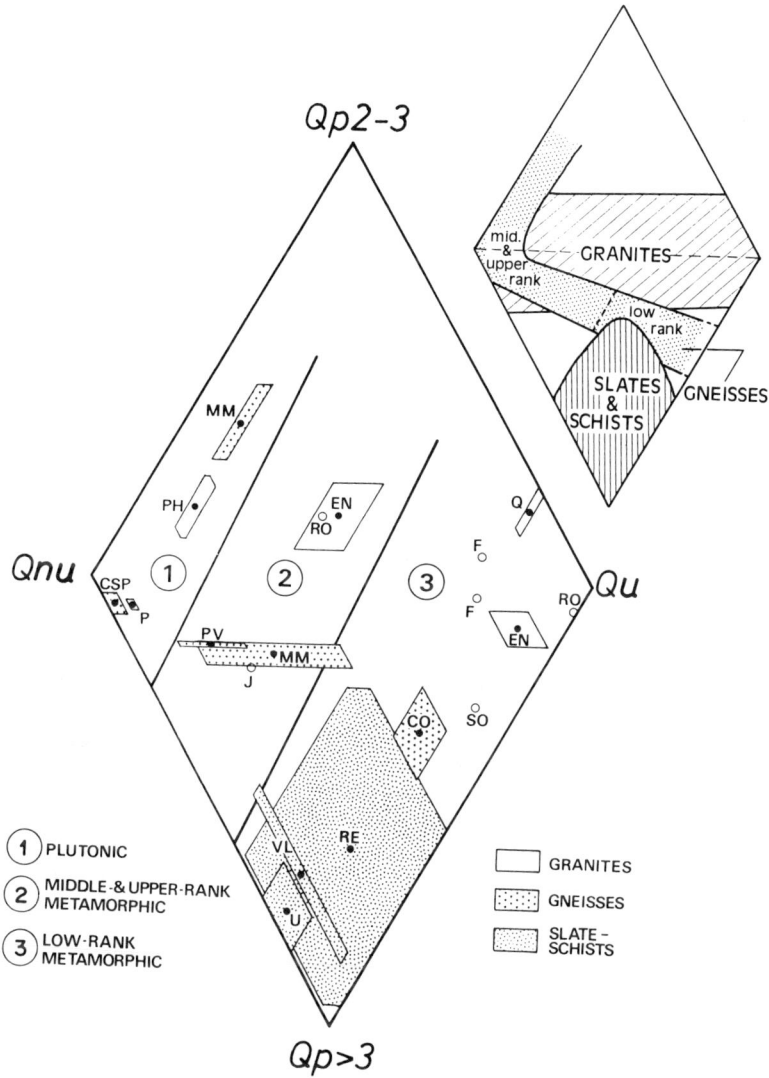

Fig. 5. Point-count data from medium-grained detrital quartz populations of Holocene sands from the Central System plotted within the diamond-shaped provenance-discrimination diagram of Basu *et al.* 1975. See Fig. 1 for key to source massifs. Little diamond-shaped diagram summarizes the projection fields of sand samples derived from granites, gneisses (low and middle-high rank metamorphic) and slates and schists (low rank metamorphic) in Central System.

tem. The grain size and source lithology are the main factors that control their relative contents. The Qp percentage decreases with decreasing grain size. The Qu percentage is very variable between the different granitic massifs, related to strain history of the plutons and crystallisation conditions. The identification of low-rank metamorphic sources on the basis of polycrystalline quartz types content is applicable to Central System deposits, although sands derived from slates and schists have larger amount of Qp than those analysed by Basu *et al.* (1975). Sands derived from gneissic areas of the diamond-shaped diagram of Basu *et al.* (1975) as a function of formation temperature rather than lithology. Granitic-derived sands plot in all fields on this diagram depending on variations in Qu content, but fall in the central area because of low contents of polycrystalline quartz types. The wide scatter of the granitic-derived sands from Central System on this diagram prevents the discrimination of plutonic sources from middle–upper-grade sources as a function of Qu content as proposed by Basu *et al.* (1975).

Therefore, this method of provenance discrimination should be used with care because of its limitations concerning granitic source rocks.

The authors are specially grateful to A. Basu and A. Morton for critically reading the manuscript. We also thank C. Casquet for helpful comments on tectonics of the Central System.

References

APARICIO, A. 1975. *Los materiales graniticos hercínicos del Sistema Central español.* Memoria del Instituto Geológico y Minero de España, **88**.

—— & GALAN, E. 1980. Las caracteristicas del metamorfismo hercínico de bajo y muy bajo grado en el sector oriental del Sistema Central (provincia de Guadalajara). *Estudios Geológicos*, **36**, 75–84.

ARZI, A. A. 1978. Critical phenomena in the rheology of partially melted rocks. *Tectonophysics*, **44**, 173–184.

BASU, A. 1976. Petrology of Holocene fluvial sand derived from plutonic source rocks: implications to paleoclimatic interpretations. *Journal of Sedimentary Petrology*, **46**, 694–709.

—— 1985. Reading provenance from detrital quartz. *In*: ZUFFA G. G. (ed.) *Provenance of Arenites*. Reidel, Dordrecht, 231–248.

——, YOUNG, S. W., SUTTNER, L. J., JAMES, W. C. & MACK, G. H. 1975. Re-evaluation of the use of undulatory extinction and polycrystallinity in detrital quartz for provenance interpretation. *Journal of Sedimentary Petrology*, **45**, 873–882.

BLATT, H. 1967*a*. Original characteristics of clastic quartz grains. *Journal of Sedimentary Petrology*, **37**, 401–424.

—— 1967*b*. Provenance determinations and recycling of sediments. *Journal of Sedimentary Petrology*, **37**, 1031–1044.

—— & CHRISTIE, J. M. 1963. Undulatory extinction in quartz of igneous and metamorphic rocks and its significance in provenance studies of sedimentary rocks. *Journal of Sedimentary Petrology*, **33**, 559–579.

CAPOTE, R., CASQUET, C. & FERNANDEZ CASALS, M. J. 1981. La tectónica hercínica de cabalgamientos en el Sistema Central español. *Cuadernos de Geología Ibérica*, **7**, 455–469.

CASILLAS, R. & PEINADO, M. 1988. Secuencias graníticas en el área de San Martin de Valdeiglesias (Sistema Central español). *In*: Libro Homenaje a L. C. García de Figuerola, Ed. Rueda, *Geologia de los granitoides y rocas asociadas del macizo hespérico*, Madrid, 281–292.

CASQUET, C., FUSTER, J. M., GONZALEZ-CASADO, J. M., PEINADO, M. and VILLASECA, C. 1988. Extensional tectonic and granite emplacement in the Spanish Central System. A discussion. *Proceeding of the fifth workshop on the European Geotraverse Project*, 65–76.

CHAYES, F. 1956. *Petrographic modal analysis.* Wiley and Sons, New York.

CONOLLY, J. R. 1965. The occurrence of polycrystallinity and undulatory extinction in quartz in sandstones. *Journal of Sedimentary Petrology*, **35**, 116–135.

DICKINSON, W. R. 1970. Interpreting detrital modes of graywacke and arkose. *Journal of Sedimentary Petrology*, **40**, 695–707.

FOLK, R. L. 1965. *Petrology of sedimentary rocks.* Hemphill, Texas.

HELMOLD, K. P. 1985. Provenance of feldspathic sandstones—the effect of diagenesis on provenance interpretations: A review. *In*: ZUFFA, G. G. (ed.) *Provenance of Arenites*, Reidel, Dordrecht, 139–164.

KRYNINE, P. D. 1946. Microscopic morphology of quartz types. *Proceedings of the 2nd Pan-American Congress of Mining Engineers and Geology*, **3**, 2nd Communication, 35–49.

LOPEZ RUIZ, J., APARICIO, A. & GARCIA CACHO, L. 1975. *El metamorfismo de la sierra de Guadarrama, Sistema Central español.* Memorias del Instituto Geológico y Minero de España, **86**.

MORTON, A. C. 1985. Heavy minerals in provenance studies. *In*: ZUFFA, G. G. (ed.) *Provenance of Arenites*. Reidel, Dordrecht, 249–278.

NAVIDAD, M. & PEINADO, M. 1981. Ortogneis y metasedimentos de la formación infrabasal al Ollo de Sapo (macizo de Hiendelaencina, Guadarrama oriental). *Cuadernos de Geología Ibérica*, **7**, 183–199.

VEGAS, R., VAZQUEZ, J. T. & MARCOS, A. 1986. Tectónica alpina y morfogénesis en el Sistema Central español: Modelo de deformación intracontinental distribuida. *Geogaceta*, **1**, 24–25.

VILLASECA, C. 1984. *Evolución metamórfica del sector centro-septentrional de la Sierra de Guadarrama.* PhD thesis, Universidad Complutense de Madrid.

YOUNG, S. W. 1976. Petrographic textures of detrital polycrystalline quartz as an aid to interpreting crystalline source rocks. *Journal of Sedimentary Petrology*, **46**, 595–603.

Reliability and application of detrital opaque Fe–Ti oxide minerals in provenance determination

ABHIJIT BASU & EMANUELA MOLINAROLI

Department of Geology, Indiana University, Bloomington, IN 47405, USA

Abstract: A survey of detrital opaque Fe–Ti oxide minerals in Holocene sands derived exclusively from known source rocks in the Rocky Mountains, USA (semi-arid climate, high relief) and the Appalachian Mountains, USA (humid climate, low relief) shows that, despite obvious differences in weathering rates, the abundance of detrital ilmenite relative to parent rocks is similar in these two regions. Because ilmenite is very common in many crystalline rocks, ilmenite is considered to be a useful Fe–Ti oxide mineral for varietal studies. About 30% to 50% of detrital Fe–Ti oxide minerals are polymineralic grains with intergrowths of two or more phases. Discriminant function analysis indicates that morphological varieties of detrital Fe–Ti oxide minerals in combination with their chemical compositions are useful guides to provenance. An igneous or metamorphic source of 96% of the Holocene samples is correctly identified by this method.

The method is used to determine the relative contribution of igneous and metamorphic source rocks to the sandstones of the part of the Tertiary Renova Formation, which was deposited within a dissected magmatic arc in southwestern Montana, USA. A separate discriminant function is calculated using the chemical compositions of detrital ilmenite and morphological properties of detrital Fe–Ti oxide minerals in Rocky Mountain Holocene sands. Classification of detrital Fe–Ti oxide minerals in Renova sandstones using this function suggests that igneous and metamorphic rocks contributed 77% and 23% respectively to the Renova sandstones in this area.

Differential weathering, hydraulic sorting, and dissimilar dissolution during diagenesis, skew the distribution of heavy minerals in sands and sandstones relative to their distribution in parent rocks. Problems caused by these processes in interpreting provenance from heavy-mineral distributions may be overcome to a large extent by characterizing the variations of morphology, internal texture, other crystallographic properties and chemical composition in a single mineral species or a single mineral group. Beginning with a study of the tourmaline group by Krynine (1946), this approach of varietal studies has been extended to garnet (Morton 1984, 1985a, b, 1987), zircon (Callender & Folk 1958; Owen, 1987), pyroxene (Cawood, 1983), ilmenite (Darby & Tsang 1987; Shukla 1988), magnetite (Grigsby 1988), and chromite (Hiscott 1978). The possibility of dating individual detrital zircon grains to decipher their provenance (Compston & Pidgeon 1986; Kober 1987; Cliff *et al.* this volume) obviously have special utility.

It has been shown that detrital opaque Fe-Ti oxide minerals (DOPQ) principally consist of magnetite, ilmenite and hematite either as monomineralic grains or as polymineralic grains with exsolution and intergrowth textures; the varieties of composition and texture of DOPQ suggest provenance (Riezebos 1979; Basu & Hood

1985). We are comparing the characteristics and distribution of DOPQ in Holocene and Tertiary sands. The purpose of our work is (a) to establish the reliability of inferring provenance from varietal studies of Holocene DOPQ, and (b) to assess the provenance of the Tertiary Renova sandstones of Montana, USA on the basis of their DOPQ characteristics.

In an ideal case:

(a) it may be possible to fingerprint each grain of one mineral group and determine its source, within some bounds of probability;

(b) the proportions of grains of one mineral group from different sources are likely to represent the relative contributions of different sources;

(c) if more than one mineral group is used, several results would be obtained, one for each mineral group. Thus, if the number of mineral groups used exceed the number of source rocks, the number of equations that may be set up to determine the relative contributions from different source rocks will exceed the number of unknowns. Such an overdetermined system may then be

From Morton, A. C., Todd, S. P. & Haughton, P. D. W. (eds), 1991,
Developments in Sedimentary Provenance Studies.
Geological Society Special Publication No. 57, pp. 55–65.

solved with least square mixing calculations. In addition, the overdetermination
could also be used to examine the internal
consistency of the data.

In this paper, we summarize our current work
and report the results of a comparative study of
DOPQ, especially ilmenite, in (a) igneous granitic rocks of the southern Appalachian Mountains of Georgia, South Carolina, North Carolina and Virginia, and the Rocky Mountains of
southwestern Montana, USA (b) the Holocene
sands derived from these rocks, and (c) the
Renova Formation (Eocene–Miocene) of southwestern Montana.

The Renova Formation

Towards the end of the Mesozoic Era an
Andean type of magmatic arc developed in the
western United States and batholiths were
emplaced below their own volcanic cover. In
southwestern Montana these igneous rocks are
represented by the Boulder Batholith and the
Elkhorn Mountain Volcanics, which intruded
and erupted through a thick pile of crustal rocks
including Precambrian metamorphic rocks at
the base and a nearly continuous sequence of
Phanerozoic sedimentary rocks. These rocks
were all involved in the late Cretaceous Laramide orogeny of the Rocky Mountains. The
Tertiary Bozeman Group of sediments were
deposited in a large basin east of this mountain
range. The Renova Formation consists of middle Eocene to early Miocene fluvial, fluvio-lacus-
trine, and alluvial fan sediments (mostly immature sandstones), which post-date the cessation of
Laramide orogenic events, lying unconformably
above older Tertiary volcaniclastic deposits or
above other pre-Tertiary sedimentary, metamorphic, and igneous rocks (Fields et al. 1985). The
Renova Formation constitutes the base of the
Bozeman Group.

Sediments that were concomitantly deposited
within the magmatic arc and in the retro-arc
areas at the western end of this basin are also
classified with the Bozeman Group (Fields et al.
1985; Monroe 1981; Kuenzi & Fields 1971). The
sandstones of the Renova Formation, which we
are investigating, were deposited in small intermontane basins. These basins owed their origin
to mostly back-arc extensional tectonics and
were surrounded by Laramide uplifts (Fig. 1).
Local igneous and metamorphic rocks supplied
detritus to this part of the Renova Formation.
Thus, the source of the sediments is predominantly a magmatic arc consisting of shallow level

granodioritic plutons under a dacite–andesite
volcanic cover, along with roof pendants and
thrust slices of Precambrian metamorphic rocks
and Phanerozoic sedimentary rocks (Fields et al.
1985).

Methodology

Sampling

Holocene sands were collected from first-order streams
draining a crystalline catchment made of a single
lithology; parent rocks of the catchments were also
sampled for comparison. Sand samples from igneous
and metamorphic source rocks were obtained in a
humid and a semi-arid climate from the Appalachian
and the Rocky Mountains, respectively. Relatively
lower relief in the Appalachians and much higher relief
in the Rocky Mountain regions, contribute to a longer
residence time of the Appalachian stream sands in
weathering zones than that in the Rocky Mountains.
The plan for sampling Holocene sands and their analysis are detailed in Basu (1976) and in Basu & Molinaroli (1989). Sandstones of the middle Eocene to early
Miocene Renova Formation were sampled from *outcrops* in the Upper Ruby River, Whitetail Creek,
North Boulder River and the Jefferson River valleys in
southwestern Montana (Fig. 1). The Holocene stream
sands were also collected from this area and it is
presumed that the source rocks of the Renova and the
local Holocene sands may be nearly identical. Sixty
one Holocene sands, 42 parent rocks, and 30 Renova
sandstones were utilized for this work.

Opaque mineral analysis

Heavy minerals were separated from the sands and
lightly crushed sandstones using standard procedures
and bromoform as the heavy liquid. Polished thin
sections of the concentrates as well as those of the
parent rocks were used for petrographic and electron
microprobe analysis. Relative abundances of monomineralic and polymineralic Fe–Ti oxide grains and
widths and directions of lamellae in polymineralic
grains were determined by point counting (8889 points
on Holocene sands, 8520 points on parent rocks, and
7382 points on Renova sandstones). Six element electron probe microanalysis (Ti, Fe, Al, Cr, Mg, Mn) of
the major phases was carried out (248 analysis of
Holocene Fe–Ti oxide grains of which 165 were ilmenite, 317 analyses of grains in parent rocks of which
233 were ilmenite, and 263 analyses of detrital ilmenite
from the Renova sandstones). Analytical methods
have been described elsewhere (Basu & Molinaroli
1989); other elements known to occur in ilmenite in
minor to trace quantities (e.g. V, Nb, Ta, Ni, etc.) are
not amenable to easy detection by the electron microprobe and were not determined. Although we analysed
mostly monomineralic detrital grains, precursors of
which might have been multiphase grains in parent
crystalline rocks, some ilmenite in multiphase detrital
grains were also analysed.

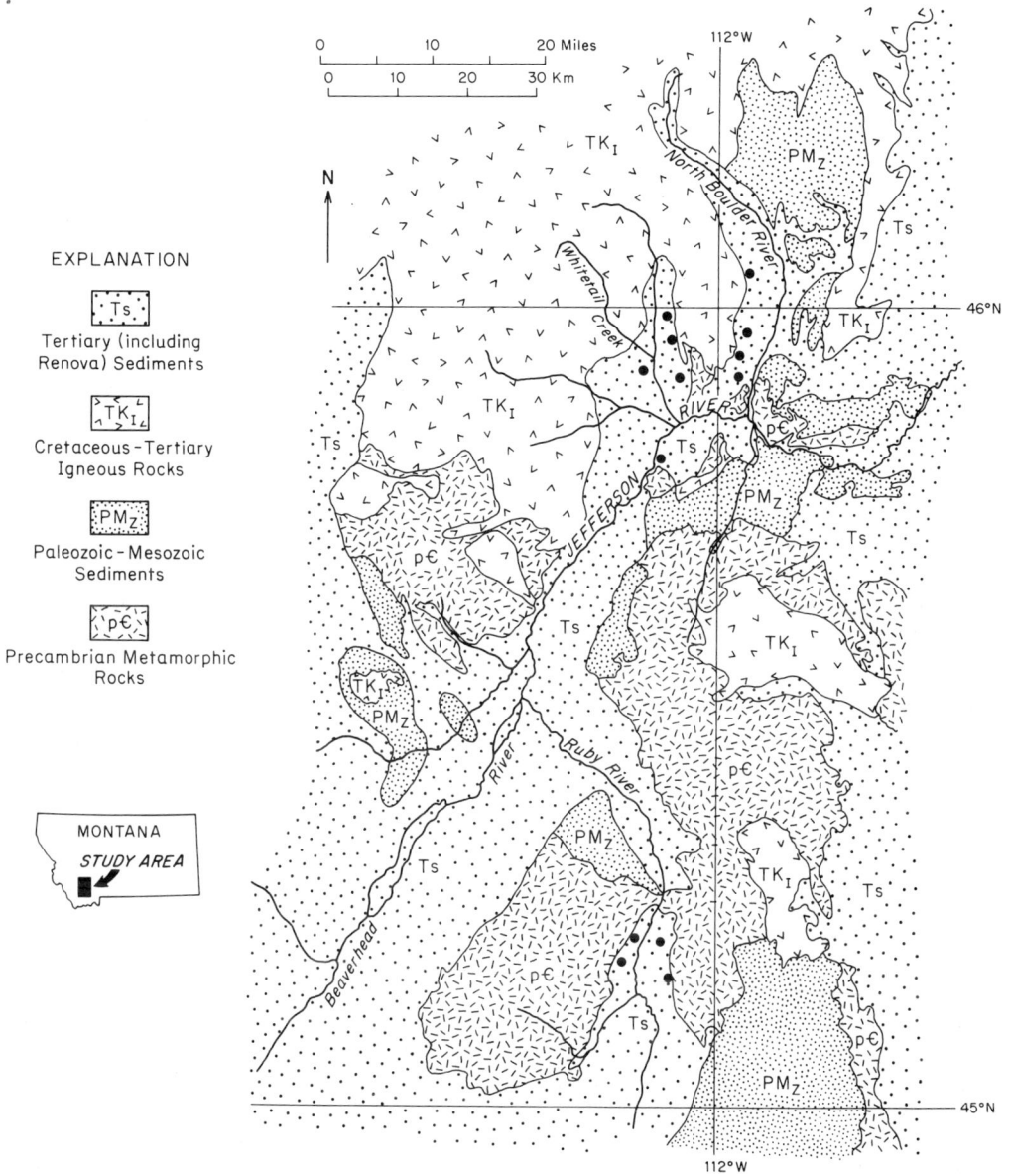

Fig. 1. Geological sketch map of a part of southwestern Montana showing the three tributaries of the Jefferson River in the valleys of which the Renova Formation (the bottom part of Ts) was sampled. Individual outcrops of the Renova Formation are not shown to avoid clutter; large black circles denote major sampling sites. Note that the Renova rests unconformably over all previous geological units.

Results and discussion

The data that we have collected are at two levels: chemical compositions represent a grain (i.e. a mineral phase), whereas the petrographic modal data represent a thin section (i.e. a sample). Therefore, representative characterization of single grains is chemical in nature. On the other hand, the characteristic of a sample is obtained from combining the modal data with the chemical compositions of many grains of different mineral phases in a sample. We examine the data separately at both levels and evaluate the reliability with which DOPQ may distinguish parent rocks and climate in source areas. In this paper we evaluate the variability of ilmenite-composition as a discriminator of source rocks. Note, however, that pure ilmenite is rare in nature; it occurs in solid solution with hematite, and replacement of major cations (Fe, Ti) by Mg, Mn, Al, and Cr is common; further, pedogenic and diagenetic alteration may affect the original chemical composition of detrital grains. Thus the interpretation of ilmenite composition is not necessarily simple. In the following discussion, we use the properties of detrital ilmenite to decipher quantitatively the provenance of Renova sandstones.

Source rock characteristics of Holocene detrital ilmenite

Chemical composition. The TiO_2 content of detrital ilmenite in Holocene sands from igneous sources (48.3% $\pm$ 3.8) is less than those from metamorphic sources (50.8% $\pm$ 3.1) and is statistically significant even at <0.01 level (Table 1). However, we are somewhat suspicious of the

geological significance of such averages of populations, the standard deviations of which overlap so much. Therefore, the statistical similarity and dissimilarity of individual variables are not discussed below. Rather, we discuss our statistical tests and procedures collectively for all data. Of the minor elements, MgO concentrations above 0.5%, Al_2O_3 concentrations above 0.4%, and MnO concentrations above 6.5% are found only in the ilmenite derived from igneous source rocks; Cr_2O_3 concentrations show no distinction between these source rocks (Fig. 2). Note, however, that concentrations of these oxides in detrital ilmenite at levels lower than those mentioned above are not suggestive of any specific source rock type. Chemical variations in ilmenite are caused by a complex combination of the activities of the elements present, other co-precipitating phases, temperature, cooling rate, oxygen fugacity, etc. during the crystallization of parent rocks. We are not aware of any petrogenetic study comparing and contrasting the specific chemistry of ilmenite occurring in igneous and metamorphic rocks. Therefore, the causes of the variations of ilmenite composition that we have observed remain poorly understood.

Petrographic properties. About 50% and 70% DOPQ grains from igneous and metamorphic sources respectively are monomineralic and do not show any intergrowth or exsolution under an optical microscope (Table 2). Polymineralic grains showing lamellar exsolution/intergrowth of more than one phase were classified on the basis of the *minimum* lamellae width in the grain. Classification on the basis of minimum lamellae width ensures objectivity, reduces grain size

Table 1. *Variability of TiO_2 content in detrital ilmenite from igneous and metamorphic source rocks and their relative distribution in daughter sands*

TiO_2 %	Igneous source (%)	Metamorphic source (%)
40–50	73.3	32.4
50–60	26.7	67.6
Average	48.2 $\pm$ 3.8	50.8 $\pm$ 3.1
$N_{analysis}$	95	70

Table 2. *Distribution of minimum lamellae widths (%) and the frequency percent of number of directions of intergrowth (exsolution) in detrital opaque oxide minerals from Holocene sands*

	Provenance			
	Igneous Rky*	Metamorphic Rky*	Igneous App*	Metamorphic App*
Grains with no exsolution				
	57.6	77.7	40.8	65.7
Minimum lamellae widths of exsolved grains				
<2 μm	42.6	4.5	29.0	11.5
2–10 μm	47.5	70.0	42.2	52.9
>10 μm	9.9	25.5	28.8	35.6
Intergrowth directions				
1	9.9	65.0	4.4	96.0
2	—	1.2	2.4	0.8
3	90.1	33.8	93.2	3.2
N_{sample}	17	19	5	20
N_{grain}	607	124	203	522

* Rky, Rocky Mountains; App, Appalachians

DETRITAL ILMENITE

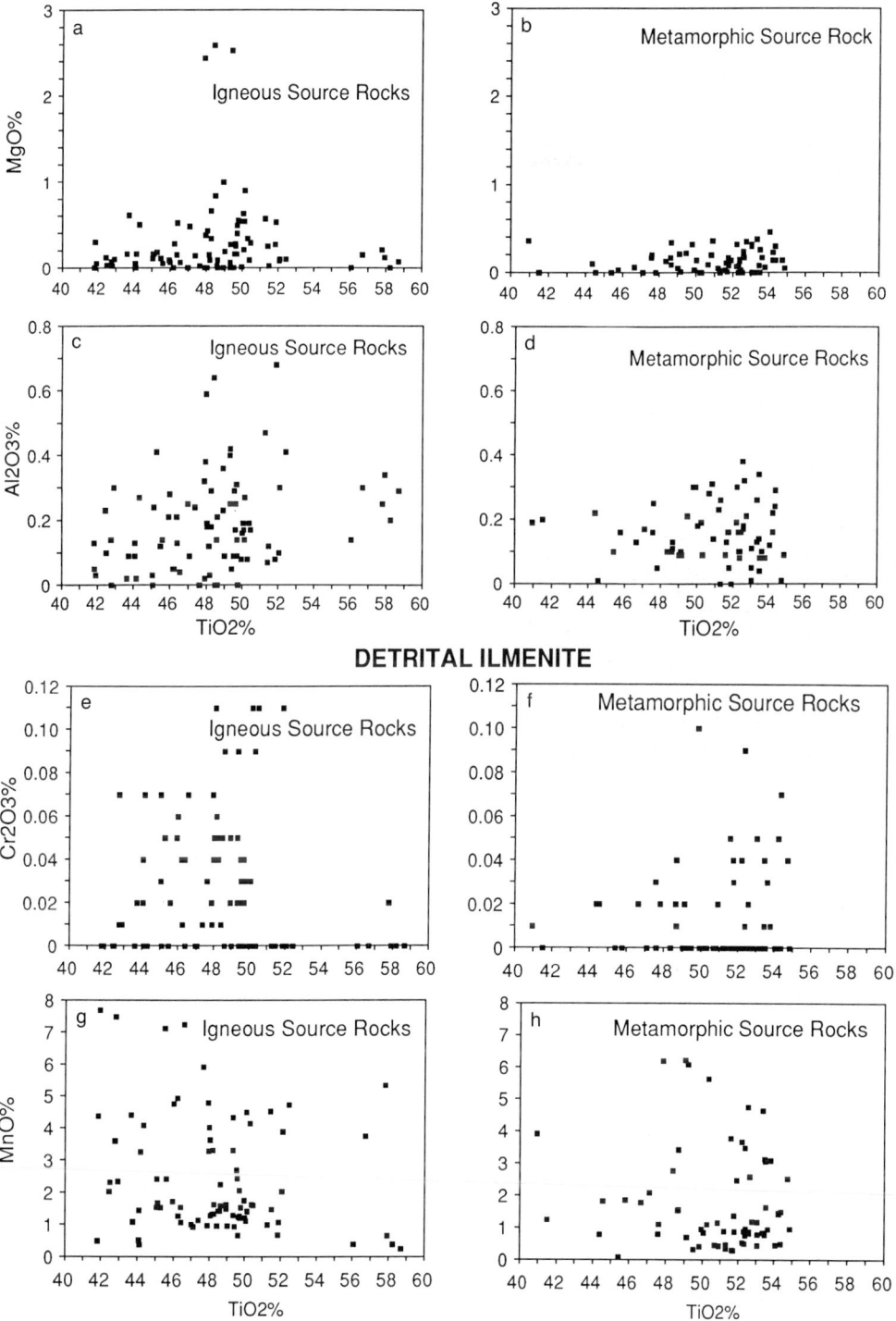

DETRITAL ILMENITE

Fig. 2. Scatter plots comparing the distribution of minor elements in detrital ilmenite in Holocene sands from igneous (**a, c, e, g**) and metamorphic (**b, d, f, h**) source rocks. Note that there is considerable similarity and overlap between the two sets of data although those from igneous sources are relatively enriched in these trace elements.

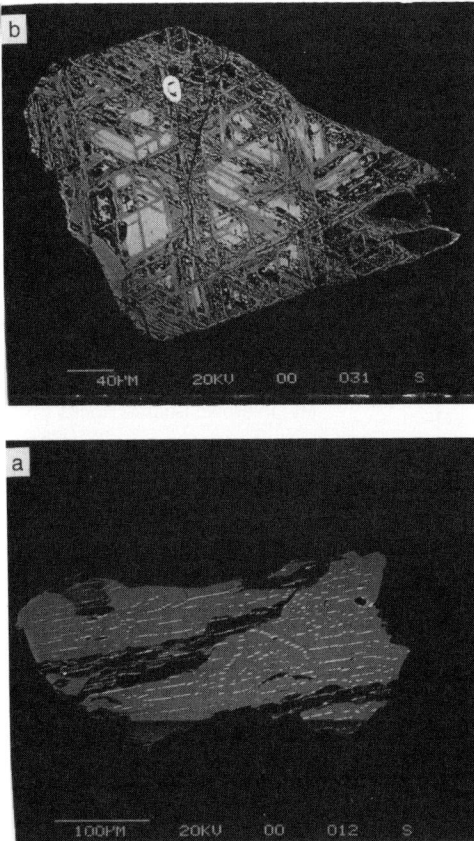

Fig. 3. Backscattered electron micrograph of detrital Fe-Ti oxide minerals in Holocene sands showing intergrowths in (**a**) one direction (presumably pinacoidal in ilmenite), and (**b**) three directions (presumably octahedral in magnetite); an inclusion of zircon appears as a very bright grain.

effects on modal data, and increases reproducibility of modal analysis. Lamellae <2 μm are more common in grains from igneous than those from metamorphic sources whereas lamellae >10 μm are more common in grains from metamorphic sources. Octahedral intergrowth in isometric phases and pinacoidal intergrowth in trigonal phases can be seen as three and one direction of lamellar structure respectively in a random plane of a thin section (Fig. 3). The frequencies of the number of directions of intergrowth lamellae in DOPQ from igneous and metamorphic rocks from the Appalachians and the Rocky Mountains are also given in Table 2. DOPQ from igneous rocks in the Rocky Mountains are dominated by three directions of lamellae whereas those from both igneous and metamorphic rocks in the Appalachian Mountains

are dominated by a single direction of exsolved lamellae. Predominance of octahedral exsolution/intergrowth patterns, especially common in DOPQ from the shallow level plutons in Montana, is expected because these are I-type magnetite-series granites; and, predictably pinacoidal exsolution/intergrowth patterns are more common in DOPQ from deep seated Appalachian plutons that are mostly S-type ilmenite-series granites.

Discriminant function analysis. Although DOPQ grains from igneous and metamorphic sources show some contrasting properties, all are somewhat overlapping and a clear-cut distinction of source rocks cannot be made directly from raw data. A simple plot cannot be constructed with so many variables for a straightforward graphical solution. We have chosen to perform discriminant function analysis simultaneously to treat the variability of six oxide components of detrital ilmenite, the numbers of directions of 'exsolution' lamellae, and their widths for ascertaining the reliability of these variables in provenance determination. Discriminant function analysis of all chemical and petrographic variables taken together provides an efficient identification of provenance with more than 95% correct determinations (Table 3). However, discriminant function analysis of the chemical data alone shows that the sources of about only 68% of the detrital grains are correctly identified; and, that of petrographic data alone shows that the sources of nearly 84% of the grains are correctly identified (Table 3). We conclude that a combination of chemical and petrographic properties of DOPQ are useful for provenance determination at the sample level.

Weathering effects. Traditionally DOPQ have been treated as a single mineral group derived from crystalline parent rocks because magnetite, ilmenite, and hematite are so intimately and ubiquitously intergrown. However, in reality these three minerals respond differently to weathering processes. Molinaroli & Basu (1987) have shown that under normal weathering conditions, detrital phases with higher TiO_2 and FeO are more durable than those with Fe_2O_3 as the dominant oxide component. A comparison of the abundances of DOPQ in igneous parent rocks and daughter sands in the humid Appalachian and the semi-arid Rocky Mountains bears this out (Table 4). The climatic contrast in these two regions may be assumed to be responsible for different degrees of weathering that the DOPQ have undergone. The data show that

Table 3. *Results of discriminant function analysis (correct identification at <0.01 significance) of petrographic and chemical properties of detrital ilmenite in Holocene sands from known sources; the number of cases are given in parentheses*

Criteria	Igneous source (%)	Metamorphic source (%)	Correctly identified (%)
Petrographic + Chemical (165 cases)	93 (88 of 95)	100 (70 of 70)	96 (158 of 165)
Petrographic only (61 cases)	68 (15 of 22)	92 (36 of 39)	84 (51 of 61)
Chemical only (165 cases)	58 (55 of 95)	81 (57 of 70)	68 (102 of 165)

Table 4. *Abundance of detrital Fe–Ti oxide minerals in igneous parent rocks and daughter sands in the Appalachian and the Rocky Mountains (in increasing order of similar durability under different weathering conditions)*

Particle type	Rocky Mountain			Appalachian Mountain		
	Sand	Rock	Sand/Rock	Sand	Rock	Sand/Rock
	%	%		%	%	
Hm	2.4	2.2	1.1	0.3	4.7	0.1
Mt-Hm	39.5	37.8	1.0	0.0	0.0	0.0
Mt	36.8	40.0	0.9	31.2	53.6	0.6
Il-Hm	11.7	8.6	1.4	35.5	23.3	1.5
Mt-Il	3.0	3.3	0.9	2.0	1.4	1.4
Il	0.5	1.5	0.3	1.1	7.1	0.2
Mt-Il-Hm	1.2	1.2	1.0	27.3	0.0	—
N_{sample}	17	13		5	22	
N_{grain}	4384	1922		1271	4870	

Hm, hematite; Mt-Hm, magnetite–hematite intergrowth; Mt, magnetite; Il-Hm, ilmenite–hematite intergrowth; Mt-Il, magnetite–ilmenite intergrowth; Il, ilmenite; Mt-Il-Hm, magnetite–ilmenite–hematite intergrowth.

(a) ilmenite and ilmenite-bearing polymineralic grains are similarly depleted or enriched in the sands relative to parent rocks from both areas, and (b) sand/rock ratios of hematite and magnetite have the highest disparity relative to ilmenite between samples from these two areas. In particular, hematite and magnetite are severely depleted in Appalachian sands. Note that some of the hematite in detrital grains is not primary; it may be a low temperature alteration product of pre-existing magnetite or Fe-rich ilmenite. Thus an original magnetite–ilmenite intergrowth may now be a hematite–ilmenite intergrowth. The hematite in grains with magnetite–ilmenite–hematite intergrowths in the Appalachian sands is probably secondary after some precursor. Ilmenite is not necessarily the most durable (resistant to weathering) of the DOPQ. But it appears to undergo similar degrees of depletion under contrasting weathering conditions dictated by not only different climates but also by different degrees of relief controlling residence

times in soil horizons. In addition, ilmenite occurs in many different kinds of rocks. Therefore, we think that ilmenite may be the most profitable DOPQ for provenance analysis.

Provenance of Renova sandstones

We are trying to determine the relative contributions of igneous and metamorphic rocks to the Renova Formation in southwestern Montana. This we attempt to do on the basis of the provenance characteristics of DOPQ as determined from the study of Holocene sands. As stated above, outcrops in the Ruby River valley, North Boulder River valley, and the Whitetail Creek valley were sampled (Fig. 1). The Holocene sands of the Rocky Mountains, which we have studied are also from the streams in these areas. Therefore, a direct comparison of DOPQ in local Holocene sands and in the Renova sandstones (outcrops) is possible.

Table 5. *Relative abundance of detrital Fe-Ti oxide minerals in Renova sandstones (tabulated according to outcrop areas)*

Particle Type	Ruby River section %	North Boulder River section %	Whitetail Creek section %
Primary			
Ilmenite	39.3	23.0	27.0
Magnetite	3.4	2.7	0.8
Hematite	1.3	0.4	0.3
Rutile	2.9	0.0	0.0
Ilmenite–hematite	8.5	6.0	34.2
Secondary			
Hematite	31.4	68.3	37.5
Rutile	13.2	0.5	0.2
Minimum lamellae width in 'exsolved' grains			
$<2\,\mu m$	78.6	92.1	85.7
2–$10\,\mu m$	21.4	7.9	14.3
$>10\,\mu m$	0.0	0.0	0.0
Number of crystallographic planes of 'exsolved' lamellae			
1 direction	42.5	22.2	1.5
2 directions	13.6	28.2	8.1
3 directions	43.9	49.2	90.4
N_{sample}	17	8	5
N_{grain}	3886	2097	1399

Table 6. *Electron probe microanalysis of representative ilmenite grains from Renova sandstones (wt% oxide)*

Analysis	MgO	Al_2O_3	TiO_2	Cr_2O_3	MnO	FeO*	Total
1	0.35	0.07	48.7	0.00	0.29	49.7	99.0
2	1.64	0.03	48.1	0.00	2.09	48.1	99.9
3	0.00	0.17	48.6	0.01	7.50	42.8	99.1
4	1.63	0.23	49.6	0.07	1.00	46.3	98.8
5	0.00	0.14	48.7	0.11	3.75	47.0	99.7
6	0.20	0.19	53.6	0.00	0.52	45.9	100.4
7	1.52	0.11	52.1	0.09	2.05	42.5	98.4
8	2.54	0.34	46.9	0.01	0.39	49.7	99.8
9	3.73	0.48	44.1	0.02	0.37	49.9	98.6
10	4.30	0.47	44.3	0.03	0.34	49.3	98.7
11	3.80	0.30	44.9	0.04	0.38	49.2	99.1
12	0.08	0.00	49.9	0.00	5.24	44.3	99.5
13	0.11	0.07	51.1	0.02	3.09	46.6	101.0
14	0.00	0.00	48.4	0.00	4.99	46.2	99.6
15	2.63	0.31	51.4	0.14	0.62	43.1	98.2
16	0.00	0.09	52.3	0.03	0.63	47.1	100.2

* All Fe is expressed as FeO

Petrographic data on the DOPQ from the Renova sandstones are summarized in Table 5, and a few representative microprobe analyses of detrital ilmenite are given in Table 6. Primary rutile is found only in the Renova samples from the Ruby River basin where large blocks of Precambrian metamorphic rocks are exposed in the surrounding potential source areas (Fig. 1). Few or only minor exposures of metamorphic rocks are presently found in the North Boulder River and Whitetail Creek basins, and primary rutile has not been found in outcrops of Renova sandstones in these areas. Rutile, a very durable heavy mineral, is derived only from metamorphic rocks (Force 1976). Therefore, we conclude that during Renova time rutile-bearing metamorphic rocks contributed little to the sediments in the two basins. It is possible that despite considerable unroofing of source rocks, the present distribution of igneous and metamorphic rocks in this area is still similar to that in the Renova time. Petrographic observations also show the presence of a significant amount of altered grains of hematite or rutile in the Renova. The origin of these grains could be diagenetic or due to outcrop weathering. However, a large fraction of the other grains retain their original intergrowth textures, although the specific phases may have undergone some diagenetic alteration (Fig. 4). In addition, microprobe analysis provides no obvious indication of

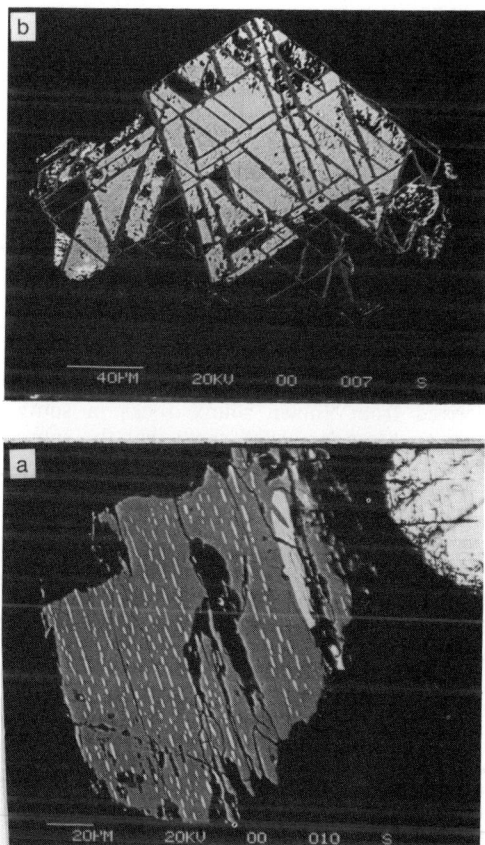

chemical modification of every ilmenite grain during diagenesis. Rather, reflected light and backscattered electron microscopy suggest that significant portions of many detrital ilmenite grains remain unaltered. Chemical compositions of detrital ilmenite in the Renova sandstones are well within range of those found in Holocene sands.

For the purpose of estimating the relative contributions of igneous and metamorphic rocks to the sediments of the Renova Formation, we have chosen to compare DOPQ of Renova sandstones *only* to those from local Holocene sands in the Rocky Mountains. A discriminant function analysis of chemical compositions of detrital ilmenite and the petrographic properties of local Holocene DOPQ derived from igneous and metamorphic rocks was performed to calculate discriminant function coefficients of the significant variables. The results show that a 100% correct determination is obtained for the Holocene samples with known source rocks (Table 7). Note that the correlation between discriminating variables for the canonical discriminant function is very good (0.969) and the Chi Squared value is very high (103.12; degree of freedom: 10) signifying a virtually 100% confidence level. These coefficients, therefore, could be used with good confidence to determine the source rocks of unknown samples. Analysis of Renova samples using the discriminant function coefficients so calculated suggests that about 77% of the DOPQ in the Renova sandstones that we have studied were derived from igneous source rocks and the other 23% were derived from metamorphic source rocks (Table 7). We have argued above, following Morton (1985a), that the population of the varieties of a single mineral species or a single mineral group in daughter sediments is more representative of their population in parent rocks than the population of many different minerals. If the DOPQ population adequately represents the clastics in the Renova, it might be concluded that igneous and metamorphic rocks contributed approximately in a 3:1 ratio to these immature sediments. The geological map of the area shows considerable outcrops of Precambrian metamorphic rocks that were the obvious source (Fig. 1). The importance of this study is not merely to infer the obvious metamorphic source in this magmatic arc (which is also independently indicated by the presence of rutile), but to estimate the relative amount of its contribution and to test the method. Our data indicate that sediments derived from an Andean type magmatic arc may consist of a very significant amount, in this case 23%, of non-igneous detritus.

Fig. 4. Backscattered electron micrographs of detrital Fe-Ti oxide minerals in Renova sandstones showing intergrowths in (**a**) one direction (presumably pinacoidal in ilmenite), and (**b**) three directions (presumably octahedral in magnetite). Note the textural similarity between the grains from Holocene sands (Fig. 3) and the Renova Sandstones.

Table 7. *Classification of DOPQ in Renova sandstones on the basis of the results of discriminant function analysis of petrographic and chemical properties of detrital ilmenite in Holocene sands in Montana; the number of cases are given in parentheses*

	Igneous Source	Metamorphic Source	Canonical Correlation	χ^2	DF*
Holocene sands (used for calculating discriminant function coefficients)	100% (14 of 14)	100% (30 of 30)	0.969	103.12	10
Renova sandstones	77% (193 of 252)	23% (59 of 252)			
Ruby River section	66% (82)	34% (43)			
N. Boulder River section	80% (60)	20% (15)			
Whitetail Creek section	98% (51)	2% (1)			

* DF, degrees of freedom.

Other heavy minerals

Although we have used ilmenite as an index mineral for our case study, other minerals such as pyroxene, garnet, zircon, etc. might as well have been used (cf. Cawood 1983; Morton 1985a; Owen 1987). Properties of these minerals in local Holocene sands derived exclusively from known source rocks would serve as the basis for setting up quantitative criteria for determining the provenance of local Tertiary sandstones e.g. the Renova. However, the results of such determinations would not be necessarily identical for each mineral. Thus, instead of the 77:23 ratio (igneous:metamorphic) indicated by detrital ilmenite, each mineral might indicate a different ratio. The result would be an array of several independently solvable equations for only one unknown i.e. the ratio of igneous and metamorphic contributions to the Renova. A truly rigorous assessment of relative contributions from igneous and metamorphic source rocks could then be obtained by solving this overdetermined system with least square mixing calculations (e.g. Wright & Doherty 1970). In addition to determining the igneous to metamorphic source rock ratio, it would also be possible quantitatively to discriminate contributions from different tectonic provenance with suitable samples.

Conclusions

(1) Combined analysis of chemical and petrographic properties (varieties of intergrowth and exsolution textures) of DOPQ is a reliable guide to determining provenance characteristics of sediments derived from igneous and metamorphic source rocks.

(2) The relative abundances of ilmenite in daughter sands and parent rocks in semi-arid and humid climates are similar; ilmenite is also ubiquitous in all crystalline common parent rocks. Therefore, of the Fe–Ti oxide minerals ilmenite is probably the most suitable for varietal studies.

(3) Discriminant function analysis of petrographic properties of DOPQ and chemical compositions of detrital ilmenite in Holocene sands derived from known source rocks in southwestern Montana, USA show that the source rocks of *all* the samples are correctly identified.

(4) The discriminant function so calculated is used to determine the provenance of DOPQ of the magmatic arc facies of the Tertiary Renova Formation in southwestern Montana. The results indicate 77% contribution from igneous source rocks and 23% from metamorphic source rocks.

The Renova samples we used were collected mostly by J. Olson. We are most grateful to her not only for the use of the samples but also for the excellent documentation she provided for each sample. The electron microprobe and the scanning electron microscope of Indiana University were obtained and then maintained with support from the NSF, NASA, and Indiana University Foundation. Reviews by D. Darby, E. Merino, B. Shukla and L. J. Suttner together with the expert editorial handling by A. Morton considerably improved the script although the authors are solely responsible for the conclusions. This research was supported in part by grants from the NSF (USA) and the NATO-CNR (Italy).

References

BASU, A. 1976. Petrology of Holocene fluvial sands derived from plutonic source rocks: implications to paleoclimatic interpretation. *Journal of Sedimentary Petrology*, **46**, 694–709.

—— & HOOD, L. 1985. Provenance significance of detrital opaque oxide minerals in Lake Erie sands near Sandusky, Ohio (abstract). *Geological Society of America, Abstracts & Program*, **17(5)**, 279.

—— & MOLINAROLI, E. 1989. Provenance characteristics of detrital opaque Fe-Ti oxide minerals. *Journal of Sedimentary Petrology*, **59**, 922–934.

CALLENDER, D. L. & FOLK, R. L. 1958. Idiomorphic zircon, key to volcanism in the lower Tertiary sands of central Texas. *American Journal of Science*, **256**, 257–269.

CAWOOD, P. A. 1983. Modal composition and detrital clinopyroxene geochemistry of lithic sandstones from the New England Fold Belt (east Australia): A Paleozoic forearc terrane. *Geological Society of America Bulletin*, **94**, 1199–1214.

CLIFF, R., DREWERY, S. & LEEDER, M. 1991 Sourcelands for Carboniferous rivers and deltas of the Permian Basin: radiometric and geological evidence. (This Volume.)

COMPSTON, W. & PIDGEON, R. T. 1986. Jack Hills, evidence of very old detrital zircons in Western Australia. *Nature*, **321**, 766–769.

DARBY, D. A. & TSANG, Y. W. 1987. Variation in ilmenite element composition within and among drainage basins: Implications for provenance. *Journal of Sedimentary Petrology*, **57**, 831–838.

FORCE, E. R. 1976. Metamorphic source rocks of titanium placer deposits—a geochemical cycle. *USGS Professional Paper*, **959-B**, 1–16.

FIELDS, R. W., RASMUSSEN, D. L., TABRUM, A. R. & NICHOLS, R. 1985. Cenozoic Rocks of the Intermontane Basins of Western Montana and Eastern Idaho: A Summary. *In*: FLORES, R. M. & KAPLAN, S. S. (eds) *Cenozoic Paleogeography of the West-Central United States*. The Rocky Mountain Section, Society of Economic Paleontologists and Mineralogists, Denver, Colorado, 9–36.

GRIGSBY, J. D. 1988. Fe-Ti oxides in provenance studies (abstract). *Geological Society of America, Abstracts & Program*, **20(5)**, 345.

HISCOTT, R. N. 1978. Provenance of Ordovician deepwater sandstones, Tourelle Formation, Quebec, and implications for initiation of the Taconic orogeny. *Canadian Journal of Earth Science*, **15**, 1579–1597.

KOBER, B. 1987. Single-zircon evaporation combined with Pb+ emitter bedding for $^{207}Pb/^{206}Pb$-age investigations using thermal ion mass spectrometry, and implications to zirconology. *Contributions to Mineralogy and Petrology*, **96**, 63–71.

KRYNINE, P. D. 1946. The tourmaline group of sediments. *Journal of Geology*, **56**, 65–87.

KUENZI, W. D. & FIELDS, R. W. 1971. Tertiary stratigraphy, structure and geologic history, Jefferson basin, Montana. *Geological Society of America Bulletin*, **82**, 3374–3394.

MOLINAROLI, E. & BASU, A. 1987. Studio di minerali opachi in sabbie fluviali oloceniche e nelle corrispondenti rocce madri di zone sottoposte a climi diversi (Montagne Rocciose e Monti Appalachi in U.S.A.). *Rendiconti della Società Italiana di Mineralogia e Petrologia*, **42**, 271–283.

MONROE, J. S. 1981. Late Oligocene–Early Miocene facies and lacustrine sedimentation, upper Ruby River basin, southwestern Montana. *Journal of Sedimentary Petrology*, **51**, 939–951.

MORTON, A. C. 1984. Stability of detrital heavy minerals in Tertiary sandstones from the North Sea basins. *Clay Minerals*, **19**, 287–308.

—— 1985a. Heavy minerals in provenance interpretation. *In*: ZUFFA, G. G. (ed.) *Provenance of Arenites*. Reidel, 249–277.

—— 1985b. A new approach to provenance studies: electron microprobe analysis of detrital garnets from Middle Jurassic sandstones of the North Sea. *Sedimentology*, **32**, 553–566.

—— 1987. Influences of provenance and diagenesis on detrital garnet suites in the Paleocene Forties Sandstone, central North Sea. *Journal of Sedimentary Petrology*, **57**, 1027–1032.

OWEN, M. R. 1987. Hafnium content of detrital zircons, a new tool for provenance study. *Journal of Sedimentary Petrology*, **57**, 824–830.

RIEZEBOS, P. A. 1979. Compositional downstream variation of opaque and translucent heavy residues in some modern Rio Magdalena sands (Colombia). *Sedimentary Geology*, **24**, 197–225.

SHUKLA, B. 1988. Provenance of Oligocene Lough Neagh Group, Northern Ireland, United Kingdom (abstract) *In*: JAMES, D. P. & LECKIE, D. A. (eds) *Sequences, Stratigraphy, Sedimentology: Surface and Subsurface*. Memoir of the Canadian Society of Petroleum Geology, **15**, 583–584.

WRIGHT, T. L. & DOHERTY, P. C. 1970. A linear programming and least square computer method for solving petrologic mixing problems. *Geological Society of America Bulletin* **81**, 1995–2008.

The role of fission track dating in discrimination of provenance

ANTHONY J. HURFORD & ANDREW CARTER

University of London Fission Track Research Group, Department of Geological Sciences, University and Birkbeck Colleges, Gower Street, London WC1E 6BT, UK

Abstract: Determination of single crystal ages by fission track dating analysis of zircons (and subordinately apatites) within a detrital heavy mineral assemblage, together with frequency distribution and radial plot analysis of the measured ages, permits identification of different age modes. Such modes may characterize the provenance of a sediment and assist in defining source areas. A limitation of the application is the thermal stability of tracks within a mineral above which tracks are lost and ages partially reset: in broad terms over geological time 200–250°C for zircon and < 100°C for apatite. Synopses are presented of published fission track provenance studies in north central Victoria, northeastern Tasminia, Barbados, the Rockall Trough, the English Wealden, Mexico, Utah and the Central Alps.

The ability of a geochronometric method to determine ages of individual mineral grains within a detrital heavy mineral population offers a tool for the discrimination of different age modes, which in turn provides a guide to possible source areas and sediment provenance. Fission track dating analysis has the power to determine single crystal ages and although there has been limited application of the method to sediment provenance studies, the method clearly has a high potential for provenance discrimination. It is the purpose of this short contribution to review briefly the fission track method and to draw together published case histories of its application to provenance studies in Northern Victoria, Tasmania, Barbados, the Rockall Trough, the English Wealden, Mexico, central Utah, and the Central Alps.

Fission track dating: a summary

Linear radiation damage tracks from the natural spontaneous fission of ^{238}U accumulate in the crystal lattice of a host mineral, and may be revealed by chemical etching, and counted using high power optical microscopy (Fleischer *et al.* 1975; Naeser & Naeser 1984; Hurford 1986a). The minerals with appropriate (ppm) trace levels of uranium and to which the fission track method has been principally applied, are zircon, apatite and subordinately titanite. As with other forms of radioactivity, ^{238}U spontaneous fission decay is exponential, with a fission half-life of *c.* 9×10^{15} a. The numerical density of spontaneous fission tracks (ρ_s) counted on a polished and etched internal surface of a mineral (Fig. 1) is function not only of the rate of spontaneous

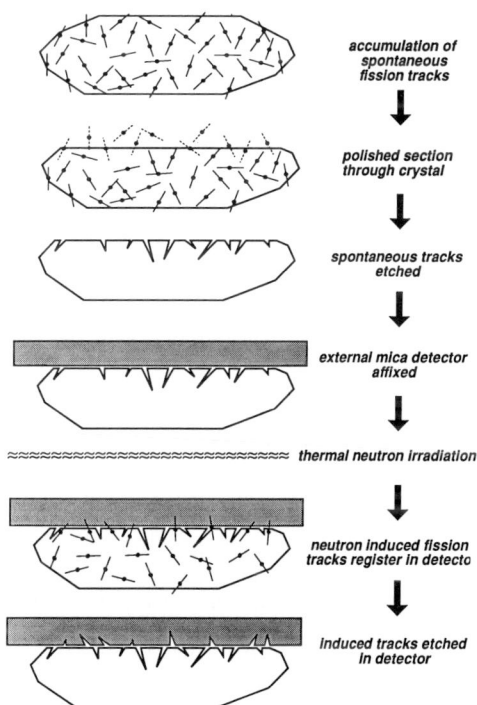

accumulation of
spontaneous
fission tracks

↓

polished section
through crystal

↓

spontaneous tracks
etched

↓

external mica detector
affixed

↓

thermal neutron irradiation

↓

neutron induced fission
tracks register in detecto

↓

induced tracks etched
in detector

Fig. 1. The external detector method of fission track dating. A grain mount of zircons is polished and etched to reveal ^{238}U spontaneous fission tracks. An external detector of low-uranium muscovite is affixed and this sandwich irradiated. Neutron-induced ^{235}U tracks pass into the detector and are subsequently etched to reveal a mirror-image impression of the crystal. The ratio of spontaneous to induced tracks counted over an identical area gives the proportion of total uranium which has fissioned and is the basis of calculating a single-crystal age.

From Morton, A. C., Todd, S. P. & Haughton, P. D. W. (eds), 1991,
Developments in Sedimentary Provenance Studies.
Geological Society Special Publication No. 57, pp. 67–78.

fission decay, but also the time during which tracks have been accumulating, and of the uranium content of the crystal: the greater the uranium content, the greater the track density.

A convenient method of empirically determining the uranium content is to irradiate the sample with a monitored fluence of thermal neutrons (i.e. neutrons with a mean energy <0.025 eV), which induces a fraction of ^{235}U atoms to fission, producing a second population of fission tracks, whose numerical density is designated ρ_i. These tracks may be recorded in an external detector of low-uranium mica held against the sample during irradiation (Fig. 1). The ratio of spontaneous to induced fission tracks (ρ_s/ρ_i) counted for each sample represents the fraction of total uranium which has undergone natural fission. In the external detector method approach (see Gleadow 1981), determination of spontaneous/induced track count ratios for each crystal in a fission track analysis yields individual ages for each grain. The neutron fluence is monitored by including in the irradiation package a dosimeter glass of known uranium content, and measuring the neutron-induced ^{235}U fission track count similarly recorded in an adjacent detector. Allowance is made for the use of different isotopes in determining spontaneous and induced track densities by inclusion of the ^{235}U/^{238}U isotope abundance ratio in the age equation (Naeser 1967).

Uncertainty over the value of the ^{238}U spontaneous fission decay constant λ_f and difficulties in absolute evaluation of the neutron fluence giving the n,f reaction with sample ^{235}U (Hurford & Green 1981, 1982) necessitates, for external detector method analyses, the use of a system calibration based on comparison with samples of known age (the zeta calibration of Fleischer & Hart 1972), as recommended by the IUGS Subcommission on Geochronology (Hurford 1990). The choice of material as age standards for the calibration baseline requires strict control and appropriate standards have been proposed and tested (see e.g. Green 1985; Hurford 1990).

Exposure of fission tracks to elevated temperatures results in a fading or annealing of the damaged lattice, with the diffusion of displaced ions back to their original lattice positions, although the exact mechanism of annealing is uncertain. Of the various natural factors which might influence annealing, temperature has been shown to be by far the most important (Fleischer et al. 1965). Various attempts have been made to establish the effective temperatures for various minerals above which fission tracks are annealed, these experiments usually taking the form of heating in the laboratory at various temperatures and for various times. Subsequent extrapolation of these data to geological time defines a 'closure temperature' (Dodson 1973) below which the fission tracks are effectively retained.

In apatite, such closure temperature estimates fall in the range between 75–130°C, these values showing a time dependency given by the rate of cooling which may vary between 1 and 100°C Ma^{-1} (Wagner & Reimer 1972; Haack 1977; Gleadow & Lovering 1978). Painstaking studies using both laboratory and natural annealing experiments (see Green et al. 1989 and references therein), have yielded highly detailed qualitative and quantitative descriptions of annealing in apatite. These studies relate fission track length and density reduction in apatite to temperature and apatite chemistry, and offer the basis of a palaeotemperature indicator and a tool for thermal history reconstruction. In contrast to the theoretical modelling of closure temperatures (see for example Dodson & McClelland-Brown 1985), Laslett et al. (1987) and Green et al. (1988) demonstrate a lack of evidence to substantiate the description of track annealing in apatite by first-order kinetics, showing rather that experiment supports a model which allows for differences in activation energy for annealing.

In comparison with apatite, the track annealing parameters for zircon are poorly defined. The extrapolation of limited laboratory annealing data to geological time, suggests an effective closure temperature of >300°C (Fleischer et al. 1965; Krishnaswami et al. 1974). Geological estimates based, primarily, on the relationship of fission track zircon ages to ages measured on other co-existing minerals, favour an effective closure temperature of 175–250°C (Gleadow & Brooks 1979; Harrison et al. 1979; Zeitler et al. 1982; Hurford 1986b). This difference is not readily explicable but may relate to the presence of fluid phases, to pressure, to the chemical composition of the zircon or simply to the current inadequate base of experimental annealing data. On the basis of geological evidence, 200–250°C represents a reasonable estimate of closure temperature for retention of tracks in zircon.

From these considerations of track retention with increased temperature, zircon is the primary candidate for use in attempting to discern provenance. This application was first suggested by Gleadow & Lovering (1974) who, in a comparison of ages from fresh syenite and from soil derived from weathered syenite, noted the stability of tracks in zircon despite extreme weathering. The following synopses of published case histories illustrate the diversity of the application.

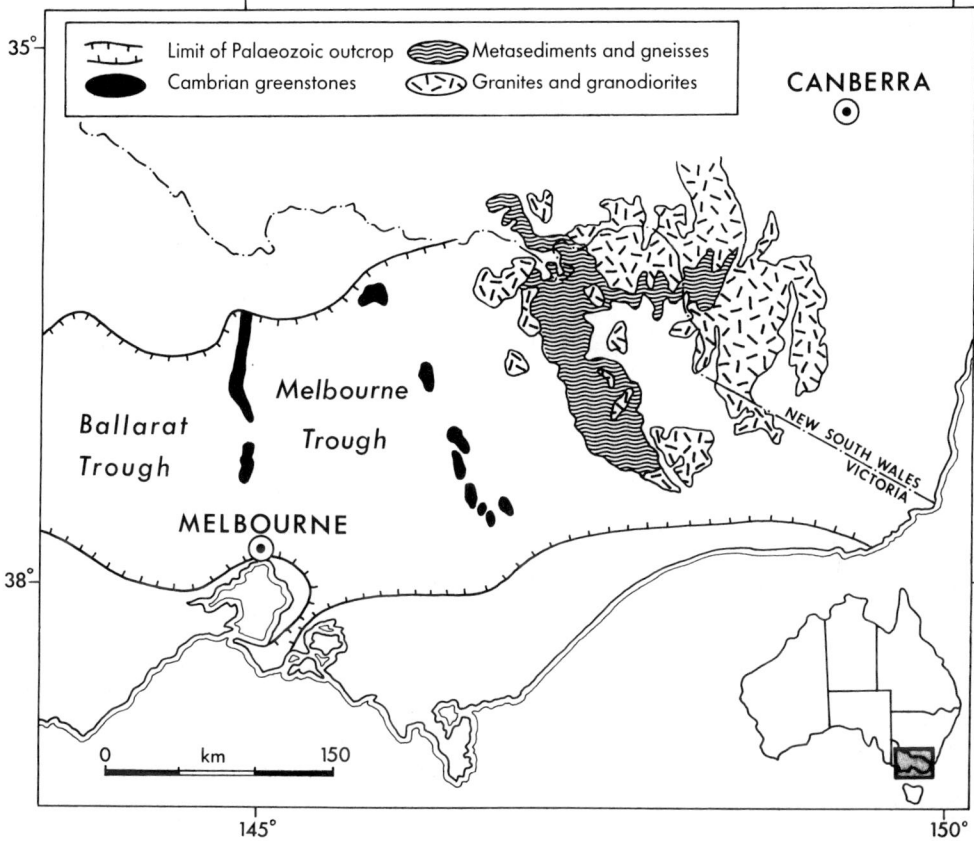

Fig. 2. Geological sketch map of eastern Victoria and New South Wales, showing sampling area and distribution of probable granitic and metamorphic source rocks; (after McGoldrick & Gleadow 1977).

Applications of fission track dating to provenance

Palaeozoic sandstones of north central Victoria, Australia

The first published application of the method to provenance studies by McGoldrick & Gleadow (1977) examined unfossiliferous sandstone samples from the area of Tatong in eastern Victoria, Australia (Fig. 2). These steeply dipping sandstones are associated spatially with the Victorian greenstones, Cambrian basalts and dolerites which have been subject to burial metamorphism to lower greenschist facies grade. Although mapped as lower Ordovician in age on the basis of lithological similarity to rocks of established age elsewhere in Victoria, the sandstones are unlike other lower Ordovician rocks in the Tatong area. Fission track ages measured

on individual zircon crystals from the sandstones varied between 300 ± 93 Ma and 718 ± 125 Ma, with a mean age of 427 ± 20 Ma, centrally placed within the spread of 390–460 Ma ages given by Rb-Sr and K-Ar analysis of granitic rocks on the Victorian–New South Wales borders (e.g. Evernden & Richards 1962) and further east in New South Wales (Williams et al. 1975; Roddick & Compston 1976), offering strong indication that the majority of dated zircons were derived from these eastern highlands granitic rocks. Provided the fission track ages have been thermally unaffected subsequent to deposition, then the mean age of 427 ± 20 Ma also provides a maximum age for timing of deposition.

Alluvial deposits, northeastern Tasmania

Crossing the Bass Strait to northeastern Tasmania, Yim et al. (1985) utilized fission track analy-

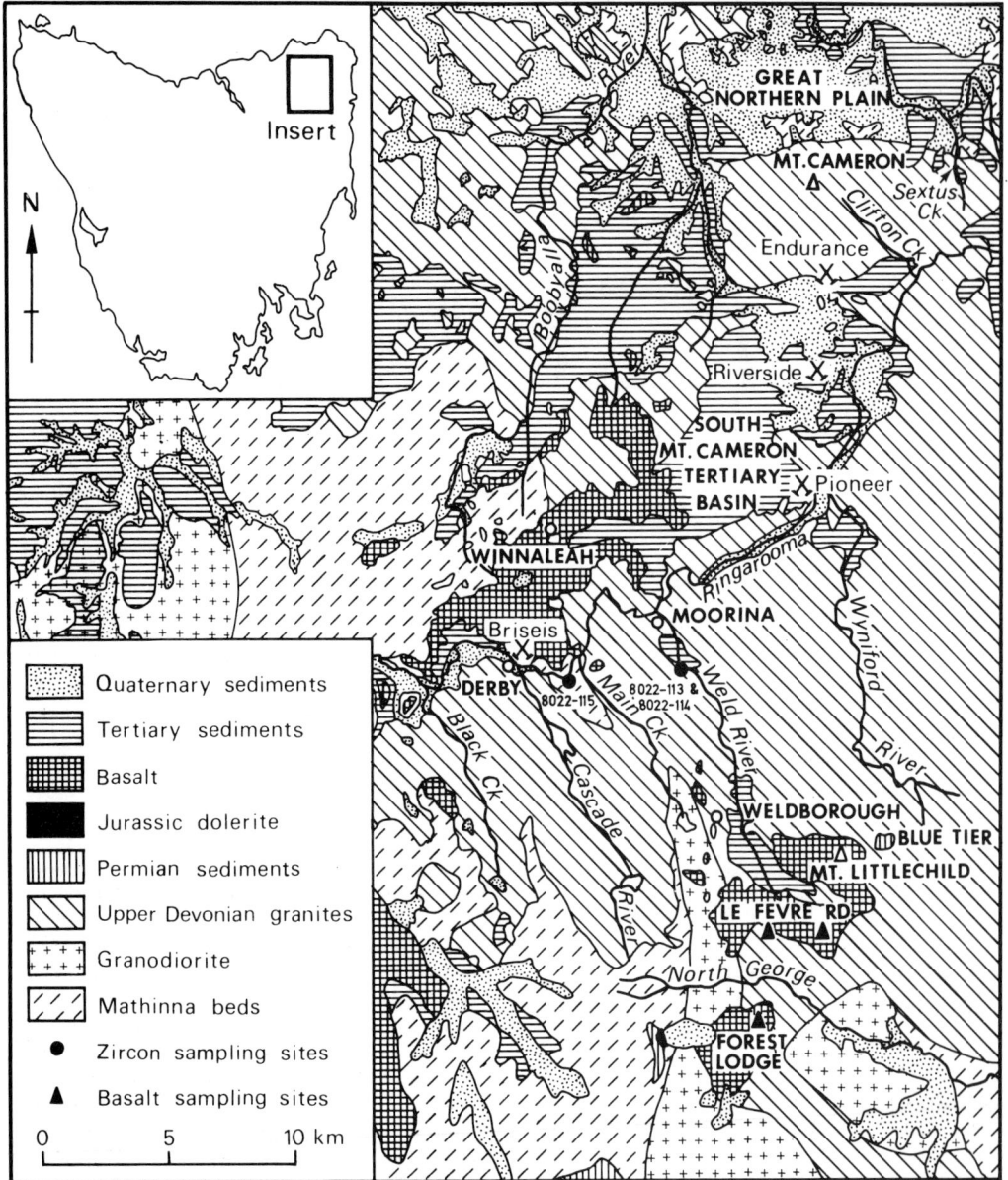

Fig. 3. Geological sketch map of NE Tasmania showing the Blue Tier and Middle Ringarooma valley, together with zircon and basalt sampling sites; (from Yim *et al.* 1985).

sis of zircons from cassiterite-bearing alluvial deposits to help differentiate zircon populations and to define probable source areas. Yim (1980) identified three heavy mineral associations within the Cenozoic sediments of NE Tasmania and noted that these contain two very distinctive zircon groups: well-formed euhedra <1 mm in

size and well- to sub-rounded zircons ranging from 2 mm to 1 cm in size. Sources for the zircons appeared to be granitic rocks of the Blue Tier batholith range (Fig. 3) or conceivably Cenozoic basaltic volcanism. Two samples were collected, with one showing both morphologies of zircon (8022-133 & 114), whilst the other

Table 1. *Single-crystal fission track ages of alluvial zircons from NE Tasmania (after Yim et al. 1985)*

Sample	Zircon size (mm)	Shape	Crystal	Spontaneous/induced track counts	Age (Ma $\pm$ 1σ)
8022-113	0.1–0.2	euhedra	♯1	426/235	405 ± 33
			♯2	297/177	375 ± 36
			♯3	186/128	326 ± 37
			♯4	430/261	369 ± 29
			♯5	221/158	314 ± 33
			♯6	290/170	381 ± 37
8022-114	2–5	rounded	♯1	281/341	48 ± 4
			♯2	158/213	43 ± 5
			♯3a	128/173	43 ± 5
			♯3b	187/235	46 ± 5
			♯4	344/422	48 ± 3
			♯5	166/206	47 ± 5
			♯6	154/194	46 ± 5
8022-115	2–5	rounded	♯1	599/774	46 ± 2
			♯2	525/622	50 ± 3
			♯3	577/693	49 ± 3
			♯4	472/610	46 ± 3
			♯5	262/369	42 ± 3
			♯6	154/184	50 ± 5

(8022-115) revealed only the larger grains. A direct correlation between fission track zircon age and crystal form (see Table 1) gives 2 clear mean age modes. The finer euhedral zircon mean age of 367 ± 15 Ma agrees well with the late Devonian age proposed by Groves *et al.* (1977) for the Blue Tier batholith, and with quantitative estimates by K-Ar biotite and Rb-Sr mineral isochron determinations ranging from 370 to 384 Ma (McDougall & Leggo 1965, IUGS constants). The early Eocene mean age of the 2 coarser zircon fractions, 46.7 ± 0.6 Ma, predates a Rb-Sr estimated age of c. 16 Ma (Brown 1977) for basalts around Winnaleah, immediately northwest of the sample sites, necessitating an additional magmatic source for the zircons.

Subsequent to the fission track study by Yim *et al.*, K-Ar ages of the Weldborough basalt at Mt Littlechild, on the present drainage pattern some 10 km upstream from the zircon sample sites, yielded ages between 46.2 and 47.4 Ma, identical with the younger mean fission track zircon age. The occurrence of similar coarse and well-rounded zircons to the north of Winnaleah in the Boobyalla Valley indicates an earlier drainage pattern with NW-flowing rivers originating in the Blue Tier mountains. Eruption of mid-Miocene basalts at Winnaleah onto a landform many of whose present-day features were already established, effectively blocked the upper drainage system, diverting it northeast to connect with Ringarooma River system, as seen today. Formation of the placer deposits is conveniently bracketed as younger than 46 Ma

by the inclusion in the placers of early Eocene zircons, and older than 16 Ma by the occurrence of coarser, rounded zircons NW of the 16 Ma-basalts and by presence of some stanniferous deposits beneath the mid-Miocene basalts.

Eocene sandstones, Scotland District of Barbados

Working in the Scotland District of northeastern Barbados, West Indies, Baldwin *et al.* (1986) analysed zircons from sandstones from the accretionary wedge of the Lesser Antilles arc, with the aim of defining the source areas of these terrigenous sediments and constraining more exactly the timing of their deposition. Paucity of fossils in the sandstones has led to uncertainty in the age and environment of their deposition, although it has been often considered that they derive from the South American continent. Allochthonous fauna in the terrigenous sediments suggested ages from Cretaceous to Eocene, although many fossils were rolled and worn. Foraminifera and radiolaria within the sediments of the Scotland District indicated a mid-Eocene age (Torrini *et al.* 1985 and references therein).

Fission track analysis of eight sandstone samples yielded a wide mixture of individual zircon crystal ages, within which Baldwin *et al.* considered that three groupings from 20 to 80 Ma, 200 to 350 Ma and >500 Ma could be discerned. Precise interpretation of these spreads in fission track zircon age is difficult in an area of such structural complexity and multiplicity of possible provenances especially considering the paucity of isotopic data in those areas. However, various possible sources for these zircons are forwarded by the authors: contemporaneous volcanic ash falls from the Aves Ridge, Lesser Antilles arc and the Netherland–Venezuelan Antilles arc, as well as mid-Tertiary uplift of the western Caribbean Mountains of Venezuela are mooted as possible suppliers of 20–80 Ma crystals. Baldwin *et al.* suggest that the youngest ages of *c.* 20 Ma may point to an Oligocene age for the Scotland District rocks rather than the mid-Eocene age indicated by microfauna. Speculation as to the source of the 200–350 Ma component includes partially annealed cratonic material, an Andean element resulting from the collision of the Americas *c.* 280 Ma ago, or a Triassic rifting event. The > 500 Ma zircons are ascribed to the South American craton, possibly the Guyana shield.

Palaeogene sediments of DSDP Leg 81, Site 555: the southwest margin of the Rockall Plateau

Examining apatite and titanite crystals recovered from six samples from 336 to 919 m sub-bottom depths, Duddy *et al.* (1984) reported the contribution of detritus from source rocks of at least 3 different age groups. Titanite crystals revealed two distinct individual crystal age groupings, a Tertiary group of 62 ± 3 Ma and four crystals which gave a Precambrian age of 1381 ± 95 Ma. Individual apatite crystal ages defined a major early Tertiary group at 56 ± 4 Ma which, with the Palaeocene titanite ages, can be attributed to either near contemporaneous volcanism associated with the opening of the NE Atlantic (Morton & Parson 1988) and/or the rapid unroofing of high level intrusives such as those in Greenland and Scotland. Cretaceous apatite ages of 100 ± 20 Ma are considered consistent with known high level microgabbroic intrusives on Helen's reef, although the authors note that such apatite ages could also be derived from the basement of ancient terranes, especially around rifted continental margins where there may have been low-level heating associated with uplift.

As we have noted above, tracks in apatite are annealed at relatively low temperatures and thus care must be taken when attempting to relate fission track ages of apatites in sediments to possible provenance. Duddy *et al.* acknowledge this by measuring the lengths of confined fission tracks in their apatites. The distribution of fission track lengths provides a sensitive record of thermal history, including possible post-depositional heating (for further description see Green *et al.* 1989 and references therein). Duddy *et al.* found that the mean track lengths and length distributions of the apatites with Tertiary and some Cretaceous ages indicate that the present temperatures are the maximum experienced since deposition, and thus those ages can be used with some confidence to identify possible provenance sites.

Early Cretaceous sandstones from the English Wealden

In the four studies summarized above, data were presented as raw crystal ages or as a form of histogram of those measured ages. Definition of modes within the spread of data from the Scotland District sandstones appears particularly speculative. In each case no regard was made for the analytical uncertainties which, of necessity, are usually large for individual crystal ages (see e.g. Table 1) since they derive from the actual numbers of tracks counted.

In a study of two Early Cretaceous sandstones from the English Wealden (Hurford *et al.* 1984), the probability distributions of each individual age and its attendant 1σ analytical uncertainty (the conventional error of Green 1981) were summed to form a more meaningful, weighted distribution of ages for the whole zircon population (Fig. 4). Such a distribution shows the frequency of occurrence of crystals within given age intervals. Note that although Hurford *et al.* suggested that more resolution of the age spectrum might be obtained by calculating the frequency curves using individual crystal uncertainties of $\pm 0.5\sigma$ such a practice is hard to justify, the logical conclusion being that each different measured age is identified as a separate age mode. At the 1σ level, the age spectra from the two Wealden samples are broadly similar with major peaks in the mid-Jurassic and pronounced tails through the Palaeozoic into the Precambrian. In addition, the Netherside Sand sample revealed a distinct late Permian subordinate age peak, only suggested in the Ashdown Sands. Clearly the zircon data indicate that the Wealden has a complex provenance, with multiple sources and/or reworking of sediments of intermediate age. The spectra admit the contribution of Precambrian, Caledonian and Cornubian components. Extensive late Permian–early Triassic volcanics occur in Aquitaine, Biscay and Iberia

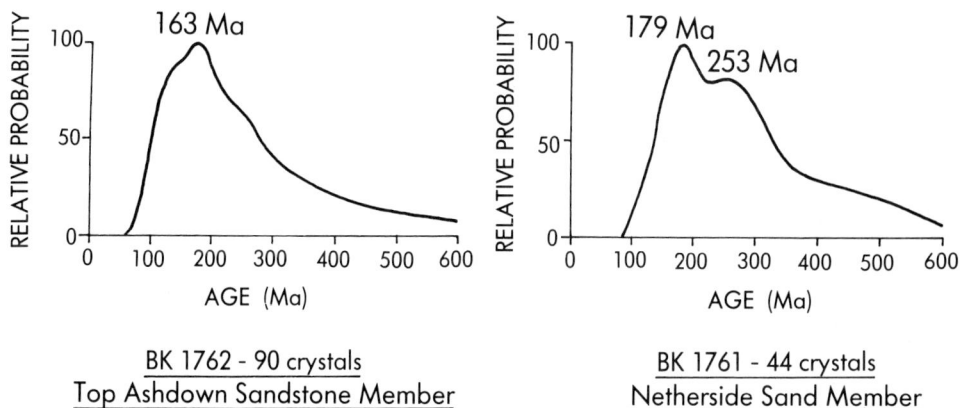

Fig. 4. Comparison of probability distribution curves ('age spectra') for the summed single crystal zircon ages from two Wealden sandstones. Age modes of *c.* 170, *c.* 260 are indicated; (after Hurford *et al.* 1984).

(Carte géologique de la France 1980), suggesting a southerly source area, whilst mid Jurassic volcanism (e.g. Dixon *et al.* 1981) is available in the North Sea as a source candidate for the 160–175 Ma zircon age peaks.

Volcanically-derived sediments from Mexico and Utah

Using the fission track dating technique on individual crystals from a single sample, the analysis of the resultant data using the age spectra probability curve concept of Hurford *et al.* (1984) permits not simply the identification of different age modes but, in volcanically-derived sediments, reveals precisely any contamination by older grains, either during eruption itself, or by subsequent reworking of the sediment. Such analysis by Kowallis *et al.* (1986) of their zircon age data sets produced probability distribution age spectra for two tuffaceous Pliocene sandstones from central Guanajuato, Mexico. Both localities are associated with the important vertebrate fossil assemblages and the study was aimed at securing the age of sediment deposition. At least two separate components with peaks at 4.6, 11.0 & 24.5 Ma (El Ocote), and 3.6 & 30.7 Ma (Rancho Viejo) (Fig. 5a) were identified, rendering meaningless the respective mean zircon ages of 17.9 and 19.6 Ma. In a search for source areas for the measured zircon age modes, Kowallis *et al.* comment that K-Ar ages of 13 and 30 Ma have been determined from nearby andesitic and rhyolitic lavas, and that numerous Pliocene extrusive rocks to the south and west of the central Mexican volcanic belt (Luhr *et al.* 1985) represent probable sources for the youngest zircons.

In a further study, Kowallis *et al.* collected a suite of bentonite samples for the late Jurassic–early Cretaceous Morrison and Cedar Mountain Formations in the Colorado Plateau of central Utah. The euhedral zircon grains were separated by hand-picking from the obviously rounded, abraded and coloured crystals, and dated, the resulting age spectra being shown in Fig. 5b. Six samples yielded results compatible with stratigraphy and other isotopic data, but (despite the hand-picking) three samples were significantly older, NTM-25 showing clear bimodality, whilst the broader skewed spectrum of NTM-11 probably represents two age populations that are too close to be resolved into separate peaks.

Bergell intrusives of the Central Alps

Wagner *et al.* (1979) reported a very different application of the fission track method to provenance. Apatite ages determined on boulders of granite, derived from the Bergell intrusive complex in the Central Alps, but now within a late Oligocene conglomerate of the Po Plain molasse at Pedrinate, northern Italy, permit reconstruction of the early uplift history of the Bergell massif. Apatite ages of 23.4 ± 1.0 to 25.9 ± 0.8 Ma for Bergell boulders substantially predate the ages measured for in situ Bergell samples, which themselves show a dependency upon sample elevation: 11.2 Ma at 310 m to 16.8 ± 0.8 Ma at 2340 m. (This direct age v. elevation relationship results from the earlier passage of the higher sample below the temperature for the retention of fission tracks during uplift). By comparing the boulder and outcrop apatite ages, Wagner *et al.* 'restored' the boulders to their original vertical position within

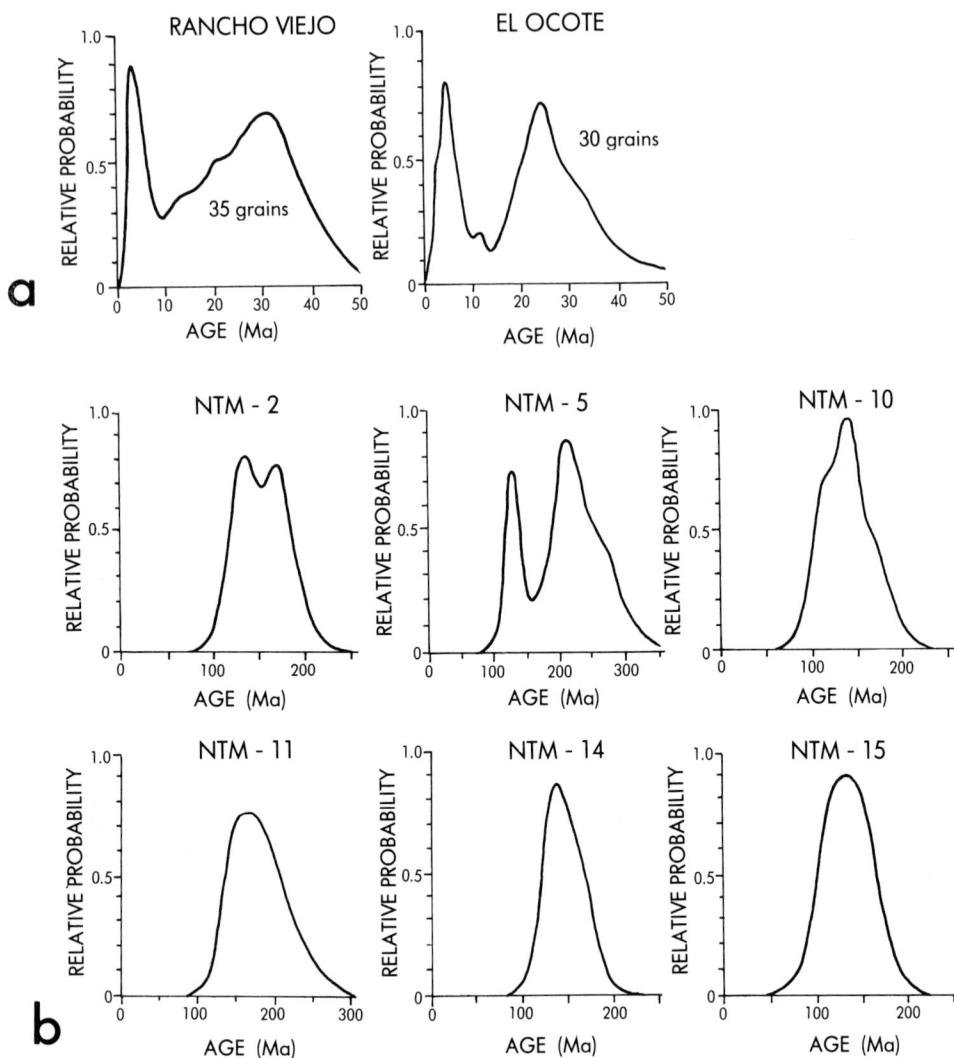

Fig. 5. (a) Age spectra for two Pliocene sandstone samples from central Guanajuato, Mexico; each reveals at least two age components demonstrating that a pooled age would be meaningless; (from Kowallis *et al.* 1986). (b) Examples of age spectra for late Jurassic and early Cretaceous bentonite samples from Utah; NTM-2, 5 and 11 reveal evidence of more than one age mode; (after Kowallis *et al.* 1986).

the Bergell intrusive: using an uplift rate of 0.7 mm a^{-1} for the Miocene and late Oligocene, the analysed boulders were replaced at altitudes of 9600 to 7800 m above present sea level (Fig. 6). The difference between these heights and the present maximum elevations indicates that some 7 km of erosion of the Bergell massif has occurred since late Oligocene times. Further, since the boulders occupied the same stratigraphic level in the molasse, there must have been at least 1800 m of palaeorelief of the Bergell granite in the late Oligocene. Wagner *et al.* conclude their report by noting that the present-

day offset of outcrop of the Bergell granite and the Pedrinate molasse indicates a dextral movement of some 30 km on the Insubric line.

Further considerations

An alternative device for analysing individual crystal data sets has been described recently by Galbraith (1988, 1989): similar to the probability distribution, his radial plot makes allowance for the differing precisions of single crystal ages, but presents them in a way which permits more

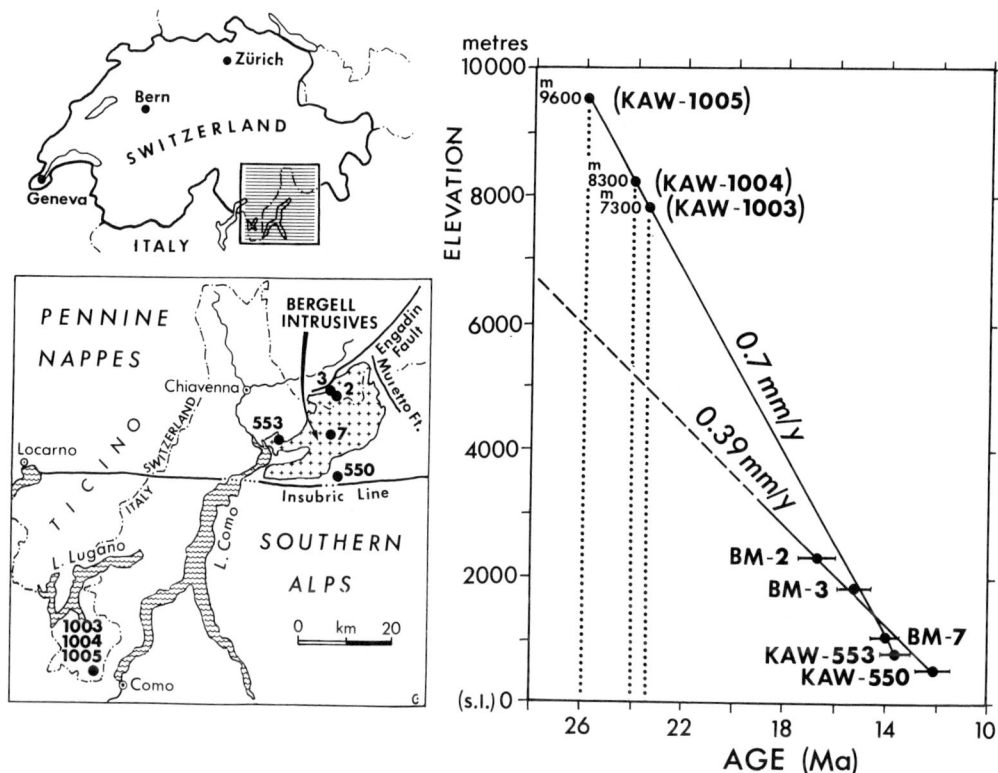

Fig. 6. Determination of the palaeo-elevation (referred to present sea level) of boulders of Bergell granite found in the late Oligocene molasse at Pedrinate. Uplift rate of 0.39 mm a^{-1} was defined from apatite age v. elevation; the earlier uplift rate of 0.7 mm a^{-1} from mid-Miocene resulted from comparison of ages from different minerals and radiometric methods. The inset shows sample localities. (After Wagner et al., 1979).

ready comparison. Figure 7 shows an example of the radial plot where the data discussed above for 44 zircon crystals from a Wealden sandstone (Hurford et al. 1984) are shown as an x,y scatter plot. Here x indicates the precision of the individual age estimates and y is a standardized estimate, such that each point has the same standard error in the y direction. Further the plot includes a circular age scale such that the age of any grain may be determined by extrapolating a line from the origin (0,0) through the crystal's (x,y) co-ordinates to intercept the age scale. The approximate relative error of each age is shown on the abscissa, whilst the standardized error on the ordinate makes the visual comparison of the precision of age estimates straightforward: points which agree with each other within error will scatter homoscedastically, with unit standard deviation, about a line through the origin. In Fig. 7 the data are consistent with three such lines. A mixture model of three ages of 168 Ma (154–183 Ma at 95% confidence level), 264 Ma (243–287 Ma) and 448 Ma (398–504 Ma) may be fitted to the data, as shown in

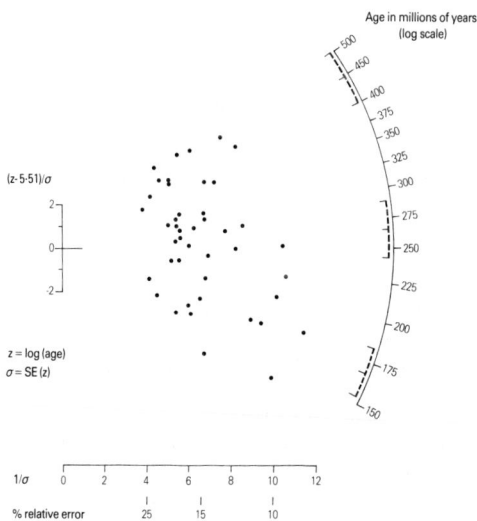

Fig. 7. Radial plot of Wealden zircon data (Hurford et al. 1984) showing three maximum likelihood estimates fitted through the points c. 168, c. 264 & c. 448 Ma; (from Galbraith 1988).

Fig. 7 (Galbraith 1988). The radial plot has a significant advantage over the probability curve approach since the latter combines two sources of variation: uncertainty in the single age estimate and variation in true ages between crystals.

As shown in the above studies, different age modes can be distinguished with relatively small numbers of crystal ages, this differentiation being enhanced by the use of graphical devices such as the radial plot (Fig. 7). Clearly larger data sets are desirable but also highly demanding of an analyst's time. In addition the possibility of unintentional bias invading a data set should be guarded against, this bias originating from inappropriate etching and counting, and (as considered in most of the above studies), from the frequent inability to count certain high uranium zircons which are often metamict. Since metamictization is caused by α-recoil damage, it is thus a function of uranium and thorium contents, and age. Baldwin et al (1986) propose that a direct comparison of the induced track counts of countable and metamict crystals irradiated together is a guide as to whether the metamict state is attributable to great uranium content or high age, and offers a means of estimating a minimum age for a metamict crystal (e.g. their > 500 Ma age group). Markedly different etch rates for zircons of different age and with differing track densities (Gleadow et al. 1976) can produce both over and under-etched crystals, presenting the possibility of erroneous track counts. Strict selection of crystals according to their orientation and degree of etching should prevent such problems, but it remains a possible source of error for a novice analyst. As underlined by experimental mixing of zircons of different ages (Naeser et al. 1987), a multi-etch procedure should be employed to optimize the revelation of tracks in crystals of differing age and uranium content within a detrital population; similarly, whilst traversing a sample during analysis, each crystal encountered which is properly orientated and etched should be counted to avoid a selection bias.

Clearly the relative sizes of different age modes detected by fission track dating of zircon cannot be used as quantitative indicators of the relative contributions of heavy minerals from different sources to a detrital mineral concentration.

Conclusions

The above case studies demonstrate that fission track dating has a significant potential as a single-crystal geochronological method for discerning ages of detrital zircons in a heavy mineral assemblage. Relating those ages to probable sources provides evidence for the route from hinterland to depocentre. Because of its relatively low temperature for fission track retention, apatite is generally unsuited for discrimination of provenance, rather providing information on the post-depositional thermal history of a sediment. However, in orogenic areas, comparison of apatite results from sediment and source can assist in reconstruction of palaeo-uplift and palaeo-relief patterns in the hinterland.

Fission track studies in London are supported by NERC Research Grant GR3/7068 and by a British Petroleum Research Award. We thank R. Galbraith and A. Gleadow for constructive comments on an earlier version.

References

BALDWIN, S. L., HARRISON, T. M. & BURKE, K. 1986. Fission track evidence for the source of accreted sandstones, Barbados. Tectonics, 5, 457–468.

BROWN, A. V. 1977. Preliminary report on age dating of basalt samples for the Ringarooma 1 : 50,000 sheet. Unpublished Report of the Department of Mines, Tasmania, 1977/25.

CARTE GÉOLOGIQUE DE LA FRANCE, 1980. Carte géologique de la France et de la marge continentale à l' echelle de 1/150,000 (including Notice Explicative) Bureau de Recherches Géologiques et Minières, Orléans.

DIXON, J. E., FITTON, J. G. & FROST, R. T. C. 1981. The tectonic significance of post-Carboniferous igneous activity in the North Sea Basin. In: ILLING, L. V. & HOBSON, G. D. (eds) Petroleum Geology of the Continental Shelf of North West Europe. Institute of Petroleum, London, 121–137.

DODSON, M. H. 1973. Closure temperature in cooling geochronological and petrological systems. Contributions to Mineralogy and Petrology, 40, 259–274.

——, & McCLELLAND-BROWN, E. 1985. Isotopic and palaeomagnetic evidence for rates of cooling, uplift and erosion. In: SNELLING, N. J. (ed.) The Chronology of the Geological Record. Geological Society, London, Memoir 10, 315–325.

DUDDY, I. R., GLEADOW, A. J. W. & KEENE, J. B. 1984. Fission track dating of apatite and sphene from Palaeogene sediments of Deep Sea Drilling Project Leg 81, Site 555. In: ROBERTS, D. G. & SCHNITKER, D. (eds) Initial Reports of the Deep

Sea Drilling Project. US Government Printing Office, Washington DC, **81**, 725–729.

EVERNDEN, J. F. & RICHARDS, J. R. 1962. Potassium-argon ages in eastern Australia. *Journal of the Geological Society of Australia*, **9**, 1–50.

FLEISCHER, R. L. & HART, H. R. 1972. Fission track dating: techniques and problems. *In*: BISHOP, W. W., MILLER, J. A. & COLE, S. (eds) *Calibration of Hominoid Evolution*. Scottish Academic Press, Edinburgh, 135–170.

——, PRICE, P. B. & WALKER, R. M. 1965. Effects of temperature, pressure and ionisation on the formation and stability of fission tracks in minerals and glasses. *Journal of Geophysical Research*, **70**, 1497–1502.

——, —— & —— 1975. *Nuclear Tracks in Solids: Principles and Applications*. University of California Press.

GALBRAITH, R. F. 1988. Graphical display of estimates having differing standard errors. *Technometrics*, **30**, 271–281.

——, 1989. The radial plot: graphical assessment of spread in ages. *Nuclear Tracks* (in press).

GLEADOW, A. J. W. 1981. Fission track dating methods: what are the real alternatives? *Nuclear Tracks*, **5**, 3–14.

—— & BROOKS, C. K. 1979. Fission track dating, thermal histories and tectonics of igneous intrusions in east Greenland. *Contributions to Mineralogy and Petrology*, **71**, 45–60.

—— & LOVERING, J. F. 1974. The effect of weathering on fission track dating. *Earth and Planetary Science Letters*, **22**, 163–168.

—— & —— 1978. Thermal history of granitic rocks from Western Australia: a fission track dating study. *Journal of the Geological Society of Australia*, **25**, 323–340.

——, HURFORD, A. J. & QUAIFE, R. D. 1976. Fission track dating of zircon: improved etching conditions. *Earth and Planetary Science Letters*, **33**, 273–276.

GREEN, P. F. 1981. A new look at statistics in fission track dating. *Nuclear Tracks*, **5**, 77–86.

—— 1985. Comparison of zeta calibration baselines for fission track dating of apatite, zircon and sphene. *Chemical Geology (Isotope Geoscience Section)*, **58**, 1–22.

——, DUDDY, I. R. & LASLETT, G. M. 1988. Can fission track annealing in apatite be described by first-order kinetics? *Earth and Planetary Science Letters*, **87**, 216–228.

——, ——, ——, HEGARTY, K. A., GLEADOW, A. J. W. & LOVERING, J. F. 1989. Thermal annealing of fission tracks in apatite: 4. Quantitative modelling techniques and extension to geological timescales. *Chemical Geology (Isotope Geoscience Section)*, **79**, 155–182.

GROVES, D. I., COCKER, J. D. & JENNINGS, D. J. 1977. The Blue Tier Batholith. *Bulletin of the Department of Mines, Tasmania*, **55**, 1–171.

HAACK, U. 1977. The closing temperature for fission track retention in minerals. *American Journal of Science*, **277**, 459–464.

HARRISON, T. M., ARMSTRONG, R. L., NAESER, C. W.

& HARAKAL, J. E. 1979. Geochronology and thermal history of the Coast Plutonic complex, near Prince Rupert, B. C. *Canadian Journal of Earth Sciences*, **16**, 400–410.

HURFORD, A. J. 1986a Application of the fission track dating method to young sediments: principles, methodology and examples. *In*: HURFORD, A. J., JÄGER, E. & TEN CATE, J. A. M. (eds). *Dating Young Sediments: Proceedings of the Workshop, Beijing, China, Setpemeber 1985*. CCOP-UNESCO Technical Publication **16**, Bangkok, 199–233.

—— 1986b. Cooling and uplift patterns in the Lepontine Alps, south-central Switzerland, and an age of vertical movement on the Insubric fault line. *Contributions to Mineralogy and Petrology*, **93**, 413–427.

—— 1990. Standardization of fission track dating calibration: recommendation by the Fission Track Working Group of the I.U.G.S. Subcommission on Geochronology. *Chemical Geology (Isotope Geoscience Section)*, **80**, 171–178.

—— & GREEN, P. F. 1981. Standards, dosimetry and the uranium-238 λ_f decay constant: a discussion. *Nuclear Tracks*, **5**, 73–75.

—— & —— 1982. A users' guide to fission track dating calibration. *Earth and Planetary Science Letters*, **59**, 343–354.

——, FITCH, F. J. & CLARKE, A. 1984. Resolution of the age structure of the detrital zircon populations of two Lower Cretaceous sandstones from the Weald of England by fission track dating. *Geological Magazine*, **121**, 269–277.

KOWALLIS, B. J., HEATON, J. S. & BRINGHURST, K. 1986. Fission track dating of volcanically derived sedimentary rocks. *Geology*, **14**, 19–22.

KRISHNASWAMI, S., LAL, D., PRABHU, N. & MACDOUGALL, D. 1974. Characteristics of fisson tracks in zircon: applications to geochronology and cosmology. *Earth and Planetary Science Letters*, **22**, 51–59.

LASLETT, G. M., GREEN, P. F., DUDDY, I. R. & GLEADOW, A. J. W. 1987. Thermal annealing of fission tracks in apatite: 2. A quantitative analysis. *Chemical Geology (Isotope Geoscience Section)*, **65**, 1–13.

LUHR, J. F., NELSON, S. A., ALLAN, J. F. and CARMICHEL, I. S. E. 1985. Active rifting in southwestern Mexico: manifestations of an incipient eastward spreading-ridge jump. *Geology*, **13**, 54–57.

McDOUGALL, I & LEGGO, P. J. 1965. Isotopic age determinations on granitic rocks from Tasmania. *Journal of the Geological Society of Australia*, **12**, 295–332.

McGOLDRICK, P. J. & GLEADOW, A. J. W. 1977. Fission track dating of Lower Palaeozoic sandstones at Tatong, North Central Victoria. *Journal of the Geological Society of Australia*, **24**, 461–464.

MORTON, A. C. & PARSON, L. M. (eds). 1988. *Early Tertiary Volcanism and the Opening of the NE Atlantic*. Geological Society, London, Special Publication No. 39.

NAESER, C. W. 1967. The use of apatite and sphene for fission track age determinations. *Geological*

Society of America Bulletin, **78**, 1523–1526.

NAESER, N. D. & NAESER, C. W. 1984. Fission Track Dating. *In*: MAHANEY, W. C. (ed.) *Quaternary Dating Methods*. Elsevier, Amsterdam.

——, ZEITLER, P. K., NAESER, C. W. & CERVENY, P. F. 1987. Provenance studies by fission track dating of zircon—etching and counting procedures. *Nuclear Tracks*, **13**, 121–126.

RODDICK, J. C. & COMPSTON, W. 1976. Radiometric evidence for the age of the Murrumbidgee Batholith. *Journal of the Geological Society of Australia*, **23**, 223–233.

TORRINI, R., SPEED, R. C. & MATTOLI, G. S. 1985. Tectonic relationships between forearc-basin strata and the accretionary complex at Bath, Barbados. *Geological Society of America Bulletin*, **96**, 861–874.

WAGNER, G. A. & REIMER, G. A. 1972. Fission track tectonics: the tectonic interpretation of fission track ages. *Earth and Planetary Science Letters*, **14**, 263–268.

——, MILLER, D. S. & JÄGER, E. 1979. Fission track

ages on apatite of Bergell rocks from the Central Alps and Bergell boulders in Oligocene sediments. *Earth and Planetary Science Letters*, **45**, 355–360.

WILLIAMS, I. S., COMPSTON, W., CHAPPELL, B. W. & SHIRAHASE, T. 1975. Rb-Sr age determinations on micas from a geologically controlled, composite batholith. *Journal of the Geological Society of Australia*, **22**, 497–505.

YIM, W. W. S. 1980. *Heavy mineral studies of stanniferous deep leads, northeast Tasmania*. Geological and Geophysical Record of the Bureau of Mineral Resources, Canberra **67**.

——, GLEADOW, A. J. W. & VAN MOORT, J. C. 1985. Fission track dating of alluvial zircons and heavy mineral provenance in Northeast Tasmania. *Journal of the Geological Society, London*, **142**, 351–356.

ZEITLER, P. K., TAHIRKHELI, R. A. K., NAESER, C. W. & JOHNSON, N. M. 1982. Unroofing history of a suture zone in the Himalaya of Pakistan by means of fission track annealing ages. *Earth and Planetary Science Letters*, **57**, 227–240.

Reworking of plant microfossils and sedimentary provenance

D. J. BATTEN

*Institute of Earth Studies, UCW-Aberystwyth, University of Wales, Aberystwyth
SY23 3DB, UK*

Abstract: Reworked plant microfossils are commonly recovered from sediments along with
palynomorphs regarded as in situ. Their potential for locating the source of terrigenous
clastics is perhaps generally underestimated. Recognition of reworking may depend on
taxonomic identification or differences in colour and preservation, and sometimes on
variations in response to staining and fluorescence microscopy. This is demonstrable in
some Mesozoic and Cenozoic sequences in NW Europe, with mixed assemblages suggesting
derivation from deposits of more than one age and/or source area and geothermal history.
Reworked palynomorphs can also indicate the presence of rocks or sediments in the vicinity
of a site of deposition that have now been completely removed by erosion.

This paper aims to discuss the significance of the occurrence of reworked palynomorphs as a basis for determining sedimentary provenance. It is divided into three main sections. The first is concerned with the criteria on which palynomorphs can be identified as reworked. Some of the literature on their dispersal and on the identification of source areas is then reviewed. Five examples follow, each of which demonstrates both the value of, and the difficulties in interpreting palynomorph assemblages of mixed origins.

Recognition of reworked palynomorphs

Spores, pollen grains, dinoflagellate cysts and other palynomorphs are eminently suitable as indicators of reworking for a number of reasons. In common with all fossils, their assemblage composition differs according to the age of the rocks from which they are recovered, but their abundance, small size and relative resistance to decay enhances their preservation potential, allowing them to withstand the ravages of recycling processes better than most other animal and plant remains. The protective walls of sporopollenin are among the most durable of organic materials (Brooks *et al.* 1987). Being silt- to sand-sized particles (mostly less than 200 μm, rarely more than 2 mm), they are more likely than larger fossils to be released from the enclosing matrix of an eroding sedimentary deposit and transported intact to a new site.

Despite their robustness, they are not, however, immune from alteration and degradation. When fresh, pollen grains may be virtually colourless or have light yellow, greenish-yellow or other hues. Most spores and phytoplankton are similarly pale or devoid of colour. Apart from usually being compressed (this will depend on the composition of the deposits in which they accumulate), once incorporated in sediments their appearance remains much the same unless they are subjected to weathering, which may ultimately lead to their destruction by oxidation, or to prolonged heating, which causes changes in colour and eventually also in size and morphology (Staplin 1969; Raynaud & Robert 1975; Batten 1980, 1981). The degree of alteration with increasing depth of burial depends on the amount of overburden and the geothermal gradient. A typical trend for a spore or pollen grain in response to rising temperatures is a gradual darkening of the wall from yellowish-orange to reddish-brown, dark brown and black, a reduction in diameter, increasing opacity and a tendency to become brittle (Combaz 1980; Batten 1980, 1981). Hence, whereas both fresh and relatively unaltered fossil palynomorphs are pliable and will fold readily if thin-walled, those from thermally mature and over-mature horizons in the petroleum-generation sense (e.g. Dow 1977; Gaupp & Batten 1985) tend to be more rigid and therefore more easily broken. The final stage in the thermal alteration process is reached when they are reduced to graphite particles, all morphological characteristics having been obliterated (cf. Batten 1982*a*).

Palynomorphs are commonly altered in other ways as well. Their walls may become pitted, generally degraded and ultimately destroyed by aerobic bacterial and fungal attack, oxidation, and prolonged exposure to an alkaline environment (Elsik 1966; Batten 1973 and others). In euxinic conditions, anaerobic bacteria can also damage their walls, and the formation of pyrite

From Morton, A. C., Todd, S. P. & Haughton, P. D. W. (eds), 1991,
Developments in Sedimentary Provenance Studies.
Geological Society Special Publication No. 57, pp. 79–90.

crystals, which is linked to the activities of sulphur-reducing bacteria can greatly alter the appearance of palynomorphs. This is most clearly seen on specimens from which the crystals have been removed leaving only impressions (relict structures) in their walls (Neves & Sullivan 1964; Batten 1985).

All of these changes in appearance of palynomorphs are also reflected in their reactions to chemical oxidation in the laboratory, to irradiation with blue or ultraviolet (UV) light, and sometimes to staining with safranin (Safranin O in Stanley 1966) or other substances. For example, well preserved spores recovered from a bituminous coal may become significantly lighter in colour on being immersed briefly in fuming nitric acid, whereas similar but poorly preserved forms that reflect deposition in oxidising conditions may be little altered by identical treatment. Palynomorphs that have never been deeply buried may show a bright yellow fluorescence when viewed under UV light and generally take up stain easily, whereas those from very mature horizons will not respond to either treatment.

These differences in appearance and reaction to physical and chemical methods of analysis can be used with varying degrees of success to distinguish between reworked and in situ palynomorphs. Reworking is easiest to detect in assemblages containing associations of species that are indicative of significantly different periods of geological time. The presence of Carboniferous spores in a Cretaceous preparation, for example, can usually be recognized without difficulty. On the other hand, it may be impossible to identify Hauterivian spores in a Barremian assemblage, or late Neogene pollen in a Quaternary sequence unless there is reliable biostratigraphic control. In both of these cases most of the taxa concerned are morphologically identical. Reworked specimens tend to be less numerous and darker in colour than those in situ but this is not always true. It depends on a number of factors including the rate of erosion, and whether the beds removed were more thermally altered than the younger host deposits. Palynomorphs that have been washed down into older deposits ('stratigraphic leakage'; Streel & Bless 1980 and references cited therein) are normally paler than the forms that are in place. Reworked specimens of *Gonyaulacysta jurassica* (Deflandre) have been found (Eaton, unpublished) that are as well preserved as the in situ dinoflagellate cysts in the Eocene Bracklesham Beds of the Isle of Wight from which they were extracted (see Eaton 1976), and I have studied Tertiary and Quaternary samples which yielded palynomorph assemblages that are dominated by, or consist only of reworked material (see below). The latter occurrences are difficult to interpret but they suggest that heating, provided it has not been excessive and led to brittleness (Muir 1967; Batten 1981), may increase the resistance of palynomorphs to oxidation (Phillips 1974) and bacterial degradation and thus enhance their chances of survival in an environment of deposition that is not otherwise conducive to the preservation of organic matter.

Significance of occurrence of reworked palynomorphs: a brief review

Palynologists commonly report the presence of reworked palynomorphs in the assemblages they have examined but, as Streel & Bless (1980) have previously indicated, seldom are they considered further. This is probably because the majority of taxonomic studies have been, and still are, aimed primarily at biostratigraphy. Although palynological data are increasingly being integrated with those derived from sedimentological analyses in order to arrive at palaeoenvironmental interpretations, even in these studies the presence of reworked specimens has been underused. Little or no attention is generally paid to the reasons for their presence. Among the exceptions are several papers by Truswell (e.g., Truswell 1982, 1983a,b; Truswell & Drewry 1984). Consideration is given in these to the geological implications of the occurrence of recycled palynomorphs in Antarctic shelf sediments, large numbers of which have been derived either from the continental landmass by glacial action or the movement of ice. Although there is commonly no clear indication of where the source beds might lie, relationships between ice drainage pattern and palynomorph distributions have been observed which in turn can be used to shed some light on the age and location of strata concealed beneath the ice cover. To this end, quantitative analyses of the density and age distribution of the dispersal data in relation to patterns of ice flow are essential. The most useful and reliable results have been obtained from areas within which processes leading to the accumulation of abundant reworked palynomorphs probably involved transport of rock by ice, followed by meltout close to the surface of deposition where bottom currents were insufficient to bring about any subsequent redistribution of released material.

These processes of reworking and redeposition are in marked contrast to those that operate in non-glacial environments. The distribution of modern pollen, dinoflagellate cysts and other palynomorphs in the marine realm depends largely on ocean currents as shown, for example, by Muller (1959), Traverse & Ginsburg (1966), Wall *et al.* (1977) and others. Reworked palynomorphs may also be widely dispersed in this way, as has been suggested by Collinson *et al.* (1985) to account for the co-occurrence of Carboniferous, Cretaceous and Tertiary megaspores in the Thanet and Reading Beds (Upper Palaeocene) of southern England. Habib (1982) reported that secondary deposition may occur as a result of displacement by turbidity or other gravity currents from depositional sites on continental shelves. An analysis by Eshet *et al.* (1988) of the palynological assemblages recovered from a Permo-Triassic sequence in the subsurface of Israel revealed that reworked material was most abundant in regressive intervals. They attribute this to changes either in the extent of exposed land or to the transportation régime.

Usually relatively local origins are invoked to account for significant numbers of recycled palynomorphs in an assemblage, such as, for example, the Devonian specimens that occur in Lower Cretaceous deposits of the Moose River Basin of Ontario, Canada (Legault & Norris 1982). Incorporation of both Devonian and Cretaceous palynomorphs in Pleistocene till in the same area is also attributed to a single event, namely of recycling Cretaceous deposits containing Devonian remains (Legault & Norris 1982).

Probably about 20% of the published records of reworking are concerned with Carboniferous spores in Jurassic and Cretaceous rocks (e.g. Muir 1967; Windle 1979; Guy-Ohlson *et al.* 1987). Windle (1979) has drawn attention to examples of unrecognised reworking which can lead to taxonomic duplications, nomenclatural inaccuracies, erroneous extensions of ranges and incorrect biostratigraphic and evolutionary conclusions. Integration of sedimentological with palynological data can bring greater rewards than reliance solely on the latter for palaeoenvironmental interpretations. Although one of the palynomorphs illustrated by Guy-Ohlson *et al.* (1987, fig. 4G) as *Tetraporina* sp. appears to be a much younger, angiospermous contaminant, some of the other spores encountered in Jurassic core samples from northwest Scania, Sweden are convincing evidence of Carboniferous reworking, which is particularly significant because rocks of this age have not so far been encountered in the region. Combined with mineralogical and sedimentological analyses the conclusion was drawn that the palynomorphs had been transported only relatively short distances from

presumed thin Namurian deposits directly overlying, or closely associated with Precambrian basement in the Fennoscandian Border Zone.

Wide variations in the preservation of reworked acritarchs recorded from Devonian assemblages extracted from core samples from a borehole in Oxfordshire, southern England, led Richardson & Rasul (1978) to the conclusion that they were derived from three or four different areas which included both metamorphosed and unmetamorphosed deposits of four ages. The Midland Shelf not far to the north, and possibly northeast, may well have been the source of the uncarbonised Tremadocian forms whereas the other three preservational types could have come from northwesterly localities in north Wales.

Major differences of another sort were encountered by Gaupp & Batten (1983) in palynological preparations of samples from Cretaceous strata in the Northern Calcareous Alps. The occurrence of palynomorphs indicative of ages ranging from Carboniferous to Early Cretaceous, the composition of heavy mineral assemblages, the lithologies of pebbles in associated conglomerates and other sedimentological characteristics all pointed to a number of different sources in this structurally complex area, and perhaps also to some multiple recycling.

Detailed analysis by Muir & Sarjeant (1978) of the taxonomic composition of palynomorph assemblages from the Langdale Beds (Middle Jurassic, Callovian) of Yorkshire revealed that these contain a substantial component of forms interpreted to have been reworked from only slightly younger (Bathonian) deposits. The most likely source was considered to be the Market Weighton block, a structurally high area on the northern margin of the East Midlands Shelf to the south and east of the depositional site.

Reworking and sedimentary provenance: further examples

In this section I briefly refer to two northwest European occurrences of palynomorph reworking and then discuss three British records in rather more detail in the context of source areas and problems connected with their identification. The first European example concerns the presence of Devonian and Carboniferous spores in Quaternary deposits containing flints derived from Upper Cretaceous chalk. Two short boreholes in the vicinity of Hombourg in northeast Belgium (Bless & Streel 1988) penetrated Pleistocene deposits, a thin Upper Cretaceous sequence

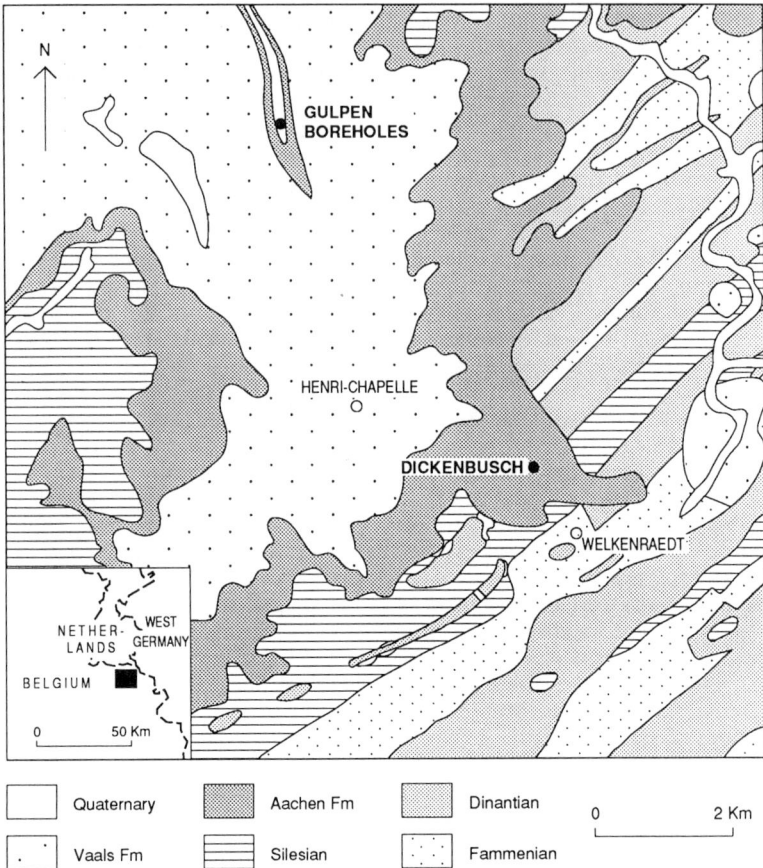

Fig. 1. Geology of the country in the vicinity of the Gulpen boreholes, Hombourg, northeast Belgium, based on, and modified from a copy of a map produced by the Belgian Geological Survey in 1897 (made available by M. Streel).

(Aachen Formation, Santonian; Batten *et al.* 1988) and Palaeozoic strata (Figs 1 & 2). The relative abundance of dark brown Devonian and Carboniferous spores in the unconsolidated sediments overlying the Cretaceous (Fig. 2) can be attributed to erosion, during the Pleistocene Epoch, of the Palaeozoic rocks of the nearby Rhenish Massif (Fig. 3). The flints must have originated from the numerous horizons in which they are abundant in the upper part of the Vaals Formation (Fig. 1) and in the younger Gulpen and Maastricht Formations which dominate the pre-Quaternary surface geology locally, particularly to the north and northwest.

A thin (15 m) lower Campanian section near Beckum in the Münster Basin is the other European example (Fig. 4). The palynomorph assemblages recovered from this rhythmically bedded limestone-marl sequence are dominated by dinoflagellate cysts. The associated miospores consist mainly of bisaccate gymnosperm pollen (*Alisporites*, *Abietineaepollenites*, *Pityosporites*), some of which, along with a few triradiate spores, may be derived from Jurassic or Lower Cretaceous deposits. Taxonomic difficulties connected with bisaccate pollen grains, and no obvious differences in colour make it difficult to determine the proportion that is likely to represent contemporaneous vegetation. The scarcity of angiosperm pollen is in marked contrast to their relative abundance in deposits of Santonian age in Limburg (Batten *et al.* 1988) and might be attributable to the depositional site being relatively distant from any source vegetation at this time. This could equally apply to the bisaccate grains, in which case it might be reasonable to argue that most are reworked. Bisaccate grains do, however, generally seem to be much more common in Campanian and Maastrichtian assemblages than in those from the Santonian. It is

GULPEN I GULPEN II

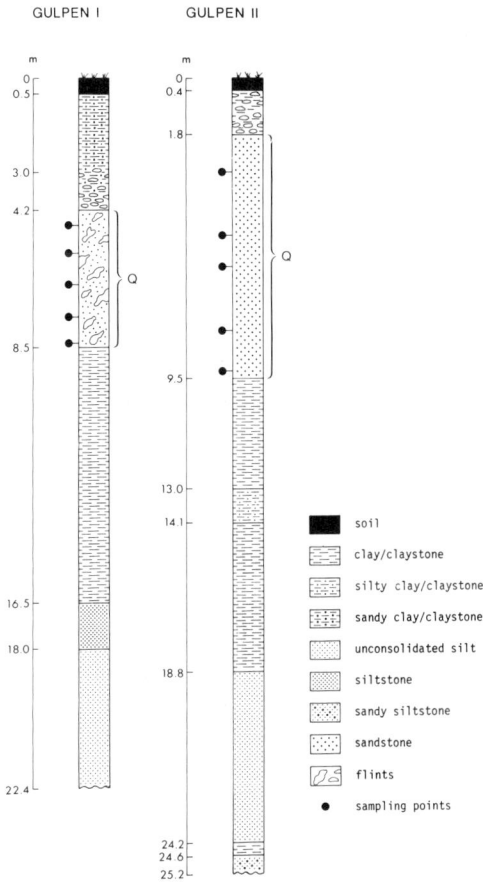

Fig. 2. Post-Palaeozoic deposits penetrated by Gulpen boreholes I and II. The beds below 8.5 and 9.5 m respectively in these holes are of Santonian age and rest unconformably on Carboniferous strata. Sampling points are shown for the overlying Quaternary part of the sequence (Q).

also well known that saccate grains can be blown vast distances from their place of origin in modern vegetation and are capable of prolonged flotation; hence, they may accumulate in sediments far removed from the trees that produced them (cf. Traverse & Ginsburg 1966). The possibility that they represent the dominant type of vegetation on the Rhenish Massif during the Campanian (Fig. 3), must, therefore, be taken seriously. This does not explain the presence of the palynomorphs that are definitely reworked. It is possible that some of these were derived from submarine erosion of older Cretaceous deposits within the Münster Basin.

This German example illustrates the difficulties that may be encountered when attempting to recognize reworked palynomorphs in the face of taxonomic, taphonomic, palaeoecological and palaeogeographical uncertainties. Other problems are encountered in trying to explain the dominance of Mesozoic miospores in lower Tertiary (Palaeocene; Curry et al. 1978) leaf beds

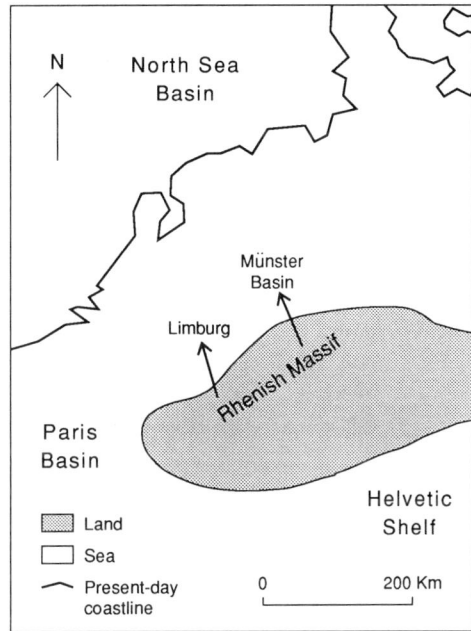

Fig. 3. Simplified Late Cretaceous palaeogeography of part of NW Europe (adapted from Ziegler 1982, enclosure 23). The western section of the mid-European island was a source of Santonian sediments in Limburg and adjacent NE Belgium and the Aachen area of West Germany, and probably also of some of the clastic component of the Upper Cretaceous deposits in the Münster Basin.

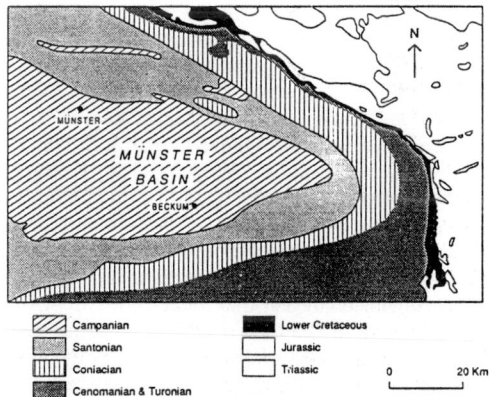

Fig. 4. Geological map of the Mesozoic of the Münster Basin and vicinity (according to Kukuk 1936 and Arnold 1964; modified after Kemper et al. 1978, fig. 4) with the location of the Campanian section of Beckum indicated; a small outlier of Tertiary deposits in the northeast is omitted.

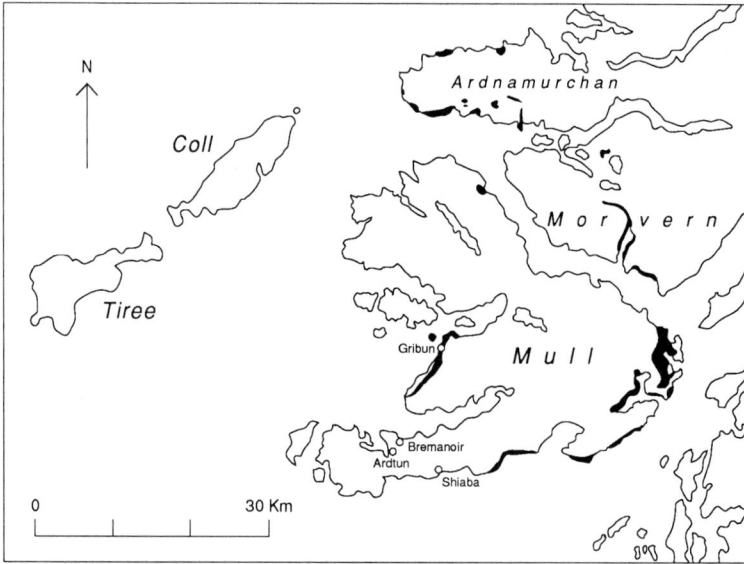

Fig. 5. Outline map of Mull, neighbouring islands, and part of the mainland to the east and north, showing the distribution (black infill) of presumed and definite Upper Triassic and Jurassic outcrop (undifferentiated; based on fig. 9 in Richey 1961 and British Geological Survey Sheets 56N 06W, Argyll, and 56N 08W, Tiree, solid geology, 1:250000). Places mentioned in the text are also indicated.

on Mull in western Scotland. Phillips (1974) reported occurrences of entirely reworked Jurassic assemblages from three samples of these beds, collected by J. S. Gardner more than 100 years ago from an outcrop near Ardtun (Fig. 5). Although her taxonomic list is short, the conclusion she drew that the assemblage suggests derivation from a mixed Jurassic source is supported by palynological data on several of my own samples from exposures in the same area (north of Ardtun, west of Bremanoir; see Macdougall *in* Skelhorn *et al.* 1969). Among the miospore taxa I recorded in addition to those Phillips listed are *Araucariacites australis* Cookson, *Converrucosisporites rariverrucatus* (Danzé-Corsin & Laveine), *Densoisporites velatus* Weyland & Krieger, *Inaperturopollenites turbatus* Balme and *Callialasporites dampieri* (Balme). Also encountered were representatives of the dinoflagellate cyst *Nannoceratopsis* and the tasmanitid *Crassosphaera*, which suggest that some of the deposits being eroded at the time originally accumulated in brackish-marine conditions.

Although not confirmed in this study, a latest Triassic (Rhaetian) contribution to the leaf beds is also a possibility. Warrington & Pollard (1985) provided palynological evidence for deposits of this age to the south of Gribun which they placed in the Penarth Group. Both these beds and the lower Lias are quite well repre-

sented in this area and easily within reworking range of Ardtun. Deposits of younger Jurassic age are poorly developed, but it is likely that they were more widespread during the early Tertiary.

The scarcity of contemporaneous flowering plant pollen in the leaf beds which are dominated by remains of angiosperm foliage is surprising; Phillips (1974) did not record any from her preparations. She suggested that either these fluvial-lacustrine, allochthonous deposits might have accumulated in flood conditions during the autumn when pollen grains were no longer being shed or the Tertiary grains were preferentially destroyed. Neither explanation is wholly satisfactory (Phillips 1974, p. 227); the first is perhaps more likely. Known autochthonous deposits in the vicinity (at Bremanoir and Shiaba; Fig. 5) which have yielded palynomorph assemblages contain only a small reworked component (Simpson 1961; Srivastava 1975).

Warrington & Pollard (1985) noted that the miospores of the Gribun sequence comprise the most northerly record of Triassic palynomorphs so far documented from onshore exposures. Upper Triassic rocks that crop out further north on Rhum and deposits that are possibly of this age on Raasay and in the Western Isles on Lewis, have so far proved unproductive (Warrington *et al.* 1980). Reworked Late Triassic forms have,

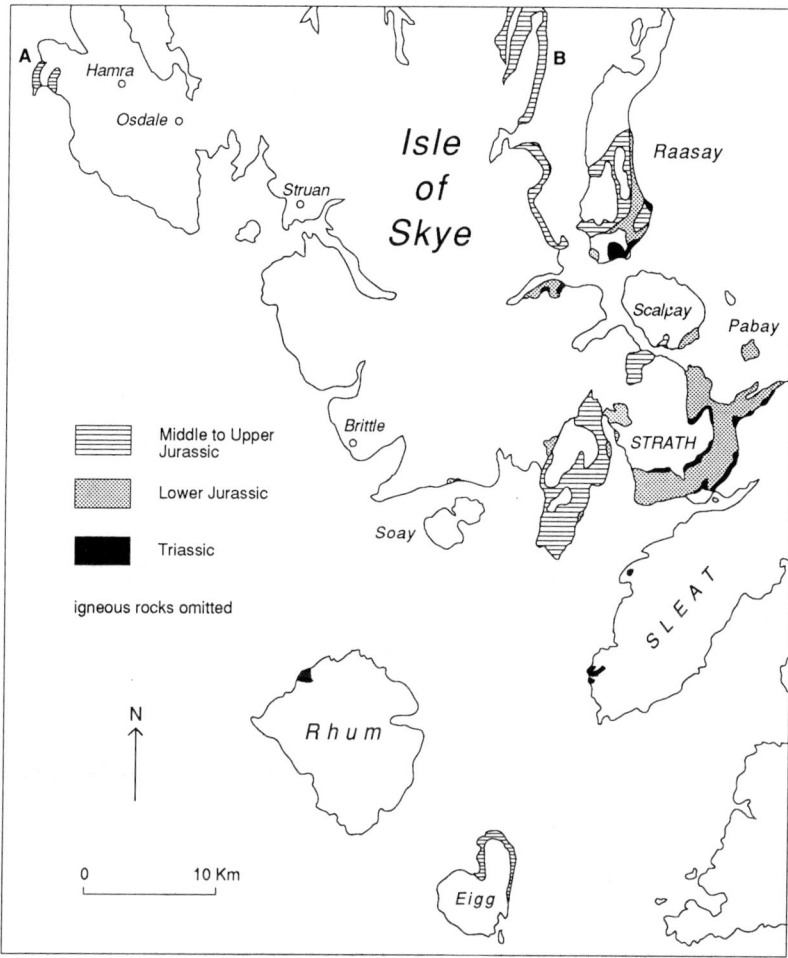

Fig. 6. Sketch map of part of the Isle of Skye, neighbouring islands and the mainland (in SE corner) showing the distribution of Triassic and Jurassic rocks (based on fig. 11 in Richey 1961, fig. 1 in Anderson & Dunham 1966, fig. 1 in Bell & Harris 1986, and British Geological Survey Sheet 57N 08W, Little Minch, solid geology, 1:250000). For explanation of A and B, see Fig. 7.

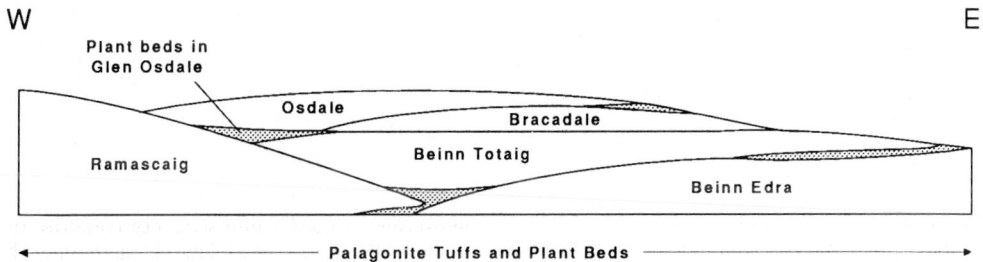

Fig. 7. Cross-section from west to east along a line from points A to B on Fig. 6 showing relationships between the main groups of lava in northern Skye (labelled). Beneath the oldest of these are palagonite tuffs which were laid down under water on a Jurassic surface. The interbasaltic plant-bearing sediments and tuffs are denoted by stippling (modified from part of fig. 13 in Anderson & Dunham 1966). The Brittle deposits may be stratigraphically equivalent to those in Glen Osdale.

however, been encountered in palynomorph assemblages recovered from sediments between Tertiary lavas on Skye. Nine samples from three of four deposits I have studied (Glen Osdale, ESE of Healaval Mhor; Sgurr à Ghobhainn, north of Loch Brittle; and northwest of Struan near Loch Bracadale) were found to contain mixed assemblages of Tertiary and Mesozoic palynomorphs. Two others from Osdale and all three from an exposure in the Hamra River, Glen Dale (Figs 6 & 7) were barren.

The interbasaltic sediments are thought to represent 'relatively long' periods of fluviatile deposition between eruptions from the major fissures that gave rise to the lava-groups (Bell & Harris 1986). The upper surfaces of these commonly show evidence of having been decomposed by contemporaneous weathering processes to red earthy (lateritic) clays which were probably vegetated. The lava fields are extensively faulted so that it is difficult to determine to which extrusive episode they belong and hence also to correlate the intervening sedimentary deposits. The north Brittle samples are from a 5 m section that is dominated by a fine-grained, red, channel-fill sandstone which grades laterally into conglomerates containing numerous pebbles of Torridonian Sandstone (75–80%) with smaller quantities of lava (15–20%), gneiss (2–3%) and acid igneous rocks (1–2%) in association (King 1976 and *in litt.*). Elsewhere, blocks of Jurassic sedimentary material and large cobbles of granite have been found. The limited distribution of conglomerates containing well-worn pebbles of lava strongly suggest that they represent river channel accumulations (King 1976) and is further evidence of prolonged erosion in between periods of basaltic extrusion. Within the sandstone, there are lenses of different lithological composition including flaggy, carbonaceous, fine-grained sandstones and silty mudstones that are similarly rich in organic matter. These are the horizons that have yielded palynomorphs.

The sedimentary sequence at Glen Osdale, which is *c.* 9 m thick, also consists of a mixture of conglomeratic, sandy and argillaceous units. The palynologically productive deposits are grey and brownish grey carbonaceous mudstones and silty mudstones. Those that have yielded leaf impressions are thought to correlate with the Ardtun beds on Mull (Bailey *et al.* 1924; Richey 1961; Bell & Harris 1986). The Struan outcrop is less than 1 m thick and consists of baked mudstones with interbedded coals.

All of the palynologically productive preparations contain bisaccate pollen, but because of their generally poor state of preservation and the

taxonomic difficulties mentioned above, it is not possible to determine whether they are of Mesozoic or Tertiary age, although the former seems more likely for the majority. The same applies to the many inaperturate pollen that are present in two of the Osdale preparations.

Unfortunately the in situ palynomorphs are neither very common nor of much stratigraphic value. The total assemblage is taxonomically impoverished. Among the pollen grains recorded are *Alnipollenites versus* (Potonié), *Fraxinoipollenites variabilis* Stanley, *Plicatopollis plicatus* (Potonié), *Quercoidites microhenricii* (Potonié), *Taxodiaceaepollenites hiatipites* (Wodehouse), *Triatriopollenites coryphaeus* (Potonié) and *Triporopollenites mullensis* (Simpson). This association, together with the absence of diagnostic latest Cretaceous (Maastrichtian) and Eocene forms, is consistent with a Palaeocene dating. Among the Mesozoic miospores are *Baculatisporites comaumensis* (Cookson), *Cerebropollenites mesozoicus* (Couper), *Chasmatosporites magnolioides* (Erdtman), *Classopollis* spp., *Contignisporites* sp., *Cycadopites* spp., *Duplexisporites problematicus* (Couper), *Eucommiidites troedssonii* Erdtman, *Exesipollenites* sp., *Lycopodiumsporites* spp., *Perinopollenites elatoides* Couper, *Vitreisporites pallidus* (Reissinger), *Ovalipollis pseudoalatus* (Thiergart), *Rhaetipollis germanicus* Schulz and *Ricciisporites tuberculatus* Lundblad. The last three of these indicate derivation from Upper Triassic rocks of probable Rhaetian age. Specimens referred to some of the other taxa may also have their origins in these deposits, but most are more likely to reflect erosion of Lower and Middle Jurassic beds. Both the presence of the dinoflagellate cyst *Rhaetogonyaulax rhaetica* (Sarjeant) and the occurrence of some foraminiferal linings and acritarchs (*Micrhystridium* sp., *Veryhachium* sp.) indicate that deposition of at least some of the Rhaetian and Early Jurassic beds occurred in brackish-marine environments.

The present-day extent of putative Triassic deposits on Skye (and on Rhum to the south) is very limited, and Jurassic outcrop is confined mainly to the northeastern–eastern and central–southern part of the island. In common with the interpretation of the reworked assemblages on Mull it can probably be assumed that deposits of these ages were more widespread during the early Tertiary, which was clearly a time of rapid erosion. In view of the extensive contemporaneous development of plateau lavas over central and western Skye, the source beds for the reworked palynomorphs were probably located mainly to the east and southeast in areas that remained exposed during this period of igneous

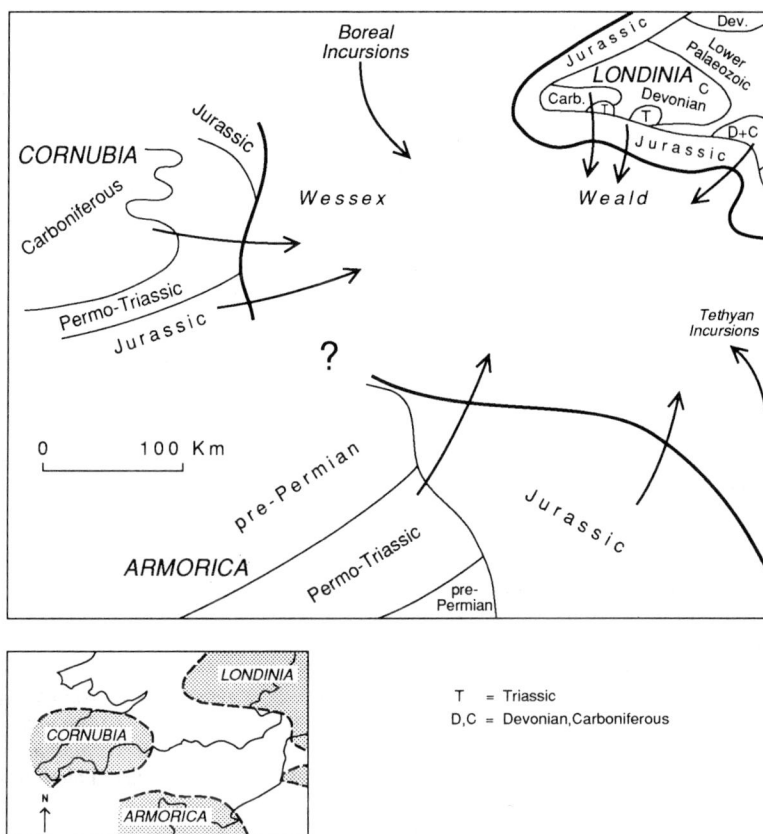

Fig. 8. Generalized palaeogeography of southern England and northern France (simplified from fig. 8 in Allen 1981) showing location of potential sources of reworked palynomorphs and directions of inflow to the Wealden outwash and meander plains *sensu* Allen (1981). Areal extent of massifs in relation to present-day coastlines is indicated on smaller map.

activity. There is, however, the possibility of minor derivation from more southerly locations on Rhum and Eigg and perhaps even further afield (e.g. from Ardnamurchan).

In the final example, I show that the occurrence of reworked palynomorphs can be used to supplement determinations of provenance based on sedimentological evidence. Although there have been problems in interpreting transport directions of the sediment that entered the Wessex-Weald basin during the Early Cretaceous, Allen's (1959, 1967a, b, 1976, 1981) conclusions based on the mineralogy and patterns of distribution of pebbles within arenaceous units of the Wealden succession are broadly supported by the palynological record. Reworked palynomorphs, particularly of Jurassic, but also of Devonian (rare), Carboniferous and Late Triassic ages, are commonly encountered in both these rocks and in the overlying Lower Green-

sand Group (e.g. Batten 1982b; Duxbury 1983; Lister & Batten 1988). It is logical to assume that for the Wealden area, these were derived principally from deposits of these ages on, and adjacent to, the eroding London-Belgium Massif (Londinia on Fig. 8) and for the Wessex subbasin, mainly from Cornubia but with perhaps some contribution of Upper Jurassic detritus from an exposed Portsdown Swell (Allen 1981; Lister & Batten 1988). The amount of reworked material contributed by Amorica to the western part of the region is difficult to gauge.

All three massifs have Palaeozoic and Permo-Triassic cores which are fringed by Jurassic rocks. It is not surprising, therefore, that the latter were a source of the majority of reworked palynomorphs. Particularly numerous in Barremian and Aptian deposits are dinoflagellate cysts of Middle to Late Jurassic age. On the basis of the composition of pebbles in Wealden deposits

Allen (1967a, 1976, 1981) has emphasized that large areas of the marginal Kimmeridgian–Portlandian scarplands were stripped of their sediments. This would also appear to be true for older Jurassic deposits, unequivocal evidence for the removal of lower Oxfordian strata being provided, for example, by the recovery of the distinctive dinoflagellate cyst *Wanaea fimbriata* Sarjeant from the Vectis Formation on the Isle of Wight, the upper Wealden in Southeast England and from the Lower Greensand in both areas. Among the numerous other reworked cysts recorded from these same sequences are *Cribroperidinium*? *longicorne* (Downie), *Dichadogonyaulax*? *pannea* (Norris), *Endoscrinium luridum* (Deflandre), *Glossodinium dimorphum* Ioannides *et al.*, *Gonyaulacysta jurassica* (Deflandre), *Hystrichodinium pulchrum* Deflandre, *Meiourogonyaulax staffinensis* Gitmez, *Nannoceratopsis pellucida* Deflandre, *Rigaudella aemula* (Deflandre), *Scriniodinium crystallinum* (Deflandre), *S. inritibile* Riley, *Stephanelytron redcliffense* Sarjeant and *Tubotuberella apatela* (Cookson & Eisenack).

Although this documentation of reworked palynomorphs in Upper Wealden and Lower Greensand sediments provides support for the general sedimentological and palaeogeographical conclusions that have been drawn previously for this period of Cretaceous deposition in southern England, there is scope for refinement. Changes in assemblage composition through a sequence may point to either the removal of old sources or the addition of new. It might be expected that, with denudation of Londinia having taken place throughout the Early Cretaceous, palynomorphs reworked from Upper Jurassic deposits would be more strongly represented in the lower part of the succession and that older forms would be more numerous at higher levels, but this has not so far been demonstrated. More rigorous quantitative palynological analyses linked precisely to the geometry of sedimentary bodies, current directions (cf. Allen 1959; Stewart 1981), clay mineralogy (cf. Sellwood & Sladen 1981; Sladen 1983; Sladen & Batten 1984), and other sedimentological parameters should enable source areas to be pinpointed, and rates of erosion to be assessed, with greater accuracy than is currently possible.

Conclusion

The basis for identifying palynological reworking and the examples discussed in this paper serve to highlight both the value of, and the difficulties associated with using plant microfossils as indicators of sedimentary provenance. Derivation from a variety of sources may be easily recognisable when the ages of the mixed assemblages of clastic and fossil detritus are significantly different (example 1), but taphonomic, taxonomic and other uncertainties may cause problems of interpretation that are difficult to resolve (example 2). Transport directions are not easy to ascertain from palynological evidence alone, but they may be inferred from the present-day distribution of outcrops representing the remains of deposits that were eroded in the past (example 3) and can sometimes indicate the former presence of sediments of a particular age in the vicinity which have since been entirely removed (example 4). They may also support previous suggestions of source areas and transport directions based only on clastic sedimentological or other analyses (example 5).

Palynologists have probably not, hitherto, made as much use of assemblages of mixed origins as they might have done. Reworked specimens have instead been seen mainly as a warning to take care with identifications in order to avoid arriving at incorrect age or palaeoenvironmental determinations; this is possible if the in situ assemblages are dominated by, or consist entirely of non-diagnostic morphotypes. On the other hand, the majority of sedimentologists do not appear to be aware of the fact that palynologists may be able to help them locate sources of clastic detritus. It is to be hoped that more routine recording of occurrences of reworked plant microfossils combined with detailed sedimentological analyses of the same sections will in future lead to an increase in their use for determining the provenance of sediments.

I thank A. C. Morton (British Geological Survey, Nottingham) for inviting me to present this paper at the Geological Society meeting on Developments in Sedimentary Provenance Studies, held in London on June 29–30, 1989. The cost of undertaking field work in northern Germany and western Scotland was met by NERC Grant GR3/3425. M. Streel (University of Liège, Belgium) kindly provided samples and slides from the Gulpen boreholes and background information on the geology of NE Belgium. The Skye study was initiated by P. King while he was a graduate student at Aberdeen University, and my analysis of the in situ Tertiary palynomorphs recovered benefited from discussion at about the same time with G. C. Wilkinson (Unocal, Sunbury-on-Thames).

References

ALLEN, P. 1959. The Wealden environment: Anglo-Paris basin. *Philosophical Transactions of the Royal Society, London,* **B242**, 283–346.

—— 1967a. Origin of the Hastings facies in north-western Europe. *Proceedings of the Geologists' Association, London,* **78**, 27–105.

—— 1967b. Strand-line pebbles in the mid-Hastings Beds and the geology of the London uplands. Old Red Sandstone, New Red Sandstone and other pebbles. Conclusion. *Proceedings of the Geologists' Association, London,* **78**, 241–276.

—— 1976. Wealden of the Weald: a new model. *Proceedings of the Geologists' Association, London,* **86** [for 1975], 389–437.

—— 1981. Pursuit of Wealden models. *Journal of the Geological Society, London,* **138**, 375–405.

ANDERSON, F. W. & DUNHAM, K. C. 1966. *The geology of northern Skye.* Memoir of the Geological Survey, Scotland.

ARNOLD, H. 1964. Zur Lithologie und Zyklik des Beckumer Campans. *Fortschritte in der Geologie von Rheinland und Westfalen,* **7**, 577–598.

BAILEY, E. B., CLOUGH, C. T., WRIGHT, W. B., RICHEY, J. E. & WILSON, G. V. 1924. *Tertiary and post-Tertiary geology of Mull, Loch Aline, and Oban.* Memoir of the Geological Survey, Scotland.

BATTEN, D. J. 1973. Use of palynologic assemblage-types in Wealden correlation. *Palaeontology,* **16**, 1–40.

—— 1980. Use of transmitted light microscopy of sedimentary organic matter for evaluation of hydrocarbon source potential. *Proceedings of the 4th International Palynological Conference,* Lucknow (1976–77), **2**, 589–594.

—— 1981. Palynofacies, organic maturation and source potential for petroleum. *In:* BROOKS, J. (ed.) *Organic maturation studies and fossil fuel exploration.* Academic Press, London, 201–223.

—— 1982a. Palynology of shales associated with the Kap Washington Group volcanics, central North Greenland. *Grønlands Geologiske Undersøgelse Rapport,* **108**, 15–23.

—— 1982b. Palynofacies and salinity in the Purbeck and Wealden of southern England. *In:* BANNER, F. T. & LORD, A. R. (eds). *Aspects of micropaleontology.* George Allen & Unwin, London, 278–308.

—— 1985. Coccolith moulds in sedimentary organic matter and their use in palynofacies analysis. *Journal of Micropalaeontology,* **4**, 111–116.

—— DUPAGNE-KIEVITS, J. & LISTER, J. K. 1988. Palynology of the Upper Cretaceous Aachen Formation of northeast Belgium. *In:* STREEL, M. & BLESS, M. J. M. (eds). *The chalk district of the Euregio Meuse-Rhine.* Natuurhistorisch Museum Maastricht, The Netherlands & Laboratoires de Paléontologie de l'Université d'Etat à Liège, Belgium, 95–103.

BELL, B. R. & HARRIS, J. W. 1986. *An excursion guide to the geology of the Isle of Skye.* The Geological Society of Glasgow.

BLESS, M. J. M. & STREEL, M. 1988. Upper Cretaceous nannofossils and palynomorphs in south Limburg and northern Liège: a review. *In:* STREEL, M. & BLESS, M. J. M. (eds) *The chalk district of the Euregio Meuse-Rhine.* Natuurhistorisch Museum Maastricht, The Netherlands & Laboratoires de Paléontologie de l'Université d'Etat à Liège, Belgium, 105–117.

BROOKS, J., GRANT, P. R., MUIR, M., VAN GIJZEL, P. & SHAW, G. (eds) 1971. *Sporopollenin.* Academic Press, London.

COLLINSON, M. E., BATTEN, D. J., SCOTT, A. C. & AYONGHE, S. N. 1985. Palaeozoic, Mesozoic and contemporaneous megaspores from the Tertiary of southern England: indicators of sedimentary provenance and ancient vegetation. *Journal of the Geological Society, London,* **142**, 375–395.

COMBAZ, A. 1980. Les kérogènes vus au microscope. *In:* DURAND, B. (ed.) *Kerogen: insoluble organic matter from sedimentary rocks,* Éditions Technip, Paris, 55–111.

CURRY, D., ADAMS, C. G., BOULTER, M. C., DILLEY, F. C. *et al.* 1978. *A correlation of Tertiary rocks in the British Isles.* Geological Society, London, Special Report, **12**.

DOW, W. G. 1977. Kerogen studies and geological interpretations. *Journal of Geochemical Exploration,* **7**, 79–99.

DUXBURY, S. 1983. A study of dinoflagellate cysts and acritarchs from the Lower Greensand (Aptian to Lower Albian) of the Isle of Wight, southern England. *Palaeontographica B,* **186**, 18–80.

EATON, G. L. 1976. Dinoflagellate cysts from the Bracklesham Beds (Eocene) of the Isle of Wight, southern England. *Bulletin of the British Museum (Natural History), Geology,* **26**, 225–332.

ELSIK, W. C. 1966. Biologic degradation of fossil pollen grains and spores. *Micropalaeontology,* **12**, 515–518.

ESHET, Y., DRUCKMAN, Y., COUSMINER, H. L., HABIB, D. & DRUGG, W. S. 1988. Reworked palynomorphs and their use in the determination of sedimentary cycles. *Geology,* **16**, 662–665.

GAUPP, R. & BATTEN, D. J. 1983. Depositional setting of middle to Upper Cretaceous sediments in the Northern Calcareous Alps from palynological evidence. *Neues Jahrbuch für Geologie und Paläontologie, Monatshefte,* **1983/10**, 585–600.

—— & —— 1985. Maturation of organic matter in Cretaceous strata of the Northern Calcareous Alps. *Neues Jahrbuch für Geologie und Paläontologie, Monatshefte,* **1985/3**, 157–175.

GUY-OHLSON, D., LINDQVIST, B. & NORLING, E. 1987. Reworked Carboniferous spores in Swedish Mesozoic sediments. *Geologiska Föreningens i Stockholm Förhandlingar,* **109**, 295–306.

HABIB, D. 1982. Sedimentation of black clay organic facies in the Mesozoic oxic North Atlantic. *Proceedings of the 3rd North American Paleontologic Convention,* **1**, 217–220.

KEMPER, E., ERNST, G. & THIERMANN, A. 1978. *Die Unterkreide im Wiehengebirgsvorland bei Lübbecke und im Osning zwischen Bielefeld und Bever-*

gern. Exkursion Al, Symposium Deutsche Kreide, Münster i. W., 1978.

KING, P. M. 1976. *The secondary minerals of the Tertiary lavas of northern and central Skye—zeolite zonation patterns, their origin and formation*. PhD Thesis, University of Aberdeen.

KUKUK, P. 1936. *Geologie des Niederrheinisch—Westfälischen Steinkohlengebietes*. Springer, Berlin.

LEGAULT, J. A. & NORRIS, G. 1982. Palynological evidence for recycling of Upper Devonian into Lower Cretaceous of the Moose River Basin, James Bay Lowland, Ontario. *Canadian Journal of Earth Sciences*, **19**, 1–7.

LISTER, J. K. & BATTEN, D. J. 1988. Stratigraphic and palaeoenvironmental distribution of Early Cretaceous dinoflagellate cysts in the Hurlands Farm Borehole, West Sussex, England. *Palaeontographica B*, **210**, 9–89.

MUIR, M.D. 1967. Reworking in Jurassic and Cretaceous spore assemblages. *Review of Palaeobotany and Palynology*, **5**, 145–154.

—— & SARJEANT, W. A. S. 1978. The palynology of the Langdale Beds (Middle Jurassic) of Yorkshire and its stratigraphical implications. *Review of Palaeobotany and Palynology*, **25**, 193–239.

MULLER, J. 1959. Palynology of Recent Orinoco delta and shelf sediments: reports of the Orinoco Shelf Expedition; volume 5. *Micropaleontology*, **5**, 1–32.

NEVES, R. & SULLIVAN, H. J. 1964. Modification of fossil spore exines associated with the presence of pyrite crystals. *Micropaleontology*, **10**, 443–452.

PHILLIPS, L. 1974. Reworked Mesozoic spores in Tertiary leaf-beds on Mull, Scotland. *Review of Palaeobotany and Palynology*, **17**, 221–232.

RAYNAUD, J. F. & ROBERT, P. 1976. Les méthodes d'étude optique de la matière organique. *Bulletin du Centre de Recherche, Pau-SNPA*, **10**, 109–127.

RICHARDSON, J. B. & RASUL, S. M. 1978. Palynological evidence for the age and provenance of the Lower Old Red Sandstone from the Apley Barn Borehole, Witney, Oxfordshire. *Proceedings of the Geologists' Association, London*, **90**, 27–42.

RICHEY, J. E. 1961. *Scotland: the Tertiary volcanic districts* (3rd edition, revised by MACGREGOR, A. G. & ANDERSON, F. W.). British Regional Geology, Her Majesty's Stationery Office, Edinburgh.

SELLWOOD, B. W. & SLADEN, C. P. 1981. Mesozoic and Tertiary argillaceous units: distribution and composition. *Quarterly Journal of Engineering Geology*, **14**, 263–275.

SIMPSON, J. B. 1961. The Tertiary pollen-flora of Mull and Ardnamurchan. *Transactions of the Royal Society of Edinburgh*, **64**, 421–468.

SKELHORN, R. R., MACDOUGALL, J. D. S. & LONGLAND, P. J. N. 1969. *The Tertiary igneous geology of the Isle of Mull*. Geologists' Association Guide **20**, Benham, Colchester.

SLADEN, C. P. 1983. Trends in Early Cretaceous clay mineralogy in NW Europe. *Zitteliana*, **10**, 349–357.

SLADEN, C. P. & BATTEN, D. J. 1984. Source-area environments of Late Jurassic and Early Cretaceous sediments in Southeast England. *Proceedings of the Geologists' Association, London*, **95**, 149–163.

SRIVASTAVA, S. K. 1975. Maastrichtian microspore assemblages from the interbasaltic lignites of Mull, Scotland. *Palaeontographica B*, **150**, 125–156, 14 pl.

STANLEY, E. A. 1966. The problem of reworked pollen and spores in marine sediments. *Marine Geology*, **4**, 397–408.

STAPLIN, F. L. 1969. Sedimentary organic matter, organic metamorphism, and oil and gas occurrence. *Bulletin of Canadian Petroleum Geology*, **17**, 47–66.

STEWART, D. J. 1981. A field guide to the Wealden Group of the Hastings area and the Isle of Wight. *In*: ELLIOTT, T. (ed.) *Field guides to modern and ancient fluvial systems in Britain and Spain*. International Fluvial Conference, University of Keele, 3.1–3.31.

STREEL, M. & BLESS, M. J. M. 1980. Occurrence and significance of reworked palynomorphs. *In*: BLESS, M. J. M., BOUCKAERT, J. & PAPROTH, E. (eds) Pre-Permian around the Brabant Massif in Belgium, The Netherlands and Germany. *Mededelingen Rijks Geologische Dienst*, **32–10**, 69–80.

TRAVERSE, A. & GINSBURG, R. N. 1966. Palynology of the surface sediments of Great Bahama Bank, as related to water movement and sedimentation. *Marine Geology*, **4**, 417–459.

TRUSWELL, E. M. 1982. Palynology of seafloor samples collected by the 1911–14 Australasian Antarctic Expedition: implications for the geology of coastal East Antarctica. *Journal of the Geological Society of Australia*, **29**, 343–356.

—— 1983*a*. Recycled Cretaceous and Tertiary pollen and spores in Antarctic marine sediments: a catalogue. *Palaeontographica B*, **186**, 121–174.

—— 1983*b*. Geological implications of recycled palynomorphs in continental shelf sediments around Antarctica. *In*: OLIVIER, R. L., JAMES, P. R. & JAEP, J. B. (eds) *Antarctic earth science*. Australian Academy of Science, Canberra, 394–399.

—— & DREWRY, D. J. 1984. Distribution and provenance of recycled palynomorphs in surficial sediments of the Ross Sea, Antarctica. *Marine Geology*, **59**, 187–214.

WALL, D., DALE, B., LOHMANN, G. P. & SMITH, W. K. 1977. The environmental and climatic distribution of dinoflagellate cysts in modern marine sediments from regions in the North and South Atlantic Oceans and adjacent seas. *Marine Micropalaeontology*, **2**, 121–200.

WARRINGTON, G. & POLLARD, J. E. 1985. Late Triassic miospores from Gribun, western Mull. *Scottish Journal of Geology*, **21**, 218–221.

——, AUDLEY-CHARLES, M. G., ELLIOTT, R. E., EVANS, W. B. *et al*. 1980. *A correlation of Triassic rocks in the British Isles*. Geological Society, London, Special Report, **13**.

WINDLE, T. M. F. 1979. Reworked Carboniferous spores: an example from the Lower Jurassic of northeast Scotland. *Review of Palaeobotany and Palynology*, **27**, 173–184.

ZIEGLER, P. A. 1982. *Geological atlas of western and central Europe*. 2 volumes; Shell Internationale Petroleum Maatschappij B. V., 130 pp + 40 enclosures.

Palaeoweathering or diagenesis as the principal modifier of sandstone framework composition? A case study from some Triassic rift-valley redbeds of eastern North America

MICHAEL A. VELBEL & MOUNIR K. SAAD

Department of Geological Sciences, Michigan State University, East Lansing, Michigan 48824-1115 USA

Abstract: Previous studies to recognize a 'palaeoweathering' signature in the framework composition of ancient sandstones have been undertaken in study areas where ancient climates are inferred to have been more humid than the present climate. Unfortunately, in studies in which the ancient climate was more humid than the modern, reduction of modal unstable-grain abundance by ancient weathering cannot be distinguished from unstable-grain destruction by burial diagenesis. Reinterpretation of data from Triassic Chatham Group sandstones of the Deep River basin (North Carolina) and modern sands derived from similar source rocks shed new light on the matter, as the Triassic sediments were generated under more arid conditions than the modern wet climate. If diagenesis were the more important factor, the sandstones would have lower contents of unstable grains. If weathering were more important, the ancient sandstones would have higher contents of unstable grains. Chatham Group sandstones have higher unstable grain contents, indicating that the compositional signature imposed on the sandstones under the Triassic climate has survived without being obliterated by diagenesis. Only 'long-term average' palaeoclimatic conditions are preserved in sandstone detrital framework modes, and several specific criteria must be met before these conditions can be inferred from sandstone framework composition. The framework-petrographic signatures of palaeoclimate can be discerned only in unglaciated sequences from mature extensional tectonic settings on continental plates, only in sedimentary basins with mild late-diagenetic histories, and only for those modern–ancient pairs where the modern climate is more humid than the inferred ancient climate.

Sandstone provenance refers to the set of factors which influenced the production of sand in its source area (Pettijohn *et al.* 1987). Because tectonism determines the nature and distribution of the source rocks, much recent research on sandstone provenance has emphasized plate-tectonic controls on sand and sandstone framework mineralogy (e.g., Dickinson & Suczek 1979; Dickinson 1988). However, the recent emphasis on tectonic controls has distracted attention from the fact that the framework composition in an ancient sandstone is a product not only of the source rocks, but also of modifying processes, both within the source area and in other compartments of the sedimentary cycle. These modifying processes include: chemical and mechanical weathering in the source area; physical modification during transportation and deposition; mineralogical alteration during diagenesis (Suttner 1974).

Weathering processes in the source area are a component of provenance (Suttner 1974; Pettijohn *et al.* 1987), because weathering modifies the source rocks' primary mineral assemblage, prior to the introduction of the weathered assemblage into the dispersal system. Such processes in the source areas of ancient sandstones are here included with related processes under the term *palaeoweathering*, to distinguish these prediagenetic near-surface alterations from weathering of sandstones in modern outcrops after diagenesis but prior to sampling. Palaeoweathering includes weathering in ancient source areas, along ancient dispersal paths (e.g., temporary storage of fluvial sediments in flood plains, etc.), or in ancient non-marine depositional environments (e.g., on flood plains immediately prior to burial).

Previous attempts to relate climate to sand composition (e.g., Young *et al.* 1975; Suttner *et al.* 1981) suggested that weathering in humid climates significantly reduces the content of unstable framework-grain types in modern sands. Topography (relief) also plays a recognized role (e.g. Suttner 1974; Pettijohn *et al.* 1987), and more recent research has elucidated the combined effects of climate and relief as important modifiers of source-rock signatures in modern

From Morton, A. C., Todd, S. P. & Haughton, P. D. W. (eds), 1991, *Developments in Sedimentary Provenance Studies.* Geological Society Special Publication No. 57, pp. 91–99.

sands (e.g., Grantham & Velbel 1988). However, applying these results from modern sands to interpreting palaeoclimatic (palaeoweathering) effects on ancient sandstones has been hampered by the fact that framework grains in ancient sandstones are vulnerable to modification during diagenesis.

It is widely accepted that source-area litholo-gical signatures can be discerned in ancient sand-stones (e.g., Dickinson 1988). It is equally well demonstrated that, in modern sands, weathering in source areas modifies the parent-rock signa-ture in systematic ways, especially with regard to climate and relief (e.g., Grantham & Velbel 1988). The purpose of this paper is to examine the 'survival' and recognition of these 'weather-ing' signatures in ancient sandstones, in a case study chosen to illustrate the specific conditions which must be met to distinguish palaeoweather-ing effects from diagenesis.

Previous work

Young *et al.* (1975), Basu (1976), and Suttner *et al.* (1981) have shown that unstable sand grains (feldspars and rock fragments) are less abundant in modern sands derived under humid climates than in modern sands derived under arid cli-mates. This difference is attributed to the greater effectiveness of weathering in wetter weathering environments. The few attempts to extend this type of analysis of sandstone detrital frame-work modes to infer palaeoclimate and palaeo-weathering in ancient source areas and/or dis-persal systems have suffered from a choice of study conditions not specifically designed to dis-tinguish palaeoweathering effects from diagene-sis. For example, the observation that unstable-grain contents of the Fountain Fm. (Pennsylva-nian) of Colorado (USA) are lower than those of modern sands derived from similar source rocks led Mack & Suttner (1977) to conclude that the climate in Fountain time was more humid than the present climate. Their argument can be summarized as follows:

> *hypothesis*—source-area weathering con-trols framework grain composition in ancient sandstones;
> *setting*—modern and ancient (Pennsylva-nian) sands of Colorado Front Range; modern climate is arid;
> *predicted consequence*—rock fragments and other unstable constituents will be less abundant in the sand produced under more humid conditions;
> *observation*—ancient sandstones have lower

abundances of rock fragments and other unstable constituents than modern sands; *conclusion*—observations are consistent with those predicted from the hypothesis; ancient sands were derived under conditions more humid than the present.

On the other hand, Walker (1978) pointed out that burial diagenesis would also result in un-stable-grain depletion in ancient sands relative to their modern counterparts. Walker's (1978) argument can be similarly summarized:

> *hypothesis*—diagenesis controls framework grain composition in ancient sandstones;
> *setting*—modern and ancient (Pennsylva-nian) sands of Colorado Front Range; modern climate is arid;
> *predicted consequence*—rock fragments and other unstable constituents will be less abundant in the sand which experienced the most diagenesis (i.e., the ancient sand-stones);
> *observation*—ancient sandstones have lower abundances of rock fragments and other unstable constituents than modern sands; *conclusion*—observations are consistent with those predicted from the hypothesis. Differences in rock fragment content of modern sands and ancient sandstones are due to diagenesis.

As can be seen when the two contrasting arguments are compared, it is impossible to distinguish ancient source-area weathering from diagenesis in the case of the Fountain Fm., because of the choice of study area and modern–ancient climate relationships. In any modern–ancient comparison where the ancient was more humid than the modern (e.g., Pennsylvanian v. modern Front Range), both hypotheses predict the observed depletion of unstable grains in ancient sediments; consequently, the two hypo-theses cannot be distinguished from one another under such conditions. However, if a setting could be found in which the ancient climate was less humid than the modern, the relative import-ance of source-area weathering and diagenesis in that specific case could be ascertained. If diage-nesis were more important than climate in des-troying unstable framework grains, ancient sandstones would have lower unstable-grain contents than their modern counterparts, regardless of the relative climates. If, on the other hand, palaeoweathering were the domi-nant modifier of sandstone framework compo-sition, ancient sands, having been derived under a drier, less vigorous weathering regime than their modern counterparts, should have higher contents of unstable grains than modern sands. Furthermore, not only would higher abundances of unstable grains in ancient sandstones be consistent with the palaeoweathering hypothesis, but such higher abundances would be inconsistent with the diagenetic-control hypothesis, because diagenesis cannot create unstable grains, especially lithic fragments. Thus, higher unstable-grain contents in (arid-climate) ancient sandstones than modern sands of otherwise identical provenance derived under a more humid climate would prove that the compositional signature imposed on the sands by the ancient climate has survived without being obliterated by diagenesis.

Study area and methods

Suttner & Dutta (1986, fig. 3) identified a number of criteria which must be satisfied for preservation of a climatic signature in ancient sandstones. These criteria are: (1) derivation from phaneritic crystalline rocks; (2) transportation over distances of less than 100 km in streams of medium to low gradient; (3) deposition in low-order stream channels and associated alluvial environments; (4) a diagenetic history dominated by early cementation at shallow burial depths, effectively 'sealing off' the framework grains from later intrastratal alteration. Suttner & Dutta (1986) concluded that the conditions stipulated for the preservation of a climatic signature in detrital framework modes

of sandstones are met in a mature extensional tectonic setting on a continental plate (e.g., an ancient rift setting, a graben bounded by basement-cored uplifts). In such a setting, the crystalline basement source rock of both modern and ancient sands is likely to be the same (or, at minimum, older than the ancient sandstones, so that the common source for both sands is at least plausible).

Based on the preceding summary, a rift-basin study area was sought, which had to satisfy not only the criteria set forth by Suttner & Dutta (1986) but several additional requirements as well. Independent evidence (e.g., caliche or other arid palaeosols, palaeobotanical data) had to indicate that the ancient climate was more arid (less humid) than the present climate acting on the same source rocks. Furthermore, the study area should be one in which Pleistocene glaciation has not modified the lithological character of the modern source materials.

Modern sands and the strata of the Chatham Group (Triassic; Upper Ladinian and Karnian; Traverse 1987) of the Deep River basin were chosen for this study on the basis of the aforementioned criteria. The Deep River basin (specifically, the subbasins commonly known as the Sanford in the south and the Durham in the north; Stuckey 1965) is a 25×240 km Triassic half-graben situated in central North Carolina, USA (Fig. 1). It is some 500 km south of the southernmost glacial deposits in eastern North America, assuring that the source materials of modern sands are not contaminated by material unrelated to the local crystalline basement rock.

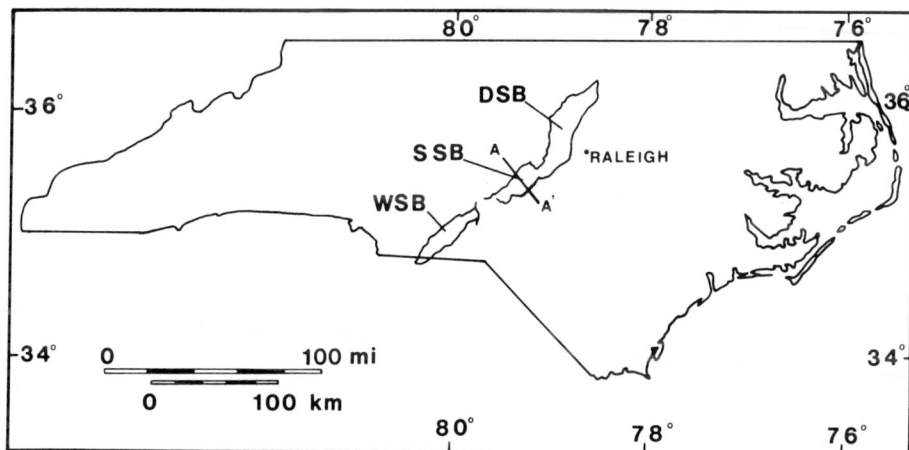

Fig. 1. Locality map of North Carolina (USA), showing the location of the Deep River basin. The basin is subdivided into the Sanford sub-basin (SSB) and the Durham sub-basin (DSB). The Wadesboro (not studied here) is shown here as a sub-basin of the Deep River, (WSB), but is often considered as a separate basin. Note location of cross-section A–A' through the Sanford sub-basin. Adapted from Gore (1986).

Basement consists of greenschist facies meta-sedimentary and metavolcanic rocks of the Carolina Slate Belt to the west; to the east, upper greenschist to amphibolite facies rocks of the Carolina Slate Belt, and crystalline rocks (pluto-nic granites, gneisses, schists, phyllites, gray-wackes, and quartzite) occur (McCarn & Mans-field 1986). Strata of the Carolina Slate Belt were deposited during the late Proterozoic to early Palaeozoic (Stuckey 1965; Secor et al. 1983), were metamorphosed during the early and poss-ibly middle Palaeozoic (Glover et al. 1983; Secor et al. 1983), and are intruded by Palaeozoic granites (Stuckey 1965; King 1977). Other crys-talline rocks of the region are also late Protero-zoic and early Palaeozoic metasediments and plutonic rocks (Stuckey 1965; King 1977).

The petrological character of all the rock-types of potential source areas was established prior to the onset of rifting along the Atlantic margin of North America (and the associated development of the Deep River basin) during the Mesozoic. No lithology-modifying episodes (e.g., metamorphism) have taken place since then. The possibility that unroofing of the uplifted blocks presently exposes rock types that were not exposed in the Triassic cannot be dis-counted, but there is no evidence to suggest that different rock types existed at shallower struc-tural levels than are exposed today. Further-more, even if they existed, they may have been eroded during late Palaeozoic and earliest Trias-sic erosion, so they could have already been

removed by the onset of rift-basin development. With these caveats in mind, we proceed on the assumption that the same potential source rock-types existed in the Triassic as today.

Clastic sedimentary rocks of the Chatham Group are red to brown poorly sorted polymic-tic conglomerates, arkosic and lithic sandstones, siltstone, mudstone, and shale (Gore 1986). In the Sanford subbasin, the Chatham Group can be divided into three formations (Fig. 2); the (basal) conglomeratic and sandy Pekin Forma-tion, overlain by the shaly, coal-bearing Cum-nock Formation, which is overlain in turn by the sandy Sanford Formation. In the adjacent Dur-ham subbasin, the undifferentiated Chatham Group consists of the same sedimentary facies, but with no distinctive stratigraphic units. With the exception of the Cumnock Formation, a shaly paludal and lacustrine unit in the Sanford subbasin, the Chatham Group consists of fluvial and alluvial fan deposits (Gore 1986). Sedimen-tology, stratigraphy, and palaeogeography of the Chatham Group are discussed in more detail by Gore (1986).

Modern rainfall in the study area averages approximately 1.2–1.4 m annually (Stuckey & Steel 1953). Plant fossil assemblages in the basal Pekin Formation of the Sanford subbasin indi-cate moist tropical upland conditions during the early history of the basin (Ladinian stage through Julian sub-stage of the Karnian stage; Gore 1986; Traverse in Gore 1986). Plant fossils are rare in the Sanford Formation (Tuvalian

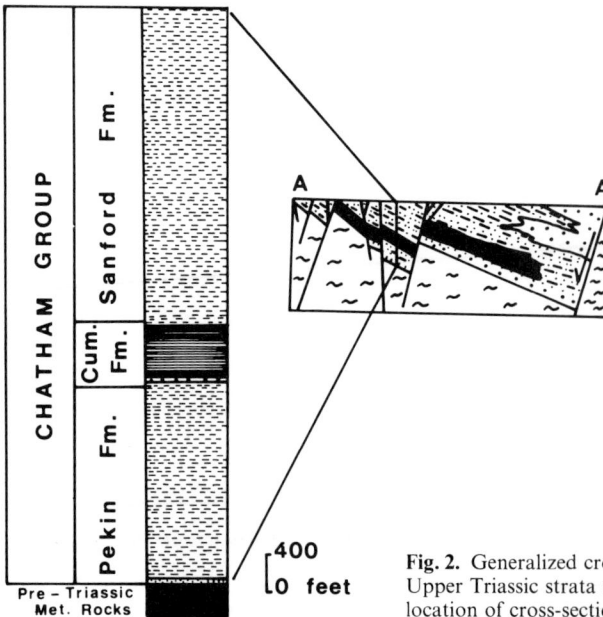

Fig. 2. Generalized cross-section and stratigraphic column for the Upper Triassic strata of the Sanford sub-basin (see Fig. 1. for location of cross-section). Adapted from Gore (1986).

sub-stage of the Karnian stage; Traverse *in* Gore 1986), but a progressive change to more xeric vegetation (drying of the climate) during the Karnian is indicated (Gore 1986; Traverse *in* Gore 1986). Calcite nodules and calcite-filled fractures interpreted as caliche palaeosols have also been reported from fluvial deposits of the Durham subbasin (Wheeler & Textoris 1978; Textoris & Holden *in* Gore 1986), also indicating relative aridity during the late Triassic.

The empirical foundation for reevaluating the palaeoweathering-diagenesis problem is based on qualitative examination of petrographic thin sections prepared from the 1986 field season, and reinterpretation of the quantitative data of McCarn (1980) and McCarn & Mansfield (1986), as summarized below.

Results

McCarn (1980) and McCarn & Mansfield (1986) studied the petrography and provenance of the Chatham Group sandstones. The Chatham Group sandstones are subarkose to sublitharenite ($Q_{50-95}F_{0-50}L_{0-50}$). Lithic fragments in the sandstones and conglomerates are similar to the pre-Triassic plutonic and metamorphic rocks of the adjacent Appalachian Piedmont;

from this, McCarn & Mansfield (1986, p. 71) concluded that 'the Chatham Group sands and gravels were likely derived from corresponding portions of these older rocks when they were exposed to weathering and erosion during the late Triassic.'

McCarn & Mansfield (1986) found that modern sands originating in the Piedmont are petrographically similar to, but somewhat more mature than, the Chatham Group sandstones. In particular, modern sands have consistently lower abundances of both total lithic fragments (Fig. 3) and metamorphic lithic fragments than the ancient sandstones. Lithic fragment content is widely regarded as the most sensitive indicator of weathering (Basu 1976; Grantham & Velbel 1988). McCarn & Mansfield (1985, 1986) concluded that the difference is due primarily to differences in climate. McCarn (1980) specifically concluded that, when interpreted according to the methods of Young *et al.* (1975) and Basu (1976), a Triassic climate more arid than the modern climate is indicated.

McCarn (1980) briefly discusses the diagenetic features of the Chatham Group sandstones, and concludes (p. 27) 'that the influence of diagenesis on framework grain composition is negligible.' Petrographic examination of Chatham Group sandstones confirms McCarn's (1980) inferences.

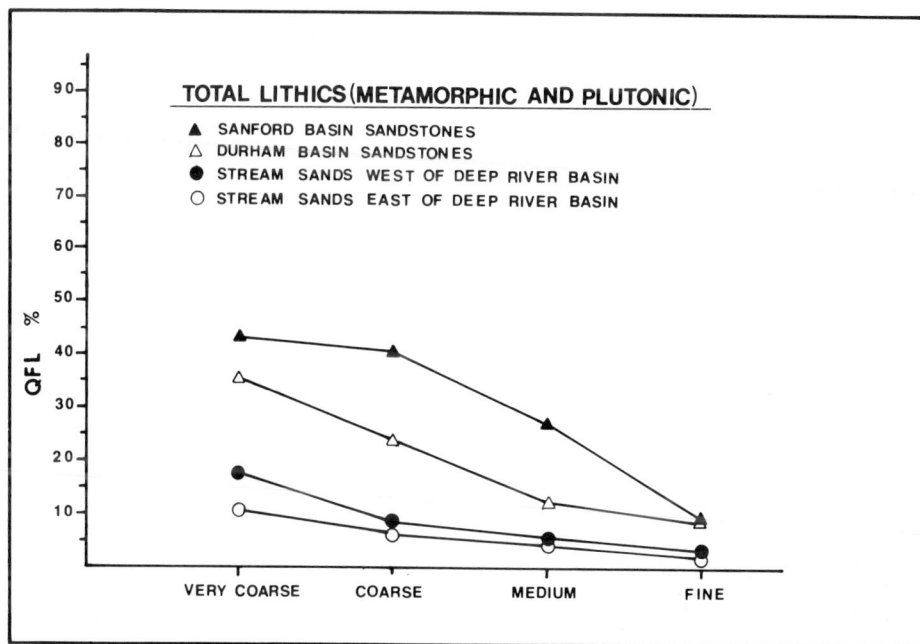

Fig. 3. Composition-size plot of total (metamorphic plus plutonic) lithic fragments in Chatham Group sandstones (triangles) and modern sands (circles). Redrawn from McCarn (1980).

We observed seven major diagenetic features: (1) partial chloritization of feldspars; (2) calcite replacement and cementation; (3) fracturing; (4) sericitization; (5) plagioclase overgrowth; (6) hematite grain coatings; (7) chalcedonic pore fillings. None of these result in major modifications of framework grains.

Plagioclase is the dominant feldspar. Diagenesis was not intense enough to destroy the detrital feldspars, or to modify them beyond recognition. The most common alteration is the replacement of feldspar by authigenic chlorite. Partial sericitization, and replacement by calcite, are locally observed. Fracturing of feldspar due to compaction is uncommon; those fractures which do occur are commonly filled with authigenic calcite or sericite. Minor authigenic feldspar overgrowths on plagioclase were locally observed. Hematite grain coatings stain most of the framework grains; calcite and chalcedonic silica are the predominant cements. Petrographic observations show that feldspars have been modified by diagenesis, but not to a degree that reduces their apparent abundance in the Chatham Group sandstones.

Metasedimentary and metaplutonic rock fragments (mainly schist, phyllite, quartzite, and gneiss) show no evidence of replacement or alteration, and are generally not even stained by hematite. Diagenesis has had no influence on the recognizability of rock fragments.

Comparison of petrographic thin-sections of Chatham Group sandstones with sandstones from other Mesozoic rift-valley basins reveals that diagenesis in the Chatham Group sandstones is manifested primarily as compaction and cementation, *without* extensive replacement of framework grains which is common in some other basins (e.g., the Hartford basin in Connecticut, USA; Saad, in prep.). Other rift-valley basins may have experienced sufficiently intense diagenesis to modify unstable detrital framework grains, but diagenesis in the Chatham Group sandstones of the Deep River basin has not been extensive enough to affect the recognizability of detrital feldspar or rock fragments.

Discussion

The full significance of McCarn's (1980) data regarding the relative importance of palaeo-weathering and diagenesis in determining the framework composition of Chatham Group sandstones becomes apparent when the two working hypotheses referred to in the 'previous work' section are recast in terms of the new study area. To see how the choice of study conditions (specifically, relative 'wetness' of modern and inferred ancient weathering environments) makes the two hypotheses distinguishable from one another, the two hypotheses are here restated, but with the climatic relations of the Deep River Basin study setting substituted for those of the Colorado Front Range, which were presented earlier. The hypothesis of diagenetic control is now:

hypothesis—diagenesis controls framework grain composition in ancient sandstones;
setting—modern and ancient (Triassic) sands of North Carolina Piedmont and Mesozoic rift-valley basins. Modern climate is humid;
predicted consequence—rock fragments and other unstable constituents will be less abundant in the ancient sandstones (which experienced diagenesis);
observation—modern sands have lower abundances of rock fragments and other unstable constituents than ancient sandstones;
conclusion—observations are inconsistent with those predicted from the hypothesis. Differences in rock fragment content of modern sands and ancient sands cannot be due to diagenesis.

The hypothesis of diagenetic control is thus rejected.

The hypothesis that source-area weathering controls sand composition can be similarly modified for the Deep River Basin study setting:

hypothesis—source-area weathering controls framework grain composition in ancient sandstones;
setting—modern and ancient (Triassic) sands of North Carolina Piedmont and Mesozoic rift-valley basins. Modern climate is humid;
predicted consequence—rock fragments and other unstable constituents will be less abundant in the sand produced under more humid conditions;
observation—modern sands have lower abundances of rock fragments and other unstable constituents than ancient sandstones;
conclusion—observations are consistent with those predicted from the hypothesis in this setting; ancient sands were derived under conditions less humid (more arid) than the present (McCarn 1980).

Thus, by careful choice of a study area, the two hypotheses predict different consequences, and are distinguishable from one another.

Summary and conclusions

Previous attempts to distinguish palaeoweathering effects from diagenesis on petrographic grounds have failed because of the choice of study conditions. For modern/ancient comparisons in which the ancient was more humid than the modern, both hypotheses predict the observed depletion of unstable grains in ancient sediments; under these conditions, the two hypotheses cannot be distinguished.

Data from Chatham Gp. (Triassic) sandstones of the Deep River Basin (North Carolina) and modern sands derived from similar (Palaeozoic) source rocks render the weathering and diagenesis hypotheses distinguishable from one another. Independent evidence (e.g., palynology, caliche palaeosols, etc.) indicates that the Triassic climate was more arid than the modern Piedmont climate. Chatham Gp. sandstones have higher unstable-grain contents than modern sands derived from the same provenance under the more humid modern climate. This observation is incompatible with the hypothesis that diagenesis is more important than climate in destroying unstable framework grains in these sandstones; diagenesis cannot create unstable detrital grains. However, the observation is compatible with the hypothesis that palaeoweathering was the dominant modifier of sandstone framework composition. The compositional signature imposed on the sandstones by weathering under the drier Triassic climate has survived, without being obliterated by diagenesis. Therefore, under appropriate circumstances, ancient/modern comparisons of detrital framework modes like those suggested by Mack & Suttner (1977) can be used to demonstrate the importance of palaeoweathering and palaeoclimate in determining sandstone framework compositions.

Several specific criteria must be met before palaeoclimate can be inferred from sandstone framework composition. Petrographic signatures of palaeoclimate can be discerned only in unglaciated settings which satisfy the requirements of Suttner & Dutta (1986). Especially important is their requirement that the sedimentary basin had a diagenetic history dominated by early cementation at shallow burial depths, effectively 'sealing off' the framework grains from later intrastratal alteration, without vigorous late-stage burial diagenesis. Furthermore, effects of weathering and diagenesis on unstable-grain abundances can only be distinguished for those modern-ancient pairs where the modern climate is more humid than the inferred ancient climate.

Finally, only 'long-term average' palaeoclimatic conditions can be inferred from sandstone framework compositions. In the southern Blue Ridge Mountains of North Carolina (400 km west of the present study area), for example, modern sands show a close relationship between source-area weathering and modal sand composition (Grantham & Velbel 1988). However, the soils whose erosion supplies these sands range from bare outcrops to deep regoliths that must have taken hundreds of thousands of years to form (Velbel 1985, 1988). Therefore, the composition of the modern sands represents a spatially (catchment-scale) and temporally averaged weathering signature representing tens or hundreds of thousands of years of accumulated mineral depletion and related weathering effects.

On the other hand, climate can vary significantly on time-scales of hundreds to thousands of years (for example, during the Pleistocene); consequently, climatically controlled early-diagenetic minerals (e.g., calcite, hydrous v. anhydrous Fe-oxides) can vary stratigraphically, often between individual strata, on a time-scale much finer than the hundreds of thousands of years represented by the detrital minerals. In the southern Blue Ridge (including the study area of Grantham & Velbel 1988), Pleistocene–Holocene alluvial-fan deposition is episodic, with a mean recurrence interval for depositional events of 3000–6000 years (Kochel & Johnson 1984; Velbel 1987). Individual strata within these alluvial-fan deposits exhibit post-depositional alteration features, including palaeosol features, many of which differ from one stratum to another (Mills 1981, 1982; Kochel & Johnson 1984; Velbel 1987). Stratigraphic variations in these post-depositional weathering characteristics, including authigenic minerals, can preserve a record of climatic variation with a temporal resolution orders of magnitude finer than that available from the 'time-averaged' weathering conditions preserved in the detrital minerals.

This suggests that detailed study of early diagenetic minerals is at least as useful for purposes of palaeoclimate reconstruction as studying detrital framework compositions, and that the relative utility of framework-grain studies diminishes as the temporal resolution required increases. For example, studies of detrital minerals and studies of authigenic minerals are equally useful if long-term climatic changes associated with plate motions are being investigated (e.g., Dutta & Suttner 1986; Suttner & Dutta 1986), but shorter-term climatic variations (e.g., those induced by Milankovich forcing) will not be discernible in the detrital minerals, because of the time-averaged nature of the detrital signature. Climatic variations on time-scales finer than the mean residence time of grains in the weathering profiles of the source areas (e.g., hundreds of thousands of years in the southern Blue Ridge) are more likely to be detected by studies of authigenic minerals (e.g., Dutta &

Suttner 1986; Suchecki *et al.* 1988) than by studies of sandstone detrital modes.

We would like to thank the students in the senior author's sandstones courses, especially J. McKee, and the fall 1985 class, for their enthusiastic discussions and comments; P. J. W. Gore, for helpful discussions; and D. S. Brandt, L. J. Suttner, and R. W. O'B. Knox for helpful comments, and reviews of this manuscript.

References

BASU, A. 1976. Petrology of Holocene fluvial sand derived from plutonic source rocks: Implications to paleoclimatic interpretation. *Journal of Sedimentary Petrology*, **46**, 694–709.

DICKINSON, W. R. 1988. Provenance and sediment dispersal in relation to paleotectonics and paleogeography of sedimentary basins. *In*: KLEINSPEHN, K. L., & PAOLA, C. (eds). *New Perspectives in Basin Analysis*. Springer-Verlag, New York, 3–25.

—— & SUCZEK, C. A. 1979. Plate tectonics and sandstone composition. *American Association of Petroleum Geologists Bulletin*, **63**, 2164–2182.

DUTTA, P. K. & SUTTNER, L. J. 1986. Alluvial sandstone composition and paleoclimate, II. Authigenic mineralogy. *Journal of Sedimentary Petrology*, **56**, 346–358.

GLOVER, L. III, SPEER, A., RUSSELL, G. S. & FARRAR, S. S. 1983. Ages of regional metamorphism and ductile deformation in the central and southern Appalachians. *Lithos*, **16**, 223–245.

GORE, P. J. W. (ed.) 1986. *Depositional framework of a Triassic rift basin: The Durham and Sanford subbasins of the Deep River Basin, North Carolina.* Society of Economic Paleontologists and Mineralogists, Third Annual Mid-year Meeting, Field Trip No. 3, Guidebook, 55–115.

GRANTHAM, J. H. & VELBEL, M. A. 1988. Influence of climate and topography on rock-fragment abundance in modern fluvial sands of the southern Blue Ridge mountains, North Carolina. *Journal of Sedimentary Petrology*, **58**, 219–227.

KING, P. B. 1977. *The Evolution of North America.* Princeton, Princeton University Press.

McCARN, S. T. 1980. *Petrology of modern stream sands and Upper Triassic sandstones in the eastern Piedmont of North Carolina: A clue to paleoclimate interpretation.* M.S. thesis, Southern Illinois University.

—— & MANSFIELD, C. F. 1985. Petrographically deduced Triassic climate for the Deep River Basin, eastern Piedmont of North Carolina. *Geological Society of America, Abstracts with Programs*, **17**, 657.

—— & —— 1986. Petrology and provenance of Upper Triassic sandstones, Deep River Basin, North Carolina. *In*: GORE, P. J. W. (ed.) *Depositional framework of a Triassic rift basin: The Durham and Sanford sub-basins of the Deep River Basin, North Carolina.* Society of Economic Paleontolo-

gists and Mineralogists, Third Annual Mid-year Meeting, Field Trip No. 3, Guidebook, 71–74.

MACK, G. H. & SUTTNER, L. J. 1977. Paleoclimate interpretation from a petrographic comparison of Holocene sands and the Fountain Formation (Pennsylvanian) in the Colorado Front Range. *Journal of Sedimentary Petrology*, **47**, 89–100.

MILLS, H. H. 1981. Piedmont-cove deposits of the Dellwood quadrangle, Great Smoky Mountains, North Carolina, U.S.A.: Some aspects of sedimentology and weathering. *Biuletyn Peryglacjalny* (Poland), **30**, 91–109.

—— 1982. Long-term episodic deposition on mountain foot slopes in the Blue Ridge province of North Carolina: Evidence from relative-age dating. *Southeastern Geology*, **23**, 123–128.

PETTIJOHN, F. J., POTTER, P. E. & SIEVER, R. 1987. *Sand and Sandstone* (2nd ed.). New York, Springer-Verlag.

SECOR, D. T., Jr., SAMSON, S. L., SNOKE, A. W. & PALMER, A. R. 1983. Confirmation of the Carolina Slate Belt as an exotic terrane. *Science*, **221**, 649–651.

STUCKEY, J. L. 1965. *North Carolina: Its Geology and Mineral Resources.* North Carolina Department of Conservation and Development, Raleigh.

—— & STEEL, W. G. 1953. *Geology and Mineral Resources of North Carolina.* North Carolina Division of Mineral Resources, Educational Series No. 3.

SUCHECKI, R. K., HUBERT, J. F. & BIRNEY DE WET, C. C. 1988. Isotopic imprint of climate and hydrogeochemistry on terrestrial strata of the Triassic-Jurassic Hartford and Fundy rift basins. *Journal of Sedimentary Petrology*, **58**, 801–811.

SUTTNER, L. J. 1974. Sedimentary petrographic provinces: An evaluation. *In*: ROSS, C. A. (ed.) *Paleogeographic Provinces and Provinciality*. Society of Economic Paleontologists and Mineralogists Special Publication **21**, 75–84.

—— & DUTTA, P. K. 1986. Alluvial sandstone composition and paleoclimate, I. Framework mineralogy. *Journal of Sedimentary Petrology*, **56**, 329–345.

——, BASU, A. & MACK, G. H. 1981. Climate and the origin of quartz arenites. *Journal of Sedimentary Petrology*, **51**, 1235–1246.

TRAVERSE, A. 1987. Pollen and spores date origin of rift basins from Texas to Nova Scotia as Early

Late Triassic. *Science*, **236**, 1469–1472.

VELBEL, M. A. 1985. Geochemical mass balances and weathering rates in forested watersheds of the southern Blue Ridge. *American Journal of Science*, **285**, 904–930.

—— 1987. Alluvial-fan origin for terrace deposits of the southeast Prentiss Quadrangle, near Otto, North Carolina. *Southeastern Geology*, **28**, 87 103.

—— 1988. Weathering and Soil Farming Processes. *In*: SWANK, W. T. & CROSSLEY, D. A., Jr. (eds) *Forest Hydrology and Ecology at Coweeta*. Springer Ecological Studies Series, **66**, 93–102.

WALKER, T. R. 1978. Discussion: Paleoclimate interpretation from a petrographic comparison of Holocene sands and the Fountain Formation (Pennsylvanian) in the Colorado Front Range. *Journal of Sedimentary Petrology*, **48**, 1011–1013.

WHEELER, W. H. & TEXTORIS, D. A. 1978. Triassic limestone and chert of playa origin in North Carolina. *Journal of Sedimentary Petrology*, **48**, 765–776.

YOUNG, S. W., BASU, A., MACK, G., DARNELL, N. & SUTTNER, L. J. 1975. Use of size-composition trends in Holocene soil and fluvial sand for paleoclimatic interpretations. *Proceedings of the IXth International Congress on Sedimentology*, Theme 1, Nice, France.

Redistribution of rare earth elements during diagenesis of turbidite/hemipelagite mudrock sequences of Llandovery age from central Wales

A. E. MILODOWSKI & J. A. ZALASIEWICZ

British Geological Survey, Keyworth, Nottingham, NG12 5GG, UK

Abstract: Geochemical studies of recently mapped Upper Llandovery turbidites in central Wales indicate localized rare earth element (REE) mobilization and fractionation in mudrock-dominated sequences. Closely associated turbidite sandstones, turbidite mudstones and laminated anoxic hemipelagites have generally similar major elemental compositions but display strongly differentiated REE distributions. Shale-normalized REE distribution patterns show that anoxic hemipelagites are consistently enriched in the light and middle REE, with greatest enrichment occurring for La–Eu. This enrichment reflects a concentration of millimetre-scale zoned REE-rich monazite nodules in the hemipelagites. In contrast, the REE patterns of the associated turbidites mirror those of the anoxic hemipelagite, being relatively depleted in the light and middle REE, with Nd and Sm generally showing greatest differentiation. The relationship between the concentration of nodules and degree of enrichment of REE in anoxic hemipelagite and turbidite unit thickness imply upward migration of REE into the overlying hemipelagite. However, mass balance considerations suggest that the turbidite-hemipelagite couplets were not necessarily closed systems.

Within the turbidites, REE were initially held in unstable volcanogenic minerals and adsorbed on to clay minerals and iron-manganese hydroxides. The REE were liberated into the sediment porewaters during early diagenesis and fractionated in the migrating porewaters as a result of the greater solubility of heavy REE relative to light REE. Expulsion of porewaters from the turbidite sediment during burial carried REE in solution through the intervening hemipelagic layers where REE were precipitated, probably in the presence of organic matter, and possibly initially as the hydrous phosphate rhabdophane which later transformed to monazite. Fractionation of the REE during this process led to preferential enrichment of the light and medium REE within the hemipelagic sediment.

The use of REE as indicators of sediment provenance in the Llandovery turbidite-hemipelagite mudrock sequences of central Wales must be treated with caution since significant REE element migration and fractionation is shown to have occurred during their diagenesis. Similar effects might be expected in other sequences comprising interbedded organic-rich/organic-poor mudrock and detailed petrographic analyses should be undertaken to evaluate the possibility of diagenetic REE fractionation and redistribution in conjunction with the use of REE patterns or Sm–Nd isotopes in provenance studies of these rocks.

The assumption that rare earth elements (REE) are essentially immobile during diagenesis and metamorphism (Cullers *et al.* 1975; Fleet 1984; Howard 1985) underlies their use in provenance studies of sedimentary rocks. In general, mudrocks display remarkably similar REE abundances and patterns (Haskin & Haskin 1966; Haskin *et al.* 1966; Wildeman & Condie 1973; Wildeman & Haskin 1973; McClennan *et al.* 1980) and this characteristic is taken to indicate that fine grained detrital material is efficiently homogenized during weathering, erosion and transport such that the REE fractionation that may have occurred during these processes (e.g. Ronov *et al.* 1967; Roaldset & Rosenqvist 1971) is effectively cancelled out. Thus, the REE abundances of these sediments are taken to represent the average REE composition of the upper continental crust (Fleet 1984). Deviations in the REE composition within individual samples of clay sediments are therefore considered to represent variations in the sediment source characteristics (Cullers *et al.* 1975; Dypvik & Brunfelt 1976; Fleet *et al.* 1976; Chaudhuri & Cullers 1979; Cullers *et al.* 1979; Howard 1985). However, modification of the REE distribution of mudrocks in the sedimentary environment during diagenesis cannot be ruled out (e.g. Girin *et al.* 1970; Dypvik & Brunfelt 1976; Roaldset 1979; Elderfield *et al.* 1981a,b; Thomson *et al.* 1984; German & Elderfield 1989) especially as a high proportion of the REE content of some clays may be readily leached (Balashov & Girin 1969; Roaldset & Rosenqvist 1971).

From Morton, A. C., Todd, S. P. & Haughton, P. D. W. (eds), 1991, *Developments in Sedimentary Provenance Studies.* Geological Society Special Publication No. 57, pp. 101–124.

Monazite nodules have recently been discovered in Lower Palaeozoic rocks of Wales. They have been recorded as locally abundant monazite grains in pan-concentrates of stream sediment samples derived from Lower Palaeozoic rocks of Caradoc to Upper Llandovery age (Cooper et al. 1983; Read et al. 1987), and observed in situ within mudrocks of Ashgill age (Kearsley, pers. comm.). Marked chemical zonation was noted with the light REE concentrated in the rim of the nodules whilst the middle to heavy REE were concentrated in the nodule cores (Read et al. 1987). The nodules are, however, not stratigraphically ubiquitous. For instance, they are rare or absent in Palaeozoic rocks of northern Britain, including those of the Lake District (Read et al. 1987).

Monazite nodules of similar chemistry and morphology have been reported in mudrocks and as placer deposits in Brittany, Belgium, Siberia, Taiwan, USA and several African states (Overstreet 1971; Donnot et al. 1973; Matzko & Overstreet 1977; Rosenblum & Mosier 1983; Burnotte et al. 1989). These monazites are characteristically enriched in Eu and most occurrences are associated with low-grade metamorphosed mudrocks. Opinions as to their mode of formation vary widely. Donnot et al. (1973) considered that the monazite 'grise' in Palaeozoic (Llanvirn and Dinantian) mudrocks of Brittany formed initally as a REE-rich gelatinous phosphate precipitated during early diagenesis in marine silts and which later transformed, via rhabdophane, to monazite. Matzko & Overstreet (1977) also suggested that the Taiwanese placer deposits were derived by erosion of coastal sediments cemented by diagenetic monazite. Overstreet (1971) and Rosenblum & Mosier (1983) proposed that these secondary monazite nodules were porphyroblastic features resulting from contact metamorphism, although they suggested that low-grade regional metamorphism may sometimes be important. Read et al. (1987) believed that the nodules in the British Palaeozoic rocks formed by in situ recrystallization and overgrowth of detrital monazites during diagenesis and low-grade metamorphism. Most recently, Burnotte et al. (1989) interpreted these features as authigenic precipitates, with REE being derived by desorption from clay minerals during diagenesis of phosphatic sediments.

REE nodule-bearing horizons are common in Lower Palaeozoic (Llandovery) turbidite sequences of central Wales. The relationship of these nodules to lithofacies and whole-rock geochemistry is described below. The results give a fresh insight into the formation of monazite nodules and the behaviour of REE within a closely-constrained sedimentary system, and show that REE patterns in mudrocks do not always give a reliable indication of provenance.

Sedimentology

Most of the nodule-bearing sedimentary rocks examined form part of a 1–2 km thick Upper Llandovery turbidite sequence in central Wales. Their location and generalised lithology are shown in Fig. 1. The turbidite units typically are 10–100 mm thick and have a thin basal sandstone (Bouma D) which grades up into a thick, apparently homogeneous turbidite mudstone (Bouma E, or the type 2 mudstone of Cave 1979). Each turbidite unit corresponds to a single depositional event. Above the turbidite unit there is commonly a layer of finely laminated mudstone typically 5–30 mm thick, comprising alternating laminae, 0.1–1.0 mm thick, of a medium grey mudstone and a darker grey, more carbonaceous, mudstone. This laminated layer is the type 3(ii) mudstone of Cave (1979) which represents a slowly accumulated hemipelagic sediment deposited under anoxic bottom conditions, where the absence of bioturbation has preserved the delicate laminae. Dimberline & Woodcock (1987) interpret similar anoxic hemipelagites developed within Wenlock mudrocks in Wales as annual in origin, with the darker laminae representing organic-rich fallout from spring plankton blooms, and the lighter laminae representing sediment brought into the basin by winter floods (cf. Santa Barbara Basin, Thornton 1984).

Material examined also includes mudstones consisting largely or almost entirely of anoxic hemipelagites from the condensed Llandovery sequence in north Wales (Fig. 1); and mudstones laid down in oxic bottom conditions, in which the hemipelagic sediment, together with the upper part of the underlying mudstone turbidite, is oxidized to a pale colour and burrowed (type 3(i) mudstone of Cave 1979). Sediment provenance was probably complex with input from turbidity currents, nepheloid plumes and pelagic fallout (Fig. 1).

Diagenetic apatite concretions a few millimetres thick are commonly present at the boundary between oxidized and unoxidized mudstones (Fig. 1; see also Cave 1979 and Smith 1987), and more rarely, a few millimetres below anoxic hemipelagite layers. They formed very early, just below the sediment-water interface. Calcareous concretions also occur sporadically in the Llandovery sequence and differential compactional relationships indicate that they also

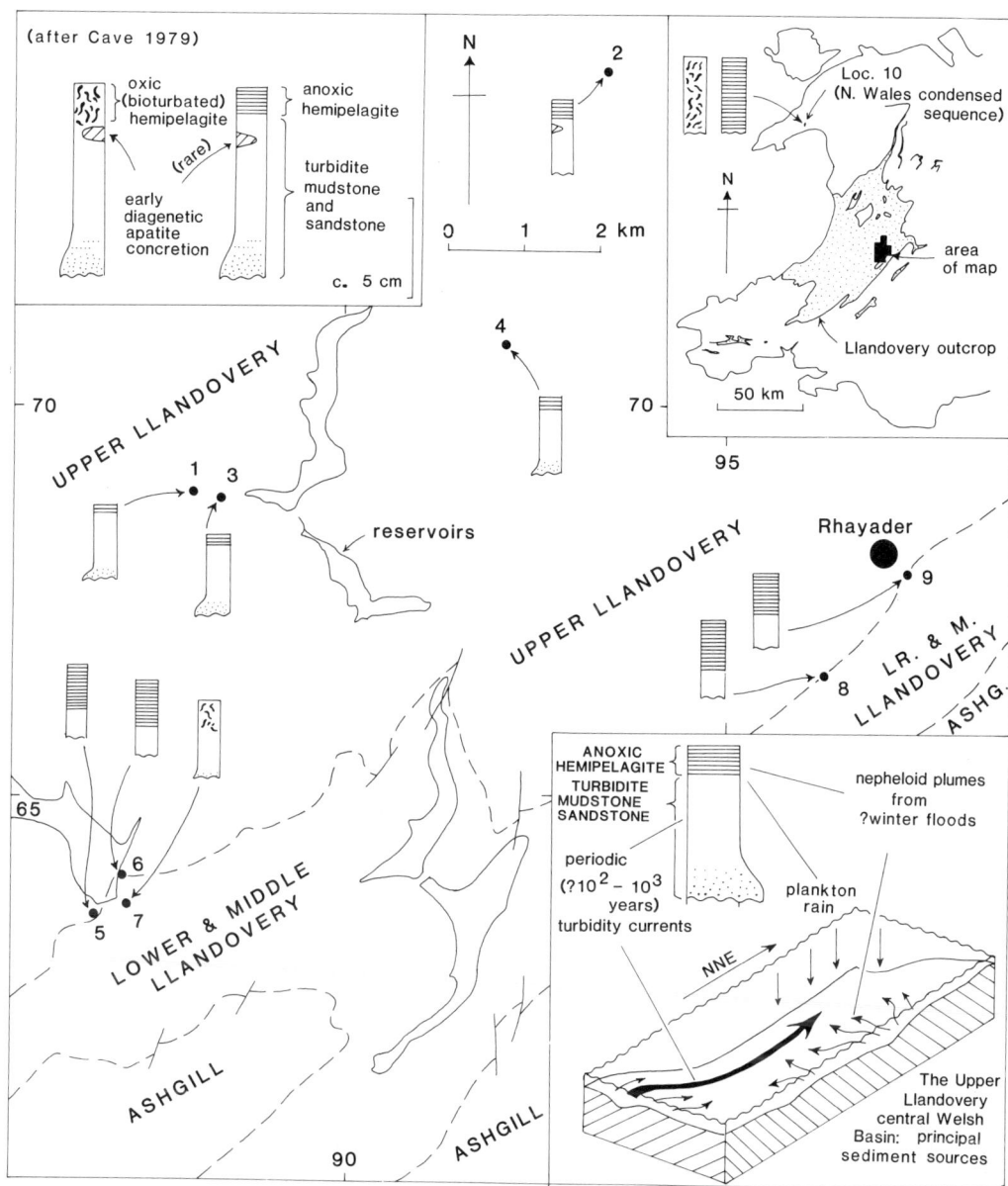

Fig. 1. Location of study area, showing sample localities with simplified lithology. Inset diagrams show generalized lithofacies scheme. Llandovery outcrops in Wales and inferred Upper Llandovery sediment provenance directions.

Localities: 1, SN 880 689; 2, SN 9342 7416; 3, SN 884 688; 4, SN 9209 7079; 5, SN 8667 6354; 6, SN 8710 6412; 7, SN 8710 6353 to SN 8718 6343; 8, SN 9625 6655; 9, SN 9745 6781; 10, Llanystumdwy outlier, N. Wales (see inset map), *turriculatus* Zone, SH 3948 4783.

formed early during diagenesis, though later than the apatite which they sometimes enclose.

Analytical techniques

The Llandovery rocks from central Wales were examined as polished thin-sections by optical petrography and by backscattered electron microscopy (BSEM). Most thin-sections were cut normal to bedding. BSEM analysis was carried out using a Cambridge Instruments Stereoscan S250 scanning electron microscope (SEM) with a KE Developments solid-state 4-element backscattered electron detector. Mineralogical identification was aided by qualitative energy-dispersive

X-ray microanalysis (EDXA) performed with a Link Systems 860 X-ray microanalyser attached to the SEM. BSEM imaging of the polished sections was carried out at an accelerating potential 20 kV after first coating the sections with a thin film of carbon (approximately 250Å).

Nodule distribution was examined by visual and X-ray methods. Visual estimates were made on large clean cleavage faces comprising several turbidite-hemipelagite alternations. Measurements were taken of the frequency and location of nodules, and of the thickness of enclosing and adjacent anoxic hemipelagites, turbidite mudstones and turbidite sandstones. X-ray radiographic examination of the rocks was conducted using cleavage slabs 10–20 mm in thickness; the frequency and location of nodules was noted, and the volume of each sedimentary unit within the slab estimated.

Chemical analyses of the nodules were obtained by energy-dispersive electron microprobe analysis (EMPA) of the polished sections. Analyses were undertaken with a Cambridge Instruments Microscan 5 instrument with a Link Systems AN10000 energy-dispersive X-ray spectrometer calibrated using REE oxide and mineral standards. X-ray spectra were recorded with an accelerating potential of 15 kV and 75 s counting livetime. The beam diameter was estimated to be of the order of 2 μm diameter. Data obtained were processed using standard ZAF4 FLS (Link Systems) software. Bulk rock chemical analyses of the rocks were determined following separation into the four lithofacies of turbidite sandstone, turbidite mudstone, anoxic hemipelagite and concretionary apatite. Major and trace elements were analysed by X-ray fluorescence (XRF) spectroscopy of pressed powder pellets (Caleb Brett Ltd). Full REE analysis of the various lithofacies were performed on selected samples by inductively coupled plasma emission spectrometry (Walsh *et al.* 1981). Organic carbon contents of selected samples were determined by evolved gas analysis (EGA) using a modification of the technique described by Morgan (1977).

Mineralogy and petrographic relationships

Host rocks

The mudrocks in which the nodules occur are cleaved, having been subjected to anchizone grade regional metamorphism during the Acadian orogeny. Both turbidites and hemipelagites comprise mainly chlorite, illite, quartz and albitized feldspar with accessory anatase and rutile. Chlorite-mica stacks (*sensu* Craig *et al.* 1982; Dimberline 1986) are common and are particularly well developed in the turbidite units where they display a graded distribution. Authigenic pyrite is present either finely disseminated throughout the matrix or as framboidal aggregates. Trace amounts of other sulphides (galena,

chalcopyrite and a Co, As-sulphide mineral) may also be present. Minor amounts of calcite are also commonly present. The turbidite sandstones have a similar mineralogy to the mudstones, though with a higher proportion of quartz, feldspar and chlorite–mica stacks.

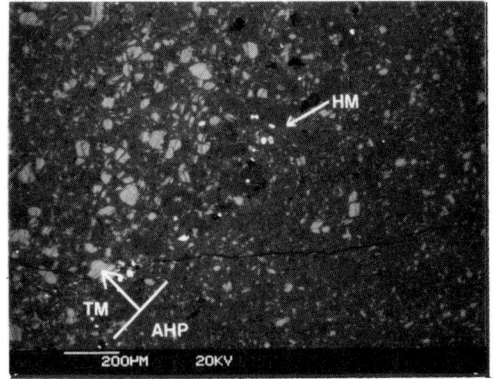

Fig. 2. BSEM photomicrograph showing concentration of detrital heavy minerals (HM) at base of turbidite mud (TM) overlying anoxic hemipelagite (AHP).

Concentrations of detrital heavy minerals are often present at the bases of turbidite units (Fig. 2). These are dominated by zircon, apatite and anatase pseudomorphs after ferromagnesian or Fe–Ti oxide minerals. Tiny grains of xenotime and thorite, commonly intergrown, may also occur and these are more uniformly distributed in the rock. Many of these heavy minerals are present as primary inclusions in some chloritemica stacks which are interpreted as altered detrital biotites. Heavy minerals are less common and of finer grain size in the anoxic hemipelagites. The presence of both rounded grains and unabraded fragments of euhedral zircon crystals may indicate a complex sediment provenance with derivation from both mature and immature sources. Within apatite concretions rare overgrowths of a secondary xenotime-like mineral were sometimes seen to be seeded on detrital zircon and replacing the adjacent authigenic apatite (Fig. 3). The morphology and texture of the anatase pseudomorphs suggests that the detrital precursors probably included magnetite (octahedral cleavages/faces preserved; Fig. 4) and ilmenite (orthogonal cleavages; Fig. 5). Apatites are often euhedral and appear to show little evidence of corrosion (Fig. 6), although in part the euhedral form may be due to recrystallization.

Detrital monazite is rare and is fine grained, generally less than 10 μm, and not especially concentrated in heavy mineral 'horizons' (the

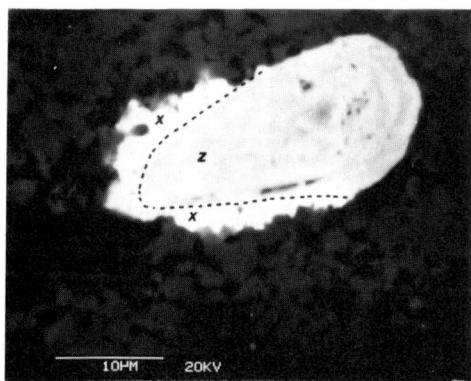

Fig. 3. BSEM photomicrograph showing growth of authigenic xenotime (x) on detrital zircon (z) in apatite-cemented early concretion. (Turbidite mudstone).

Fig. 4. BSEM photomicrograph showing skeletal anatase pseudomorph after probable detrital magnetite. Note preservation of octahedral crystallographic features. (Turbidite mudstone).

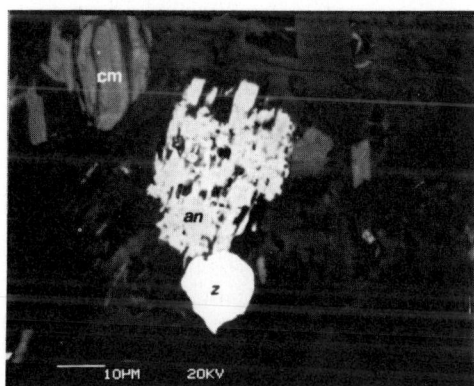

Fig. 5. BSEM photomicrograph showing anatase pseudomorph (an) after possible prismatic Fe–Ti detrital grain, associated with zircon (z) and chlorite-mica stack (cm) at base of turbidite unit. (Turbidite mudstone).

Fig. 6. BSEM photomicrograph showing detrital apatite (ap) with euhedral crystal faces. (Turbidite mudstone).

small grain size probably limited any effective hydrodynamic separation during deposition). The thorium content (estimated to be of the order 5–10% ThO_2 by comparison of qualitative EDXA spectra with known monazites) clearly distinguishes these grains from the larger thorium-deficient secondary monazite nodules. Despite their fine grain size these grains do not appear to exhibit corrosion and are euhedral (Fig. 7). In addition, corroded grains of an extremely rare Th–Ca–Si–P-rich phase, possibly cheralite (a Th-rich variety of monazite), and a La–Ce–Nd phase, possibly a carbonate (e.g. bastnaesite) or fluoride (e.g. fluocerite) were identified as trace detrital constituents.

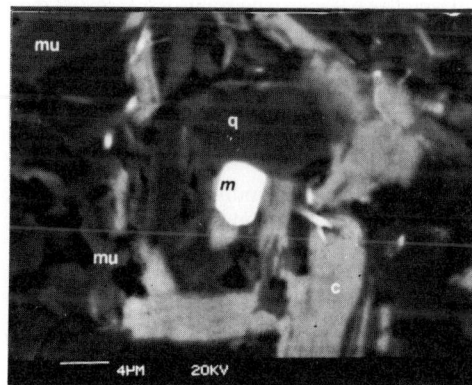

Fig. 7. Small detrital monazite grain (m) in matrix of quartz (q), mica (mu) and chlorite (c). Note lack of evidence for dissolution of monazite. (Turbidite mudstone).

Monazite nodules

Well-developed monazite nodules possess typically discoid to ovoid morphology (Fig. 8) and

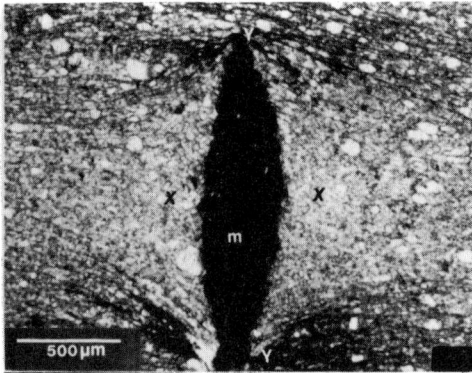

Fig. 8. Optical photomicrograph (plane polarized light) showing ellipsoidal nodule of opaque monazite (m). Note development of pressure shadow (X) either side of nodule and distortion of cleavage at ends of nodule (Y). (Turbidite mudstone).

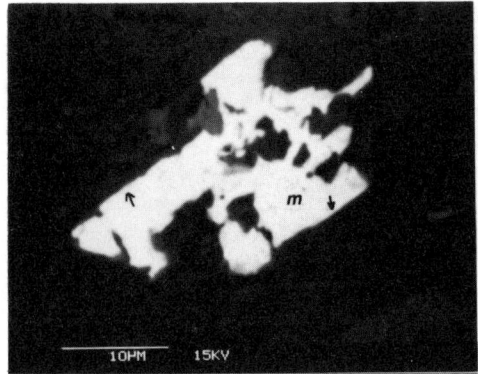

Fig. 9. BSEM photomicrograph showing early stage of monazite nodule development as tiny replacive skeletal poikiloblastic single monazite crystal (m). Prominent developments of crystal faces arrowed. (Anoxic hemipelagite).

range in size from 0.05 mm to 2 mm, although most are of the order of 0.5 mm to 1 mm in diameter. They are identical in morphology to nodules found in stream sediments draining Welsh Palaeozoic rocks (Cooper *et al.* 1983; Read *et al.* 1987) and to nodules described from other areas (e.g. Donnot *et al.* 1973; Matzko & Overstreet 1977; Rosenblum & Mosier 1983; Burnotte *et al.* 1989). They occur in both turbidite mudstone and anoxic hemipelagite layers although they are considerably more abundant in the latter. In anoxic hemipelagites monazite may also be rarely developed as discontinuous thin lamellae parallel to bedding, less than 0.2 mm thick but up to 5 mm in length.

In thin section the nodules are dark brown to opaque (Fig. 8) sometimes surrounded by more translucent rims. In most cases the monazite comprising individual nodules appears to be largely in optical continuity but some polycrystalline monazite nodules were also seen. The more elongated nodules generally exhibit inclined length-slow extinction, typically 10–15°. However, some sections may display length-fast characteristics. In the majority of examples studied extinction is often indistinct or incomplete because of the dark to opaque nature of the monazite, the abundance of strongly birefringent inclusions (micas) and the intrinsically high birefringence of the monazite. Some nodules display a diffuse wavy extinction.

In any given sample there is a continuum between these larger nodules and early growth stages represented by tiny skeletal, poikiloblastic single crystals of the order of 10 μm in size (Fig. 9). The monazite is occasionally seen to have nucleated around or include thorium-rich detrital monazite grains which may have acted as seeds

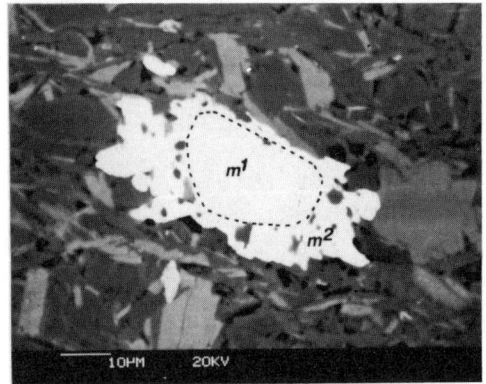

Fig. 10. BSEM photomicrograph showing localised secondary low-Th monazite (m^2) nucleated on Th-rich detrital monazite (m^1). (Anoxic hemipelagite).

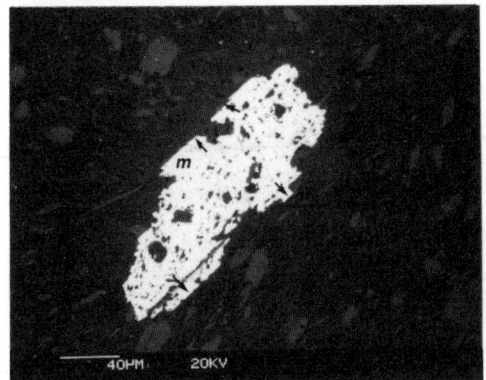

Fig. 11. BSEM photomicrograph showing intermediate stage of monazite nodule development as subhedral to euhedral poikiloblastic monocrystals (m) with development of prominent tabular faces (arrowed). (Anoxic hemipelagite).

for growth (Fig. 10). Intermediate growth stages are developed as subhedral to euhedral poikiloblastic monocrystals with prominent tabular faces (Fig. 11). Similar tabular faces can be recognized at the margins of the larger nodules (Fig. 12).

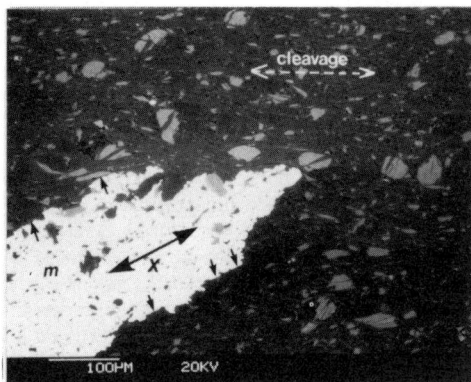

Fig. 12. BSEM photomicrograph of end of monazite nodule (m) showing relationship between nodule elongation (X), orientation of prominent tabular crystal faces (arrowed) and cleavage direction. (Anoxic hemipelagite).

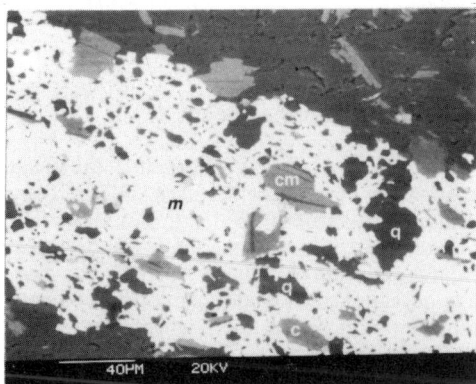

Fig. 13. BSEM photomicrograph showing detail of variably replaced and corroded inclusion fabric of quartz (q), chlorite (c) and poorly developed chlorite-mica stacks (cm) within monazite nodule (m). (Anoxic hemipelagite).

Further examination shows that elongation of the nodules was controlled by growth parallel, or subparallel, to this crystallographic plane. The optical properties described above indicate that this prominent crystal face may be the [100] plane (Deer *et al.* 1962). At their terminations, the nodules may display other well-developed crystal faces. Monazite growth is clearly replacive (Figs 9–12) and the nodules contain an inclusion fabric of variably replaced and corroded low-grade metamorphic minerals identical

to those of the host rock assemblage (Fig. 13). The finer grained (phyllosilicate) minerals and to a lesser extent the coarser quartz silt have been replaced preferentially in comparison to chlorite (Fig. 13). Early diagenetic framboidal pyrite inclusions do not appear to have suffered any replacement.

The relationship between the orientations of the nodules and inclusion, bedding and metamorphic fabrics is complex. The inclusion fabric displays a strong sub-parallel alignment of micaceous and elongate detrital mineral grains orientated parallel to the direction of nodule elongation (Figs 14 & 15). This is particularly well-developed within the anoxic hemipelagite units and is interpreted as a relict bedding fabric.

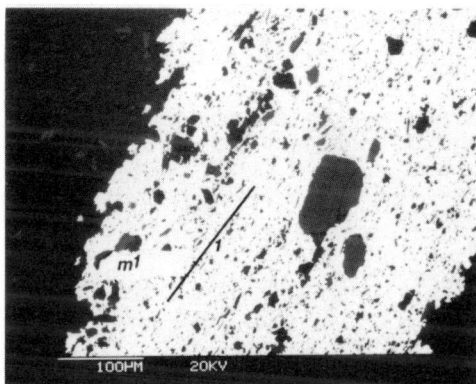

Fig. 14. BSEM photomicrograph showing relation of strong parallel orientation of phyllosilicate inclusion fabric to direction of nodule elongation (1). Inclusion of detrital Th-rich monazite grain (m1) also shown. (Anoxic hemipelagite).

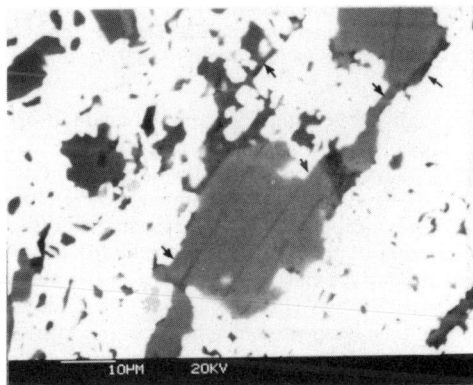

Fig. 15. BSEM photomicrograph showing a corroded relict of detrital chlorite with very limited chlorite-mica stack development. Prominent tabular faces of replacive monazite arrowed. (Anoxic hemipelagite).

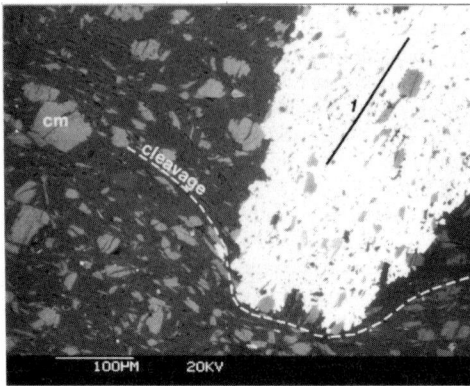

Fig. 16. BSEM photomicrograph showing relationship between nodule elongation (1) and cleavage orientation. Note distortion of cleavage around end of nodular monazite and lack of nodule deformation. Chlorite-mica stacks (cm) shown. (Turbidite mudstone).

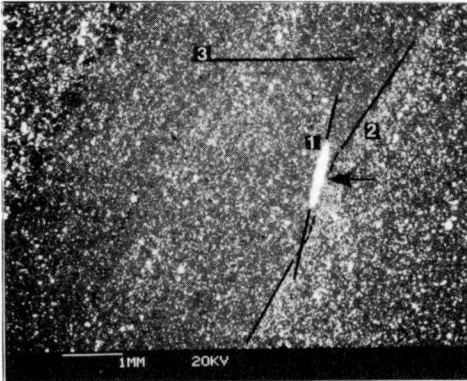

Fig. 17. BSEM photomicrograph (low magnification) showing relationship between orientations of nodule elongation (1), bedding fabric (2) and cleavage direction (3). Note slight distortion of bedding fabric adjacent to nodule. (Anoxic hemipelagite).

Although the orientation and elongate morphology of the inclusions may be exaggerated by the replacive development of the dominant tabular [100] crystal face of the monazite (Fig. 15), their orientation contrasts markedly with the orientation of phyllosilicates in the matrix of the host rock which have recrystallised in the plane of slaty cleavage (Fig. 16). Thus the nodules appear to have protected relict bedding fabrics from cleavage-oriented recrystallization during metamorphism. However, within any one sample monazite nodules tend to be similarly orientated with elongation at a small angle to bedding, between 5° and 20° (Fig. 17) but orientations up to 30° from bedding were recorded.

Further evidence for pre-cleavage nodule formation is implied by the distortion of cleavage planes around the ends of the nodules and the formation of pressure fringes above and below the nodules (Figs 8, 16). The monazite nodules do not appear to have suffered deformation or grain breakage during cleavage formation (e.g. Figs 8, 12, 16). The development of chlorite-mica stacks is also very limited within the monazite nodules (Fig. 15).

Secondary monazite was not found included within early authigenic apatite concretions. It is also absent from the cores of early diagenetic carbonate concretions which preserve relatively uncompacted sediment structures. Very rare traces of monazite (similar in development to that in Fig. 9) were found replacing relicts of the clay mineral matrix trapped within later cone-in-cone calcite developed around the outside of carbonate concretions. There is a distinct lack of differential compaction of the fine bedding laminae within and around monazite nodules from anoxic hemipelagites which implies that nodule growth occurred after the main sediment compaction. This observation is consistent with the absence of monazite nodules in the early (pre-compaction) apatite and calcite concretions.

Nodule distribution

Nodule distribution was determined by visual examination of cleavage surfaces and by X-ray radiography. Errors in estimating nodule content are inherent in both methods. For example, visual recognition of nodules depends upon cleavage development; nodules cannot be observed within turbidite sandstones where cleavage is poorly developed. Using X-ray radiographs, it is difficult to distinguish pyrite framboids and small fragments of pyritised graptolites from monazite nodules. Both methods indicate only numbers of nodules present which, given the observed (see above) and reported (Read *et al.* 1987) variation in nodule size, gives only a rough indication of total monazite content. Nevertheless, BSEM qualitatively shows that the same continuum of nodule sizes occurs in both turbidite mudstone and anoxic hemipelagite and by comparing nodule distribution to sedimentary parameters (Figs 18 & 19), the following are apparent.

(a) Nodules are markedly concentrated within anoxic hemipelagites, where they are between one and two orders of magnitude more frequent than in the associated turbidite mudstones (Fig. 18).

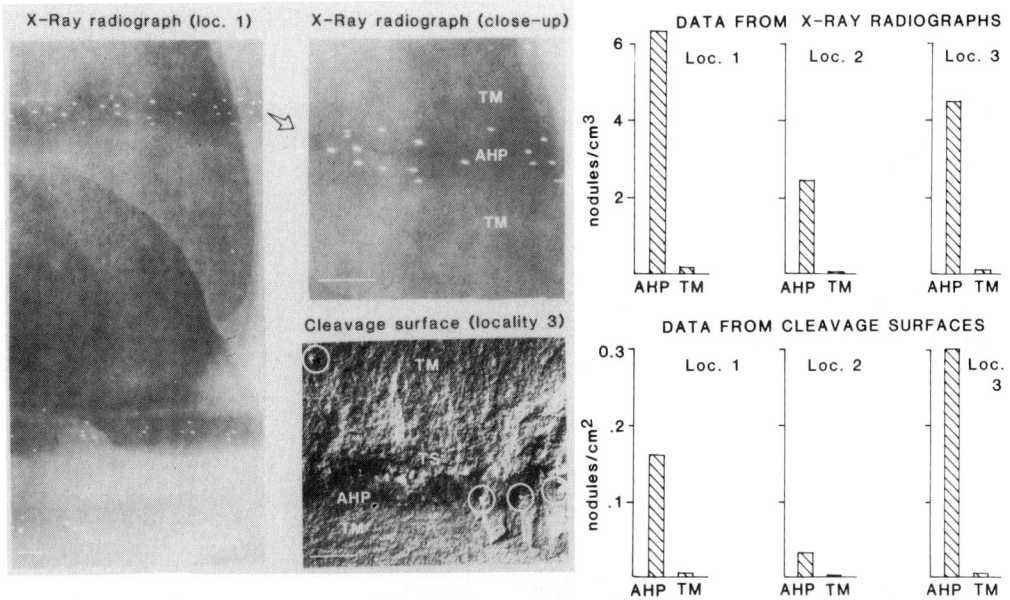

Fig. 18. Comparison of monazite nodule content of anoxic hemipelagites (AHP) and turbidite mudstone (TM). TM, turbidite mudstone; TS, turbidite sandstone; AHP, anoxic hemipelagite. Scale bar on photographs is 10 mm. Nodules circled on photographs of cleavage surface.

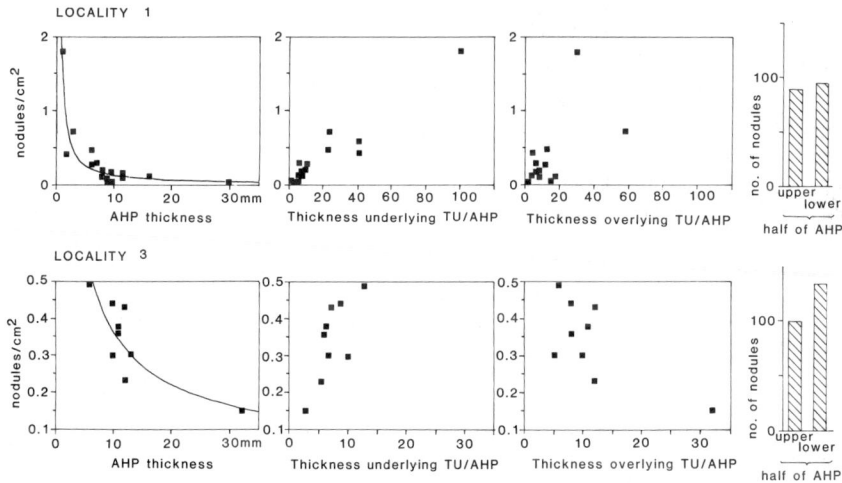

Fig. 19. Distribution of monazite nodules at localities 1 and 3 in relation to lithology (data from visual examination of cleavage surfaces). Nodule content of anoxic hemipelagites in relation to: the thickness of the enclosing anoxic hemipelagite; the thickness of the underlying turbidite unit (relative to that of the anoxic hemipelagite); and the thickness of the overlying turbidite unit (relative to that of the anoxic hemipelagite). On right, distribution of nodules within anoxic hemipelagite layers. AHP, anoxic hemipelagite; TU, turbidite unit.

(b) There may be a slight tendency for nodules to be concentrated in the bottom half of anoxic hemipelagite layers (Fig. 19).

(c) Nodule frequency is inversely proportional to the thickness of the enclosing anoxic hemipelagite (Fig. 19). This is particularly well-defined at Locality 1 where the relationship is asymptotic rather than linear.

(d) A positive correlation exists between nodule frequency and the relative thickness of the underlying turbidite unit. This correlation is still maintained when outlying data points are

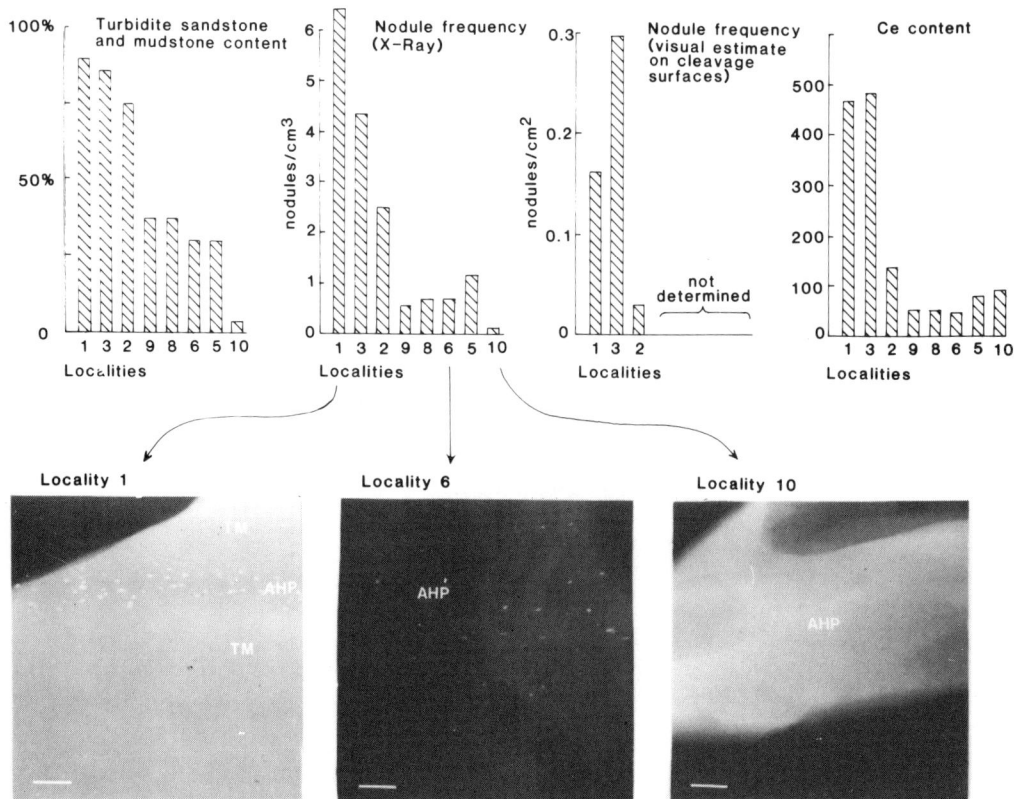

Fig. 20. The relation of turbidite/hemipelagite ratios to the monazite nodule content and the Ce content of anoxic hemipelagites. X-ray radiographs show decreasing nodule content in AHPs from left to right. AHP, anoxic hemipelagite; TM, turbidite mudstone. Scale bar is 10 mm.

excluded. There is little or no relationship between nodule frequency and the relative thickness of the overlying turbidite unit (Fig. 19).

The average nodule frequency of the anoxic hemipelagite in relation to the overall proportion of turbiditic sediment (i.e. turbidite sandstone and turbidite mudstone) in the sequence at each locality studied is shown in Fig. 20. The levels of Ce (as a guide to the variation in REE levels) are also shown for each locality, where available. Within anoxic hemipelagites, there is a strong positive correlation between nodule frequency, Ce content and the relative amount of associated turbidite sediment. Thus anoxic hemipelagites that are interbedded with a high proportion of turbidite sediment have higher contents of REE and exhibit a greater development of monazite nodules than those in turbidite-poor sequences.

Geochemistry

Major and trace element compositions for a representative suite of turbidites and anoxic

hemipelagites are given in Table 1. These samples were selected for detailed REE analysis on the basis of results of a parallel, more extensive regional geochemical study involving in excess of 80 samples from 53 localities covering the entire Llandovery sequence (T. K. Ball, pers. comm.). Detailed REE analyses of the same samples together with REE analyses of anoxic and oxic hemipelagite from the condensed Llandovery sequence (Locality 10) and from a sequence containing apatite concretions (Locality 2) are presented in Table 2. Samples from Locality 3 represent material taken from individually separated turbidite units (100–200 g) and their associated overlying anoxic hemipelagites. All other samples consist of material bulked (100–200 g) from several separated turbidite or anoxic hemipelagite units from a small stratigraphic thickness at a single locality.

Major and trace elements

Little variation is seen in most of the elemental compositions of the turbidite sandstone, turbi-

Table 1. *XRF analyses of major and trace elements for Llandovery mudrocks*

	Locality 1 AHP	Locality 1 TM	Locality 3 AHP(1)	Locality 3 TM(1)	Locality 3 TS(1)	Locality 3 AHP(2)	Locality 3 TM(2)	Locality 3 TS(2)
All values %								
SiO_2	64.40	58.81	64.86	58.22	55.55	64.00	60.31	52.91
Al_2O_3	21.14	23.04	22.54	23.42	24.40	22.90	23.37	24.39
TiO_2	1.10	1.23	1.11	1.25	1.43	1.18	1.27	1.42
Fe_2O_3	5.17	10.29	4.58	10.78	10.99	4.64	8.92	14.06
MgO	0.79	1.40	0.86	1.30	1.30	0.85	1.14	1.45
MnO	0.06	0.12	0.06	0.14	0.14	0.05	0.11	0.16
CaO	0.14	0.12	0.12	0.10	0.21	0.12	0.10	0.16
Na_2O	1.24	1.07	1.24	1.10	1.22	1.21	1.13	1.17
K_2O	3.70	3.75	3.77	3.70	3.95	3.86	3.79	3.89
P_2O_5	0.08	0.04	0.05	0.03	0.12	0.07	0.03	0.11
Total	97.81	99.87	99.19	100.03	99.32	98.89	100.16	99.72
All values ppm								
Th	13	12	13	13	12	17	17	14
U	<1	<1	3	<1	3	3	1	2
Y	42	35	33	32	43	44	33	42
Zr	160	174	168	173	292	177	163	299

AHP, anoxic hemipelagite; TM, turbidite mudstone; TS, turbidite sandstone

dite mudstone and anoxic hemipelagite. The turbiditic rocks (in particular the turbidite sandstones) have significantly lower SiO_2 and higher Fe_2O_3 and MgO contents than the anoxic hemipelagites. This reflects the greater abundance of iron-magnesium rich chlorite-mica stacks within the turbidite units which are concentrated towards the base of the turbidite units. The apatite band (Locality 2), apart from much higher P_2O_5 contents, is also broadly similar and the lower levels of other major elements can be accounted for as a consequence of dilution by apatite.

Th and U show no real variation between turbiditic and anoxic hemipelagic mudrocks. Th contents (12–17 ppm) are similar to those expected for most shales and within the range expected for modern-day pelagic clays (Wedepohl 1978). U contents (<1–3 ppm) are similar to modern-day oceanic muddy sediments but low in comparison to typical black shales (Wedepohl 1978). Zr levels are similar in both turbidite mudstones and anoxic hemipelagites (*c.* 170 ppm) but are approximately doubled in the basal turbidite sandstones (*c.* 300 ppm) reflecting the concentration of heavy minerals (dominated by zircon) at the base of turbidite units (e.g. Fig. 2). Y (Tables 1 & 2) tends to be slightly enriched in anoxic hemipelagites relative to turbidite mudstones. Here, the higher levels of Y (which behaves like the heavy REE) may be associated with the secondary monazite. Y is also enriched in the turbidite sandstones relative to the turbidite mudstones, but less so than Zr. This may

reflect a heavy mineral control, with Y being associated with the more evenly distributed xenotime rather than with zircon.

Anoxic hemipelagites contain more organic carbon than do turbidite mudstones. At Locality 1, average organic carbon contents of the anoxic hemipelagite and turbidite mudstones are 0.17% and 0.09% respectively; at Locality 3 (Triplet 2) they are 0.21% and 0.06% respectively.

REE geochemistry

The REE data for the various lithologies analysed (Table 2) normalised against the North American Shales Composite (NASC), which is taken to represent an 'average shale' composition, are plotted in Fig. 21. REE distribution patterns in the turbidite mudstones and turbidite sandstones are essentially similar, with slightly higher REE contents within the sandstones (Fig. 21, Table 2). However, there is a striking contrast between the REE compositions of the anoxic hemipelagites and of the associated turbidite units (Fig. 21, Table 2). Relative to 'average shale' (NASC), most anoxic hemipelagites (except at Locality 10) are highly enriched in the light rare earths (LREE) and middle rare earths (MREE), in particular La to Gd. The enrichment of LREE and MREE in the anoxic hemipelagites contrasts with a general depletion of these elements in the turbidite units relative to NASC (Fig. 21). Normalisation of the anoxic

Table 2. *REE analysis of Llandovery mudrocks*

(a) Composite samples (all values ppm)

	Loc. 1 AHP	Loc. 1 TM	Loc. 2 AHP	Loc. 2 TM	Loc. 2 AP	Loc. 4 AHP	Loc. 4 TM	Loc. 10 AHP	Loc. 10 OHP
La	217.30	38.20	67.50	36.40	20.60	183.70	11.20	48.85	41.96
Ce	469.43	61.76	138.53	57.26	65.74	386.98	19.58	97.27	94.12
Pr	56.68	6.37	15.64	6.06	9.80	45.13	2.32	10.90	9.35
Nd	220.80	23.10	58.40	20.00	58.70	172.30	9.20	41.30	36.10
Sm	42.33	4.31	10.58	3.29	27.37	31.87	1.97	7.84	6.90
Eu	7.32	0.97	2.05	0.76	8.08	5.55	0.54	1.50	1.41
Gd	24.34	5.08	7.73	3.80	39.47	18.53	3.06	6.68	5.86
Dy	10.11	5.66	6.08	4.51	22.07	8.39	4.03	6.67	5.65
Ho	1.77	1.14	1.24	0.95	3.36	1.46	0.85	1.23	1.04
Er	4.30	3.53	3.69	3.04	7.51	3.61	2.78	3.37	2.90
Yb	3.60	3.53	3.69	3.11	4.29	3.10	2.90	3.39	3.02
Lu	0.56	0.54	0.59	0.50	0.59	0.48	0.46	0.52	0.47
Y	37.0	31.0	31.0	25.0	89.0	31.0	23.0	32.0	26.0
Sm/Nd	0.192	0.187	0.181	0.165	0.466	0.185	0.214	0.190	0.191

Volume of lithotype comprising sequence:
Locality 1　: AHP = 11%; TM = 75%; TS = 14%
Locality 2　: AHP = 26%; TM = 74%; TS = <1%; AP = <1%
Locality 4　: AHP = 15%; TM = 75%; TS = 10%
Locality 10 : AHP + OHP = >95%

(b) Individual units (Locality 3)

	AHP(1)	TM(1)	TS(1)	AHP(2)	TM(2)	TS(2)
La	134.90	21.30	27.80	319.20	16.40	27.70
Ce	299.29	33.32	52.21	678.26	22.77	45.97
Pr	36.50	3.33	6.29	80.33	2.58	4.90
Nd	143.10	11.80	21.70	305.70	8.90	17.90
Sm	27.73	2.26	4.34	55.42	1.91	3.69
Eu	4.97	0.63	1.09	9.58	0.58	0.92
Gd	16.54	3.91	6.13	30.10	3.69	5.29
Dy	7.96	5.06	6.71	12.00	4.82	6.24
Ho	1.44	1.06	1.38	2.04	1.03	1.28
Er	3.84	3.36	4.16	4.70	3.26	4.07
Yb	3.44	3.43	4.31	3.81	3.30	4.18
Lu	0.53	0.54	0.69	0.58	0.52	0.65
Y	32.0	28.0	36.0	40.0	27.0	34.0
Sm/Nd	0.194	0.192	0.200	0.181	0.215	0.206

Thickness of lithotype in "triplet"
AHP(1) = 32 mm; TM(1) = 78 mm; TS(1) = 15 mm
AHP(2) = 10 mm; TM(2) = 70 mm; TS(2) = 30 mm

AHP, anoxic hemipelagite; OHP, oxic hemipelagite; TM, turbidite mudstone; TS, turbidite sandstone; AP, apatite band.

hemipelagite against its associated turbidite mudstone (Fig. 22) shows that the peak differentiation of the REE between the two lithologies occurs between Ce and Sm, where levels may be over 30 times higher than in the corresponding turbidite mudstone.

The heavy REE (HREE), Er to Lu show little or no differentiation between anoxic hemipelagite and turbidite lithologies, and may be present in slightly higher concentration in the coarser basal turbidite sandstones (e.g. Locality 3). These elements are probably influenced by the greater abundance of detrital zircon at the base of turbidite units, this mineral tending to concentrate HREE (Clark 1984).

The anoxic and oxic hemipelagites from

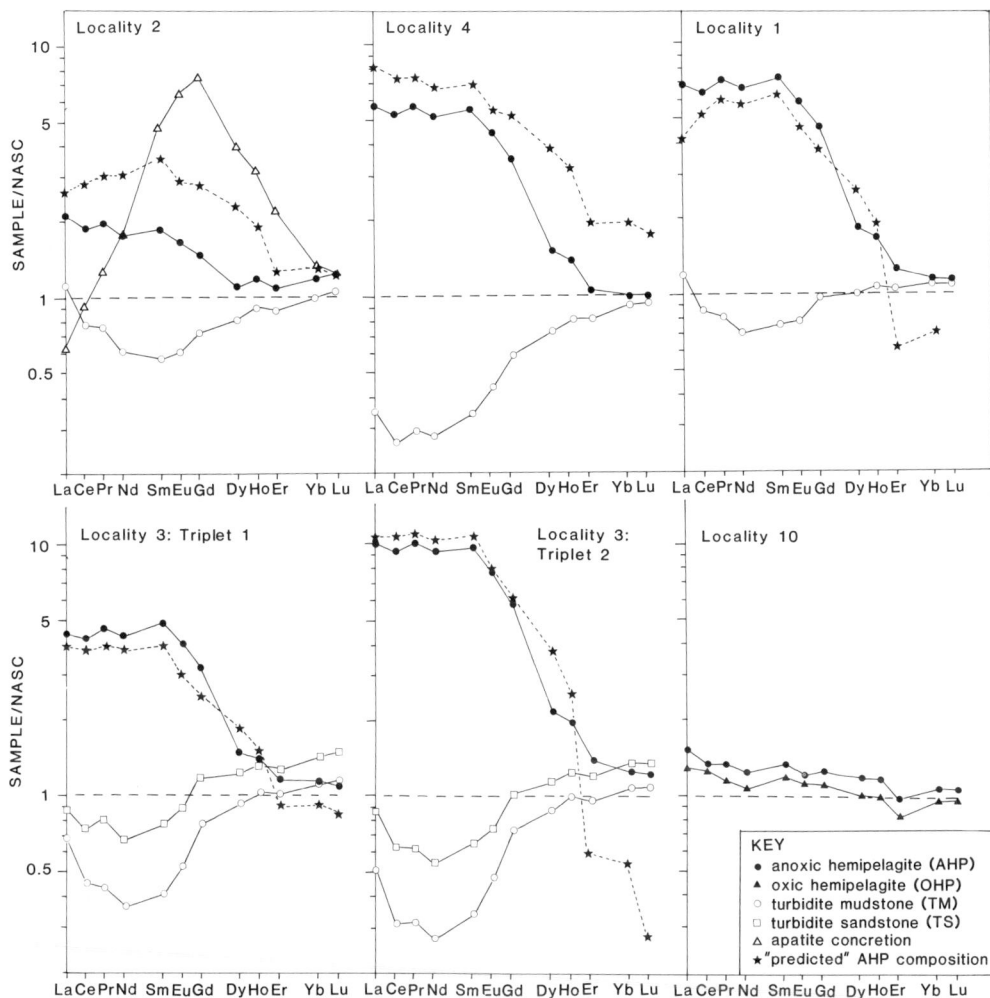

Fig. 21. Shale-normalized (North American Shale Composite) REE abundances for Llandovery mudrocks. Predicted REE abundances for anoxic hemipelagites assuming REE enrichment derived by depletion from association turbidite unit, are also shown. Value for normalization based on Haskin *et al.* (1968) with data for Dy taken from Gromet *et al.* (1984). Thickness values used in calculated curves taken from Table 2.

Locality 10 have similar REE levels to NASC although they appear to be slightly more enriched in MREE and LREE (Fig. 21). The anoxic hemipelagite has slightly higher overall levels of REE in comparison to the oxic hemipelagite but basically the patterns of both lithologies are similar. This largely turbidite-free condensed Llandovery sequence appears to show little of the REE enrichment or depletion exhibited by mixed turbidite-anoxic hemipelagite assemblages. In general, the degree of enrichment in REE displayed by the anoxic hemipelagites increases with either the proportion of turbidite units in the sequence (e.g. Fig. 21, Table 2: Location 1 > Location 4 > Location

2 > Location 10) or the thickness of the underlying turbidite unit (e.g. Fig. 21: Location 3; triplet 2 > Location 3; triplet 1).

The single apatite concretion analysed also shows a REE enrichment relative to 'average shale' (Locality 2; Fig. 21, Table 2). The most marked effect is seen from Nd to Er with peak enrichment occurring at Gd. However, the apatite concretion appears to be depleted in La relative to NASC and La to Nd in comparison to the associated anoxic hemipelagites.

Limited Ce and La analyses (T. K. Ball, pers. comm.) available for oxic hemipelagite/turbidite mudstone couplets from Locality 7 present a more complicated picture. Two of these couplets

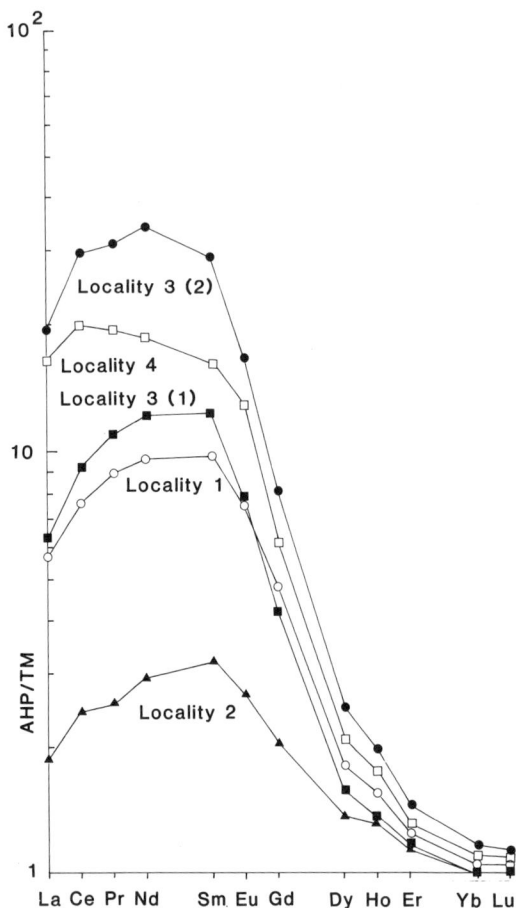

Fig. 22. REE abundances in anoxic hemipelagites normalized against associated turbidite mudstones.

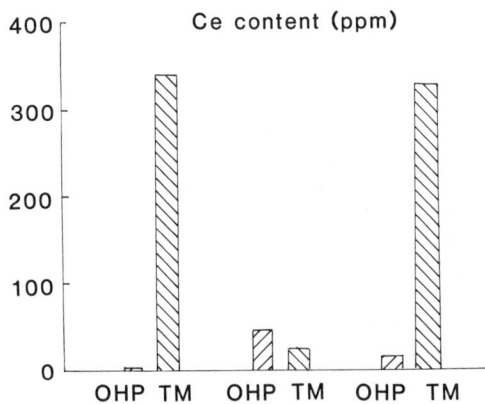

Fig. 23. Distribution of Ce within oxic mudrock sequences (see Fig. 1). OHP, oxic hemipelagite; TM, turbidite mudstone. All three sample pairs from locality 7.

show marked enrichment of Ce and La in the turbidite mudstones relative to the oxic hemipelagites, whilst in another there is a slight enrichment in the oxic hemipelagite (Fig. 23).

Nodule composition

Energy-dispersive EMPA data for typical nodules from both anoxic hemipelagite and turbidite mudstone are given in Table 3. Monazite nodules from both lithologies display a very strong REE differentiation from core to rim (Table 3). Nodule cores are enriched in MREE and heavier LREE (Pr to Dy); shale-normalized ratios in the cores increase from La to a maximum at Sm then decrease rapidly for heavier elements (Fig. 24). In contrast, the outer regions of the nodules are preferentially enriched in La and Ce, shale-normalized ratios showing a maximum at La and decreasing rapidly towards MREE and HREE, Eu and Dy were generally at or below detection limit (approximately 0.5 weight %) for energy-dispersive EMPA in most analyses. The REE zonation displayed by these monazite nodules is symmetrical from rim to rim but the La concentrations increase rapidly above 10% La_2O_3 in the outer 25% (normally outer 100–200 μm) of the nodules (Table 3). As a consequence of this zonation Nd/Sm ratios vary considerably across nodules from approximately 3 in the core to over 10 at the margins (Table 3). Th distribution tends to be erratic and BSEM-EDXA indicated that this is, in some cases, due to the presence of primary Th-rich inclusions, such as thorite or rare detrital monazite. The composition and zonation of the monazites is similar to that described by Read et al. (1987) for monazites found in stream sediments from central Wales and Exmoor. A shale-normalized composition of monazite from stream sediment concentrates (based on data from Read et al. 1987) is also shown in Fig. 24 and probably represents a reasonable average monazite composition.

No significant chemical difference was discerned between monazite nodules from anoxic hemipelagites and turbidite mudstones from the same localities, nor between nodules from different localities.

Discussion

REE abundance and distribution of secondary monazite

The development of monazite nodules and the enrichment of REE within the Llandovery mud-

Table 3. *Composition of typical monazite nodules from Llandovery mudrocks (samples taken from nodules in anoxic hemipelagite (AHP) and turbidite mudstone (TM) at Locality 1)*

	AHP 1	AHP 2	AHP 3	TM 1	TM 2	TM 3
SiO_2	1.04			1.80		
CaO		0.26				0.28
Y_2O_3	1.18					
La_2O_3	5.36	11.51	23.11	4.92	10.62	19.39
Ce_2O_3	26.86	33.76	33.44	23.56	34.44	34.08
Pr_2O_3	3.66	3.13	2.24	4.11	3.25	3.59
Nd_2O_3	22.40	13.44	7.43	22.30	15.58	9.56
Sm_2O_3	4.37	2.08	0.86	6.85	2.62	0.95
Eu_2O_3						
Gd_2O_3	2.65	0.88	0.75	2.11	0.99	0.92
Dy_2O_3				1.23		
ThO_2		0.84	0.80	2.49	1.09	0.74
P_2O_5	31.60	31.16	31.82	30.19	31.18	30.88
Total	99.12	97.06	100.45	99.56	99.77	100.39

Number of ions on basis of 4[O]

	AHP 1	AHP 2	AHP 3	TM 1	TM 2	TM 3
Si^{4+}	0.040			0.069		
Ca^{2+}		0.011				0.012
Y^{3+}	0.024					
La^{3+}	0.075	0.167	0.325	0.070	0.152	0.276
Ce^{3+}	0.375	0.485	0.467	0.333	0.488	0.482
Pr^{3+}	0.051	0.045	0.037	0.058	0.046	0.051
Nd^{3+}	0.304	0.188	0.101	0.307	0.216	0.132
Sm^{3+}	0.057	0.028	0.011	0.091	0.035	0.013
Eu^{3+}						
Gd^{3+}	0.033	0.011	0.009	0.027	0.013	0.012
Dy^{3+}				0.015		
Th^{4+}		0.007	0.007	0.022	0.010	0.007
P^{5+}	1.017	1.035	1.027	0.986	1.023	1.011
Nd/Sm^{3+}	5.33	6.71	9.18	3.37	6.17	10.15

AHP 1, core; AHP 2, 3/4 from rim to core; AHP 3, rim; TM 1, core; TM 2, 3/4 from rim to core; TM 3, rim.

rocks of central Wales are clearly linked. Nodules show a greater abundance within the anoxic hemipelages (Fig. 18) and this is reflected in a corresponding enrichment of REE, and in particular the LREE and MREE, within the anoxic hemipelagites relative to turbiditic units (Fig. 20, Table 2). Shale-normalized LREE and MREE distributions of average nodular monazite (Fig. 24) and anoxic hemipelagites (Fig. 21), within which the nodules are abundant, are similar. This indicates that the REE content of these rocks is dominated by the presence, and composition, of monazite nodules, Burnotte *et al.* (1989) similarly concluded that the bulk rock REE composition was related to the monazite content in the Palaeozoic of Belgium. These observations contrast with the conclusion by Read *et al.* (1987) that the monazite nodules do not exert a major control over REE abundances in their host rocks. The discrepancy probably

arises from a lack of appreciation by the earlier authors of the lithological control on nodule distribution in the host formation, since, previously, monazite nodules had only been found as reworked grains in stream sediments.

The monazite nodules are not enriched in HREE (Read *et al.* 1987) and consequently would not be expected to influence strongly the HREE spectrum of the host rocks. This is consistent with the observation that anoxic hemipelagites and turbidite mudrocks have similar levels of HREE (in particular, Yb and Lu). By comparison, the basal turbidite sandstones often display slightly higher levels of HREE than either anoxic hemipelagites or turbidite mudstones (Fig. 21, Table 2). This can be explained by the presence of detrital zircon and lesser amounts of xenotime. These minerals preferentially incorporate HREE (Clark 1984) and zircon, in particular, dominates the heavy mineral

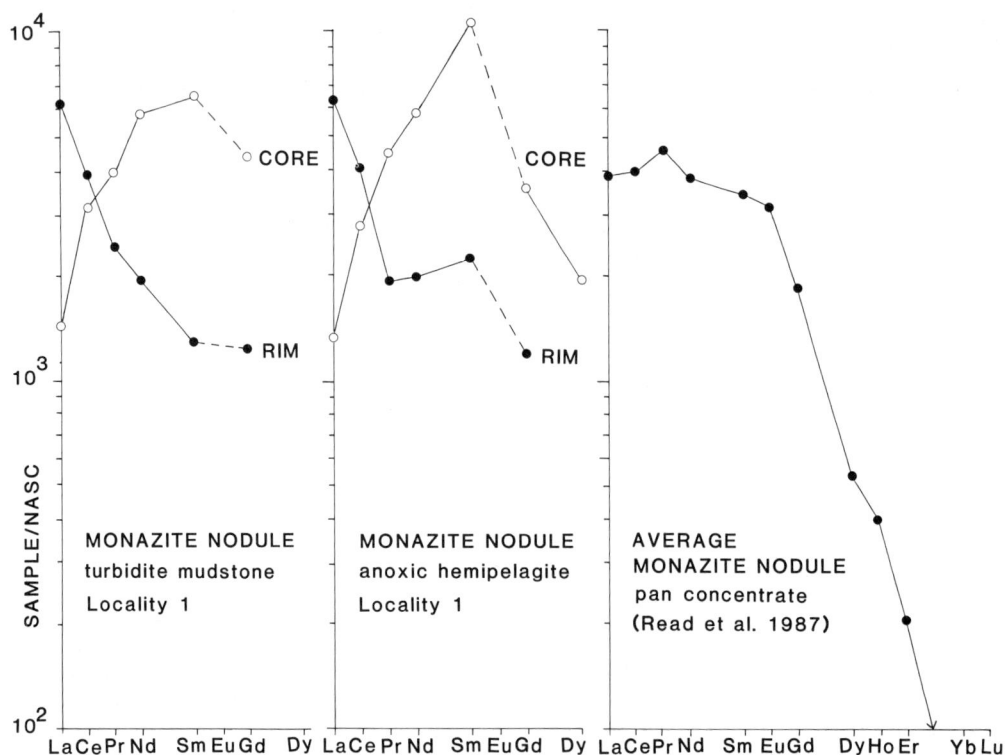

Fig. 24. Shale-normalized (North American Shale Composite) REE abundances of typical monazite nodules from anoxic hemipelagite and turbidite mudstone in Llandovery mudrocks (Locality 1; this study) and from stream sediment pan concentrates from central Wales (Read *et al.* 1987).

assemblage which is concentrated towards the base of turbidite units. Zircon also appears to have acted as a seed for rare minor secondary xenotime precipitation in some cases. The HREE patterns of both the anoxic hemipelagite and turbidite units are therefore considered to be dominated by detrital minerals.

Origin of the monazite nodules

Published hypotheses on the origin of such nodules have been largely based on derived nodule concentrations within placers, and so far have been largely based on nodule structure and chemistry (e.g. Matzko & Overstreet 1977; Cooper *et al.* 1983; Rosenblum & Mosier 1983; Read *et al.* 1987). The examination of the distribution of in situ nodules within a sequence of rhythmically alternating turbiditic lithologies allows greater constraints to be placed on their origin.

A contact metamorphic or metasomatic origin as porphyroblasts (cf. Overstreet 1971; Rosen-

blum & Mosier 1983) is not applicable to the monazite nodules of central Wales as no igneous rocks or contact metamorphic aureoles are associated with their occurrence. Read *et al.* (1987) also point out that nodules are not recorded from areas containing similar Lower Palaeozoic sedimentary sequences, such as the Lake District and southern Scotland which have been thermally metamorphosed by Caledonian intrusions. Neither are the monazite nodules essentially altered granite-derived monazites that have recrystallised and overgrown during diagenesis and low-grade metamorphism, as Read *et al.* (1987) proposed; their concentration within the fine-grained hemipelagites, lack of association with detrital heavy mineral concentrates at the base of turbidite units, and the identification of rare, apparently unaltered and chemically distinct (Th-rich) detrital monazite as a separate entity, preclude this hypothesis. Donnot *et al.* (1973) proposed that similar monazite nodules in the Ordovician of Brittany were formed by transformation of an early diagenetic gelatinous marine phosphate precipitate, via a rhabdophane precursor, during progressive diagenesis.

However, this fails to explain the absence of monazite (or rhabdophane) inclusions within demonstrably early diagenetic carbonate and apatite concretions (cf. Cave 1979; Smith 1987), and the absence of a negative Ce anomaly which although not a reliable criterion, might be expected for some marine phosphate precipitates (Fleet 1984). It also fails to explain the absence of monazite nodules within pelagic mud-dominated sequences (e.g. Locality 10) which would have had long contact times with marine waters.

This study indicates that the formation of the monazite nodules and REE enrichment in the anoxic hemipelagites is diagenetic in origin. But unlike Donnot *et al.* (1973), we would argue that the monazite is a later (post-compactional) diagenetic phase and that the REE were largely derived from the associated turbidite sediments rather than from seawater. Burnotte *et al.* (1989) also considered that the REE were derived from within the sediments rather than from extraneous sources. A number of observations support this. Firstly, some nodules are present within the turbidite sediments, well below and unrelated to the contemporaneous sediment–water interface. Secondly, there is the strong tendency for nodules to be more abundant, and REE levels to be higher, where there is a high proportion of associated turbidite sediment; hemipelagite-dominated sequences contain few nodules and do not have elevated REE levels (Fig. 4). Most strikingly, the hemipelagite-dominated sediments from north Wales (Locality 10), which would have been expected to have maintained the longest contact with contemporaneous seawater, have neither elevated nor depleted REE contents with respect to average values for mudrocks (e.g. Fleet 1984), neither do they show anomalous behaviour with respect to Ce. The REE content of the distal north Wales sequence is therefore best interpreted as a detrital signature.

Textural relationships demonstrate that monazite nodule formation post-dates the formation of early diagenetic carbonate or apatite concretions and also that they grew after the main sediment compaction. However, they are included to a minor extent within cone-in-cone carbonate concretions, generally considered to be later burial diagenesis features (Marshall 1982). The relationship of both REE enrichment and nodule abundance in the anoxic hemipelagites to the thickness of the adjacent underlying turbidite unit suggests upward movement of porewater and REE migration. This is consistent with the direction of water movement expected during burial and compactional dewatering.

There is no evidence to indicate or refute the formation of a hydrous rhabdophanic precursor which might be expected to form under diagenetic conditions (Jonasson *et al.* 1985). Dehydration of rhabdophane (Bowles & Morgan 1984) to produce monazite during late diagenesis or metamorphism would seem feasible. Experimental studies (Jonasson & Vance 1986) tend to indicate that much higher temperatures are required than would be expected during diagenesis or anchizone (greenschist) metamorphism but this may be an artefact of slow reaction kinetics at low temperature. It is evident from the relationship of the nodules with respect to bedding fabric, lack of nodule deformation and local disruption of cleavage fabrics, that the nodules behaved as rigid particles during metamorphism and cleavage development (cf. Ramsey 1967). Therefore, the nodules were lithified prior to metamorphism.

Nodule growth appears to be related to the original sedimentary fabric, as there was replacement of bedding laminae and inclusion of relict bedding fabrics, which were protected from later metamorphic recrystallization. Burnotte *et al.* (1989) considered that the bedding-parallel nodule elongation was due to crystallization in a rock displaying a pronounced unidirectional anisotropy (i.e. a laminated sediment). This may be true in the case of the finely laminated anoxic hemipelagites but bedding-parallel or sub-parallel elongated monazite nodules are also found in the turbidite mudstones where there is often no pronounced anisotropy. The oriented development of prominent monazite crystal faces ([100]) with the direction of nodule elongation parallel to original bedding (as given by the inclusion fabric) may be more indicative that growth of monazite or its precursor crystal was stress-controlled by burial compaction.

Rare earth element diagenesis

The relationship between REE enrichment in anoxic hemipelagite and REE depletion in turbidite sediments is best seen by referencing both lithologies to an unaltered background mudrock composition. Locality 10 was considered to provide a background REE composition for the Llandovery mudrocks, since this sequence appeared to show no evidence for marine or diagenetic enrichment of REE, and although distant from the main area of study it was felt that it might be more representative of local pre-diagenetic muddy sediment than the commonly accepted 'European Palaeozoic Shale Composite' standard (Haskin & Haskin 1966). Burnotte *et al.* (1989) similarly adopted a local reference

Table 4. *Calculated proportion (%) of REE depleted from turbidite units (assuming initial composition equivalent to 'background' anoxic hemipelagite from Locality 10. Table 2)*

| | Loc. 1 | Loc. 2 | Loc. 3 | | Loc. 3 | | Loc. 4 |
	TM	TM	TM(1)	TS(1)	TM(2)	TS(2)	TM
La	22	26	56	43	66	43	77
Ce	37	41	66	46	77	53	80
Pr	42	44	69	42	76	55	79
Nd	44	52	71	47	78	57	78
Sm	45	58	71	45	76	53	75
Eu	35	49	58	27	61	39	64
Gd	24	43	42	8	45	21	54
Dy	15	32	24	(+0.6)	28	6	40
Ho	7	23	14	(+12)	16	(+4)	31
Er	(+5)	10	0.3	(+23)	3	(+21)	18
Yb	(+4)	8	(+1)	(+27)	3	(+23)	15
Lu	(+4)	4	(+4)	(+33)	0	(+25)	12
Y	3	22	13	(+13)	16	(+6)	28

TM, turbidite mudstone; TS, turbidite sandstone.
(Figures in parentheses represent relative enrichments rather than depletions.)

by which they compared their monazite and host rock compositions.

Assuming that Locality 10 is representative of the composition of pre-diagenetic muddy sediment of the Welsh Llandovery, the predicted shale-normalised REE abundances for the anoxic hemipelagites are presented in Fig. 21. This has been calculated assuming that the REE depleted from the turbidite units, relative to the reference Llandovery sediment, were quantitatively fixed in the anoxic hemipelagite. The relative thickness of the lithologies at each locality (Table 2) were taken into account in constructing this mass balance as outlined below:

$$Ln_{AHP}(\text{calculated}) = Ln_{REF} +$$
$$\left((Ln_{REF} - Ln_{TM}) \times (d_{TM}/d_{AHP}) \right)$$

where Ln_{AHP} = calculated concentration of individual REE in anoxic hemipelagite unit; Ln_{REF} = concentration of individual REE in reference analysis Locality 10 (anoxic hemipelagite); Ln_{TM} = concentration of individual REE in turbidite unit, d = thickness (or % proportion) of anoxic hemipelagite or turbidite unit. Where turbidite mudstone and turbidite sandstone were separately analysed their individual contributions to the calculated anoxic hemipelagite were calculated. The calculated proportions of REE depleted from the turbidite units relative to the reference anoxic hemipelagite are also given in Table 4.

The results of this exercise show a close resemblance between the shape of the actual shale-normalized REE abundances and that predicted by the proposed model for most elements from La to Ho (Fig. 21). Poorer fits were given by the heavy REE Er to Lu and in some cases the distribution calculated by the above method gave depleted values for the theoretical anoxic hemipelagite relative to the turbidite mudstone (this is also reflected in Table 4 as an apparent enrichment rather than depletion for these elements in the turbidite unit relative to the reference anoxic hemipelagite). This calculated depletion from Er to Lu is artificial and probably arises from a higher initial content of these elements in the turbidite units due to the greater abundance of detrital heavy minerals (principally xenotime and zircon). In most cases the actual calculated enrichment or depletion in Er, Yb and Lu is very small and insignificant in comparison to the errors in the assumptions made in the above calculations. If the assumption of compositional similarity between the original turbidite sediment and the reference sediment (Locality 10) is correct then calculations suggest that up to 80% of the LREE may have been mobilised from the turbidite units. This effect progressively decreases through the MREE, becoming less significant for the HREE.

In most cases there is an imbalance between predicted and actual values of REE enrichment in the anoxic hemipelagites (Fig. 21). This imbalance appears mainly to relate to the proportions of anoxic hemipelagite and turbidite present in the couplet or sequence. In sequences or couplets in which the relative proportion of anoxic hemipelagite is low, the predicted REE content is higher than actual values. The converse is true for sequences which have relatively high proportions of anoxic hemipelagite. The relationship between REE enrichment in the anoxic hemipelagite and the thickness of the underlying turbidite unit (Fig. 19) indicates that the REE migration was a largely localized effect occurring within adjacent turbidite-anoxic hemipelagite couplets. However, the discrepancy in the overall mass-balance suggests that the system was not completely closed with respect to REE migration and may have depended to some extent on the capacity of the overlying anoxic hemipelagite to 'mop-up' the REE released. Thin units would have had less capacity than thick units, which may also have had capacity to incorporate additional REE 'leaking' through the sequence. This localized REE redistribution between adjacent lithologies is similar to that observed from shale-limestone sequences in the Newland Formation of Montana (Schieber 1988).

The worst-fit between predicted and actual REE enrichment in the anoxic hemipelagite was found for Locality 2 (Fig. 21). Here the situation may have been complicated by the presence of early diagenetic apatite concretions within the material sampled. During later diagenesis the apatite may have scavenged REE from the migrating pore fluids by exchange for Ca (e.g. Jonasson et al. 1985) or acted as a barrier inhibiting the migration of pore fluids and REE through the sediments. The effect is most marked for MREE, which are enriched in the apatite.

Dissolution of detrital heavy minerals is not considered to have been the major primary source of REE in the turbidite sediments. Traces of fine-grained detrital monazite are still present and do not appear to have dissolved or corroded during diagenesis. Corroded primary possible REE carbonate or fluoride have been noted and may represent a minor potential source. A more important source is likely to have been the original clay minerals and iron-manganese oxides or oxyhydroxides; the bulk of the REE in eroded material is contained in these phases (see Dypvik & Brunfelt 1979; Flett 1984). A significant proportion of this is loosely bound (adsorbed) on the mineral surfaces and readily leachable (Balashov & Girin 1969; Roaldset & Rosenqvist 1971) and is therefore available to take part in diagenetic reactions. Altered (?smectitic) volcanoclastic debris is likely to have been a major original component of the turbidites and subsequent transformation to chlorite or illite during burial diagenesis could also have released REE to sediment porewaters.

Fixation of REE in the anoxic hemipelagites appears to be related to their higher organic carbon content. Further support for REE accumulation being controlled by the distribution of organic carbon is found in the reversal of the enrichment pattern seen between oxic hemipela-

gite (generally organic poor) and turbidite units (Fig. 23, Locality 7). Optimum conditions for REE release, migration and fixation are likely to have been achieved during early diagenesis. Recent observations have demonstrated that sediment porewaters in anoxic sediments become significantly enriched in REE during early diagenesis due to reductive liberation from iron-manganese oxides (Elderfield et al. 1985; Elderfield & Sholkovitz 1987; German & Elderfield 1989). Analogous early release of REE in the Llandovery turbidites could be envisaged. This would also have coincided with the production of phosphate during the microbial breakdown of organic matter in the anoxic hemipelagites. Upward migration of REE from the turbidite sediment into the overlying anoxic hemipelagite, enhanced by compactional dewatering of the sediment pile could have facilitated the coprecipitation of highly insoluble REE phosphates largely within the anoxic hemipelagite. However, the nucleation of small, widely-spaced monazite nodules is unlikely to have accounted for the efficient fixation of REE and in any case the nodule growth appears to be a later diagenetic feature. A more plausible explanation is that during early reductive diagenesis REE were efficiently filtered by the finely-dispersed REE-phosphate or complexed with the organic matter (perhaps as REE–organophosphate complexes).

If the early porewaters of the Llandovery sediments showed similar levels of REE enrichment to those seen in modern sediments (e.g. Elderfield & Sholkovitz 1987; German & Elderfield 1989) then the volume of water required to transport the REE in order to produce the observed effects can be estimated for a simple advective model. Reported porewater concentrations for Ce have maxima up to 1.9×10^{-9} moles kg^{-1}; thus, to produce the observed depletion of 74 ppm Ce in turbidite mudstone TM(2) from Locality 3, relative to the background sediment, would require a flux of 2.8×10^5 litres of porewater per kilogram of sediment. Calculations assuming an original sediment water content of about 80% by volume (cf. Singer & Muller 1983, p. 188) suggest that this equates to approximately 2×10^5 pore volumes. It is unlikely that sufficient volume of water could be produced by compactional dewatering alone. Therefore it is likely that REE migration was enhanced by diffusion along chemical gradients towards the anoxic hemipelagites.

Nucleation of the monazite nodules occurred as a result of later diagenetic recrystallization with nodules forming at the expense of finely dispersed REE phosphate (or organic complexes). The compositional zonation of the monazite seems to imply increasing LREE/MREE as porewaters evolved during later diagenesis. The REE most probably migrated as bicarbonate or organic complexes during diagenesis; MREE and HREE form more stable (soluble) bicarbonate and organic complexes than LREE and consequently would have been more mobile and selectively desorbed (Girin *et al.* 1970; Humphris 1984; see also discussion by Fleet 1984). Therefore, the heavy REE might be expected to be initially present at higher concentrations in the porewaters during migration of the REE to the sites of secondary nucleation. Alternatively, as suggested by Burnotte *et al.* (1989) the zonation may be the result of differential solubilities of the REE phosphates. Carron *et al.* (1958) showed that Sm phosphate will precipitate preferentially to all the other lanthanides from La to Gd and therefore would be expected to compete effectively for phosphate during the earlier stages of nodule growth, until the supply of Sm diminished. The results of both these processes are similar and it is not possible to distinguish between their relative importance in this study.

The process described above has profound implications for the use of REE patterns as indicators of sediment provenance, particularly if the anoxic hemipelagite-turbidite couplets do not act as perfectly closed systems. The changing Nd/Sm ratio from core to rim of the nodules (Table 3) indicates that Nd and Sm were undergoing significant fractionation in the porewater during later diagenesis when the monazite nodules were forming. However, the overall mobilization of REE from the turbidite units and their reconcentration within the anoxic hemipelagites appears to have had little effect on whole rock Sm/Nd ratio (Table 2) and both turbidite and anoxic hemipelagite units display Sm/Nd ratios largely within the range of average upper crustal rock values (Faure 1986). Only in the case of the turbidite units showing the most extreme depletion of REE (Locality 3, TM(2) and Locality 4; Table 4) are the Sm and Nd perhaps fractionated enough to have a significant effect on the Sm/Nd ratio.

Conclusions

This study has revealed that marked redistribution of REE occurred during the diagenesis of Upper Llandovery mud-dominated turbidite sequences from central Wales. Lithology exerts a major control on the redistribution of REE; LREE and MREE were mobilized during early

diagenesis from turbidite units and redeposited largely within adjacent anoxic hemipelagite units, where immobilization of REE was related to the high content of organic material. This either fixed the REE directly and/or provided phosphate for coprecipitation as REE phosphate during diagenetic breakdown. The reverse REE enrichment trend may be seen where the relative distribution of organic carbon is reversed, as in the case of oxic hemipelagite and turbidite sequences. The REE were subsequently incorporated into authigenic monazite nodules during later recrystallization.

A pre-cleavage or diagenetic origin is ascribed to the monazite nodules on textural evidence. Textural relationships also indicate that the nodules post-date very early diagenetic apatite and calcite and also formed after the main sediment compaction. However, they may be earlier than cone-in-cone carbonate structures generally ascribed to later burial diagenesis and dewatering processes. The strong positive correlation between the volume of turbidite and the REE enrichment in the overlying anoxic mudstone indicates upward migration of porewaters transporting REE and is consistent with an origin during dewatering. However, advective transport alone is unlikely to have been sufficient and must have been enhanced by diffusion processes. The major REE source within the turbidite units is considered to have been REE sorbed on the surfaces of clay minerals and Fe–Mn oxides. Much of the clay mineral fraction may have been derived from altered (smectitic) volcanoclastic material. Liberation of the REE to porewaters occurred probably as the result of clay mineral transformations and reduction of Fe–Mn oxides during diagenesis. During this process the REE were probably complexed with bicarbonate or organics and fractionation occurred due to the greater mobility of the more stable complexes with MREE and HREE. The results of this process are reflected in the marked zonation of the monazite concretions; the cores being enriched in MREE and the rims enriched in LREE. In general, the turbidite–anoxic hemipelagite couplets approximate to closed systems with respect to the REE although significant loss of REE may have occurred from systems with relatively small proportions of anoxic hemipelagite. No anomalous behaviour was found for either Eu or Ce; both behaved as the trivalent REE, indicating that although reducing conditions prevailed during nodule formation (Ce behaved as Ce^{3+} not as Ce^{4+}) they were not extreme enough to reduce Eu to the divalent species.

No evidence could be found for a precursor

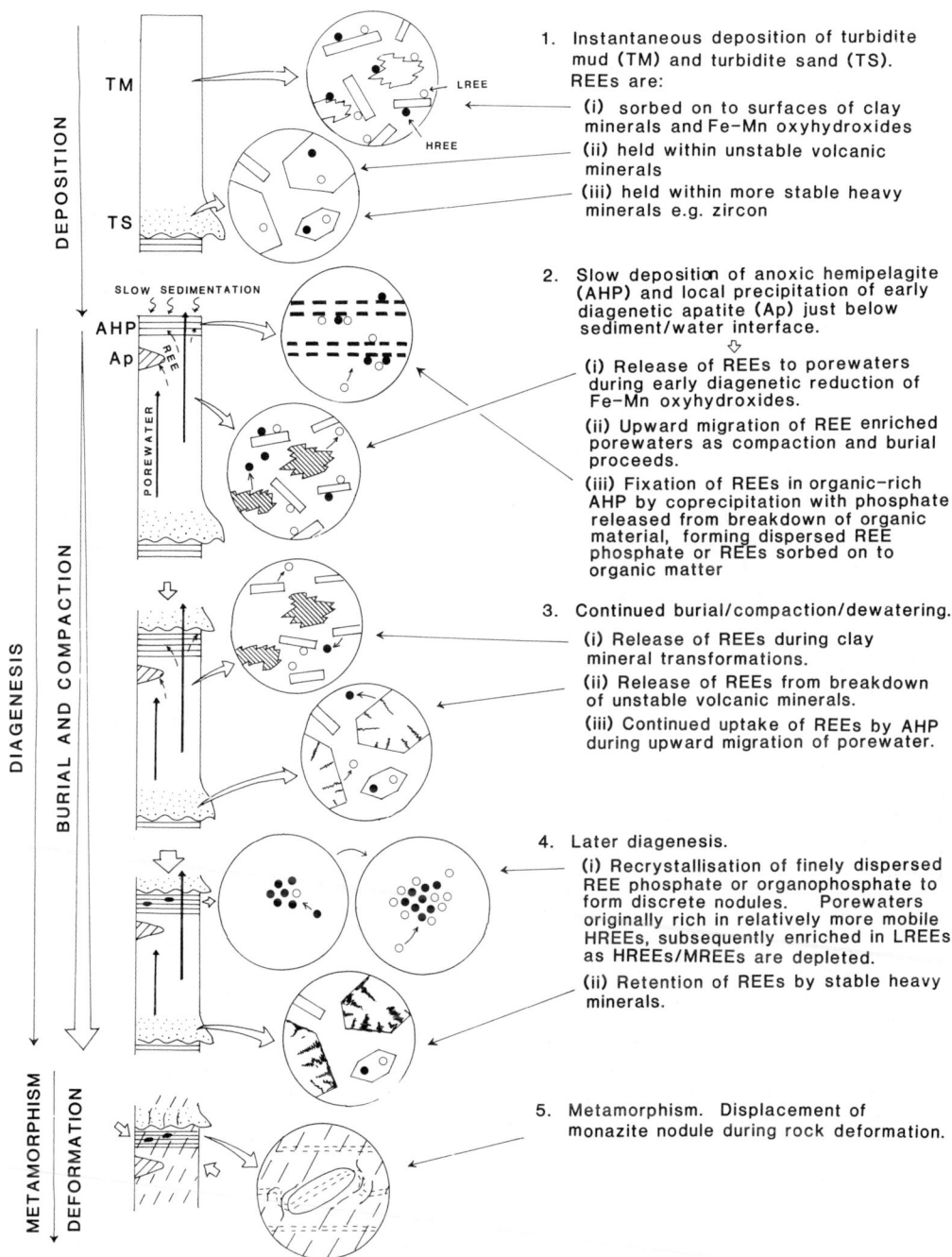

Fig. 25. Summary model for the origin of monazite nodules in Llandovery mudrocks from central Wales.

phase but it is possible that rhabdophane or hydrous REE phosphate could have formed initially, transforming to monazite later during diagenesis or metamorphism. The formation of authigenic monazite in the Llandovery sediments is summarized in Fig. 25.

The use of REE geochemistry in the study of sediment provenance relies on the assumption that these elements are relatively immobile during diagenesis and the REE composition of fine grained sediments represents an average source composition. However, this study shows that

REE can be significantly mobile during diagenesis and may be associated with enrichment or depletion in adjacent sediment units dependent upon local geochemical environments. Significant REE fractionation and/or loss of REE from the system can accompany this process resulting in a major modification of the REE pattern of the sediment. Ce, La and Y data from a parallel more extensive but less detailed regional geochemical study (with respect to REE) covering the entire Llandovery sequence (T. K. Ball, pers. comm.) indicate that the features described in this paper are not localized phenomena. Therefore the use of REE data to infer characteristics of sediment provenance must be treated with caution and should be supported by detailed petrographic analysis to evaluate the possible effects of REE mobility. In the case of the Llandovery of central Wales care must also be taken during sampling of mudstone-pelagic mudstone sequences so that results are not erroneously biased by the presence of authigenic monazite nodules. In most cases the REE redistribution and monazite authigenesis appears to have had little effect on the Sm/Nd ratio of the sediments. However, Sm and Nd may be fractionated in the most extremely depleted turbidite units and this may have an effect on the interpretation of Sm/Nd isotope ratios in Llandovery mudrocks. Previous studies of derived monazite nodules in stream sediments suggest that authigenic monazite nodules are widespread within the Lower Palaeozoic Welsh basin and not just restricted to Llandovery mudrocks.

The authors express their gratitude to their colleagues at the British Geological Survey (BGS) for help and encouragement during this study. In particular, we would like to thank D. Savage and J. Evans for advice and useful discussions on REE geochemistry; T. J. McEwan and S. T. Horseman who made helpful comments with regard to structural aspects of nodule morphology; and C. J. N. Fletcher, T. K. Ball, A. H. Bath, D. Cooper, R. J. Merriman, A. J. Reedman and M. F. Howells for encouragement and help as the work was progressing and for improving the manuscript. The authors are grateful to T. K. Ball for providing the analytical data on oxic hemipelagite/turbidite mudstone couplets used in this paper and to S. Inglethorpe for undertaking evolved gas analyses. The study was undertaken as part of the BGS Central Wales research programme. This paper is published by permission of the Director, British Geological Survey (NERC).

References

BALASHOV, Y. A. & GIRIN, Y. P. 1969. On the reserve of mobile rare earth elements in sedimentary rocks. *Geochemistry International*, **7**, 649–659.

BOWLES, J. F. W. & MORGAN, D. J. 1984. The composition of rhabdophane. *Mineralogical Magazine*, **48**, 146–147.

BURNOTTE, E., PIRARD, E. & MICHEL, G. 1989. Genesis of gray monazites: evidence from the Paleozoic of Belgium. *Economic Geology*, **84**, 1417–1429.

CAVE, R. 1979. Sedimentary environments of the basinal Llandovery of mid-Wales. *In*: HARRIS, A. L., HOLLAND, C. H. & LEAKE, B. E. (eds) *The Caledonides of the British Isles—reviewed*. Geological Society, London, Special Publication, **8**, 517–526.

CARRON, M. K., NAESER, C. R., ROSE, H. J. & HILDEBRAND, F. A. 1958. Fractional precipitation of rare earths with phosphoric acid. *Bulletin U.S. Geological Survey*, **1036-N**, 253–275.

CHAUDHURI, S. & CULLERS, R. L. 1979. The distribution of rare-earth elements in deeply buried Gulf Coast sediments. *Chemical Geology*, **24**, 327–338.

CLARK, A. M. 1984. Mineralogy of the rare earth elements. *In*: HENDERSON, P. (ed.) *Rare Earth Element Geochemistry*. Developments in Geochemistry Series, **2**, Elsevier, 33–61.

COOPER, D. C., BASHAM, I. R. & SMITH, T. K. 1983. On the occurrence of an unusual form of monazite in panned stream sediments in Wales. *Geological Journal*, **18**, 121–127.

CRAIG, J., FITCHES, W. R. & MALTMAN, A. J. 1982. Chlorite-mica stacks in low-strain rocks from central Wales. *Geological Magazine*, **119**, 243–256.

CULLERS, R. L., CHAUDHURI, S., ARNOLD, B., LEE, M. & WOLFE, C. 1975. Rare earth distributions in clay minerals and in the clay-sized fraction of the Lower Permian Havensville and Eskridge shales of Kansas and Oklahoma. *Geochimica et Cosmochimica Acta*, **39**, 1691–1703.

——, ——, KILBANE, N. & KOCH, R. 1979. Rare-earths in size fractions and sedimentary rocks of Pennsylvanian-Permian age from the mid-continent of the U.S.A. *Geochimica et Cosmochimica Acta*, **43**, 1285–1301.

DEER, W. A., HOWIE, R. A. & ZUSSMAN, J. 1962. *Rock-Forming Minerals: volume 5; Non-Silicates*. Longman.

DIMBERLINE, A. J. 1986. Electron microscope and microprobe analysis of chlorite-mica stacks in the Wenlock turbidites, mid-Wales, U.K. *Geological Magazine*, **123**, 299–306.

—— & WOODCOCK, N. H. 1987. The southeast margin of the Wenlock turbidite system, mid-Wales, *Geological Journal*, **22**, 61–71.

DONNOT, M., GUIGUES, J., LULZAC, Y., MAGNIEN, A., PARFENOFF, A. & PICOT, P. 1973. Un nouveau type de gisement d'europium: la monazite grise à europium en nodules dans les schistes paléozoïques de Bretagne. *Mineralium Deposita*, **8**, 7–18.

DYPVIK, H. & BRUNFELT, A. O. 1976. Rare earth

elements in Lower Palaeozoic epicontinental and eugeosynclinal sediments from the Oslo and Trondheim regions. *Sedimentology*, **23**, 363–378.

—— & —— 1979. Distribution of rare earth elements in some North Atlantic Kimmeridgian black shales. *Nature*, **278**, 339–341.

ELDERFIELD, H. & SHOLKOVITZ, E. R. 1987. Rare earth elements in the pore waters of reducing nearshore sediments. *Earth and Planetary Science Letters*, **82**, 280–288.

——, HAWKESWORTH, C. J. & GREAVES, M. J. 1981*a*. Rare earth element geochemistry of oceanic ferromanganese nodules and associated sediments. *Geochimica et Cosmochimica Acta*, **45**, 513–528.

——, —— & —— 1981*b*. Rare earth element zonation in Pacific ferromanganese nodules. *Geochimica et Cosmochimica Acta*, **45**, 1231–1234.

——, KENNEDY, H., KLINKHAMMER, G. & SHOLKOVITZ, E. R. 1985. Rare earth element distributions in marine pore waters and associated sediments. *Terra Cognita*, **5**, 188.

FAURE, G. 1986. *Principles of Isotope Geology*, second edition. John Wiley & Sons, New York.

FLEET, A. J. 1984. Aqueous and sedimentary geochemistry of the rare earth elements. *In*: HENDERSON, P. (ed.) *Rare Earth Element Geochemistry*. Developments in Geochemistry Series 2, Elsevier, 343–373.

——, HENDERSON, P. & KEMPE, D. R. C. 1976. Rare earth element and related chemistry of some drilled southern Indian Ocean basalts and volcanogenic sediments. *Journal of Geophysical Research*, **81**, 4257–4268.

GERMAN, C. R. & ELDERFIELD, H. 1989. Rare earth elements in Saanich Inlet, British Columbia, a seasonally anoxic basin. *Geochimica et Cosmochimica Acta*, **53**, 2561–2571.

GIRIN, Y. P., BALASHOV, Y. A. & BRATISHKO, R. K. 1970. Redistribution of the rare earths during diagenesis of humid sediments. *Geochemistry International*, **7**, 438–452.

GROMET, L. P., DYMEK, R. F., HASKIN, L. A. & KOROTEV, R. L. 1984. The "North American shale composite": its compilation, major and trace element characteristics. *Geochimica et Cosmochimica Acta*, **48**, 2469–2482.

HASKIN, L. A., HASKIN, M. A., FREY, F. A. & WILDEMAN, T. R. 1968. Relative and absolute terrestrial abundances of the rare earths. *In*: AHRENS, L. H. (ed.) *Origin and Distribution of the Elements, 1.* Pergamon, 889–911.

——, WILDEMAN, T. R., FREY, F. A., COLLINS, K. A., KEEDY, C. R. & HASKIN, M. A. 1966. Rare earths in sediments. *Journal of Geophysical Research*, **71**, 6091–6105.

HASKIN, M. A. & HASKIN, L. A. 1966. Rare earths in European shales: a redetermination. *Science*, **154**, 507–509.

HOWARD, J. J. 1985. Influence of shale fabric on illite/smectite diagenesis in the Oligocene Frio Formation, south Texas. *In*: SCHULTZ, L. G., VAN OLPHEN, H. & MUPTON, F. A. (eds) *Proceedings of the International Clay Conference, Denver, 1985.*

HUMPHRIS, S. E. 1984. The mobility of the rare earth elements in the crust. *In*: HENDERSON, P. (ed.) *Rare Earth Element Geochemistry*. Developments in Geochemistry Series 2, Elsevier. 317–342.

JONASSON, R. G. & VANCE, E. R. 1986. DTA study of the rhabdophane to monazite transformation in rare earth (La-Dy) phosphates. *Thermochimica Acta*, **108**, 65–72.

——, BANCROFT, G. M. & NESBIT, H. W. 1985. Solubilities of some hydrous REE phosphates with implications for diagenesis and seawater concentrations. *Geochimica et Cosmochimica Acta*, **49**, 2133–2139.

MARSHALL, J. D. 1982. Isotopic composition of displacive fibrous calcite veins: reversals in pore-water composition trends during burial diagenesis. *Journal of Sedimentary Petrology*, **52**, 615–630.

MATZKO, J. J. & OVERSTREET, W. C. 1977. Black monazite from Taiwan. *Proceedings Geological Society of China*, **20**, 16–35.

McCLENNAN, S. M., NANCE, W. B. & TAYLOR, S. R. 1980. Rare earth element–thorium correlations in sedimentary rocks, and the composition of the continental crust. *Geochimica et Cosmochimica Acta*, **44**, 1833–1839.

MORGAN, D. J. 1977. Simultaneous DTA-EGA of minerals and natural mineral mixtures. *Journal of Thermal Analysis*, **12**, 245–263.

OVERSTREET, W. C. 1971. *Monazite from Taiwan.* U.S. Geological Survey Open-File Report.

RAMSEY, J. G. 1967. *Folding and Fracturing of Rocks.* McGraw Hill.

READ, D., COOPER, D. C. & MCARTHUR, J. M. 1987. The composition and distribution of nodular monazite in the Lower Palaeozoic rocks of Great Britain. *Mineralogical Magazine*, **51**, 271–280.

ROALDSET, E. 1979. Rare earth elements in different size fractions of a marine quick clay from Ullensaker, and a till from upper Numedal, Norway. *Clay Minerals*, **14**, 229–239.

—— & ROSENQVIST, I. T. 1971. Unusual lanthanide distribution. *Nature Physical Science*, **231**, 153–154.

RONOV, A B., BALASHOV, Y. A. & MIGDISOV, A. A. 1967. Geochemistry of the rare earths in the sedimentary cycle. *Geochemistry International*, **4**, 1–17.

ROSENBLUM, S. & MOSIER, E. L. 1983. *Mineralogy and occurrence of europium-rich dark monazite.* U.S. Geological Survey Professional Paper, **1181**.

SCHIEBER, J. 1988. Redistribution of rare-earth elements during diagenesis of carbonate rocks from the Mid-Proterozoic Newland Formation, Montanna, USA. *Chemical Geology*, **69**, 111–126.

SINGER, L. & MULLER, G. 1983. Diagenesis in argillaceous sediments. *In*: LARSEN, G. & CHILINGAR, G. V. (eds) *Diagenesis in Sediments and Sedimentary Rocks, 2.* Developments in Sedimentology, **25B**, Elsevier, 115–212.

SMITH, R. D. H. 1987. Early diagenetic phosphate cements in a turbidite basin. *In*: MARSHALL, J. D. (ed.) *Diagenesis of Sedimentary Sequences.* Geological Society, London, Special Publication, **36**, 141–156.

THOMSON, J., CARPENTER, M. S. N., COLLEY, S. & WILSON, T. R. S. 1984. Metal accumulation rates in northwest Atlantic pelagic sediments. *Geochimica et Cosmochimica Acta*, **48**, 1935–1948.

THORNTON, S. E. 1984. Basin model for hemipelagic sedimentation in a tectonically active continental margin: Santa Barbara Basin, California Continental Borderland. *In*: STOW, D. A. V. & PIPER, D. J. W. (eds) *Fine-Grained Sediments: Deep-Water Processes and Facies*. Geological Society, London, Special Publication, **15**, 377–394.

WALSH, J. N., BUCKLEY, F. & BARKER, J. 1981. The simultaneous determination of the rare-earth elements in rocks using inductively coupled plasma source spectrometry. *Chemical Geology*, **33**, 141–153.

WEDEPOHL, K. H. 1978. *Handbook of Geochemistry, Vol II/5: Elements La (57) to U (92)*. Springer-Verlag.

WILDEMAN, T. R. & CONDIE, K. C. 1973. Rare earths in Archean graywackes from Wyoming and from the Fig Tree Group, South Africa. *Geochimica et Cosmochimica Acta*, **37**, 439–453.

—— & HASKIN, L. A. 1973. Rare earths in Precambrian sediments. *Geochimica et Cosmochimica Acta*, **37**, 419–438.

Selective alteration of arkose framework in Oligo-Miocene turbidites of the Northern Apennines foreland: impact on sedimentary provenance analysis

R. VALLONI, D. LAZZARI & M. A. CALZOLARI

Istituto di Petrografia, Università di Parma, Area delle Scienze, 43100 PARMA, Italy

Abstract: Upper Oligocene–Middle Miocene turbidites of the Macigno and Cervarola Formations fill post-collisional elongate basins of the Northern Apennines foreland; they crop out axially for distances up to 300 km and reach a total thickness of 4000 m. Sandstones are arkosic with quartzose grains averaging 50% and fine-grained lithic clasts from 5 to 25%; K-feldspar and plagioclase abundances are similar, and rock fragments are dominated by igneous and metamorphic types. The Macigno sandstones reach a much higher diagenetic grade than the younger Cervarola sandstones. The latter (mature stage of diagenesis), have a clay mineral assemblage with ordered I/S interlayers and illite crystallinity (I.K.) averaging from 8.0 to 10.1, while in the Lower Unit of the Macigno Formation clay crystallinity averages 3.8 (Kübler's epizone). The evolution of diagenetic processes is unusual because of the abundance of unstable grains. Physical compaction is the main cause of pore reduction. Post-depositional mineralogical changes are relevant in both formations; ignoring authigenic overgrowths and late calcite replacements, they commonly affect 10–20% of the framework grains.

Several mineralogical–textural changes influence provenance determinations, including: dissolution processes, which almost totally eliminate unstable species, such as amphibole and pyroxene, originally dominating the heavy mineral population; replacement processes, mainly represented by albite after plagioclase, and by clay or mica materials after unstable grains; intense deformation plus replacement of ductile or brittle grains with the authigenesis of chlorite, white mica and pseudomatrix.

Arkosic arenites, with a quartzose grain content of about 50%, occur very frequently in nature, in many different tectonic settings (Dickinson *et al.* 1983). In the post-depositional phase, several detrital components, such as plagioclases and fine-grained lithic clasts are unstable and may undergo significant changes (McBride 1985). The type and amount of these unstable grains are especially important as they affect the diagenetic evolution that ultimately influences the modal composition (Helmold 1985).

The turbiditic sandstones of the Macigno and Cervarola Formations are characterized by a $Q_{40-60} F_{25-50} L_{5-25}$ detrital grain proportion (Fig. 2), with 5–15% of mica and chlorite grains and a similar amount of altered framework (pseudomatrix and/or secondary matrix). Both formations were deposited during Oligo-Miocene times in the foreland basins of the Northern Apennines (Ricci Lucchi 1986). The deposits have experienced mainly physical compaction processes (cf. Fontana *et al.* 1986), the effects of which are controlled by the abundance of grains susceptible to deformation (Rittenhouse 1971).

The diagenetic evolution of arkosic sandstones has been more extensively investigated in extensional basins (e.g. North Sea, Burley *et al.* 1985), where the applied lithostatic pressure is a function of the burial depth, compared to orogenic settings in which tectonic compression is important (Hoffman & Hower 1979). For turbidites in the latter setting the careful study under a polarizing microscope of the unstable-grain textural relations is of particular importance.

In this paper an attempt is made at quantifying the effects of dissolution, replacement, deformation and recrystallization processes which have modified the original detrital mineralogy of turbiditic deposits. Because of the inadequacy of current analytical procedures some operative definitions are introduced in this paper. In order to relate the degree of framework alteration to specific diagenetic stages, the microscopic data on sandstones are integrated with the diffractometric data on pelites.

Geological setting

The northern segment of the Apennines is traditionally separated into two principal palaeogeographic domains, namely the Ligurian and

From Morton, A. C., Todd, S. P. & Haughton, P. D. W. (eds), 1991,
Developments in Sedimentary Provenance Studies.
Geological Society Special Publication No. 57, pp. 125–136.

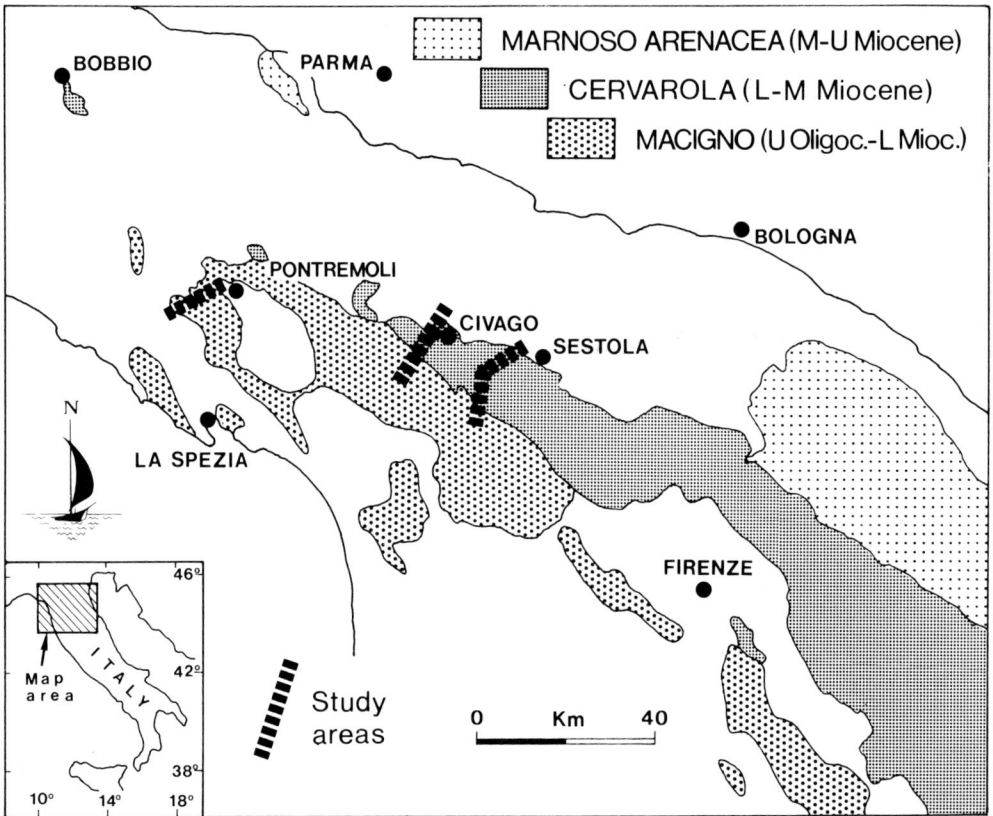

Fig. 1. Geographical distribution of the Oligo-Miocene turbidite formations of the Northern Apennines foreland and location of the study areas.

the Tuscan-Umbrian, with oceanic-quasioceanic crust and continental crust basement respectively. The latter domain is divided into two main sequences, the Tuscan to the west and the Umbrian to the east, both of which were deposited on the Adria continental margin. From the onset of marine conditions, in the Late Triassic, until the Oligocene, the Tuscan and the Umbrian sequences developed with a similar stratigraphy. Because of the progressive shifting of compression from west to east during Miocene times, a pile of Ligurian nappes was emplaced upon the Tuscan–Umbrian sequences, which were progressively deformed from southwest to northeast (Sagri & Zanzucchi 1975). From Middle Oligocene to Miocene times elongate turbidite basins, about 300 km long, formed in front of the nascent mountain belt on top of the Tuscan (Macigno Formation) and Umbrian (Marnoso-arenacea Formation) sequences. During the Late Miocene, continuing tectonic compression thrusted the clastic bodies themselves toward the

foreland. The degree of superposition and deformation decreases from the internal Macigno Formation to the external Marnoso-arenacea Formation (Ricci Lucchi 1986).

With reference to the Oligo-Miocene foreland deposits, the northwestern portion of the Northern Apennines (NW of Firenze) comprises a structurally uniform section in which the thrusted turbidite bodies constitute the backbone of the mountain belt. Two main turbidite formations are recognized: the Macigno and the Cervarola sandstones. Guenther & Reutter (1985) subdivided the deposits of the Cervarola Formation into five tectonic elements (A to E, from internal to external areas). The Cervarola Formation is traditionally considered to be part of a distinct tectonic unit, named Unità Modino–Cervarola, which occupies an intermediate position between the Macigno and the Marnoso-arenacea Formations.

Figure 1 shows the geographic distribution of the Macigno and Cervarola Formations. For

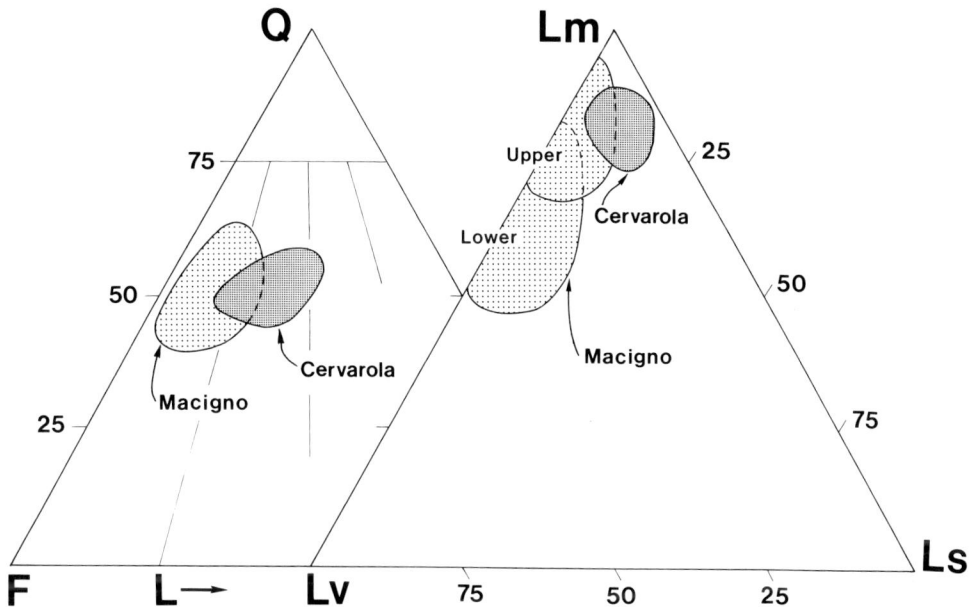

Fig. 2. QFL and lithic grain proportions of the Macigno (Lower and Upper Unit) and Cervarola sandstones. Q, quartzose grains; F, feldspathic grains; L, fine-grained rock fragments; Lm, metamorphic L; Lv, volcanic L; Ls, sedimentary L (including limeclasts). Number of samples is Macigno, 30 and Cervarola, 21.

reasons of simplicity, the Upper Oligocene Modino Formation (up to 700 m thick), whose palaeogeographic relationships with the Macigno and Cervarola Formations are not clear, has been ignored. The Macigno Formation is of Upper Oligocene–Lower Miocene age and has a thickness of 2000–2200 m while the Cervarola Formation is of Lower-Middle Miocene age and ranges from 700–1400 m in thickness (Iaccarino 1975).

The turbidite depositional facies of the study areas (Fig. 1) have been investigated by Ghibaudo (1975, 1980) for the Pontremoli section of the Macigno Formation and the Civago section of the Cervarola Formation. Sandy turbidite facies dominate the succession. Muddy turbidites attain a significant thickness in the central portion of the Macigno Formation which can thus be subdivided into two (Upper and Lower) sandstone units of similar thickness (about 700 m). Although Macigno Formation sandstones show some evidence of minor channelling, an outer fan (lobe–interlobe) depositional environment is likely for both formations (Ghibaudo 1975, 1980).

Palaeocurrents and petrography indicate that Macigno and Cervarola basins were fed from the northwest (Valloni & Zuffa 1984). The clastic supply probably travelled along the front of the collided plates and entered the basins from their northern apex. This source was also active during deposition of the Marnoso-arenacea Formation where the 'alpine' provenance (Gandolfi et al. 1983) was dominant.

Sandstone petrography

The spectrum in mineral composition of the Northern Apennine sandstones was discussed by Valloni & Zuffa (1984) and Denke & Gunther (1981). The Oligo-Miocene turbidites of the Macigno, Cervarola and Marnoso-arenacea Formations are arkosic with quartzose grains averaging 50% and fine-grained lithic grains ranging from 5 to 25%.

The key oucrops of the Macigno and Cervarola Formations, where sandstone detrital modes (51 samples) and heavy minerals (39 samples) have been studied (Valloni 1978; Mezzadri & Valloni 1981; Lazzari 1987; Valloni et al. 1990), are indicated in Fig. 1. In this area, some 100 km long, detrital modes are very uniform along strike and vary only a little in the stratigraphic succession.

The Macigno and Cervarola sandstones have similar bulk compositions. In both cases essential framework grains account for 70–80%, while micas and altered or deformed framework grains vary from 5 to 15%. Intrabasinal grains (e.g. fossils) are consistently less than 1% and the intergranular phase consists of fine-grained

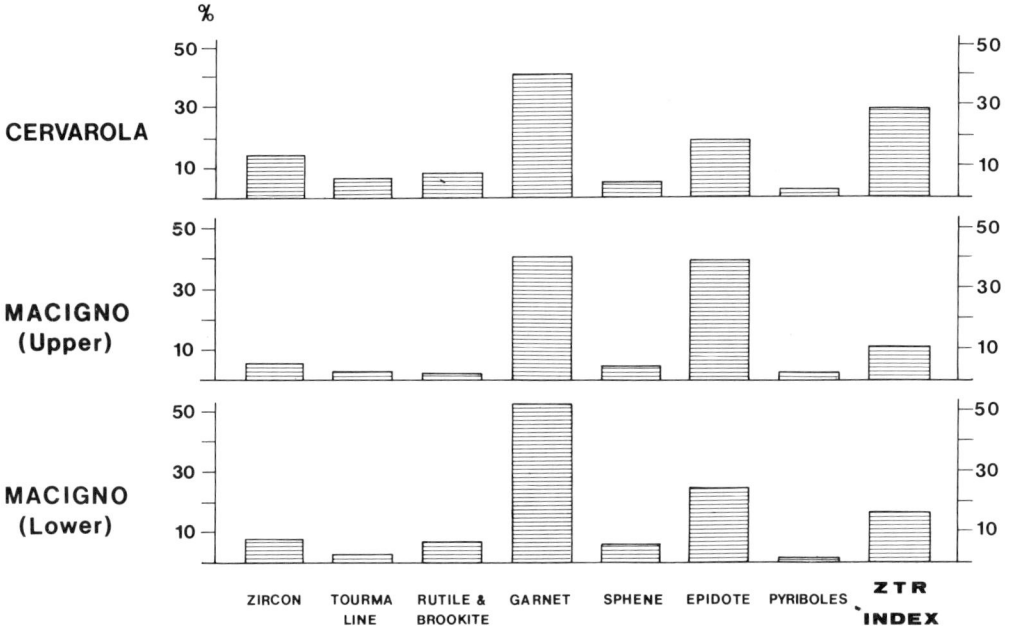

Fig. 3. Percent frequency in the 2–4 φ size fraction of transparent heavy mineral species for the Macigno (Lower and Upper Unit) and Cervarola sandstones. The ZTR index is the sum of the three left hand side columns. Pyriboles = Pyroxene + Amphibole. Number of samples is Macigno, 25 and Cervarola, 14.

detritus (reorganized into epimatrix), a characteristic of outer-fan sandstone (Chan 1985; Fontana *et al.* 1986).

Plotting the essential framework grains and the key lithic types on ternary diagrams allows the distinction of the sandstones of the Macigno and Cervarola Formations (Fig. 2). The former are feldspathic with a metamorphic–volcanic lithic assemblage while the latter are lithofeldspathic with an essentially metamorphic lithic assemblage. The Lower and Upper sandstone units of the Macigno Formation are distinguished by their lithic grain association; the Lower Unit is Lv bearing, while the Upper Unit is Lm rich.

Heavy-mineral proportions (Fig. 3) are calculated separately for the three petrofacies described above. Garnet and epidote consistently dominate; titanite (approaching 5%) and ZTR (ranging 10–30%) are also common. The ZTR index, defined as the ratio of the sum of zircon, tourmaline and rutile percentages with the total transparent nonmicaceous grains (Hubert 1962), is higher in the Cervarola Formation. This is not consistent with the general diagenetic trend of the Macigno and Cervarola Formations discussed in the following sections. However, in the two units of the Macigno Formation, stable (e.g.

garnet) and ultrastable (ZTR) grain frequencies increase with stratigraphic depth.

The small variations in mineral composition described above should not control the development of diagenetic processes (Loucks *et al.* 1984). Detrital modes and heavy mineral proportions of Figs 2 and 3, however, are not very reliable for provenance determinations since the alteration of unstable grains, which is discussed in the next sections, affects 10–20% of the framework.

Sandstone textural relations

During diagenesis, sandstone textures evolve toward closer packing. Compaction processes are traditionally divided into physical and chemical (Bjørlykke 1988). The turbidites discussed here lack an early cement and their porosity has been occluded mainly as a result of physical compaction. In contrast, chemical processes, being mainly replacive, do not alter rock volume. Chemical and physical processes often occur together, both contributing to the total diagenetic transformations. The rate of diagenetic transformation diminishes from the Lower Unit of the Macigno Formation to the Cervarola Formation (Fig. 2) and will be discussed in the following sections. The main post-depositional events are now discussed and grouped into textural classes; the sites of occurrence consist of framework grains, interstitial fines, and pore spaces (cf. Wilson & Pittman 1977).

Interstices

Authigenic clay rims and, most commonly, quartz overgrowths cement primary pore spaces. Residual primary and secondary pores as well as fractures are filled by authigenic quartz, K-feldspar and albite (Fig. 4 E & F); later fibrous illite and chlorite rosettes also occur.

The component resulting from the diagenetic growth in open interstices and/or from the diagenetic reorganization of the original fines is locally abundant. This material, which has been called epimatrix by Dickinson (1970), retains its interstitial textural position and exhibits somewhat inhomogeneous mineralogical features (Fig. 4 A).

Framework and interstices

Irregular patches of calcite are present in the form of a late replacive cement on framework and intergranular components.

Framework

(a) Chemical events: Some framework grains undergo selective changes that may result in a pseudomorph replacement. The most striking phenomena are the albitization of K-feldspar and plagioclase (Fig. 4 E & F) and the crystallization of white mica (Fig. 4 C & D), usually on altered biotite (Fig. 4 E & F).

(b) Physico-chemical events: These compression processes do not cause a significant grain deformation. A modest degree of pressure solution, generating concavo-convex contacts, may occur between quartz grains. The alteration of the more unstable grains, e.g. plagioclase, may also take place forming an aggregate of clay minerals and sericite, possibly with the addition of silicate, carbonate and opaque minerals, which may make the original grains unrecognizable; these polymineralic aggregates retain their textural relations which characterize framework components and are here termed secondary matrix (Fig. 4 B) because of their resemblance to interstitial fines.

(c) Physical events: When mechanical action on unstable framework grains causes their deformation, the grains are squeezed into the interstices (Fig. 4 C). The brittle or ductile (clayey, glauconitic, micaceous, carbonate and other) grains may be fractured and/or undergo plastic deformation, in some cases without any conspicuous change in their mineralogical features. By contrast, when the chemical and physical processes act to produce an aggregate of chlorite and/or sericite which totally replaces and even partly recrystallizes the original grains (Fig. 4 C & D), the result is a pseudomatrix (Dickinson 1970).

Sandstone components lacking a crystalline habit are summarized in Fig. 5. Criteria for the

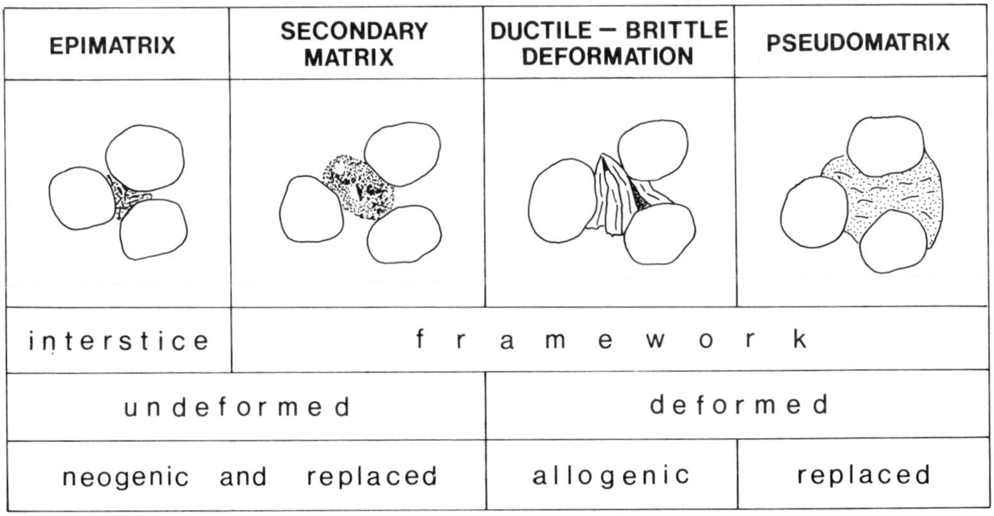

EPIMATRIX	SECONDARY MATRIX	DUCTILE – BRITTLE DEFORMATION	PSEUDOMATRIX
interstice	framework		
undeformed		deformed	
neogenic and replaced		allogenic	replaced

Fig. 5. Criteria for distinction of 'matrix'.

distinction of the 'matrix' are essentially the textural position and the degree of deformation and authigenesis.

Diagenetic processes

Although the issue is a complex one, a measure of the variability of the diagenetic transformations which occur in the sandy layers of turbidites is now given. The processes discussed below are those traditionally considered most significant: (1) intrastratal solution of unstable minerals, (2) conditions of sandstone compaction, and (3) modifications of the clay mineral assemblage.

Heavy minerals

The more unstable components of arenites can be dissolved during burial and this is especially true for some types of heavy minerals (Morton 1984). Gazzi (1965) has shown that in the turbidites of the Northern Apennines the dissolution of heavy minerals is extremely marked and is related to the age of the deposits.

Heavy mineral proportions in the Macigno and Cervarola Formations are shown in Fig. 3. The modification of the relative abundance of the various mineral types in the Macigno Formation caused by diagenesis is discussed considering only the stable and unstable mineral types (i.e. ignoring the ultrastable ZTR). In the diagrams of Fig. 6, the percentages found in the Pontremoli section (mean of the eighteen samples of the lower and upper unit) are compared to (1) those calculated considering the heavy minerals present in the sandstone rock fragments (determined by point-counting on thin-section) and (2) those determined from coastal sands of a modern basin in southwestern Calabria, Italy (unpublished data of authors), which have a similar provenance to the Macigno Formation sandstones. It should be noted that the two procedures used for the evaluation of the original heavy mineral association provide similar results (Fig. 6).

The strongest elimination occurs in pyriboles (pyroxene + amphibole; cf. Milliken 1988), which are reduced from a calculated incidence of 32% to an incidence of 2%, followed by other unstable types (principally kyanite, andalusite, sillimanite, titanite and barite) which vary from a calculated incidence of 29% to an incidence of 6%. Epidote group minerals have been partly dissolved during diagenesis, since their incidence increases much less over the calculated percent-

age (from 25% to 37%) than the more stable garnet, which increases from 14% to 55%, out of the total of the stable + unstable types.

The lack of comparable data on modern sediments makes it difficult to assess the absolute weight loss of heavy minerals during diagenesis; however, the loss is not the same in the two sandstone units of the Macigno Formation, with the Lower Unit having a lower heavy-mineral content in the 2–4 φ fraction (0.26% by weight) suggesting stronger intrastratal solution.

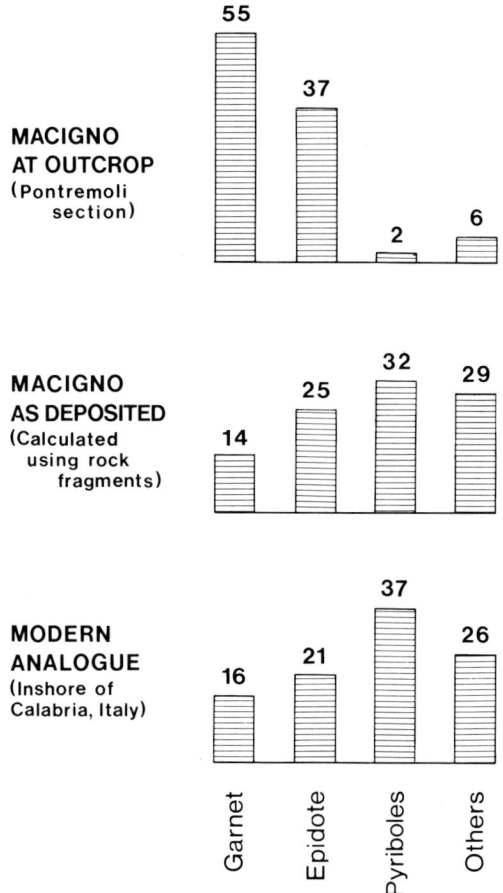

Fig. 6. Percent frequency in the 2–4 φ size fraction of heavy minerals on total stable plus unstable transparent species (i.e. ZTR excluded). Proportions of the middle histogram are calculated on the basis of the minerals present in the point-counted rock fragments.

Sandstone compaction

Arkosic arenites contain high amounts of minerals susceptible to mechanical deformation, such as mica and chlorite grains, which usually

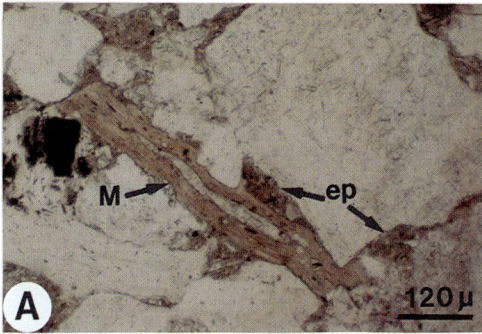

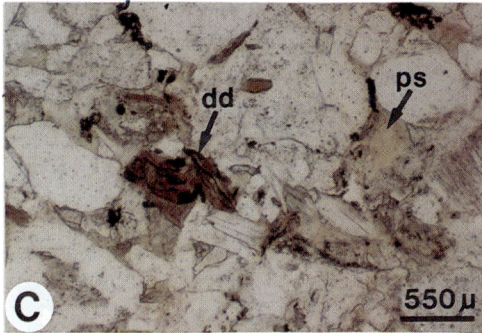

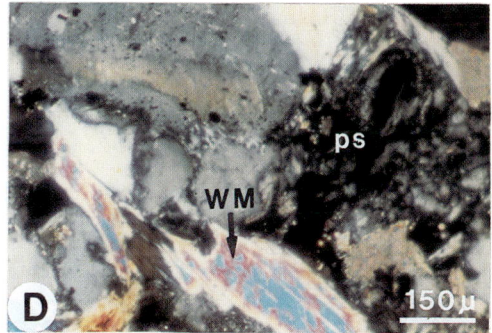

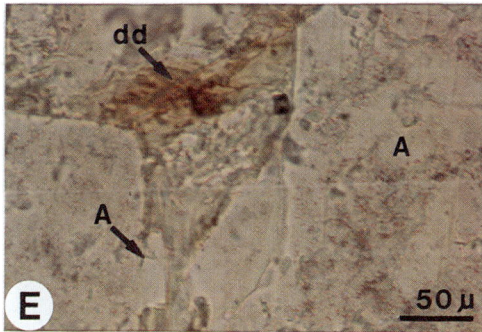

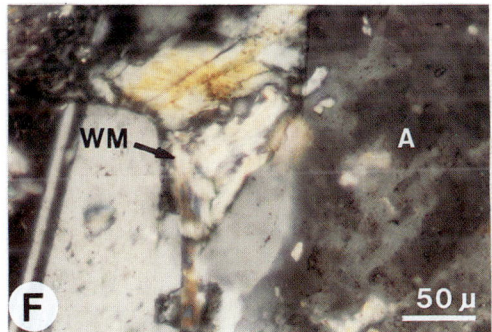

Fig. 4. (**A**) Cervarola, packing is low and biotite grain is little deformed (PPL); (**B**) Macigno, mineral aggregate (sm) after framework grains (XPL); (**C**) Macigno, packing is high, detrital ductile grains are strongly deformed or mashed (PPL); (**D**) enlarged view of the centre-right portion of the previous photo in XPL, undeformed authigenic white mica on altered framework; (**E**) Macigno, plagioclase on the left (with authigenic overgrowth) and K-feldspar on the right (with authigenic replacement) (PPL); (**F**) previous photo in XPL.

A, authigenic albite; WM, authigenic white mica; M, detrital mica grain. ep, epimatrix; sm, secondary matrix; ps, pseudomatrix; dd, deformed ductile grain.

Table 1. *Range of grain contact values (percent frequency)*

| Along traverse stable framework grain is in contact with: | STABLE GRAIN | | ALTERED GRAIN | DEFORMED GRAIN | | Epimatrix and authigenic rim & overgrowth | Authigenic mineral replacement | Authigenic patchy replacement |
	Concavo-convex (1)	Linear and tangential (2)	Secondary matrix (3)	Ductile and brittle (4)	Pseudomatrix (5)	(6)	(7)	(8)
Cervarola	3	51	19	11	—	5	1	10
Lower Macigno	5	19	—	35	22	3	9	7

Column numbers are given in parentheses

account for 5–15% of the total. This means that the framework records a sequence of gradual changes toward closer packing (Smosna 1989), and for equal applied pressures, pore space is more reduced than in a stable framework (Meade 1966; Nagtegaal 1978).

Packing can be measured on thin section counting contacts along a traverse (Pettijohn *et al.* 1987, p. 85). The values for the textural extremes, which in this study are represented by the Macigno Lower Unit and the Sestola section of the Cervarola Formation (4 + 4 representative samples), are indicated in Table 1 (100 counts per sample).

Table 1 records the mineralogical–textural classes discussed in the previous section together with their percentages of contacts with stable grains of columns 1 and 2 (i.e. undeformed, unaltered, rigid or flexible grains) on total contacts. Columns 1 to 5 represent framework units the sum of their contact percentages providing an indication of the actual packing. Column 6 represents the interstitial materials, both detrital (orthomatrix, rarely present) and authigenic (epimatrix, cement); in theory, it also contains the pores, but these have been reduced to zero. Columns 7 and 8 represent authigenic replacive minerals occurring on single framework grains (also pseudomorphs) and on two or more of the remaining mineralogical-textural classes, respectively. The patchy replacement in Column 8, represented by calcite, is an equal-volume process which takes place at a later stage, when a rigid fabric had been produced by compaction.

Table 1 also shows that in both formations grain-to-grain contacts predominate and that interstitial components are found only in a few cases (3–5%). In the Cervarola sandstones, most contacts occur between stable grains, although significant numbers (19%) of stable grains are in contact with altered grains; the interstitial materials are represented by both epimatrix and authigenic overgrowths. In the Macigno sandstones, most contacts occur between deformed grains, which in some cases have recrystallized and spread around rigid grains (22% pseudomatrix); interstitial materials consist mainly of authigenic overgrowths, with a significant proportion (9%) of authigenic mineral replacements.

In conclusion, the two formations differ because of a different degree of deformation induced by physical compaction, which is particularly marked in the Lower Unit of the Macigno Formation and is only modest in the Cervarola Formation, in which the altered framework (secondary matrix) appears to be more significant.

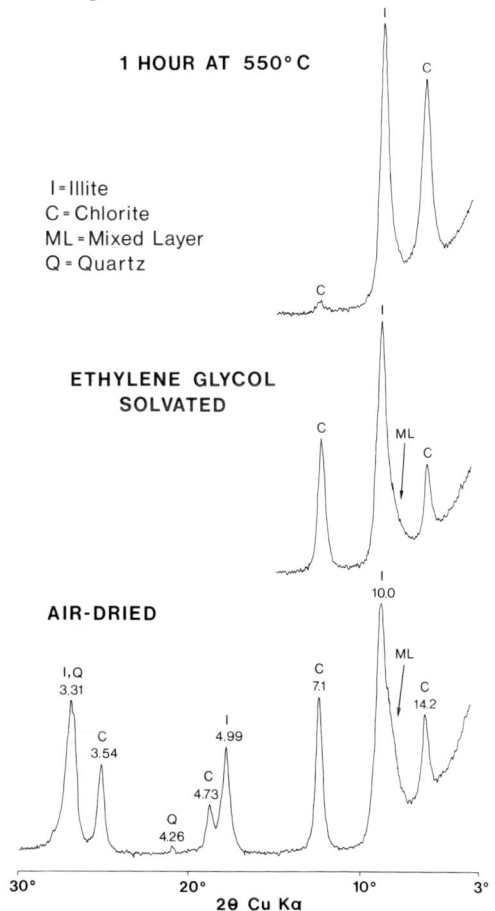

Fig. 7. X-ray diffractograms for the less than 2 μm fraction of an argillaceous sample of the Cervarola Formation outcropping in the Civago area. ML are ordered illite/smectite interlayers.

Clay mineral assemblage

·The analysed shales have been taken from both the pelitic top of sandy turbidites and the body of muddy turbidites, and are therefore closely associated with the sandstones discussed previously. The data utilized here are partly original (Civago section) and partly taken from Bonazzi *et al.* (1984) on the Macigno of the Pontremoli area and Calzolari (1984) on the Cervarola of the Sestola area.

All data have been obtained with a standardized ·procedure of preparation and analysis. The less than 2 µm size fraction was separated by sedimentation from a suspension in deionized water. After filtering, an oriented clay aggregate was prepared by pipetting the dispersed slurries onto a glass slide. Air-dried, ethylene-glycol solvated (24 hours at 60°C) and heated (1 hour at 550°C) mounts were analysed using a Philips X-ray diffractometer with Ni filtered Cu Kα radiation (40 kV, 20 mA) at a scanning speed of 1° (2θ) per minute.

Semiquantitative measurements of the relative abundance of clay minerals were done according to the Biscaye (1965) method. The area under the peaks was measured for determining the illite/chlorite ratio, the average value of which is reported in Table 2 together with the illite crystallinity index (Kübler 1967), given by the width of the first order basal reflection at half height, measured in millimetres from traces of air-dried mounts.

In the Cervarola Formation, the crystallinity index of illite ranges from 8.0 to 10.1 (Kübler's diagenetic zone) and the illite/chlorite ratio is 5/1, while in the Lower Unit of the Macigno Formation the values are 3.8 (Kübler's epizone) and 1/1, respectively.

Figure 7 depicts the diffractograms from a sample of Cervarola sandstones, indicative of the comparatively lower diagenetic stage. The diagrams show a clay mineral assemblage dominated by illite and chlorite, the presence of moderate amounts of the I/S mixed-layers (less than 10% in the Cervarola Formation), and the absence of kaolinite. In the Macigno Formation, where the highest diagenetic stage is reached, the I/S mixed layers disappear and chlorite increases conspicuously, even becoming in some cases more abundant than illite. This trend is well known in the literature (e.g. Boles & Franks 1979).

Diagenetic stage

The previous discussion on post-depositional events has stressed the importance of deformation (cf. Frey *et al.* 1980). The unstable components of the arkosic sandstones of the Macigno and Cervarola Formations show a sequence of mineralogical-textural transformations. It is then possible to determine the diagenetic stage of turbidite sandstones also using some parameters that can be measured on thin-sections. Such parameters will be discussed here by merely considering the variability of their values, with no attempt at identifying formal diagenetic stages. For correlation purposes, the data for the medium- to coarse-grained sandstone samples are compared with those traditionally obtained on X-ray diffractometry of clay minerals. The whole data set, ordered according to the geographic and stratigraphic positions of the formations under study, is reported in Table 2.

Table 2. *Diagentic stages*

	% Epimatrix and Secondary Matrix	% Chlorite and Sericite Pseudomatrix	Intact Biotite Grains (% total mica)	Heavy Minerals (weight %)	Illite/ Chlorite	% Ordered Interlayered I/S	Illite Crystallinity I.K.
CERVAROLA (top)	7.0	little	3.0	?	5/1	5.0	10.1
CERVAROLA	15.0		3.0	.41	5/1		8.0
MACIGNO Upper Unit	little	9.0	0.0	.95	3/1	little	5.8
MACIGNO Lower Unit	0.0		0.0	.26	1/1	0.0	3.8
	med.-coarse base of sandstone			*2-4 φ fraction*	*<2µm fraction of pelite*		

One of the parameters indicative of the diagenetic stage is the epimatrix, i.e. original fines under recrystallization, and the secondary matrix, i.e. altered framework, which predominate in the Cervarola sandstones, accounting for 7% of the Sestola area and 15% of the Civago area, respectively; the epimatrix and secondary matrix have almost disappeared from the Upper Unit of the Macigno Formation, as they succumbed to increased alteration and physical compaction.

By contrast, the chlorite and sericite pseudomatrix, which is virtually absent in the Cervarola Formation, is an important component of the Macigno Formation, where it accounts on average for 9% and occasionally up to 15% of the total rock. Clayey, micaceous and volcanic grains are probably the framework components that are more deeply involved in the alteration into pseudomatrix.

Among the framework grains, micas appear especially susceptible to mechanical deformation. They are more abundant in the Macigno Formation (averaging 10.1% in the Pontremoli area) than in the Cervarola Formation, where they average 5.5% in the Civago area, with comparable amounts of muscovite and biotite. The latter exhibits an interesting trend through increasing alteration and deformation to total masking and recrystallization. As shown in Table 2, no intact biotite grains remain in the Macigno Formation, while in the Cervarola Formation they account for 3% of total micaceous grains.

To integrate the data on sandstones with a quantitative index of the intrastratal solution processes, the percentage weight of the mineral grains (of the 2–4 φ fraction insoluble in 1 N HCl) with density > 2.97, is reported in Table 2. The elimination of the unstable heavy minerals (compare Fig. 6) appears especially marked in the Lower Unit of the Macigno Formation, where they account on average for 0.26% of the total, varying between 0.1 and 0.8%.

The diffractometric data for the clay minerals in Table 2 have been obtained with samples taken from the sequence studied microscopically. In both formations, the clay minerals are dominated by illite and chlorite, their ratio varying gradually from 5/1 in the Cervarola Formation to 1/1 in the Lower Unit of the Macigno Formation.

Such minerals may occur in the presence of small amounts, i.e. less than 10%, of mixed-layers illite/smectite (ordered I/S) which show significant changes. They account on average for 5% of total clay minerals ($< 2\,\mu m$) in the Cervarola Formation and are entirely lacking in the Lower Unit of the Macigno Formation. Here, only illite and chlorite are present, with an illite crystallinity index averaging 3.8 suggesting epimetamorphic conditions. This value increases to 5.8 (anchizone) in the Macigno Upper Unit and up to 10.1 in the Cervarola.

The data in Table 2 can be compared with vitrinite reflectance (Reutter et al. 1983). In the three areas discussed here, only eight (3 + 3 + 2) measurements are available, which provide average Rm values of 1.8% for the Macigno Formation in the Pontremoli area and 2.0% and 1.2% for the Cervarola Formation in the Civago and Sestola areas, respectively.

It should be noted that in the range of modifications reported in Table 2, the Lower Unit of the Macigno sandstones is in a stage of incipient metamorphism characterized by the authigenesis of white mica crystals due to selective replacement of altered and squashed phyllosilicates, e.g. biotite. This event is recognized on a textural basis with the authigenic white micas occurring in totally undeformed single grains, on a background of mashed ductile grains.

Conclusion

The post-depositional history of arkosic turbidites of the Oligo-Miocene Macigno and Cervarola Formations has been greatly affected by the abundance of labile grains and by the marked predominance of physical processes during compaction. The unstable framework grains undergoing mechanical deformation and chemical reaction show a selective and very gradual record of diagenetic effects, which can be distinguished under the polarizing microscope (Fig. 4).

For the genetic interpretation of a sandstone with mechanically unstable grains, a series of operative definitions has proved useful to discriminate between various types of 'matrix' (Fig. 5) produced by the framework grains, separating the highly deformed components (pseudomatrix) from the basically undeformed ones (secondary matrix). The sandstones of the Cervarola Formation differ considerably from those of the Macigno Formation: considering the percentage of contacts with deformed grains (out of the total contacts, Table 1) as an index of deformation, the resulting values are 11 (with no generation of pseudomatrix) for the Cervarola Formation and 57 (including two-fifths pseudomatrix) for the Macigno Formation.

Data from sandstone labile grains and shale clay minerals of the Macigno and Cervarola Formations cover the range of the mature and supermature textural stages of Schmidt & McDonald (1979), but in general the Cervarola Formation of the Sestola and Civago areas is at a mature diagenetic stage, whereas the Macigno Lower Unit of the Pontremoli area is at an incipient metamorphic stage (Table 2). The two stages are characterized as follows.

Mature diagenetic stage: Porosity reduced to zero, with moderate framework deformation (11% of deformed contacts). Preservation of intact grains among the ductile–flexible ones (3% biotite). The interstitial fine material (epimatrix) are still recognizable while pseudomatrix is virtually absent. The illite/chlorite ratio is 5/1 with a crystallinity index for illite (I.K.) up to 10 and ordered I/S interlayers less than 10%.

Incipient metamorphic stage: Strong framework deformation (22% of contacts with squashed grains) with total deformation of the ductile-flexible grains (no intact biotite) and removal of the primary and secondary matrix. Diffuse chloritization of the unstables with generation of considerable amounts of pseudomatrix (5–13%) and authigenesis of replacive white mica above mashed or altered phyllosilicate grains. Clay minerals are represented only by illite and chlorite in similar proportions (1/1 ratio) with ordered I/S interlayers reduced to zero and a crystallinity index for illite <4.

The loss of provenance information, due to the selective alteration of framework grains, is demonstrated by several sandstone components.

(1) The transparent fraction of the heavy mineral population undergoes a conspicuous dissolution of the unstable pyroxene and amphibole grains plus other minerals of higher stability so that, in the Lower Unit of the Macigno Formation, even the moderately stable epidote dissolves. As a consequence, heavy minerals consist of a relict association of garnet and epidote which is totally different from the original unstable-dominated association.

(2) Patchy calcite is a late diagenetic replacement whose occurrence strongly varies between samples and may even be absent.

(3) Authigenic albite, white mica and chlorite occur as replacive minerals in plagioclase and other unstable grains especially in the Macigno Formation where they account for a low percentage.

(4) Pseudomatrix after mashed ductile grains is an important constituent of the Macigno Formation sandstones, averaging 9% of the framework.

(5) Secondary matrix replacements of labile feldspar and lithic grains are particularly abundant in the Cervarola Formation sandstones ranging from 5–10% of the total framework.

References

BISCAYE, P. E. 1965. Mineralogy and sedimentation of Recent deep-sea clay in the Atlantic Ocean and adjacent seas and oceans. *Geological Society of America Bulletin*, **76**, 803–831.

BJØRLYKKE, K. 1988. Sandstone diagenesis in relation to preservation, destruction and creation of porosity. *In*: CHILINGARIAN, G. V. & WOLF, K. H. (eds) *Diagenesis, I. Developments in Sedimentology*, **41**. Elsevier, Amsterdam, 555–588.

BOLES, J. R. & FRANKS, S. G. 1979. Clay diagenesis in Wilcox Sandstones of Southwest Texas: implications of smectite diagenesis on sandstone cementation. *Journal of Sedimentary Petrology*, **49**, 55–70.

BONAZZI, A., SALVIOLI MARIANI, E. & VERNIA, L. 1984. Diagenesi e metamorfismo dedotti dalla cristallinità dell'illite in formazioni sedimentarie affioranti tra Pontremoli e Salsomaggiore (Appennino Tosco-Emiliano). *Mineralogica Petrographica Acta*, **27**, 123–138.

BURLEY, S. D., KANTOROWICZ, J. D. & WAUGH, B. 1985. Clastic diagenesis. *In*: BRENCHLEY, P. J. & WILLIAMS, B. P. J. (eds) *Sedimentology: Recent Developments and Applied Aspects*. Geological Society of London, Special Publication, **18**, 189–226.

CALZOLARI, M. A. 1984. *Diagenesi e metamorfismo nell'Unità di M. Modino–M. Cervarola, nell'area compresa tra la Val Trebbia e il Passo della Futa (Appennino Settentrionale): Dati di Cristallinità dell'illite*. Tesi di Laurea, Facoltà di Scienze M. F. N., Università di Parma. Italia.

CHAN, M. A. 1985. Correlations of diagenesis with sedimentary facies in Eocene sandstones, Western Oregon. *Journal of Sedimentary Petrology*, **55**, 322–333.

DENEKE, E. & GUNTHER, K. 1981. Petrography and arrangement of Tertiary graywacke and sandstone sequences of the Northern Apennines. *Sedimentary Geology*, **28**, 189–230.

DICKINSON, W. R. 1970. Interpreting Detrital Modes of Graywacke and Arkose. *Journal of Sedimentary Petrology*, **40**, 695–707.

——, BEARD, L. S., BRAKENRIDGE, G. R., ERJAVEC, J. L., FERGUSON, R. C., INMAN, K. F., KNEPP, R. A., LINDBERG, F. A. & RYBERG, P. T. 1983. Provenance of North American Phanerozoic sandstones in relation to tectonic setting. *Geological Society of America Bulletin*, **94**, 222–235.

FONTANA, D., MCBRIDE, E. & KUGLER, R. 1986. Diagenesis and porosity evolution of submarine-fan and basin-plain sandstones, Marnoso-arenacea Formation, Northern Apennines, Italy. *Bulletin of the Canadian Association of Petroleum Geologist*, **34**, 313–328.

FREY, M., TEICHMÜLLER, M., TEICHMÜLLER, R., MULLIS, J., KÜNZI, B., BREITSCHMID, A., GRUNER, U. & SCHWIZER, B. 1980. Very low-grade metamorphism in external parts of the Central Alps: Illite crystallinity, coal rank and fluid inclusion data. *Eclogae Geologicae Helvetica*, **73**, 173–203.

GANDOLFI, G., PAGANELLI, L. & ZUFFA, G. G. 1983. Petrology and dispersal pattern in the Marnosoarenacea Formation (Miocene, Northern Apennines). *Journal of Sedimentary Petrology*, **53**, 493–507.

GAZZI, P. 1965. On the heavy mineral zones in the geosyncline series. Recent studies in the Northern Apennines, Italy. *Journal of Sedimentary Petrology*, **35**, 109–115.

GHIBAUDO, G. 1975. Remarks on the Cervarola Sandstone at Torre degli Amorotti Locality. *In: Examples of Turbidite-Facies and Facies Associations from Selected Formations of the Northern Apennines.* IX International Congress of Sedimentology, Nice, Field Trip, **A11**, 68–70.

—— 1980. Deep-sea fan deposits in the Macigno Formation (Middle-Upper Oligocene) of the Gordana Valley, Northern Apennines, Italy. *Journal of Sedimentary Petrology*, **50**, 723–742.

GUENTHER, K. & REUTTER, K. J. 1985. Il significato delle strutture dell'unità di M. Modino–M. Cervarola tra il Passo delle Radici e il M. Falterona in relazione alla tettonica dell'Appennino settentrionale. *Giornale di Geologia*, **47**, 15–34.

HELMOLD, K. P. 1985. Provenance of feldspathic sandstones—the effect of diagenesis on provenance interpretations: a review. *In:* ZUFFA, G. G. (ed.) *Provenance of Arenites*. Reidel, Dordrecht, 139–163.

HOFFMAN, J. & HOWER, J. 1979. Clay mineral assemblages as low grade metamorphic geothermometers: application to the thrust faulted disturbed belt of Montana, U.S.A. *In:* SCHOLLE, P. A. & SCHLUGER, P. R. (eds) *Aspects of Diagenesis*. Special Publication of the Society of Economic Paleontologists Mineralogists, **26**, 55–79.

HUBERT, J. F. 1962. A zircon-tourmaline-rutile maturity index and the interdependence of the composition of heavy mineral assemblages with the gross composition and texture of sandstones. *Journal of Sedimentary Petrology*, **32**, 440–450.

IACCARINO, S. 1975. Biostratigraphic correlations of the Oligocene and Miocene turbidite formations of the Tuscan, Modino-Cervarola and Umbrian sequences. *In: Examples of Turbidite Facies and Facies Associations from Selected Formations of the Northern Apennines.* IX International Congress of Sedimentology, Nice, Field Trip, **A11**, 14–20.

KÜBLER, B. 1967. La cristallinité de l'illite et les zones tout à fait supérieures du métamorphisme. *Etages tectoniques, Colloque Neuchâtel*, 105–122.

LAZZARI, D. 1987. *Studio composizionale delle torbiditi arenacee del Macigno (Oligomiocene) affioranti sull'allineamento Abetone—Monti del Chianti.* Tesi di Laurea, Facoltà di Scienze M. F. N., Università di Parma, Italia.

LOUCKS, R. G., DODGE, M. M. & GALLOWAY, W. E. 1984. Regional Controls on Diagenesis and Reservoir Quality in Lower Tertiary Sandstones along the Texas Gulf Coast. *In:* McDONALD, D. A. & SURDAM, R. C. (eds) *Clastic Diagenesis*. American Association of Petroleum Geolgists Memoir, **37**, 15–45.

McBRIDE, E. F. 1985. Diagenetic Processes that Affect Provenance Determinations in Sandstone. *In:* ZUFFA, G. G. (ed.) *Provenance of Arenites*. Reidel, Dordrecht, 95–113.

MEADE, R. H. 1966. Factors Influencing the Early Stages of the Compaction of Clays and Sands—Review. *Journal of Sedimentary Petrology*, **36**, 1085–1101.

MEZZADRI, G. & VALLONI, R. 1981. Studio di Provenienza delle arenarie di M. Cervarola (Torre degli Amorotti, Reggio E.). *Mineralogica Petrographica Acta*, **25**, 91–102.

MILLIKEN, K. L. 1988. Loss of provenance information through subsurface diagenesis in Plio-Pleistocene sandstones, northern Gulf of Mexico. *Journal of Sedimentary Petrology*, **58**, 992–1002.

MORTON, A. C. 1984. Stability of detrital heavy minerals in Tertiary sandstones from the North Sea Basin. *Clay Minerals*, **19**, 287–308.

NAGTEGAAL, P. J. C. 1978. Sandstone-framework instability as a function of burial diagenesis. *Journal of Sedimentary Petrology*, **135**, 101–105.

PETTIJOHN, F. J., POTTER, P. E. & SIEVER, R. 1973. *Sand and Sandstone*. Springer-Verlag, Berlin.

REUTTER, K. J., TEICHMÜLLER, M., TEICHMÜLLER, R. & ZANZUCCHI, G. 1983. The coalification pattern in the Northern Apennines and its paleogeothermic and tectonic significance. *Geologisches Rundschau*, **72**, 861–894.

RICCI LUCCHI, F. 1986. The Oligocene to Recent foreland basins of the northern Apennines. *Special Publication of the International Association of Sedimentologists*, **8**, 105–139.

RITTENHOUSE, G. 1971. Mechanical compaction of sands containing different percentages of ductile grains: a theoretical approach. *Bulletin of the American Association of Petroleum Geologists*, **55**, 92–96.

SAGRI, M. & ZANZUCCHI, G. 1975. Geologic setting of the Northern Apennines. *In: Examples of Turbidite Facies and Facies Associations from Selected Formations of the Northern Apennines.* IX International Congress of Sedimentology, Nice, Field Trip, **A11**, 4–11.

SCHMIDT, V. & McDONALD, D. A. 1979. The Role of Secondary Porosity in the Course of Sandstone Diagenesis. *In:* SCHOLLE, P. A. & SCHLUGER, P. R. (eds) *Aspects of Diagenesis*. Special Publication of the Society of Economic Paleontologists and Mineralogists **26**, 175–207.

SMOSNA, R. 1989. Compaction Law for Cretaceous Sandstones of Alaska's North Slope. *Journal of Sedimentary Petrology*, **59**, 572–584.

VALLONI, R. 1978. Provenieza e storia post-deposizionale del Macigno di Pontremoli (Massa). *Bollet-*

tino della Societa Geologica Italiana **98**, 317–326.

—— & ZUFFA, G. G. 1984. Provenance changes for arenaceous formations for the Northern Apennines, Italy. *Geological Society of America Bulletin* **95**, 1035–1039.

——, BELFIORE, A., BERSELLI, M., CALZETTI, L. & CALZOLARI, M. A. 1990. Moda detritica e stadio diagenetico delle arenarie del Macigno di M. Modino e di M. Cervarola affioranti sulla trasversale Sassorosso (Lucca)–Civago (Reggio E.). (In press).

WILSON M. D. & PITTMAN, E. D. 1977. Authigenic clays in sandstones: Recognition and influence on reservoir properties and paleoenvironmental analysis. *Journal of Sedimentary Petrology*, **47**, 3–31.

Sourcelands for the Carboniferous Pennine river system: constraints from sedimentary evidence and U-Pb geochronology using zircon and monazite

R. A. CLIFF, S. E. DREWERY & M. R. LEEDER

Department of Earth Sciences, University of Leeds, Leeds LS2 9JT, UK

Abstract: Petrographic and U–Pb isotopic data are presented for Carboniferous sandstones from the Pennine basin ranging from Chadian to Westphalian B in age. Variations in the petrography of these samples, particularly the greater abundance of feldspar in Namurian sandstones reported qualitatively by previous authors, are confirmed by modal analyses. The influx of coarse feldspathic sandstones into the Pennine and Southern North Sea basins in the early Namurian is attributed to a marked increase of sediment yields and discharge due to increased precipitation in the hinterlands. The same detrital heavy mineral species occur throughout the Carboniferous. Zircons are petrographically variable both in colour and degree of rounding. Two major sub-populations are recognized; pink–purple grains and colourless–brown grains. U–Pb isotopic analyses of multi-grain fractions of zircon from five of the samples show very similar age patterns; the sixth sample, a Yoredale sandstone appears to be distinct.

Single grain analyses of zircons were used to investigate one sample of late Namurian (Rough Rock) sandstone in detail. Concordant or slightly discordant results demonstrate the presence of: rounded brown grains >3.5 Ga, rounded purple grains 2.8–3.2 Ga, euhedral purple grains $\geqslant$ 2.9 Ga and clear euhedral grains *c*. 0.41–0.43 Ga.

Monazites are shown to be very suitable for single grain analysis, yielding concordant results more often than zircon. Most of the monazite in the Rough Rock and a Westphalian sandstone are Caledonian (0.41–0.43 Ga) but two individual grains have ages of 0.947 and 2.70 Ga.

The isotopic results appear to exclude large areas of the North Atlantic region as potential sourcelands and no presently exposed rocks provide a close match with the inferred sourceland geology; rocks now under the northern North sea and/or between Scotland and Greenland are possible, as yet untested, sources.

Hypotheses for the initiation, origin and evolution of sedimentary basins and crustal terranes must involve knowledge of the location, extent, nature and age of sourcelands for detritus. Provenance studies should also try to deduce the route by which detritus reaches a depositional basin from the drainage basin. It is well known that many factors other than hinterland geology may affect the modal composition of clastic rocks, particularly sandstones. These factors include the nature of contemporaneous hinterland weathering, relief, transport, diagenetic dissolution/replacement and subsequent deformation. Determination of the provenance of single detrital and sand-size grains is also made very difficult when problems of recycling are considered, such problems increasing with decreasing age of the sedimentary sequence.

Isotopic dating of detrital minerals is a valuable tool for assessing the age structure of potential sourcelands (e.g. Allen 1972; Hurford *et al.* 1984; Drewery *et al.* 1987) but the problems noted above still remain. One additional problem of radiometric dating is the interpretation of dates derived from populations of detrital minerals as distinct from single grains. Current analytical techniques enable single crystal dating, either using ion probe techniques or by conventional methods, and it is this development that will undoubtedly revolutionise the study of provenance. Results of single crystal dating of both monazite and zircon carried out in the Leeds laboratory are reported in this contribution and this new data is discussed in the light of previous results from multi-grain samples.

Throughout the Carboniferous in northern England large amounts of clastic sediment were deposited in extensional sedimentary basins by fluvio-deltaic systems. Despite considerable study of these deposits, beginning with Sorby's 1859 paper, the sourcelands for, and the evolution of, these drainage systems continues to be a major problem of North Atlantic palaeogeographical development. Figure 1 shows the present-day extent of Carboniferous deposits at outcrop, together with the large extent of pre-Permian

From Morton, A. C., Todd, S. P. & Haughton, P. D. W. (eds), 1991,
Developments in Sedimentary Provenance Studies.
Geological Society Special Publication No. 57, pp. 137–159.

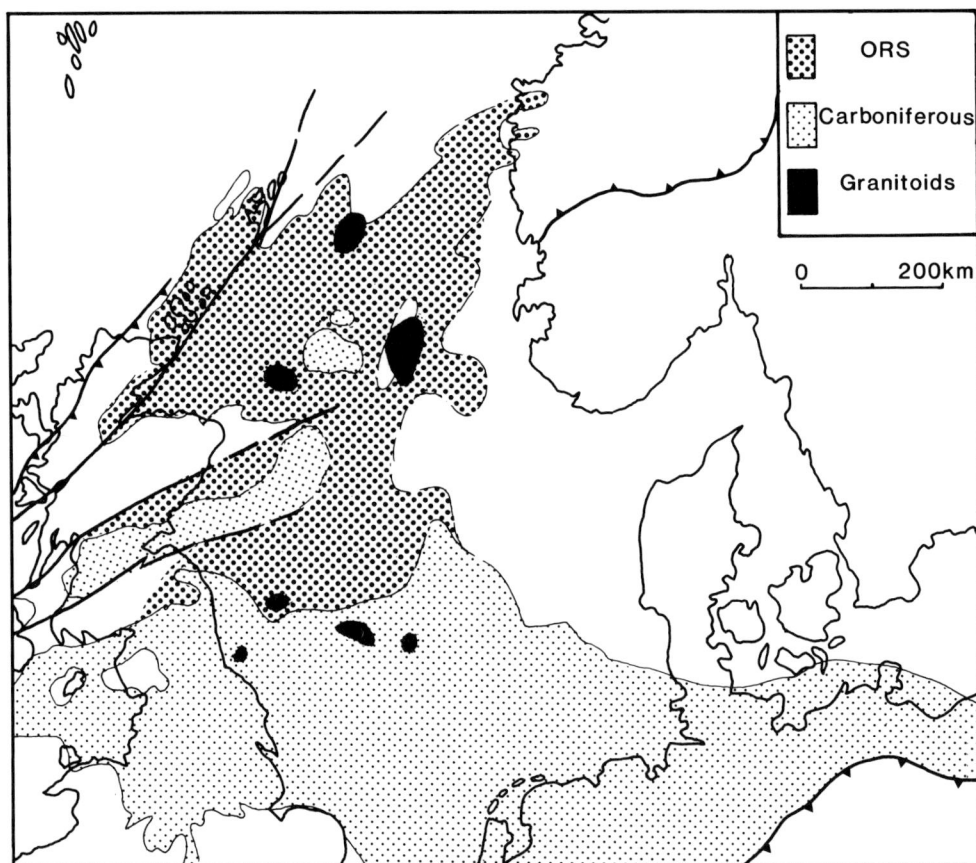

Fig. 1. Map to show the extent of Carboniferous outcrop and pre-Permian subcrop in the Pennine and North Sea basins (after Ziegler 1982; Leeder & Boldy 1990; Leeder & Hardman 1990). Also shown is the outcrop and subcrop of Old Red Sandstone (after Ziegler 1982), although it should be noted that the distribution is highly uncertain in the area on and to the east of the Shetland Platform. The distribution of buried ?Caledonian granitoid plutons largely follows the results of Donato *et al.* (1982), Zervos (1987), Leeder & Hardman (1990).

Carboniferous subcrop in the North Sea Mesozoic basin. In the present study we have sampled a total of 16 sites at outcrop in northern Britain, chosen to establish a wide geographic and stratigraphic coverage (Figs 2 & 3). Isotopic results from multigrain heavy mineral separates are presented for 7 samples from 6 of these sites (D2, D4, N4, N8, N9, W2, W3), together with single grain data from two of them. In addition petrographic data from all 16 sites are integrated with the mostly qualitative results of previous workers.

Petrographic studies

Pebble suites

Dinantian pebbly sandstones are not common and no systematic study has been undertaken of

pebble compositions. Our field observations on the Fell Sandstone of Northumberland indicate a predominantly quartzose assemblage. Namurian pebbly sandstones are widespread in the Central Pennine Basin. Although no systematic or statistical study has been carried out, the extensive observations and collecting by Gilligan (1920; confirmed by our reexamination of the Gilligan petrographic collection in the University of Leeds) reveals that quartz pebbles of various colours dominate the assemblages, with mylonitized fabrics, undulose extinction, fluid inclusions, rutile inclusions and blue/opalescent varieties commonly recorded. Less common are pebbles of (1) lustrous pink fresh microcline or microperthitic feldspars, (2) acid pegmatite, (3) acid igneous rocks including 2-mica granites, quartz and feldspar porphyries, (4) mica schists and quartz–mica schists, sometimes garnetiferous with zircon and tourmaline, (5) granitoid gneisses showing textural intergradations to (3), (6) sedimentary rocks including cherts and silicified oolites.

Westphalian pebble suites are less common than in the Namurian and no systematic study has been undertaken, although field observations indicate a predominance of quartz pebbles with a much-reduced variety of other types. Large pink feldspars occur locally.

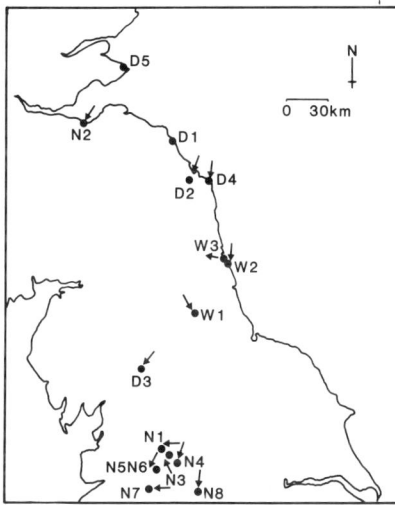

Fig. 2. Sketch map to show the location of sampling sites of fine-grained sandstones used in the present study, together with the local mean palaeocurrent direction obtained from large scale cross stratification in facies of channel origins (see also data in Table 1). D, Dinantian samples: D1 Cementstones, Burnmouth, Berwickshire (NT 962606); D2 Fell Sandstones, Belford Moor, Northumberland (NU 071326); D3 Ashfell Sandstone, Ashfell, Cumbria (SD 736427); D4 Middle Limestone Group, Beadnell, Northumberland (NU 235294); D5 Calciferous Sandstone Series, Cambo Ness, Fife (NO 612114). N, Namurian samples: N1 Skipton Moor Grits, Jenny Gill Quarry, Yorkshire (SE 00355090); N2 Joppa Sandstone, Prestonpans, Lothians (NT 383744); N3 Middleton Grit, Silsden, Yorkshire (SE 052478); N4 Earl Crag Grit, Keighley, Yorkshire (SD 990430); N5, N6 Woodhouse Grit, Haworth, Yorkshire (SE 022364); N7 Haslingden Flags, Haslingden Garth, Yorkshire (SD 758235); N8 Rough Rock, Elland Bypass, Yorkshire (SE 103213). W, Westphalian samples: W1 Quickburn Quarry Sandstone, Quickburn, Durham (NZ 080427); W2 Hartley Bay Sandstone, Tyne and Wear (NZ 345757); W3 Seaton Sluice Sandstone, Seaton Sluice, Tyne and Wear (NZ 339768).

Sandstones

Although Carboniferous sandstones have been examined petrographically for over 140 years there is very little quantitative data in the literature on composition, particularly the degree to which this is dependent upon grain size. The feldspathic nature of Namurian samples is well known, as is the low to very low feldspar content of most Dinantian sandstones. There may be a strong influence of burial history upon feldspar content since recent results from deep Southern North Sea boreholes indicate that extensive kaolinization of feldspar occurs in many samples (unpublished results of the authors). Feldspar contents can thus be very seriously underestimated, since loss of framework support renders estimates of original feldspar content highly difficult (Leeder & Hardman 1990).

Data from samples of fluviatile sandstones collected from outcrop are presented in Fig. 4. The fine-grained samples are moderately to well sorted, those from the Dinantian being generally better sorted and having more rounded grains. Quartz, including polycrystalline grains and chert, always constitutes more than 75% of the rock. The average feldspar content is 10%, never rising to above 30%. Rock fragments are rare. The sandstones are mostly sub-arkosic, exceptions being the two Dinantian samples (Fell Sandstone (D2) and Middle Limestone Group (D4)) which are quartz arenites. The quartz grains are angular to subrounded grains with the majority showing undulose extinction and abundant mineral (zircon, apatite, mica, tourmaline) and fluid inclusions. Polycrystalline quartz is rare. K-feldspar and plagioclase are present in all samples, the former being most altered and abundant (>63% total feldspar). Microline is present in only half of the samples and perthite is rare.

Data from coarse-grained Namurian samples (Fig. 4a) shows a significant increase in the proportion of polycrystalline quartz and in the proportion of microcline feldspar compared to the fine-grained samples. However, microcline is only rarely more abundant than untwinned K-feldspar, although it is usually more abundant than plagioclase. Examination of the coarse polycrystalline grains reveals schistose, grain rolled, sutured, undulose and annealed fabrics (Bristow 1987). The predominance of sutured and annealed textures indicates high temperature but low stress conditions. On average, Namurian sandstones from the Central Pennine Basin are significantly more feldspathic than either Dinantian sandstones from Northumberland or Westphalian sandstones from Durham/Northumberland (Fig. 4b).

SUB SYSTEM	SERIES	STAGES	Goniatite Zones	DIVISIONS N.ENGLAND	DIVISIONS SCOTLAND	SAMPLE NAMES	SAMPLE NOS
SILESIAN	WESTPHALIAN	WESTPHALIAN D		Upper Coal Measures	Upper (Barren) Coal Measures		
		WESTPHALIAN C		Middle Coal Measures			
		WESTPHALIAN B			Middle Coal Measures	Seaton Sluice sst	W3
						Hartley Bay sst	W2
		WESTPHALIAN A	G_2	Lower Coal Measures	Lower Coal Measures	Quickburn Quarry sst	W1
	NAMURIAN	YEODONIAN	G_1		Passage Group	Rough Rock	N8
						Haslingden Flags	N7
		MARSDENIAN	R_2			Woodhouse Grit II	N6
						Woodhouse Grit I	N5
		KINDERSCOUTIAN	R_1	Millstone Grit 'Series'		Earl Crag Grit	N4
		SABDENIAN	H			Middleton Grit	N3
		ARNSBERGIAN	E_2		Upper Limestone Group	Joppa sst	N2
		PENDLEIAN	E_1		Limestone Coal Group	Skipton Moor Grit	N1
DINANTIAN	VISEAN	BRIGANTIAN	P_2	Middle Limestone Group	Lower Limestone Group	Yordales sst – Fife	D5
			P_1	Lower Limestone Group	Upper Oil-Shale Group	Yordales sst – Beadnell	D4
		ASBIAN	B	Scremerston Coal Group			
		HOLKERIAN		Fell Sandstone Group	Lower Oil-Shale Group		
		ARUNDIAN				Ashfell sst	D3
		CHADIAN				Fell sst	D2
	TOURNASIAN	COURCEYAN	Pe — Ga	Cementstone Group	Cementstone Group	Cementstone sst	D1

Fig. 3. Summary stratigraphic chart of the north British Carboniferous to show the stratigraphic position of samples used in the present study.

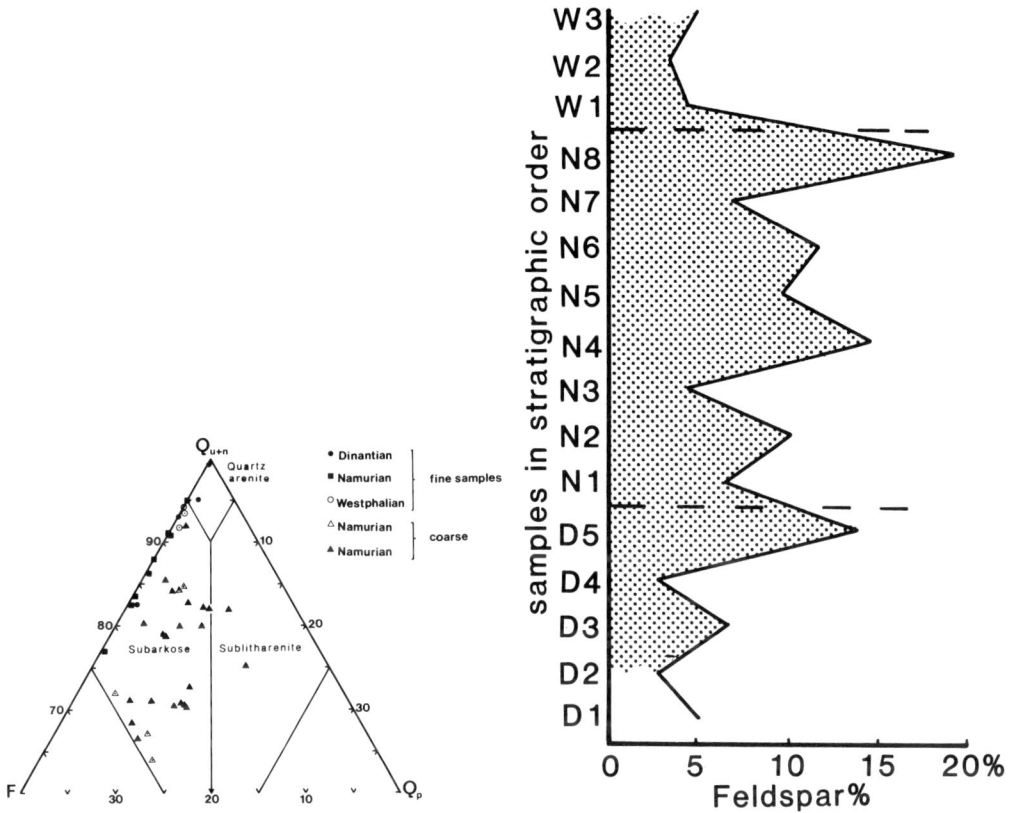

Fig. 4. (a) Part of the apex of the ternary diagram for undulose + normal quartz (Qu + n), polycrystalline quartz (Qp) and feldspar (F), together with the plots of 16 fine-grained samples from the present study, five coarse-grained samples from the Gilligan/Walker collections, University of Leeds and 21 coarse-grained samples of Rough Rock collected and analysed by Bristow (1987). The fields within the ternary diagram are those conventionally defined (Pettijohn *et al.* 1972), with Qp being part of total Q and the relevant apex being other lithic fragments. The majority of samples are clearly subarkosic in composition. The form of the present plot clearly separates the fine and coarse samples, the latter enriched in polycrystalline quartz and in feldspar.

(b) The stratigraphical distribution of feldspar abundance for the 16 fine-grained samples in the present study. Note the clear but irregular trend towards feldspar enrichment in the Namurian samples.

Heavy minerals

Previous workers, chiefly Gilligan (1920), have given much attention to the heavy minerals of Namurian sandstones and their pebbles, although the data is of a qualitative nature only. The granite pebbles noted above were found to contain rutile; euhedral, commonly zoned, zircon; monazite and apatite. The schistose pebbles contained garnet, zircon and tourmaline. Separations from the sand-sized fractions yielded garnets (usually almandine and commonest in the coarser sandstones), tourmaline and monazite. The latter mineral is uncommon but ubiquitous through the Namurian according to Gilligan (1920).

In our current studies we have undertaken detailed studies of heavy minerals in six fine-grained samples, two each from the Dinantian, Namurian and Westphalian. Mineral mounts of the total heavy population (after separation using bromoform) were counted for the most common heavy minerals (zircon, tourmaline, rutile, garnet, apatite and monazite). The data are shown in Table 1, but definite conclusions concerning the quantitative geographical and stratigraphical distribution of these species cannot be made at the present time because of the small sample population and because of the wide variations in relative and absolute abundance that evidently characterize even closely spaced stratigraphic sections (see also Gilligan's remarks on this subject, 1920, p. 268). Local hydraulic sorting may explain some of these

Table 1. *Point counted data for the relative abundance of the main heavy mineral species. Figures are a percentage of the total translucent species present minus mica and unidentified minerals*

Sample No.	% of total (−mica–opaques–others)						ZTR index	Wt % heavies*
	Zr	To	Gt	Ru	Ap	Mo		
D2	27.5	57.5	0	12.6	0.6	1.8	0.98	0.03
D4	27.0	4.1	51.3	8.6	7.5	0.4	0.39	0.18
N4	75.6	5.0	0.4	16.3	0.8	0.4	0.97	0.07
N8	70.8	3.4	12.9	7.6	4.2	tr	0.82	0.09
W2	41.8	6.2	13.0	10.7	26.6	0.6	0.59	0.13
W3	0.6	4.6	88.3	4.8	1.3	tr	0.10	0.12

* Weight % total heavy mineral population in sample; tr, trace.

local variations, particularly wave-inducing sorting of mouth bar beach sands and sorting in channel-floor lag deposits (Myers & Bristow 1989).

Since zircon, and monazite are the main mineral phases used for radiometric dating we present more detailed notes on their characteristics below.

Zircon. Similar zircon types occur throughout the Carboniferous. Most grains are small, being <105 μm and many are <75 μm. Each population includes grains from two colour groupings: (1) 'colourless'–pale yellow–yellow-brown–pale brown; (2) 'colourless'–pale pink–purple (hyacinth). Group 1 are usually commonest, particularly in the small fractions. In sample D2, and to a lesser extent W2, yellow grains are comparatively more common and may be considered as a separate group from the brown zircons. In the Yoredale sample (D4) Group 2 zircons are rarer.

Regarding shape, the majority of Group 1 zircons are euhedral, particularly the finer fractions, whilst the majority of Group 2 are well rounded. The distinctive yellow grains of D2 are generally well rounded. Grains with outgrowths, parallel growths and internal zoning are comparatively rare although distinct cores are quite common among Group 1. Minute inclusions are particularly common within group 1 grains. Euhedral zircon crystal shapes (see Pupin 1980) in our samples resemble subtypes Q3 (or ?S11 and ?S18) and imply crystallization temperatures of 750°C and 800°C respectively.

Monazite. This mineral was originally discovered by Gilligan (1920) in many samples from the Namurian of the Pennines. The mineral is characteristically present as a primary phase in acid plutonic rocks, having been found in 85 out of 178 specimens of granites from Scotland by

Mackie (1928). It has been identified optically in all of our samples and confirmed by electron microprobe analyses of samples D2, N4, N8, N9 & W3. It occurs as small, well-rounded yellow grains but is never abundant (<10% of total heavy fraction. However, recent results of detailed logging with a handheld natural gamma ray spectrometer (Myers & Bristow 1989) at locality N8 has revealed extremely high thorium concentrations from one horizon. Heavy mineral separations from the horizon (sample N9), probably a wave-reworked spit-beach facies, has revealed a high concentration of zircon, monazite and other heavy minerals.

Petrographic evidence for provenance

From his petrographic studies, Gilligan (1920, p. 281) deduced that the sourcelands for the Pennine river and delta lay 'in what is now the North of Scotland, and its extension lay east within a larger Scandinavia, thus forming part of a great continental land, the limits (northern and western) of which it is not at present possible to define'. These hinterlands were deduced to have been composed of acid granite-gneiss intruded by pegmatites, unaltered granites, quartz–mica–garnet schists, chertified limestones and various other clastic sedimentary rocks.

Gilligan's palaeogeographic map (1920, p. 280; reproduced in our Fig. 5) shows a large delta issuing into the Pennine province from the northeast, with tributaries draining named areas of hinterland in the Scottish Caledonides; in the Torridonian of the NW of Scotland and the Christiana region of southern Norway. Gilligan deduced that a lengthy period of chemical weathering in Lower Carboniferous times 'would result in a thick cover of rotted rock being left over the low ground and along the river-courses of the old continent' (Gilligan 1920, p. 282). This leached material was stripped off by the late Dinantian Yoredale rivers and the

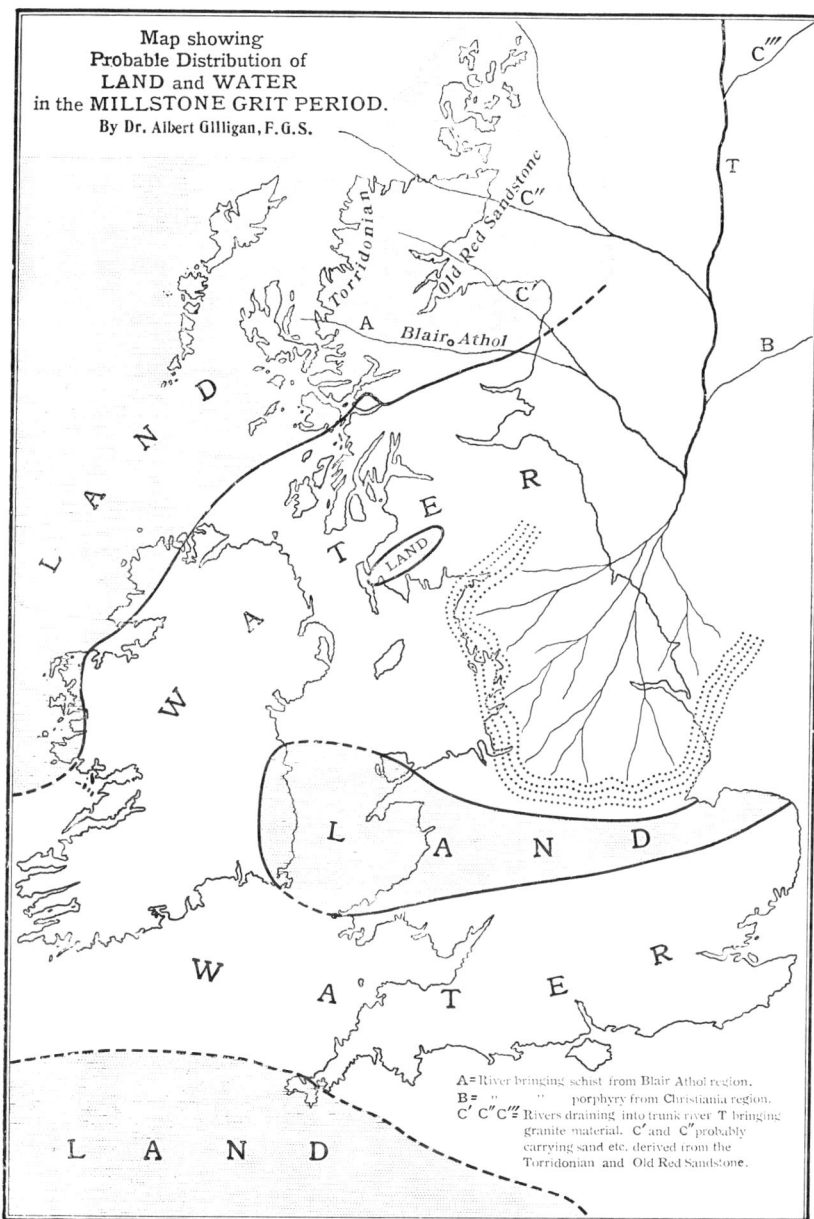

Map showing
Probable Distribution of
LAND and WATER
in the MILLSTONE GRIT PERIOD.
By Dr. Albert Gilligan, F.G.S.

A = River bringing schist from Blair Athol region.
B = " " porphyry from Christiania region.
C′ C″ C‴ Rivers draining into trunk river T bringing
granite material. C′ and C″ probably
carrying sand etc. derived from the
Torridonian and Old Red Sandstone.

Fig. 5. Gilligan's (1920) pioneering map showing the postulated course of the great Pennine river, its delta and certain exactly located tributaries draining parts of northern Scotland and Scandinavia. Note that the trunk river carries on northwards towards the Shetland/Scandinavian 'gap', scarcely diminishing in emphasis. It is this latter intuition which finds support from the isotopic results presented in this paper.

remaining 'fresh supply of unleached rock' was subject to great regional uplift to form the sourceland for the Millstone Grit. In Gilligan's words 'it would indeed seem to me that the relief of the land must have been of Himalayan grandeur' (1920, p. 283). This postulated erosion of fresh hinterland was thought to be aided by rapid mechanical weathering and by a monsoonal climate.

Gilligan's hypothesis for the provenance of Carboniferous sediments was amplified by Greensmith (1965) who noted that the Lower

Carboniferous fluvio-deltaic deposits of the eastern Midland Valley were derived from a sediment-mantled provenance to the north and north-east, as witnessed by the stable, sub-rounded heavy mineral population and the low percentage of feldspar (< 3%). The Yoredale sandstones of the northern Pennine province are generally fine grained, have a relatively high percentage of unstrained quartz (Butterfield 1938) and have a similar heavy mineral composition to the Midland Valley sandstones. The occurrence of a few garnets indicated to Greensmith (1965, p. 379) that the provenance was still sedimentary, but with an increasing proportion of low-grade metamorphic and igneous rocks being uncovered. The great change observed in Namurian clastic compositions discussed previously was attributed to uplift at source, probably a phase of the Sudetic earth movements. By Westphalian times Greensmith deduced that hinterland relief was reduced considerably, but still with a metamorphic and igneous source for most of the clastic materials, but with the proportion of feldspar much reduced. Greensmith deduced that the same drainage system existed throughout the Carboniferous, sourced in the present-day Central North Sea area, and his palaeogeographic sketches showing a gradual southerly progradation of Gilligan's great fluvio-deltaic system with time.

Our petrographic studies of both sandstones and heavy mineral separates add little to the general conclusions reached by Gilligan (1920) concerning the palaeogeology of the Carboniferous hinterlands. However, our isotopic results presented and discussed below suggest new age constraints upon the geographic location of these hinterlands.

Our early Dinantian sample (Fell Sandstone, D2) is a mature quartz arenite with a relatively impoverished suite of heavy minerals and includes a high proportion of rounded zircon grains compared to the other samples. The immediate source rocks for at least some part of the detritus are thus sedimentary rocks, a conclusion similar to that reached by Greensmith (1965, p. 380). The relative abundance of heavy minerals within the mature Yoredale sandstones is highly variable. Thus our sample D4 has abundant garnets (Table 1), whereas previous studies have indicated generally low abundances (Butterfield 1939).

On balance the significantly higher percentage of feldspar in the Namurian sandstones, together with a greater proportion of subangular to angular grains implies that these deposits are relatively immature. The increase in the proportion of feldspars in the Namurian could be due to:

(1) a decrease in the intensity of hydrolytic weathering during the Namurian, due perhaps to a climatic or relief change;

(2) a change of source to an area of rocks with a higher feldspar content and/or a higher content of more stable feldspar;

(3) a change in the relative proportions of different source rock types in the same source area;

(4) a change in the transport or depositional environment;

(5) a decrease in intensity of diagenetic alteration.

We have no reason to think that alternatives (4) and (5) are likely since our sampling strategy was designed to avoid such problems. Our isotopic data (see below) indicate that there is no change in the age of source during the Carboniferous. The Westphalian samples in our study, and previous results summarised by Greensmith (1965), contain significantly less common feldspar than the Namurian. However, the common apatite and garnet in samples W2 and W3 respectively suggest that this decrease is probably not due to a significantly more 'mature' hinterland being exposed to erosion at this time. A change in the relative contributions of different source rocks in the same area(s), combined with a change of climate is a more attractive alternative. This conclusion is amplified below after we have presented our isotopic results.

Palaeocurrents and Palaeoclimates

Palaeocurrent studies

Regional palaeocurrent data obtained from large-scale bedform structures within fluvio-deltaic channel facies gives clear insights into palaeoslopes within depositional areas of sedimentary basins. However, such data may provide misleading indications of drainage basin provenance and the transport route taken by the detritus if extrapolated 'upcurrent' since large river systems rarely flow from source to sea in a linear fashion.

The majority of onshore Carboniferous palaeocurrent data (Figs 2 & 6) indicates flow within the depositional area from the quadrant 0–90°. We have no palaeocurrent data from offshore wells that penetrate the Carboniferous section, but the persistence of characteristic facies over the Mid-North Sea High and Southern North Sea Basin may indicate that these trends persist out to at least longitude 3°E (Leeder & Hardman 1990). Secondly, it is apparent that there was a gradual southward spread of fluvio-deltaic sedimentation with time (Fig. 6). Thus early- to mid-Dinantian rivers dominated sedimentation only in the northeast, in the Northumberland Basin, mostly flowing as axial drainage from the N and NE (e.g. Fell Sandstone). Many locally sourced drainage systems (e.g. Basement Series of Alston/Cumbria; Whita/Annan Sandstones of the Southern Uplands/Scottish Borders) may be related to local tectonic slopes in the footwalls and hangingwalls of rotated tilt blocks (Gawthorpe et al. 1989). Later S to SE flowing Dinantian rivers and delta lobes periodically advanced across the Alston/Stainmore/Askrigg Blocks, depositing the characteristic Yoredale cycles.

During Namurian times there is clear evidence both in the onshore and offshore successions for the gradual southerly spread of particularly coarse grained fluvio-deltaic sediments as the syn-rift sites of marine basinal deposition were gradually infilled during the thermal 'sag' phase of subsidence. Studies in the offshore (Leeder & Hardman 1990) and palaeocurrent data from the East Midlands and Derbyshire (Jones 1980; Steele 1987) strongly indicate that the large rivers responsible for these deposits, particularly in the late Namurian, flowed in from the east. By Westphalian A/B times a pattern of palaeoflow from the N to NE was widely established in the Pennine depositional basin. There are strong indications from several wells in the Southern North Sea Basin that more proximal fluviatile sandstones occur eastwards out to the eastern side of Quadrant 44, close to the median British/Netherlands sector line.

Size of drainage basin and palaeoclimate

Hydraulic geometry requires that the size of a river channel should be adjusted to the magni-

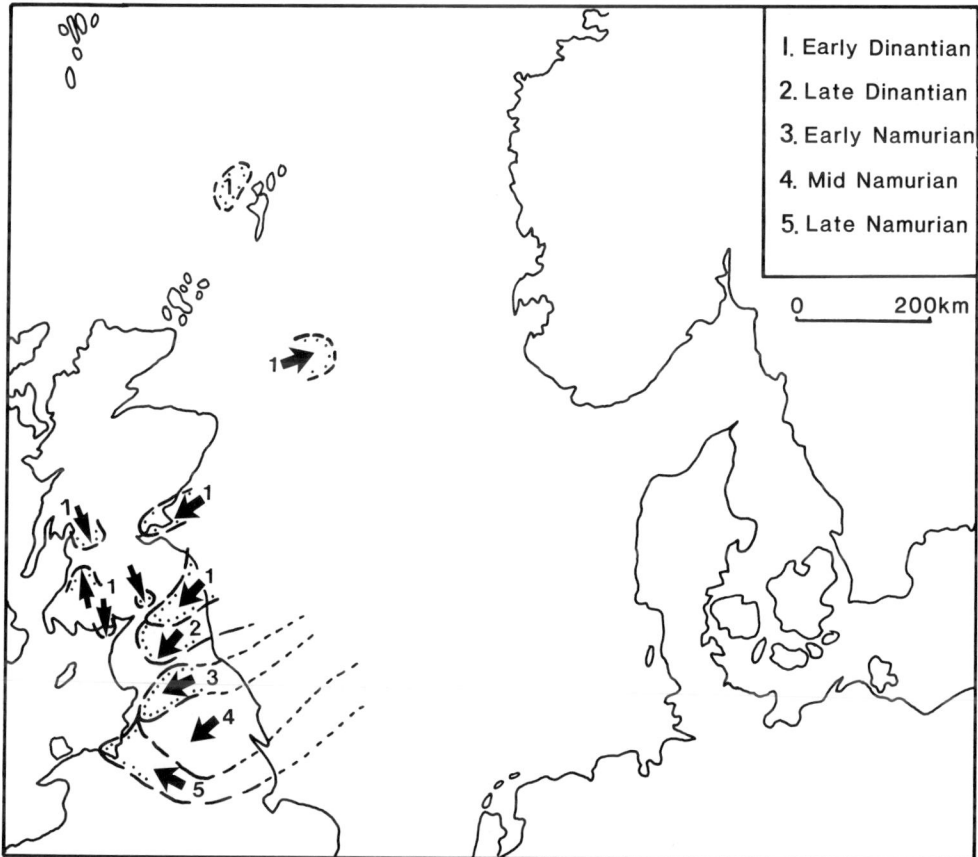

Fig. 6. Map to show the location of successive Carboniferous fluvio-deltaic 'lobes' in northern Britain, the ornament shown corresponding to areas newly affected by fluvio-deltaic deposition. Note the gradual southerly migration of the locii of fluvio-deltaic activity. The arrows represent a regional generalization of the dominant palaeocurrent trends for the area and interval in question.

tude of the discharge that passes through the channel. Since this discharge is a direct function of the drainage basin area then it follows that large channels should have large drainage basin areas. The regression equations that express these relationships have relatively large standard errors and differ according to climate and other independent variables (Leopold *et al.* 1964). One way forward, as applied by Simon & Bluck (1982) to Old Red Sandstone river deposits, is to make use of the empirical relationship (Hack 1957; Leopold *et al.* 1964) that expresses the relationship between channel length (L) to drainage basin area (Ad) by

$$L = 1.4Ad^{0.6}$$

This approach depends on independent evidence for L. In the present case it would seem that this was at least 750 km (see discussion below). Values for Ad of at least $3.5 \times 10^4 \, \text{km}^2$ seem possible.

Although some early progress was made in the study of the palaeohydrology of Upper Palaeozoic channels in the UK (Leeder 1973; Elliott 1976) there has been little attempt to apply such results to the problem of provenance. Several problems arise. One is the difficulty of deciding whether a sandbody was deposited by a fluvial trunk stream or a distributary. Another is the identification of channel morphology and magnitude from outcrop data. Thus a channel can have a large cross sectional area in two ways, either by being very wide and relatively shallow or very deep and relatively narrow. These alternatives are exceedingly difficult to distinguish, particularly in the kinds of multistorey and multilateral sandbodies that characterise much of the Carboniferous of the Pennine Basin. Finally, even supposing that maximum and minimum estimates of drainage basin size could be made then the problem of determining the direction and shape of the tributive network still remains.

Despite the above problems it is still possible to contemplate a comparison of certain Namurian palaeochannels with modern rivers, using the analysis of McCabe (1977) who describes in-channel bars from low-sinuosity channels with slip face deposits that imply channel depths of over 30 m. Assuming McCabe's interpretation is correct, and there are grounds for thinking this is not so (A. Simms pers. comm. 1989), such very large channel systems seem to be restricted to the Namurian, those of the Dinantian and Westphalian being smaller, usually a maximum of around 10 m deep. Thus along with increased grain size and feldspar content (see previous discussion)

it would appear that channel size and hence discharge also reached a maximum in the Namurian.

It is tempting to attribute the above changes to a major climatic shift in the hinterlands of the Pennine drainage system. Support for this inference comes from evidence for a major expansion of the Gondwanan ice sheet in the early Namurian (see Powell & Veevers 1987) which would certainly have influenced climatic conditions at low latitudes. There is abundant evidence for semi-arid to subhumid climatic conditions in Dinantian depositional areas of the British Isles (Besly 1987). An increase in precipitation in early Namurian times would have led to an increase in sediment yields from the unvegetated hinterland uplands as predicted by Schumm's (1968) general model of pre-Mesozoic drainage basin behaviour. It is possible that a significant geomorphic 'threshold' (Schumm 1977) may have been triggered by such a climatic change, leading to a 'flushing-out' of the coarse feldspathic detritus that is so characteristic of the Namurian sandstones of the Pennine basin. It must be assumed that the increase of precipitation was sufficient to cause a marked increase of erosion rates without increasing the degree of feldspar dissolution by hydrolytic weathering.

Isotopic studies

Among the detrital heavy minerals in Carboniferous sediments, many are potentially able to provide clues to their provenance through radiogenic isotope studies. The work reported here focusses on U–Pb measurements on zircons and to a lesser extent monazite; the same system can potentially be applied to rutile and possibly garnet and the latter mineral also has potential for Sm–Nd analysis.

The interpretation of U–Pb isotopic data is generally complicated by discordance between the ages calculated using the two independent decay schemes of ^{238}U to ^{206}Pb and ^{235}U to ^{207}Pb. For detrital samples such discordance is a combination of the effects of mixing of, possibly concordant, zircon of different ages and disturbance of the uranium-lead isotope systems at some time after crystallisation, usually resulting in loss of radiogenic daughter lead. The systematics of discordant uranium/lead systems are reviewed by Gebauer & Grünenfelder (1979). It is important to recognize that even single zircon crystals may represent mixtures reflecting crystallization from granitic magmas onto inherited nucleii or multiple crystallization during polymetamorphism (Pidgeon & Aftalion 1978; Black *et al.* 1986).

Analytical methods

Analytical details for the multi-grain zircon analyses are given in full in Drewery (1987); these samples were totally spiked with a $^{202}Pb-^{235}U$ spike in the early stages of the work and later with a $^{202}Pb-^{236}U-^{233}U$ spike, which was also used in the single-grain zircon and monazite analyses. Samples were decomposed in batches of 3, each batch accompanied by a blank, using 60% HF for zircons and 6M NCl for monazites.

Separations were made using a miniaturization of the method of Krogh (1973). For the multi-grain samples a 50 μl resin bed was used whereas single grains were processed on 15 μl columns. Zircon solutions were separated in HCl while the monazites were processed using HBr. Normal blank levels were 50–100 pg but were subject to major fluctuations; however the blank compositions were measured with each batch of three samples and their uncertainty is fully reflected in the propagated analytical errors. Lead (loaded with silica gel and phosphoric acid) and uranium (loaded with colloidal graphite) were analysed on a VG Micromass 30 mass spectrometer using Faraday Cup or Daly collectors for beams greater or less than 0.5 picoamps, respectively. Mass fractionation for lead was monitored by frequent analysis of SRM 981 and SRM 983 and the raw data were corrected by 1.3 ± 0.5‰ per mass unit for data measured on the Faraday Cup and 4 ± 1‰ per mass unit for data measured on the Daly Detector. Use of the uranium double spike allowed direct correction for mass fractionation in the uranium runs. The common lead was assumed to be completely dominated by the blank and the measured blank composition for each batch of samples was used to calculate the radiogenic lead isotopic compositions for that batch. Ages were calculated using the constants recommended in Steiger & Jäger (1977).

Analytical uncertainties and correlations in the concordia parameters were calculated taking full account of individual error contributions and their correlations, using a numerical error propagation program written by R. Powell. Errors calculated using this procedure were cross-checked against a less complete analytical error propagation procedure courtesy of F. Oberli.

Results

Data for milligram-sized zircon fractions from two Namurian samples have been reported previously (Drewery et al. 1987). Here we extend the stratigraphic coverage of the Carboniferous by presenting data on zircons from Dinantian and Westphalian sandstones (from Drewery 1987). In addition a second sample from the Rough Rock has been investigated, making use of further technical developments which allow the analysis of individual zircon grains. Some of these grains have been abraded in the laboratory (Krogh 1982) in order to minimize discordance of the analysed grains. Monazite has particular potential for provenance studies. It tends to have

uranium concentrations at the high end of the zircon range and in contrast to zircon it is generally concordant. Thus monazites are highly suited to provide reliable ages on single detrital grains. Data, in part on single grains, from three samples are presented here. Monazite crystallizes during both magmatic and metamorphic events and, for detrital grains, it will inevitably remain uncertain which alternative applies to a particular grain.

Zircon

The results of the zircon isotopic analyses are presented in Table 2 and also presented on concordia diagrams in Figs 7–10. A wide range of ages has been found, ranging from Lower Palaeozoic to early Archaean. The interpretation of the data is best considered first in relation to the Rough Rock for which both single-grain and multi-grain data are available. For the multi-grain analyses (Drewery et al. 1987) nine 0.4–1.9 mg fractions were hand-picked to give homogeneous samples representing distinct colours, morphologies and sizes from the total heterogeneous population. The dominant colourless-pale brown, generally euhedral, zircons show a 'reverse' discordance pattern extending away from a Lower Palaeozoic lower intercept (Fig. 7a), interpreted by Drewery et al. (1987) in terms of Caledonian granites containing zircons with an inherited PreCambrian component. Pink–purple zircons, although highly discordant, have Archaean $^{207}Pb/^{206}Pb$ ages with a pattern of discordance that led Drewery et al. to suggest a primary age close to 3.0 Ga.

The new data presented here comes from 17 individual grains from the second sample of Rough Rock, described above. Six of these grains belong to the pink–purple subpopulation, three euhedral and three rounded. All six were abraded in the laboratory before analysis. The three rounded grains (analyses 35–37 in Table 2) are all more concordant than any of the multi-grain analyses (Fig. 7b): two are almost concordant at 2.8 Ga whilst the third has a significantly older $^{207}Pb/^{206}Pb$ age of 3.2 Ga but is 18% discordant. The euhedral grains are no less discordant than the multi-grain fractions; taking data for all three grains together, they are compatible with a single episode of Palaeozoic lead-loss from 3.0 Ga zircons. Analyses 41–43 are on turbid brown rounded grains, unlike other zircons in the population. Grain 43 is concordant at 2.8 Ga whereas the other two are much older, with $^{207}Pb/^{206}Pb$ ages of 3.52 and 3.55 Ga, grain 41 plotting just above concordia and grain 42 being 7% discordant.

Table 2. *Isotopic analytical data*

Sample Size (µm)	Wt mg	^{238}U	^{206}Pb	^{206}Pb/^{204}Pb[†]	^{207}Pb/^{206}Pb	^{207}Pb/^{235}U	^{206}Pb/^{238}U
Dinantian							
D2, Fell Sandstone, Belford Moor							
Zircons, purple, rounded:							
1 125–105	1.32	1409	397	3333	0.1346	5.173	0.2787
2 150–125	1.22	1082	373	3188	0.1573	7.472	0.3445
3 210–150	1.11	1031	393	4042	0.1715	8.994	0.3804
euhedral:							
4 105–75	1.32	1686	609	13928	0.1840	9.099	0.3586
Zircons, brown, rounded:							
5 125–105	1.29	1247	276	1429	0.1249	3.761	0.2184
elongate, subhedral:							
6 105–75	1.42	1754	379	3724	0.1575	4.632	0.2133
euhedral Type I:							
7 105–75	1.05	1522	285	2486	0.1613	4.189	0.1884
euhedral:							
8 125–105	1.52	1986	398	14663	0.1613	4.429	0.1991
9 150–125	1.01	1044	216	1180	0.1606	4.471	0.2019
Zircons, yellow, rounded:							
10 210–150	1.46	1012	319	2284	0.1646	7.050	0.3106
D4, Yoredale Sandstone, Beadnell							
Zircons, purple, rounded:							
11 105–75	0.96	1154	252	4277	0.1052	3.178	0.2191
12 125–105	1.08	1366	294	5131	0.1094	3.254	0.2157
13 125–105 m	0.61	1097	207	3337	0.1029	2.649	0.1867
Zircons, colourless, euhedral							
14 105–75	1.28	1850	211	2706	0.0887	1.384	0.1132
Zircons, brown, euhedral							
15 125–105	0.92	1059	174	2709	0.1000	2.260	0.1639
Namurian							
N4, Earls Crag Grit, Keighley Moor							
Zircons, purple, rounded:							
16 105–75	1.22	1394	438	3600	0.1501	6.482	0.3132
17 125–105	1.71	1775	581	3846	0.1563	7.049	0.3271
18 150–125	1.37	1209	452	6068	0.1729	8.828	0.3700
euhedral Type I:							
19 125–105	1.10	1548	587	3321	0.1916	9.989	0.3774
Zircons, colourless euhedral Type I:							
20 <75	1.04	1691	141	2924	0.0826	0.951	0.0835
brown euhedral, Type I:							
21 125–105	1.98	2529	279	2053	0.1147	1.730	0.1094
22 150–125	1.38	1705	194	3438	0.1083	1.705	0.1142
N8, Rough Rock, Elland							
Zircons, purple rounded:							
23 105–75	1.54	1741	516	253	0.1489	5.67	0.2760
24 125–105	1.90	1621	526	1432	0.1523	6.77	0.3226
25 210–150	1.03	766	275	5590	0.1658	8.18	0.3580
26 250–210	0.39	192	87	3954	0.1981	12.32	0.4510
Zircons, purple euhedral Type I:							
27 125–105	0.45	459	190	5942	0.1997	11.36	0.4125
Zircons, colourless, subhedral:							
28 150–125	1.02	1016	109	857	0.1001	1.446	0.1049
Zircons, brown euhedral, Type I:							
29 105–125	1.76	1028	94	270	0.0603	0.690	0.0830
30 210–150	1.12	1355	196	3820	0.1352	2.701	0.1449

Table 2. Continued

Sample Size (μm)	Wt mg	^{238}U	^{206}Pb	$^{206}Pb/^{204}Pb$†	$^{207}Pb/^{206}Pb$	$^{207}Pb/^{235}U$	$^{206}Pb/^{238}U$
Zircons, brown euhedral, Type II:							
31 210–150	0.74	1110	187	2270	0.1578	3.603	0.1656
Monazite							
32 yellow	0.05	2757	35	45	0.07061	0.727	0.07467
33 yellow	0.02	466	36	359	0.05661	0.547	0.07006
34 turbid	0.05	282	26	105	0.05363	0.517	0.06997
N9, Rough Rock, Elland							
Zircons							
35 pu.ro.,abr.S		7.8	5.11	92	0.1979	14.58	0.5344
36 pu.ro.,abr.S		6.7	4.99	59	0.1966	14.32	0.5282
37 pu.ro.,abr.S		2.02	2.36	32	0.2544	17.99	0.5127
38 pu.e.,abr.S		6.7	4.33	61	0.2105	13.46	0.4638
39 pu.e.,abr.S		10.0	3.55	1325	0.1922	9.315	0.3515
40 pu.e.,abr.S		3.88	1.80	97	0.2066	11.18	0.3925
41 br.S		20.3	16.6	126	0.3140	32.19	0.7436
42 br.S		18.0	12.1	3088	0.3150	29.02	0.6680
43 br.S		11.9	0.707	95.2	0.1940	14.02	0.5241
44 cl.e.S		9.0	0.69	179	0.0569	0.541	0.0691
45 cl.e.S		10.4	0.88	76	0.0426	0.382	0.0650
46 cl.e.S		22.7	1.62	224	0.0525	0.475	0.0657
47 cl.e.S		2.28	0.298	53	0.0535	0.417	0.0564
48 cl.e.S		33.0	2.26	452	0.0552	0.498	0.0655
49 cl.e.S		19.7	1.46	982	0.0552	0.514	0.0675
50 cl.e.S		15.9	1.11	1083	0.0568	0.535	0.0683
51 cl.e.R		5.34	0.36	686	0.0517	0.467	0.0655
52 cl.e.C		2.22	0.26	156	0.0944	1.35	0.104
Monazite							
53 S		59	30.5	6299	0.1869	13.43	0.5210
54 S		309	21.6	5789	0.05532	0.5345	0.07008
55 S		442	30.1	29440	0.05494	0.5199	0.06864
Westphalian							
W2, Hartley Bay							
Zircons, purple, rounded							
56 <75 nm2	1.19	1550	468	4307	0.1475	6.119	0.3009
57 105–75 nm2	0.84	1094	298	3565	0.1403	5.274	0.2727
Zircons, purple, euhedral							
58 105–75 nm3	0.52	846	230	3973	0.1690	6.270	0.2691
Zircons, colourless, euhedral:							
59 <75 nm2	0.75	1147	113	428	0.0952	1.254	0.0955
60 <75 nm3	0.81	1273	122	579	0.0912	1.174	0.0934
W3, Seaton Sluice							
Zircons, purple, rounded							
61 105–75 nm10	1.19	1560	462	2750	0.1631	6.598	0.2934
62 125–105 nm10	1.35	1469	452	6676	0.1661	7.015	0.3063
Zircons, purple, euhedral							
63 105–75 nm10	1.12	1981	560	1252	0.1911	7.323	0.2779
Zircons, brown, rounded							
64 105–75	1.21	3244	429	3017	0.1364	2.481	0.1319
euhedral, Type I							
65 125–105	1.85	2892	341	2628	0.1254	2.020	0.1168
Monazites							
66 clear, 15 grains		1964	146	3294	0.06006	0.6134	0.07406
67 30 grains		3159	228	1984	0.05817	0.5735	0.07150
68 S		975	155	5875	0.07066	1.543	0.1584

m, magnetic grains; abr., abraded grain; S, single grain; C,R, core and rim of a single grain; † raw ratio; pu, purple; br, brown; cl, colourless; e, euhedral; ro, rounded.

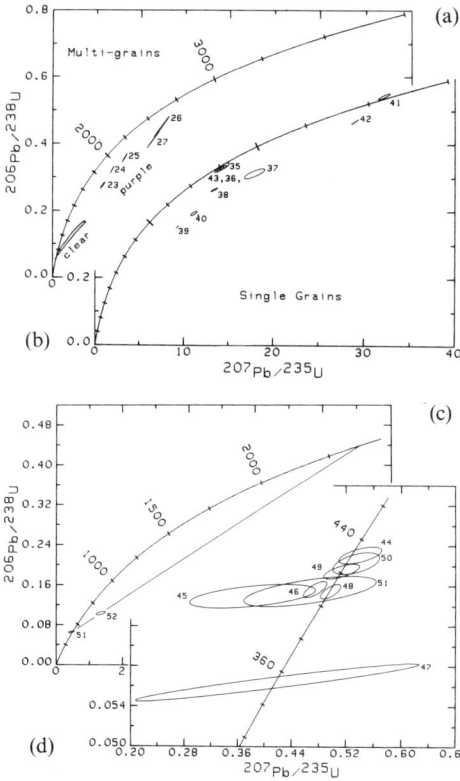

Fig. 7. Concordia diagram summarising U–Pb isotopic data for zircons from the Rough Rock at Elland. Numbers refer to the analysis number in Table 2.

(a) Multi-grain data from Drewery *et al.* (1987); only a general data field is shown for the clear–pale brown fractions, individual data points are shown in Fig. 9.

(b) Single grain data for purple and turbid brown zircons.

(c) Data for the separated core and rim of a single clear zircon, demonstrating the presence of inherited cores in some Caledonian zircons.

(d) Single grain data for clear–pale brown euhedral zircons. Data in Figs 7–10 are shown as 2σ error ellipses.

The new data on the pink–purple sub-population confirm its Archaean age but, in detail, suggest a somewhat different interpretation of the earlier, more discordant data. It is now evident that the purple zircons include grains of different primary age ranging at least from 2.8 to 3.2 Ga.

The remaining single zircon data are for grains selected from the dominant clear–pale brown euhedral zircons (grains 44–52). These zircons generally give Caledonian ages and consequently they contain much less radiogenic

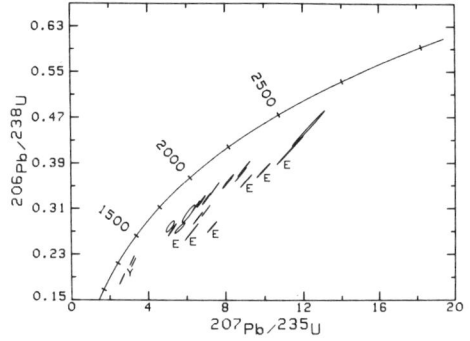

Fig. 8. Concordia diagram summarising U–Pb isotopic data on multi-grain fractions of purple zircons from six Carboniferous sandstones ranging in age from Dinantian (Chadian) to Westphalian B, including data from Drewery *et al.* (1987). The euhedral fractions are labelled 'E' while the distinctive Yoredale samples are labelled 'Y'.

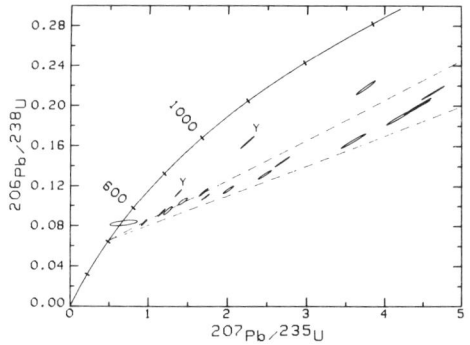

Fig. 9. Concordia diagram showing U–Pb isotopic data on colourless–brown zircons from the same six samples as Fig. 8. Two fractions from the Yoredale sandstone (D4) are labelled 'Y' and appear to have a younger, Proterozoic inherited component.

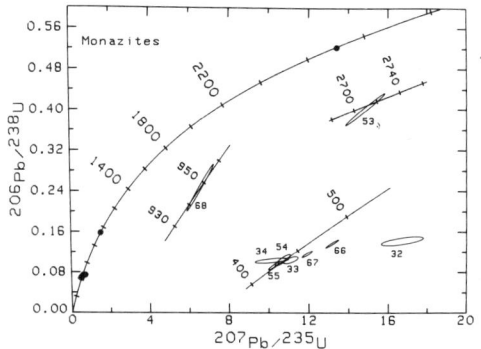

Fig. 10. Concordia diagram summarizing U–Pb isotopic data on monazites from two Carboniferous sandstones, the Rough Rock (N8, N9) and the Seaton Sluice sandstone (W3). The enlarged insets show more clearly the concordance of most of the monazite data.

lead than the Archaean grains discussed above, down to 0.3 picomoles ^{206}Pb. As a result the errors on the ages are greater; the ^{206}Pb/^{238}U ages still have useful precision but it is not possible to determine the degree of discordance with the usual precision. Four grains (44, 48, 49, 50) plot within error on concordia (Fig. 7c) between 410 and 435 Ma. Grains 45 and 46 have ^{206}Pb/^{238}U ages close to 410 Ma but plot above concordia; in the former case the apparent ^{207}Pb/^{206}Pb ratio is physically unattainable (negative age) indicating a problem with the common lead correction of this sample. These data are tentatively grouped with the other 410–435 Ma grains.

During hand-picking a single $c.$ 100 μm grain was noted with an obvious zoned structure. A central core was surrounded by an outer rim, part of which had broken away to reveal the core. This grain was broken further, using tweezers and a fine needle under the microscope, to produce three small fragments from the rim and a larger fragment volumetrically dominated by the core. The three small fragments were combined for analysis 51 while the core provided analysis 52. Both contained extremely small amounts of lead (0.36 and 0.26 picomoles ^{206}Pb respectively) but the analyses are sufficiently precise to demonstrate a marked difference in isotopic composition between core and rim (Fig. 7d). The rim is concordant within error and plots with the 410 Ma grains already discussed. Analysis 52 is strongly discordant with a ^{207}Pb/^{206}Pb age close to 1.5 Ga. Bearing in mind that this fragment was not free of rim material the age of the core must be even greater. Construction of a discordia through the two results yields an upper intercept age of 2.3 ± 0.2 Ga which is a minimum for the age of the core of this grain. Grains with inherited cores were noted visually in many of the multi-grain fractions of clear–pale brown zircons which also show reverse discordance to varying degrees. This led Drewery et $al.$ (1987) to interpret the multi-grain data in terms of zircons derived from Caledonian granites containing inherited cores of older, apparently Archaean, zircons. The results of zircon microsurgery reported above lend support to that interpretation. The possibility remains that some of the discordance seen in the multi-grain data may result from admixture of discrete grains of Archaean age although clear–pale brown grains of that age have not so far been analysed.

Grain 47 was a $c.$ 350 μm euhedral grain dominated by pyramid faces. It has a low uranium content of $c.$ 180 ppm U using a visual estimate of the grain volume and the resulting low radiogenic lead content leads to exceptionally high errors on the ^{207}Pb/^{206}Pb age. The ^{206}Pb/^{238}U age of 354 Ma is younger than any of the other samples analysed. This could be the result of lead loss from an older grain but in view of the pristine euhedral character and low uranium content it is possible this grain was derived from a Lower Carboniferous volcanic rock.

It is important to note that the single-grain data-set has provided no evidence for crystallization of zircons in the Carboniferous sourcelands between 2.7 and 0.5 Ga, though the number of analyses is limited.

The zircon data from the other samples comprise multi-grain analyses exclusively and their interpretation must make due allowance for mixing of grains with different primary ages, as revealed in the single grain measurements on the Rough Rock. With the notable exception of the Yoredales sample, discussed separately below, a similar pattern of zircon ages is found throughout the stratigraphic range from Chadian to Westphalian B. Pink–purple grains are present throughout, the majority of them rounded and plotting close to the more discordant Rough Rock data. The purple euhedral fractions plot slightly below the rounded fractions, showing an apparent difference in lead-loss history from rounded grains.

Interpretation of the discordance pattern of the multi-grain analyses of the purple zircons can be only tentative, the present position of any point representing the integrated effect of possible lead loss at any time from the Proterozoic to the present. Post-Carboniferous lead-loss does not appear to be a major influence, given the concordance of unabraded colourless zircons discussed above. This is in accordance with the tendency for uranium-rich, discordant zircon to be destroyed during the sedimentary cycle, leading to the lower average uranium content of detrital zircons. Lead-loss in the mid-Palaeozoic is implied by the single grain data on euhedral grains and could explain the multi-grain euhedral fractions as well. The rounded purple grains present a more complex pattern. Accepting the apparent absence of Proterozoic zircons as real, their general alignment along a discordia chord from 3.02 to 1.19 Ga implies at least some Precambrian lead-loss. This pattern could be produced by mid-Proterozoic loss combined with Palaeozoic lead-loss from zircons with the range of Archaean ages recorded in the single grain data. Alternatively the lower intercept can be taken to approximate the actual time of a single major lead-loss event; such an interpretation gains some support from the presence of late Proterozoic monazite reported below.

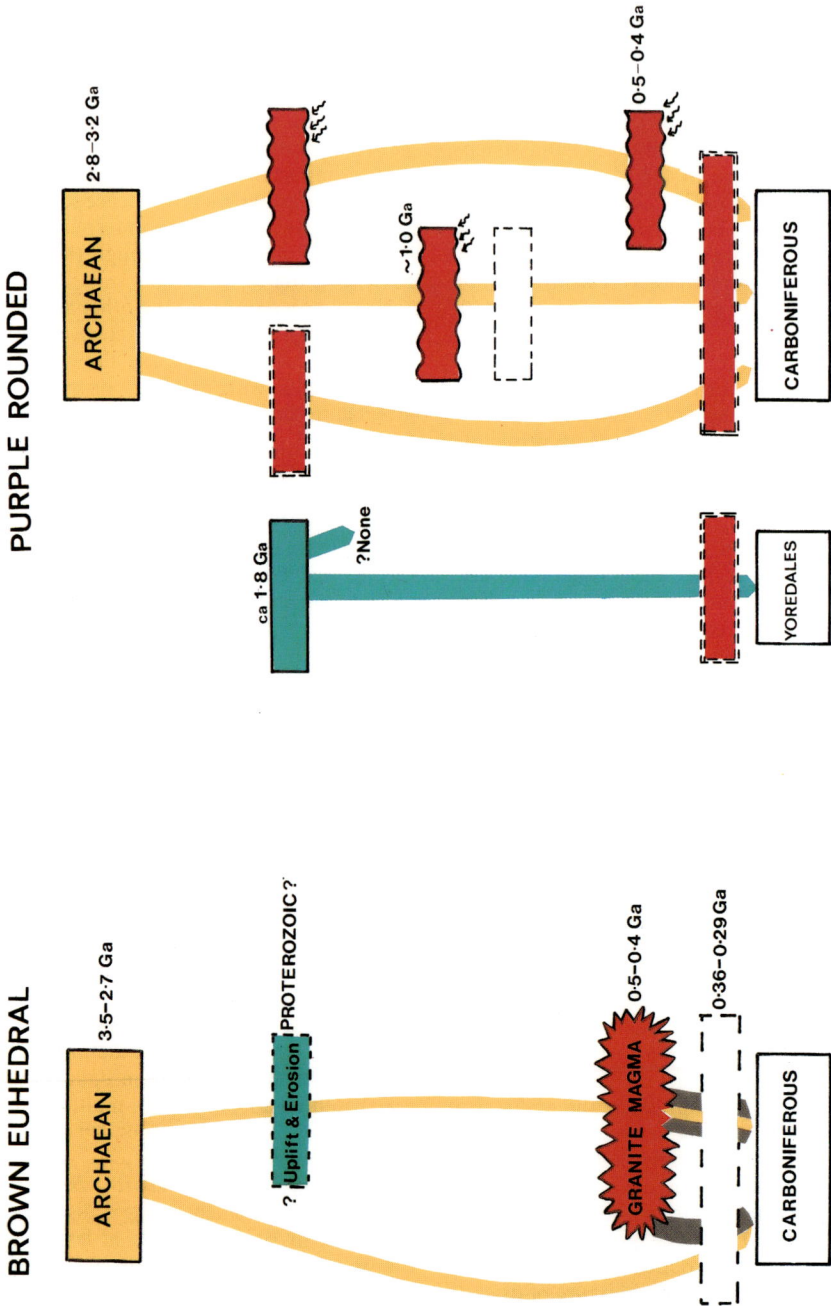

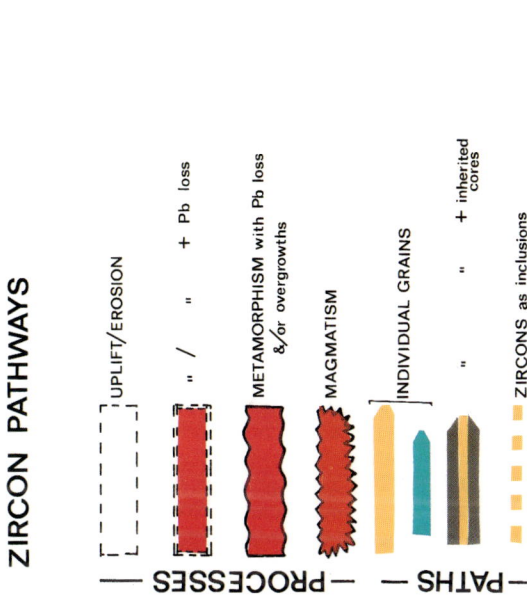

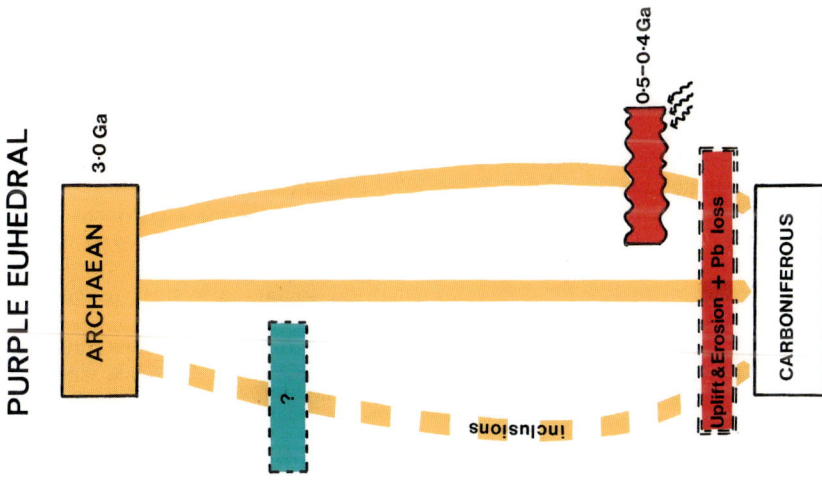

Fig. 11. (a) Possible pathways for the incorporation of colourless–brown euhedral zircons from source to Carboniferous sediments.
(b) Possible pathways for the incorporation of rounded pink–purple zircons from source to Carboniferous sediments.
(c) Possible pathways for the incorporation of euhedral pink–purple zircons from source to Carboniferous sediments.

As with the Rough Rock samples the colour-less–brown zircons scatter about a chord extending from *c.* 0.4 Ga toward Archaean ages between 2.75 and 3.00 Ga. A high proportion of these grains are euhedral/subhedral and some of them contain optically distinct cores. Incorporation of inherited zircon into zircons crystallized in Lower Palaeozoic granites again seems to offer an explanation for much of the discordance. Similar patterns are widely observed, in the Caledonian granites of Scotland for example (Pidgeon & Aftalion 1978). Unlike the Scottish granites, however, the mean age of the inherited component was Archaean, rather than Protero-zoic. To our knowledge the only British granitic rocks with this characteristic are minor granite sheets cutting Moine/Lewisian rocks on the Black Isle (Cliff, unpubl. data).

The Yoredale sample (analyses 11–15) presents a distinctly different picture. Both colour-less–brown and purple zircons are again present. Three fractions of rounded purple zircons have been analysed: they are all discordant and have lower $^{207}Pb/^{206}Pb$ ages than any of the other purple zircons, ranging from 1.69–1.79 Ga. On the concordia plot in Fig. 8 they show a crude alignment which is compatible with Palaeozoic lead-loss from mid-Proterozoic (1.8–2.0) zircons. The other two fractions comprise euhedral grains, one brown and one colourless. These are both strongly discordant, plotting on the same general alignment as the purple grains but closer to the lower intercept at *c.* 400 Ma. In the absence of single grain data the interpretation of this pattern can only be tentative. However, it is clear that the zircons in the source rocks for this Yoredale sample are distinct from those in the other samples analysed. The difference could be due to more severe late Proterozoic lead-loss from Archaean zircons similar to those in the other samples. Alternatively the apparent 1.8–2.0 Ga upper intercept age could be the true age of these zircons and indicate the presence of mid-Proterozoic rocks in the sourcelands, at least briefly during the Lower Carboniferous.

Monazite

Monazite data for two Namurian (Rough Rock) and one Westphalian sample are plotted in Fig. 10. In all cases the monazites are rounded; most are transparent yellow grains although a few are turbid and rare grains are a darker, browner colour. As predicted the results plot very close to concordia.

For the Rough Rock, early multi-grain analyses have very low $^{206}Pb/^{204}Pb$ ratios, caused by a high blank. Within the resulting high errors both samples plot just on the concordia at 436 Ma. This result is supported by the more precise single grain analyses (54, 55) with ages of 436 and 426 Ma. The other single grain analysis (53) is also concordant but with an Archaean age of 2.72 Ga. Multi-grain fraction 32 is the most discordant monazite analysed but it also has an unusually high apparent common lead content. It is possible that this common lead had an isotopic composition significantly different from the measured blank which was used to calculate the radiogenic lead composition and its uncertainty. It would be unwise to place much weight on this single result although the type of discordance observed has been reported by Copeland *et al.* (1988) and interpreted as due to incorporation of inherited monazite in a granitic intrusion.

Two multi-grain fractions comprising 30 and 15 grains from a Westphalian sample gave slightly discordant ages, whilst a single large grain (68) from this sample yielded a concordant 0.947 Ga age. The discordant samples are aligned close to a lower concordia intercept at 420 Ma on a discordia extending toward the concordant sample 68.

Implications of the isotopic data

The isotopic data presented above clearly show that the Carboniferous rivers drained a source-land which provided detritus from rocks with a wide range of ages. Possible zircon pathways from sourcelands of different ages are presented in Fig. 11.

Lower Palaeozoic granitoid rocks appear to be the source of the dominant, mainly euhedral, colourless-pale brown zircons (Fig. 11a) as suggested previously by Drewery *et al.* (1987). These source rocks contained a distinctive Archaean inherited component which distinguishes them from the typical Caledonian granites of Scotland. Most of the monazites analysed so far have also yielded Caledonian ages.

The second group of zircons comprises purple grains. Except in the Yoredale sample these are Archaean, with $^{207}Pb/^{206}Pb$ ages in single grains from 2.8–3.2 Ga. Most of these are well rounded and, together with the very old (> 3.5 Ga) rounded brown grains from the Rough Rock, have almost certainly passed through more than one sedimentary cycle. This suggests that the sourcelands contained significant outcrops of Proterozoic or Lower Palaeozoic sediments derived from an Archaean source area. In contrast

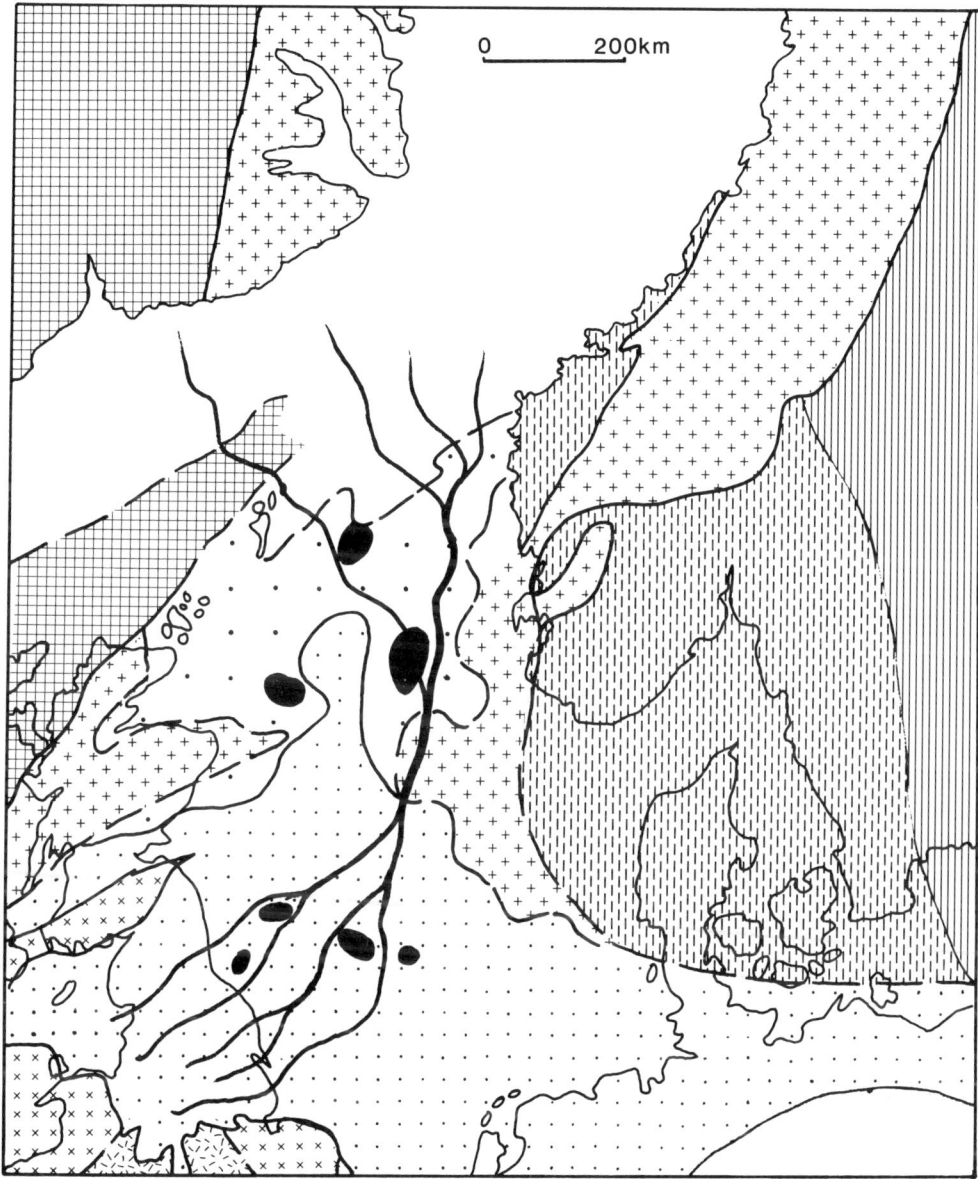

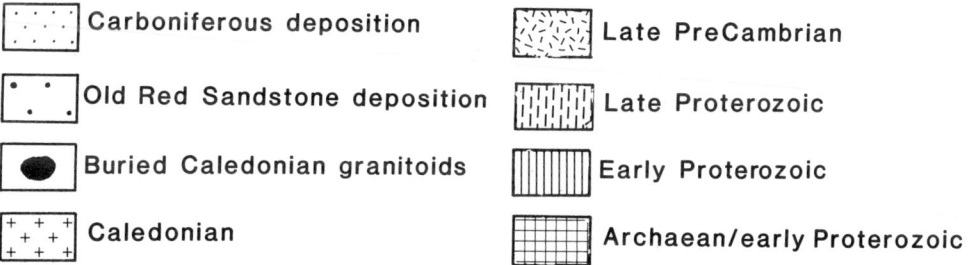

Fig. 12. Map to show the age provinces of the NE Atlantic borderlands and a tentative location for the drainage basin for the Pennine River and its depositional area, the latter shown for its maximum extent in late Namurian/early Westphalian times. See text for full discussion.

a small proportion (*c.* 5–10%) of the purple zircons are well preserved euhedral crystals; this suggests that primary detritus from Archaean basement may also have been involved although it is also possible that some euhedral grains may survive sedimentary recycling as inclusions in larger sand grains or pebbles. The occurrence of Archaean monazite grains, documented so far by isotopic data from a single concordant 2.7 Ga grain, may also point to exposure of Archaean basement in the Carboniferous sourcelands.

The time interval between 2.7 Ga and the Caledonian orogeny has apparently left little imprint in the isotope systematics investigated here. With the exception of the Yoredale sandstone, there is no evidence for zircon crystallization between 2.7 and 0.6 Ga. The discordance pattern of the purple rounded zircons shows a crude alignment with a lower intercept with concordia near 1 Ga. This could reflect an important Grenville/Sveconorwegian lead-loss at that time, or alternatively could be the net effect of multiple lead-loss during the mid-Proterozoic and Lower Palaeozoic. The single 0.95 Ga monazite grain in one of the Westphalian sandstones lends additional credibility to the first alternative.

Potential sourcelands

In comparing the age pattern just reviewed with the pre-Carboniferous geology of the North Atlantic region, the identification of possible sourcelands is necessarily tentative. The data-set from the Carboniferous sediments themselves is still very limited. Equally important, even those potential sourcelands that are exposed today are inadequately documented from the geochronological point of view. Figure 12 summarizes key features on a pre-Atlantic drift reconstruction.

Areas with a dominant Caledonian influence/imprint are widespread and include much of the Scottish Highlands, western Scandinavia and East Greenland. Of these Scotland appears unsuitable because of the strong mid-Proterozoic signature in the inherited zircons in its Caledonian granites (Pidgeon & Aftalion 1978) and it is difficult to envisage Scandinavia providing Caledonian granitic zircons without abundant Proterozoic grains as well (e.g. Tucker *et al.* 1987; Schaerer 1980, Patchett *et al.* 1987). The inherited zircons in Caledonian granites from East Greenland appear to be similar to those of Scotland. However, there are several unexposed granitoids beneath the Central and Northern North Sea (Donato *et al.* 1983; Zervos 1987) and limited geochronological data (Frost *et al.* 1981)

suggests that these are also Caledonian. The occurrence of older inherited zircons in granites from the Black Isle noted above emphasizes the possibility that suitable source rocks may be available offshore, off the east coast of Scotland, in the area of the Viking Graben and northwards towards the Greenland continental margin.

Potential Archaean source areas are much more restricted. The classical suggestion is the Lewisian of NW Scotland but as previously discussed by Drewery *et al.* (1987) available Lewisian zircon ages are generally younger than 2.8 Ga whereas the euhedral purple zircons in the Rough Rock are in the range 2.8–3.2 Ga. Furthermore, although there is little data (van Breemen *et al.* 1971), it is likely that the Lewisian contains considerable amounts of mid-Proterozoic zircon, related to Laxfordian metamorphism and granite emplacement. However, there is a large area of Precambrian extending northward from the Lewisian, offshore (Ziegler 1982) which may contain suitable rocks. Indications of a further area of Pre-Cambrian basement are to be found in the contamination signatures seen in Tertiary basalts from the Faroes (Gariepy *et al.* 1983).

Also in Scotland, the Torridonian is a potential source of recycled Archaean detritus although Rb–Sr isotope studies on pebbles (Moorbath *et al.* 1967) and Sm–Nd model ages (O'Nions *et al.* 1983) both indicate the importance of Proterozoic detritus in the Torridonian, in conflict with the inferred characteristics of the Carboniferous source material.

East Greenland provides a second alternative source of Archaean detritus, both primary and recycled. There is very little published geochronological information on the Pre-Cambrian basement southwest of the Caledonian Front but there are Archaean rocks within the Caledonian Belt (Higgins & Borchardt 1987) and the late Proterozoic sediments of the Eleonore Bay Group. Here zircons are preserved at chlorite-grade and are petrographically and isotopically (Peucat *et al.* 1985) similar (but not identical) to the purple rounded grains from the Carboniferous. The metamorphic history of this area is complex, including proposed pre-Caledonian events at 1.1 and 1.9 Ga (Hansen *et al.* 1987). Once again the lack of mid-Proterozoic zircons in the Carboniferous appears problematical.

Overall it appears that none of the presently exposed areas of the North Atlantic region provides a close match to the sourceland geology inferred from this petrographic and isotopic study. It is perhaps more likely that these sourcelands are to be found buried in the northern North Sea or between northern Scotland and

Greenland. Fig. 12 attempts to locate a possible drainage basin in this area taking account of the isotopic constraints, regional facies trends and palaeogeographical constraints discussed in this paper. The large hinterlands to the Pennine River are predicted to have included:

(1) common Caledonian granitoids that were partly sourced in Archaean crust;
(2) Proterozoic or Lower Palaeozoic sediments and low-grade metasediments derived from an Archaean source area;
(3) relatively minor outcrops of Archaean and Grenville/Sveconorwegian basement.

It is possible that climatic changes in these hinterlands were responsible for the massive influx of coarse, immature, Namurian clastic rocks discussed above. There is no independent evidence of regional uplift in the hinterlands, the effect of heterogeneous lithospheric stretching postulated by Leeder (1982) to account for the influx being an order of magnitude too small to explain the volumes of sediment involved. Neither is there any change in the pattern of zircon ages from our sample suites that might indicate a regional unroofing event at this time.

Conclusions

This investigation has shown that despite systematic variations in the feldspar content of sediment delivered to the Pennine Basin at different times in the Carboniferous, some features of the sourceland geology persisted throughout. In general the same types and ages of zircon are found in Dinantian, Namurian and Westphalian samples. They are dominated by grains from Caledonian rocks but also include Archaean grains, both rounded and euhedral. Mid-Proterozoic zircons appear to be absent except in the sample from a Yoredale sandstone; this provides the strongest constraint on potential source areas. No close match between the inferred sourceland geology and presently exposed rocks in the North Atlantic region has been found; submarine outcrops and basement subcrops beneath younger strata in the northern North Sea basin and in the area between Scotland and Greenland remain to be tested.

In a more general context we have shown that it is practicable to determine strict constraints on sedimentary provenance using modern radiogenic isotope techniques. The progression from multi-grain to single-grain analyses in the course of this work has greatly reduced the ambiguity of the results. It is also evident that careful selection techniques, in some cases coupled with laboratory abrasion, can yield concordant zircon grains in sediments as well as in igneous rocks. The great potential of monazite, which usually gives concordant ages, to provide valuable information on provenance has also been demonstrated.

References

ALLEN, P. 1972. Wealden detrital tourmaline: implications for northwestern Europe. *Journal of the Geological Society, London*, **128**, 273–294.

BESLY, B. M. 1987. Sedimentological evidence for Carboniferous and early Permian palaeoclimates of Europe. *Annales de la Société géologique Nord.* **106**, 131–143.

BLACK, L. P., WILLIAMS, I. S. & COMPSTON, W. 1986. Four zircon ages from one rock: the history of a 3930 Ma old granulite from Mount Sones, Enderby Land, Antarctica. *Contributions to Mineralogy and Petrology*, **94**, 427–437.

BRISTOW, C. S. 1987. *Sedimentology of large braided rivers ancient and modern*. PhD thesis, University of Leeds.

BUTTERFIELD, J. A. 1939. A petrological study of some Yoredale sandstones. *Transactions of the Leeds Geological Association*, **5**, 264–284.

COPELAND, P., PARRISH, R. R. & HARRISON, T. M. 1988. Identification of inherited radiogenic Pb in monazite and its implications for U-Pb systematics. *Nature*, **333**, 760–765.

DONATO, J. A., MARTINDALE, W. & TULLY, M. 1983.

Buried granites within the Mid North Sea High. *Journal of the Geological Society, London*, **140**, 825–837.

DREWERY, S. E. 1987. *Provenance of Carboniferous sandstones: geochronological and petrographic studies*. PhD thesis, University of Leeds.

——, CLIFF, R. A. & LEEDER, M. R. 1987. Provenance of Carboniferous sandstones form U-Pb dating of detrital zircons. *Nature*, **325**, 50–53.

ELLIOTT, T. E. 1976. The morphology, magnitude and regime of a Carboniferous fluvio-deltaic channel. *Journal of Sedimentary Petrology*, **46**, 70–76.

FROST, R. T. C., FITCH, F. J. & MILLER, J. A. 1981. The age and nature of crystalline basement of the North Sea Basin. *In*: ILLING, L. V. & HOBSON, G. D. (eds) *Petroleum Geology of the Continental Shelf of NW Europe*. Heyden & Son, 43–57.

GARIEPY, C., LUDDEN, J. & BROOKS, C. 1983. Isotopic and trace element constraints on the genesis of the Faroes lava pile. *Earth and Planetary Science Letters*, **63**, 257–272.

GAWTHORPE, R. L., GUTTERIDGE, P. & LEEDER, M. R. 1989. Late Devonian and Dinantian basin evolution in northern England and North Wales. *In*: ARTHURTON, R. S., GUTTERIDGE, P. & NOLAND, S. C. (eds) *The Role of tectonics in Devonian and Carboniferous sedimentation in the British Isles*. Special Publication of the Yorkshire Geological Society, 1–24.

GEBAUER, D. & GRÜNENFELDER, M. 1979. U-Th-Pb dating of minerals. *In*: JÄGER, E. & HUNZIKER, J. C. (eds) *Lectures in Isotope Geology*, Springer, Berlin, 105–131.

GILLIGAN, A. 1920. The petrography of the Millstone Grit of Yorkshire. *Quarterly Journal of the Geological Society of London*, **75**, 251–293.

GREENSMITH, J. T. 1965. *The petrology of the sedimentary rocks*. 4th Ed. Murby, London.

HACK, J. T. 1957. *Studies of longitudinal stream profiles in Virginia and Maryland*. United States Geological Survey Professional Paper. **294B**.

HANSEN, B. T., STEIGER, R. H. & HENRIKSEN, N. 1987. U-Pb and Rb-Sr age determinations on Caledonian plutonic rocks in the central part of the Scoresby Sund region, East Greenland. *Gronlands Geologiske Undersøgelse Rapport*, **134**, 5–18.

HIGGINS, K. & BORCHARDT, B. 1987. Archaean U-Pb zircon ages from the Scoresby Sund region, East Greenland, *Gronlands Geologiske Undersøgelse Rapport*, **134**, 19–24.

HURFORD, A. J., FITCH, F. T. & CLARKE, A. 1984. Resolution of the age structure of the detrital zircon populations of two Lower Cretaceous sandstones from the Weald of England by fission track dating. *Geological Magazine*, **121**, 263–296.

JONES, C. M. 1980. Deltaic sedimentation in the Roaches Grit and associated sediments (Namurian R2B) in the south-west Pennines. *Proceedings of the Yorkshire Geological Society*, **43**, 39–67.

KROGH, T. E. 1982. Improved accuracy of U-Pb zircon ages by the creation of more concordant systems using an air abrasion technique. *Geochimica et Cosmochimica Acta*, **46**, 637–649.

LEEDER, M. R. 1973. Fluviatile fining-upwards cycles and the magnitude of palaeochannels. *Geological Magazine*, **110**, 265–276.

—— 1982. Upper Palaeozoic basins of the British Isles: Caledonide inheritance versus Hercynian plate margin processes. *Journal of the Geological Society, London*, **139**, 479–491.

—— & HARDMAN, M. 1990. Carboniferous geology of the Southern North Sea Basin and controls on hydrocarbon prospectivity. *In*: HARMDMAN, R. F. P. & BROOKS, J. (eds) *Tectonic Events Responsible for Britain's Oil and Gas Reserves*. Geological Society, London, Special Publication, (In press).

—— & BOLDY, S. A. R. 1990. Carboniferous of the Outer Moray Firth basin, Quadrants 14, 15, Central North Sea. *Marine and Petroleum Geology*, (In press).

LEOPOLD, L. B., WOLMAN, M. G. & MILLER, J. P. 1964. *Fluvial Processes in Geomorphology*. Freeman, San Francisco.

MACKIE, W. 1928. The heavier accessory minerals of the granites of Scotland. *Transactions of the Edinburgh Geological Society*, **12**, 22–40.

MCCABE, P. J. 1977. Deep distributary channels and giant bedforms in the Upper Carboniferous of the Central Pennines, England. *Sedimentology*, **24**, 271–290.

MOORBATH, S., STEWART, A. D., LAWSON, D. E. & WILLIAMS, G. E. 1967. Geochemical studies on the Torridonian sediments of NW Scotland. *Scottish Journal of Geology*, **3**, 389–412.

MYERS, K. J. & BRISTOW, C. S. 1989. Detailed sedimentology and gamma-ray log characteristics of a Namurian deltaic succession II: Gamma-ray logging. *In*: WHATELEY, M. K. & PICKERING, K. T. (eds) *Deltas: Sites and Traps for Fossil Fuels*. Geological Society, London, Special Publication, **41**, 81–88.

O'NIONS, R. K., HAMILTON, P. J. & HOOKER, P. J. 1983. A Nd-isotope investigation of sediments related to crustal development in the British Isles. *Earth and Planetary Science Letters*, **6**, 229–240.

PATCHETT, P. J., TODT, W. & GORBATSCHEV, R. 1987. Origin of continental crust of 1.7–1.9 Ga age: Nd isotopes in the Svecofennian Orogenic Terrane of Sweden. *Precambrian Research*, **35**, 145–160.

PETTIJOHN, F. J., POTTER, P. E. & SIEVER, R. 1972. *Sand and Sandstone*. New York, Springer.

PEUCAT, J. J., TISSERANT, D., CABY, R. & CLAUER, N. 1985. Resistance of zircons to U-Pb resetting in a prograde metamorphic sequence of Caledonian age in East Greenland. *Canadian Journal of Earth Sciences*, **22**, 330–338.

PIDGEON, R. T. & AFTALION, M. 1978. Cogenetic and inherited zircon U-Pb systems in granites: Palaeozoic granites of Scotland and England. *In*: BOWES, D. R. & LEAKE, B. E. (eds) *Crustal Evolution in NW Britain and adjacent regions*. Seel House Press, Liverpool, 183–220.

POWELL, C. MCA. & VEEVERS, J. J. 1987. Namurian uplift in Australia and South America triggered the main Gondwanan glaciation. *Nature*, **326**, 177–179.

PUPIN, J. P. 1980. Zircons and granite petrology. *Contributions to Mineralogy and Petrology*, **73**, 207–220.

SCHAERER, U. 1980. U-Pb and Rb-Sr dating of a polymetamorphic nappe terrane: the Caledonian Jotun Nappe, Norway. *Earth and Planetary Science Letters*, **49**, 205–218.

SCHUMM, S. A. 1968. Speculations concerning palaeo-hydrologic controls of terrestrial sedimentation. *Geological Society of America Bulletin*, **79**, 1573–1588.

—— 1977. *The Fluvial System*. Wiley, New York. 338pp.

SIMON, J. B. & BLUCK, B. J. 1982. Paleodrainage of the southern margin of the Caledonian mountain chain in the northern British Isles. *Transactions of the Royal Society of Edinburgh*, **73**, 11–15.

SORBY, H. C. 1859. On the structure and origin of the Millstone Grit in South Yorkshire. *Proceedings of the Yorkshire Geological & Polytechnic Society*, **3**, 673–674.

STEELE, R. P. 1987. The Namurian sedimentary history of the Gainsborough Trough. *In*: BESLY, B. M. &

KELLING, G. (eds) *Sedimentation in a synorogenic basin complex: the Upper Carboniferous of NW Europe*. Blackie, Glasgow & London, 102–113.

STEIGER, R. H. & JÄGER, E. 1977. Sub-commision on geochronology: conventions on the use of decay constants in geo- and cosmochronology. *Earth and Planetary Science Letters*, **36**, 359–362.

TUCKER, R. D., RAHEIM, A., KROGH, T. E. & CORFU, F. 1987. Uranium–lead zircon and titanite ages from the northern portion of the Western Gneiss Region, south-central Norway. *Earth and Planetary Science Letters*, **82**, 201–211.

VAN BREEMEN, O., AFTALION, M. & PIDGEON, R. T. 1971. The age of the granite injection complex of Harris, Outer Hebrides. *Scottish Journal of Geology*, **7**, 139–152.

ZERVOS, F. 1987. A compilation and regional interpretation of the northern North Sea gravity map. *In*: COWARD, M. P., DEWEY, J. F. & HANCOCK, P. L. (eds) *Continental Extensional Tectonics*. Geological Society, London, Special Publication, **28**, 477–493.

ZIEGLER, P. A. (1982). *Geological Atlas of western and central Europe*. Shell, Hague.

Isotopic characteristics of Ordovician greywacke provenance in the Southern Uplands of Scotland

JANE A. EVANS[1], PHILIP STONE[2] & JAMES D. FLOYD[2]

[1] NERC Isotope Geoscience Laboratories, Keyworth, Nottingham NG12 5GG, UK

[2] British Geological Survey, Murchison House, West Mains Road, Edinburgh EH9 3LA, UK

Abstract: Interbedded Ordovician greywackes in the Southern Uplands are of markedly different composition. An integration of Sm-Nd isotope data, petrography of detrital clasts, and palaeocurrent flow analyses allows the likely provenance character and distribution to be deduced. The Portpatrick and Galdenoch formations contain detritus from volcanic provenances: andesite dominates the composition of Caradoc–Ashgill greywacke from the Portpatrick Formation which have εNd_{445} (time of deposition) between -2.3 and -2.9, the Llandeilo–Caradoc Galdenoch Formation is richer in hornblende and has a slightly lower range of εNd_{445} between -3.4 and -4.4. These values contrast with the more isotopically depleted signature of the Llandeilo–Caradoc Kirkcolm Formation greywackes which contain mainly quartz and feldspar clasts and have εNd_{445} as low as -11.2. The source of the volcanic rocks, deduced from palaeocurrent analysis, lay on side of the depositional basin (south in terms of modern geography) and the isotopic composition of the volcanic rocks is consistent with a calc-alkaline arc or back-arc assemblage, founded on continental crust. A Proterozoic terrane, represented by detritus from the Kirkcolm Formation, lay on the opposite side of the depositional basin. Dalradian metamorphic rocks currently exposed to the north of the Southern Uplands display a less radiogenic signature than Kirkcolm Formation greywackes and seem unlikely to have acted as their source. This may support interpretations of the Southern Upland Fault as a locus of major sinistral strike-slip movement.

The Ordovician greywacke sequence of the Southern Uplands of Scotland is notable for the marked compositional contrast between the component formations (Floyd 1982). These have an elongate outcrop parallel to the NE–SW regional strike with either faulted or interbedded relationships with adjacent formations (Fig. 1). The oldest such lithostratigraphic unit (March-burn–Tappins) crops out to the NW of the Glen App and Carcow faults in the NW of the region and overall the fault-bounded formational packets become sequentially younger towards the SE (Fig. 2). Petrographical characteristics of the detrital grains forming the greywackes have been widely described both qualitatively and quantitatively (e.g. Kelling 1961; Floyd 1982; Floyd & Trench 1989; Styles et al. 1989). They are summarized in Table 1 and clearly demonstrate that a variety of provenance areas are required, through time, to supply the sediment now forming the greywacke sequence. The interbedding of markedly contrasting greywackes, suggesting the contemporaneous availability of contrasting provenances, must also be stressed (Fig. 3; cf. Stone et al. 1987).

A mixed ophiolitic and continental basement (plutonic–metamorphic) source area was available in gracilis times, from which the March-burn Formation was derived, with the ophiolitic provenance contributing to the remarkably high magnetic susceptibility of the Marchburn Formation greywackes (Floyd & Trench 1989). Subsequently two contrasting provenances were available virtually simultaneously; one consisted dominantly of mature, quartzo-feldspathic continental basement and deeply-dissected magmatic arc, perhaps still with ophiolitic vestiges, whereas the other consisted of a similar basement together with voluminous fresh andesitic lithologies. A detailed study of the andesitic grains, together with the abundance detrital pyroxenes (Styles et al. 1989) has confirmed their provenance as a calc-alkaline ensialic island arc. These contrasting continental and island arc provenances were available in the gracilis and peltifer graptolite zones as the sources of the interbedded Kirkcolm and Galdenoch Formations (Fig. 3). At this time, the dominant or 'background' sediment source was continental (Kirkcolm Formation), with subordinate amounts of island arc material (Galdenoch Formation). Slightly later, during the clingani and linearis graptolite zones, the reverse situation pertained with the bulk of the sediment of obvious island arc derivation (Portpatrick Formation) and only relatively minor amounts of

From Morton, A. C., Todd, S. P. & Haughton, P. D. W. (eds), 1991,
Developments in Sedimentary Provenance Studies.
Geological Society Special Publication No. 57, pp. 161–172.

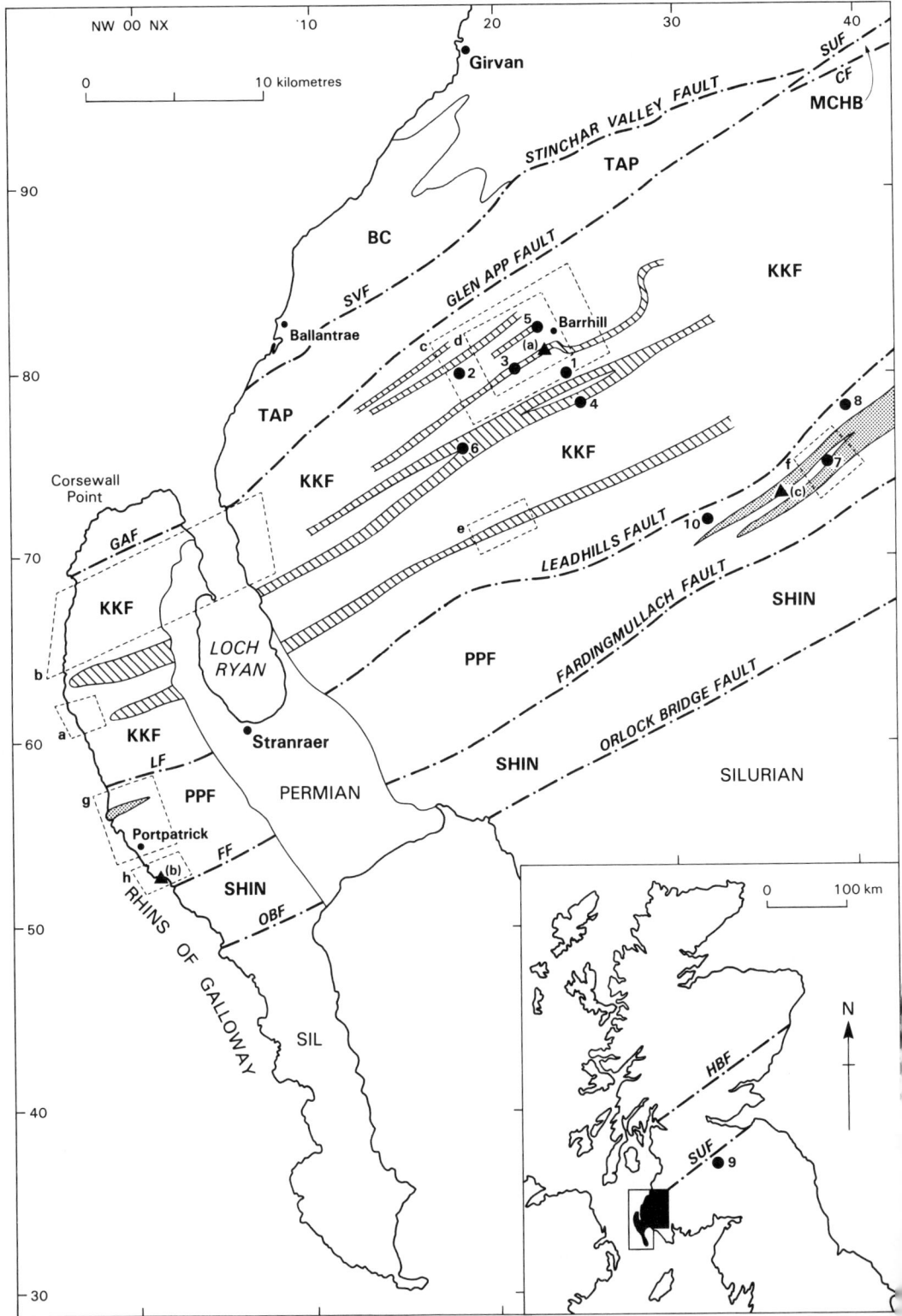

● 1–10   Location of analysed samples   a–h ⌐ ¬ ⌐ ¬   Sub-area for palaeocurrent analysis
                                                      └ ┘ └ ┘
▲ (a)–(c) Location of measured sections

89 PS 52

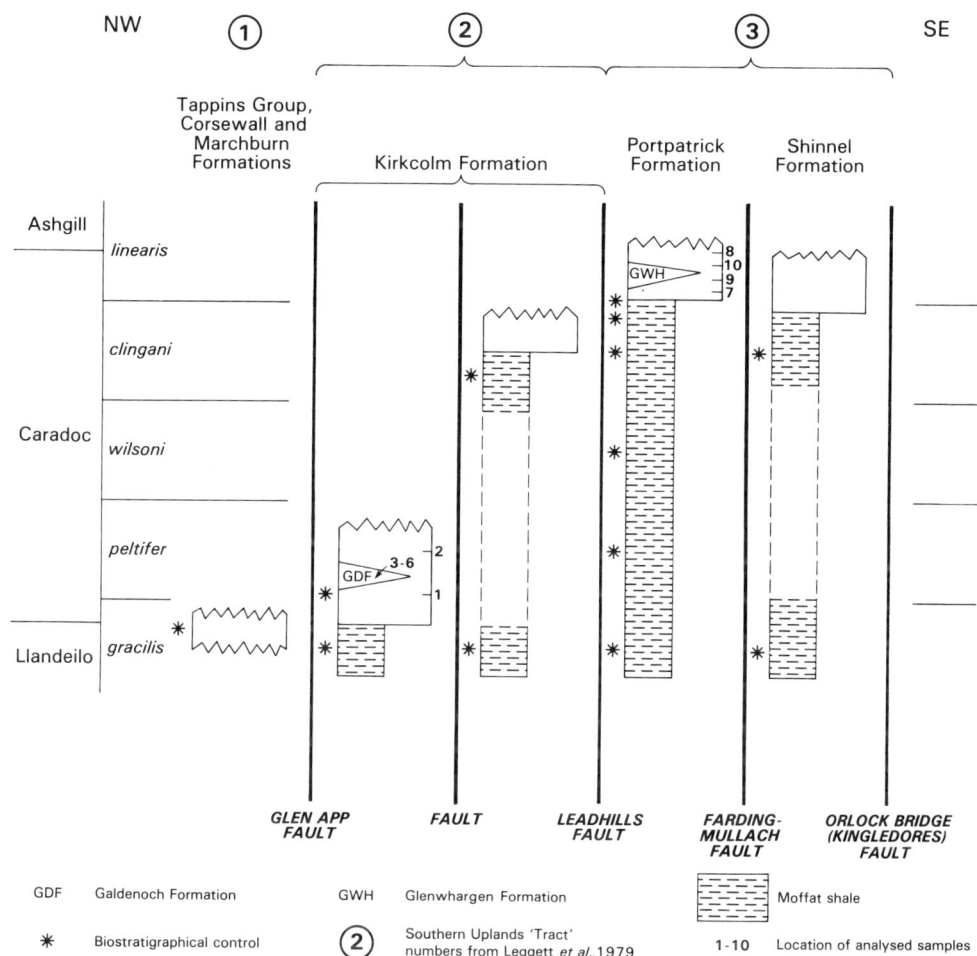

Fig. 2. Biostratigraphical summary of the Ordovician formations forming the SW Southern Uplands.

interbedded sediment of continental derivation (Glenwhargen Formation). The Shinnel Formation, as far as is known, was derived from a mature continental source and no obviously volcaniclastic interbeds have yet been reported. However, there is abundant evidence that the volcanic island arc provenance was still available in the earliest Llandovery (Barnes *et al.* 1987) although thereafter, the younger (Llandovery and Wenlock) greywackes of the Southern Uplands are exclusively cratonic.

Palaeocurrent evidence for provenance location

The marked compositional variation in the greywackes of the Southern Uplands was funda-

Fig. 1. Location and outline geology of the southwestern Southern Uplands. BC, Ballantrae ophiolite complex; CF, Carcow Fault; HBF, Highland Boundary Fault; KKF, Kirkcolm Formation; MCHB, Marchburn Formation; PPF Portpatrick Formation; SHIN, Shinnel Formation; SUF, Southern Upland fault; TAP, Tappings Group. The Galdenoch Formation is shown cross hatched and the Glenwhargen Formation is shown stippled. Filled circles annotated with numbers indicate locations of Southern Uplands samples in Table 2. Note that Sample No. 9 is shown only on the inset map as it is from the Portpatrick Formation approximately 90 km along strike to the NE of the main area of the figure. Filled triangles annotated (a), (b) and (c) indicate the locations of the 3 measured sections in Fig. 3. Outlined boxes annotated with letters a–h indicate the areas where the palaeocurrent data in Fig. 4 were obtained.

mental to the back-arc depositional models proposed by Morris (1987) and Stone *et al.* (1987) both of which envisaged a mature continental

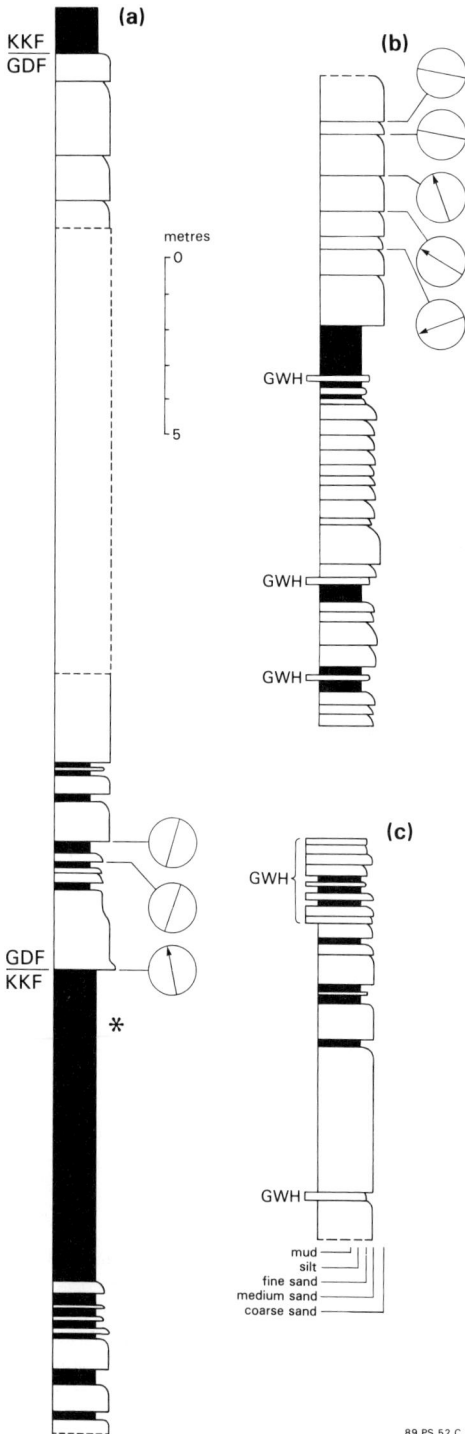

margin with ophiolite fragments to the N and a volcanic (andesitic) arc to the south. Further, Stone *et al.* (1987) and Styles *et al.* (1989) considered the arc to be ensialic, founded on basement similar to that exposed to the north of the basin. In its original form the widely accepted accretionary prism model for the Southern Uplands (Leggett *et al.* 1979) did not accommodate a southerly provenance but a later modification (Kelling *et al.* 1987) envisaged an active arc briefly juxtaposed against the northern fore-arc region by strike-slip faulting. This modification is difficult to reconcile with the enormous volume of volcanic material available over a long time period (Llandeilo–Llandovery) and the likely ensialic nature of the arc.

The location of the contrasting provenance areas is thus of great importance in terms of the regional geotectonic debate. A broad control on this problem can be gained from a consideration of palaeocurrent flow patterns, some details of which are shown in Fig. 4. Deduction of original palaeocurrent flow directions is complicated in areas of complex structure and this study utilizes only data from relatively simple structural zones. Thus, within the area shown on Fig. 1 palaeocurrent observations were only made in areas of uniform bedding attitude where the rare minor folds had sub-horizontal hinges. All strikes (within the range 045–085°) were corrected to a regional trend of 060° before unfolding to enable an overall comparison to be made. The Marchburn Formation, Tappins Group and Corsewall Formation, all likely correlatives, were probably derived predominantly from the NW with a minor trend from the NE (Kelling 1962). The slightly younger Kirkcolm Formation shows a much more varied flow trend (Fig. 4 a–c) with variation both geographically and stratigraphically. The simplest flow pattern appears in the farthest SW part of the area studied (Fig. 1), in the Rhins of Galloway, where an axial flow from the northeast is predominant (Fig. 4a). However,

Fig. 3. Graphic lithostratigraphic sections illustrating interbedding of contrasting greywacke types. Numbers 1–10 indicate approximate position of samples in Table 2. (**a**) GDF, Galdenoch Formation inter-bedded with KKF, Kirkcolm Formation. * indicates location of *gracilis* Zone graptolite fauna. Cross Water, Barrhill [NX 2297 8148]. (**b**) GWH, Glenwhargen Formation interbedded with the Portpatrick Formation. Port of Spittal Bay [Nx 0190 5219]. (**c**) Glenwhargen Formation interbedded with Portpatrick Formation. Knockville Moor, Newton Stewart [NX 3548 7323].

Table 1. *The principal detrital components in the Ordovician greywackes of the SW Southern Uplands*

Formation	Lithic fragments	Mineral grains
Marchburn and Corsewall (=Tappins Group) in part	granodiorite, quartz porphyrite, chlorite schist, microgranite, rhyolite, felsite, andesite, gabbro, spilite, serpentinite, chert	quartz (14%), feldspar, apatite, clinopyroxene, hornblende, spinel, epidote, biotite
Kirkcolm	spilite, quartzite, garnet schist, biotite schist	quartz (45%), feldspar, garnet, zircon
Galdenoch	andesite, dacite, rhyolite, spilite, chlorite schist	quartz (18%), feldspar, hornblende, augite, epidote, apatite
	Total 'andesitic' may exceed 25%	
Portpatrick	hornblende andesite, dacite, rhyolite, spilite, glaucophane schist, garnet schist, shale, gabbro, diorite, granite	quartz (15%), feldspar, clinopyroxene, hornblende, garnet, epidote, glaucophane, spinel
	Total 'andesitic' may exceed 25%	
Glenwhargen	quartzite, mica schist, spilite, chert	quartz (67%) feldspar
Shinnel	granodiorite, granophyre, rhyolite, felsite, spilite, quartzite, shale	quartz (57%), feldspar, zircon, apatite

farther north in the Rhins outcrop and on the eastern side of Loch Ryan the axial flow is both equally important transverse flow from the south (Fig. 4b). Farther northeast, in the Barrhill area, axial flow is again predominantly from the northeast but is there subordinate to transverse flow from both the north and south (Fig. 4c). One possible interpretation, in view of the likely structural imbrication within the formation, is of an early phase of variable turbidite flow settling down through time to give a fairly regular dispersal pattern from the northeast. Palaeocurrent evidence is sparse in the Galdenoch Formation, the volcanic-rich greywacke unit interbedded with the Kirkcolm Formation, but there is nevertheless a suggestion of bimodality in the turbidite flow regime. Within the Loch Ryan to Barrhill area palaeocurrents from the more northerly exposed horizons flowed from the south (Fig. 4d) whereas palaeocurrents from the more southerly exposed horizons flowed from the north (Fig. 4e). Kelling (1962) reported flow from the south on the Rhins of Galloway but the examples cited are in a complex structural zone and, depending upon the structural interpretation, could equally well be unfolded to indicate flow from the northeast.

A structural and stratigraphical complexity along the southern margin of the Kirkcolm Formation outcrop is the presence of a fault-bounded unit which has the Kirkcolm-type lithostratigraphic characteristics but is probably of *clingani* zone age (Fig. 3; biostratigraphy, from the northeast and south west with an A. W. A. Rushton pers. comm.). Conglomerates within the Kirkcolm Formation which are de-

rived from the northwest (Kelling *et al.* 1987, fig. 6c) probably lie within this ambiguous sequence but, therein, more general palaeocurrent flow is from the SW (Kelling 1962, fig. 16). This contrasts, on the Rhins of Galloway, with the generally NE-dominated regime developed in the adjacent *gracilis–peltifer* Kirkcolm Formation greywackes. A dominant flow from the SW does however match that present in the other *clingani–linearis* succession, the Glenwhargen Formation (Fig. 4f) and the Portpatrick Formation (Fig. 4g). Such an interpretation from the Portpatrick Formation is supported by the more comprehensive sedimentological modelling of Kelling *et al.* (1987) who describe the Portpatrick Formation as the product of several small coalescing submarine fans prograding from a source to the south and southwest. The most consistent evidence for a southerly source comes from the structurally isolated Portpatrick Formation sequence at the southern limit of its outcrop on the Rhins of Galloway (Figs 3 and 4h). The interbedded, exceptionally quartz-rich Glenwhargen Formation (Fig. 3) also appears to be derived from the southwest but, nevertheless, the dramatic compositional contrasts with the Portpatrick Formation greywackes requires provenances with some considerable spatial separation. Palaeocurrent evidence is sparse in the Shinnel Formation; in the eastern part of the area shown in Fig. 1 there is limited evidence of flow from the north (Fig. 4) but Kelling *et al.* (1987) were able to deduce, on the Rhins of Galloway, that sheet flow sequences were cut by ephemeral channels, all derived from a source to the south and/or southeast.

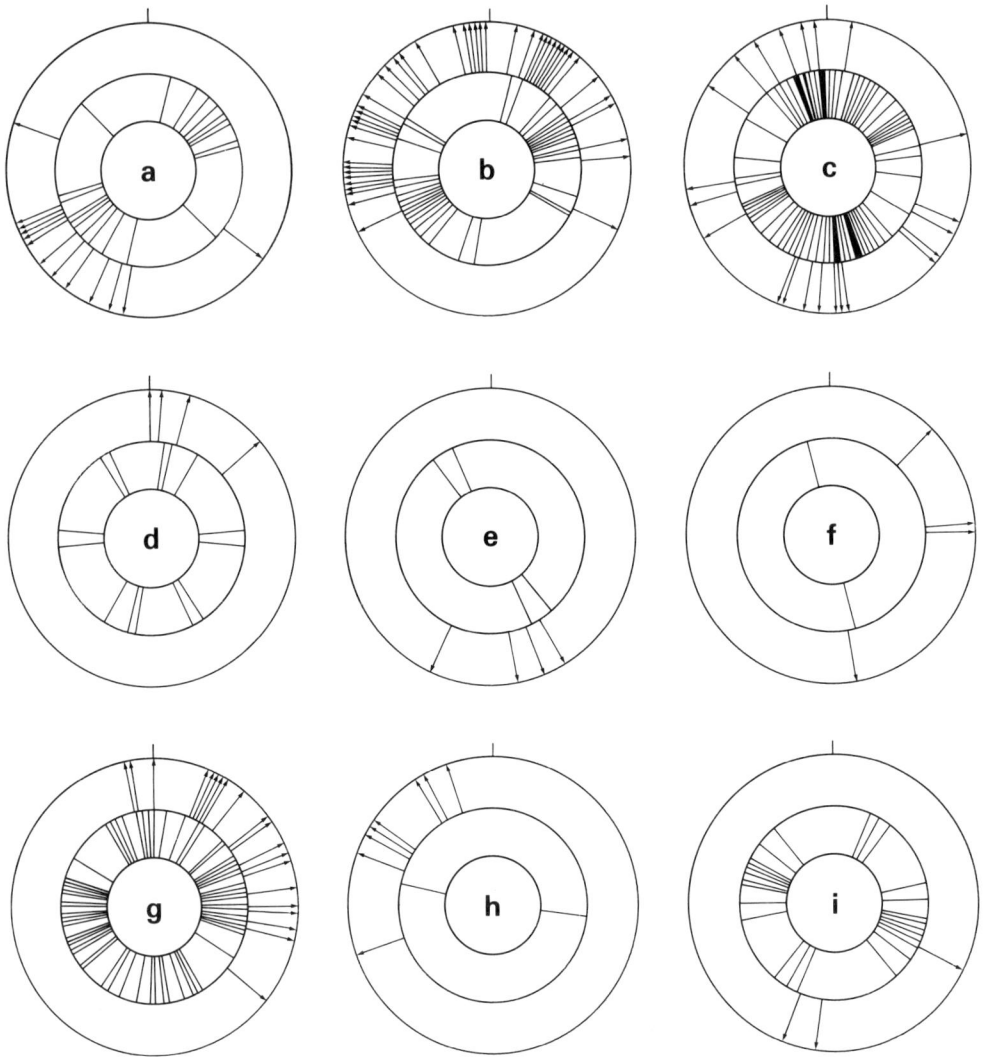

Fig. 4. A summary of palaeocurrent data for the Ordovician formations of SW Scotland; inner circle, groove casts; outer circle, flute casts; (**a**) Kirkcolm Formation (*gracilis–peltifer*), southern part of outcrop, Rhins of Galloway. (**b**) Kirkcolm Formation (*gracilis–peltifer*) northern part of outcrop, Rhins of Galloway and the eastern side of Loch Ryan. (**c**) Kirkcolm Formation (*gracilis–peltifer*) Barrhill area. (**d**) Galdenoch Formation, northern outcrops. (**e**) Galdenoch Formation, southern outcrop. (**f**) Glenwhargen Formation. (**g**) Portpatrick Formation, main part of outcrop. (**h**) Portpatrick Formation, southern margin of outcrop on the Rhins of Galloway. (**i**) Shinnel Formation. Palaeocurrent observations were only made in areas of uniform bedding attitude where the rare minor folds had sub-horizontal hinges. All strikes (within the range 045–085°) were corrected to a regional trend of 060° before unfolding to enable an overall comparison to be made.

There is now no obvious presently exposed candidate for the southern, volcanic provenance of the Southern Uplands greywackes. To the north of the Southern Uplands the crystalline basement of the Laurentian margin is believed to underlie the Midland Valley (e.g. Bamford *et al.* 1977) and the Ballantrae ophiolite complex may represent a once more extensive ophiolitic ter-

rane accreted at the Laurentian margin in the late Arenig. A potentially suitable northern provenance therefore exists nearby but most recent interpretations have preferred to look farther afield. A Nankai Trough-type arrangement was proposed by Leggett (1987) in which all of the Southern Uplands greywackes were derived from a considerable distance away in the northeast by axial flow along the elongate depositional basin. Even allowing for considerable basin-floor channel meandering such an analogy is difficult to reconcile with the palaeocurrent data (Fig. 4), particularly when the specific associations of flow pattern and greywacke composition are considered. Strike-slip faulting has also been invoked to remove the Southern Uplands from its provenance area, with Elders (1987) sourcing Corsewall Formation conglomerates in Newfoundland and suggesting 1500 km of strike-slip displacement. Even in its absence, the southern arc terrane also provokes controversy. An active arc contemporaneous with sedimentation (?Llandeilo–Llandovery) was envisaged by Kelling et al. (1987), Morris (1987) and Stone et al. (1987) influenced by the fresh angular nature of the andesitic detritus. However, Kelly & Bluck (1989) presented Ar–Ar laser probe dates obtained from detrital grains which support an early Cambrian age for andesite eruption.

Previous isotopic studies of Southern Uplands sedimentary rocks

O'Nions et al. 1983 suggested that Southern Uplands greywackes and mudrocks were derived by mixing of relatively young ocean floor volcanic debris with the detritus of Dalradian metamorphic rocks to give sediments with crustal residence ages between 1.1 and 1.7 Ga. The previous view, that the mudrocks were derived essentially from erosion of the Dalradian was not supported by isotopic evidence because samples of the currently exposed Dalradian gave much older model ages of 2.0–2.7 Ga. Further, basic volcanic rocks have Sm/Nd ratios above those of average crust so that a mixture of Dalradian rocks and volcanic rocks penecontemporaneous with sedimentation, would show correlation between model ages and Sm/Nd ratios. Thus, shales dominated by volcanics would have younger ages and higher Sm/Nd ratios, whereas shales dominated by Dalradian would have older ages and lower Sm/Nd ratios. This correlation is not apparent in the O'Nions et al. (1983) data set which has a wide range of Sm/Nd ratios but a relatively narrow range of crustal residence ages. Miller & O'Nions (1984) put forward an alternative model in which two old, Proterozoic sources, one geochemically evolved and one of basaltic composition, mixed to produce the observed results. This model was later shown to be unnecessary because crustal residence ages are relatively insensitive to mixing

of a MORB type and older upper-crustal material (Haughton 1988).

Southern Uplands grewackes contain a varied clast assemblage (Table 1; cf. Floyd 1982; Floyd & Trench 1989) and their resulting heterogeneous composition has not been fully considered in previous isotope studies. In this paper we combine palaeocurrent information and petrography with the isotopic data from stratigraphically well-constrained samples so that their provenance may be better defined.

Constraints on Sm–Nd isotope studies

The application of Sm–Nd isotope studies to the source of sedimentary rocks is dependent upon the assumption that no fractionation of Sm from Nd occurred during the sedimentary process. It is assumed that sedimentary rocks are disaggregates of older igneous, metamorphic and sedimentary precursors and that the analysis of clastic sedimentary rocks provides an average composition of the source region that they sample. A common observation in support of this theory is that sedimentary rocks display a range of Sm/Nd ratios which is comparable with continental rocks (McCulloch & Wasserburg 1978). Further, Frost & Winston (1987), who examined a sequence of both coarse- and fine-grained rocks and found that the fine-grained samples gave uniform crustal residence ages, which represented the average signature of a large catchment area of detritus, whereas the coarse-grained sediments were isotopically less uniform and provided data on proximal detrital sources.

The Nd-isotopic value of sedimentary rocks generally correlates well with petrographic discriminants of source, but the isotopes are more sensitive to small components not easily distinguishable petrographically (Nelson & De Paolo 1988). Sorting and preferential concentration of heavy mineral grains in sandy layers can lead to a biased sample of the source material in coarser grained rocks, a problem carefully avoided when choosing samples for this study.

Nd Isotope analytical details

The concentrations of Sm and Nd, determined by isotope dilution using ^{149}Sm and ^{150}Nd enriched isotopic tracers, and the ^{143}Nd/^{144}Nd isotope ratios were measured on a VG-354 mass spectrometer which gave a value of 0.511843 ± 0.000016 (2-sigma) for the La Jolla international ^{143}Nd/^{144}Nd standard during

Table 2. *Sm and Nd abundance and isotopic data*

Sample	Sm ppm	Nd ppm	$^{147}Sm/^{144}Nd$	$^{143}Nd/^{144}Nd$	εNd_{445}	T_{DM}
Portpatrick Formation						
S 62373	3.796	17.69	0.1297	0.512325	−2.3	1.30
S 62379	4.008	18.67	0.1298	0.512296	−2.9	1.35
S 62420	5.004	23.74	0.1274	0.512317	−2.4	1.28
S 73805	4.930	23.67	0.1259	0.512293	−2.7	1.30
Galdenoch Formation						
S 73800	7.717	35.78	0.1303	0.512274	−3.4	1.39
S 76598	12.12	64.85	0.1130	0.512182	−4.2	1.30
S 73807	7.364	34.41	0.1294	0.512220	−4.4	1.47
S 73801	9.799	52.21	0.1135	0.512221	−3.4	1.25
Kirkcolm Formation						
S 73826	7.223	38.81	0.1125	0.511946	−8.8	1.62
S 76514	6.314	33.35	0.1144	0.511829	−11.2	1.82

analyses. The decay constant taken for $^{147}Sm = 6.54 \times 10^{-12}$ a^{-1}. The following values were used in the calculation of depleted mantle model ages and epsilon values; $^{143}Nd/^{144}Nd_{CHUR} = 0.512640$, $^{143}Nd/^{144}Nd_{DM} = 0.513114$, $^{147}Sm/^{144}Nd_{CHUR} = 0.1967$ and $^{147}Sm/^{144}Nd_{DM} = 0.222$; depleted mantle values from Michard *et al.* (1985). Data are presented in Table 2.

The Kirkcolm Formation

The Kirkcolm Formation is a Llandeilo–Caradoc, greywacke sequence derived largely from the NE but with a very varied palaeocurrent pattern (Fig. 4a–c). The rocks are dominated modally by the minerals quartz and feldspar with lithic fragments of acid and basic (spilitic) igneous and metamorphic rocks generally typical of a 'mature' sediment from a continental source (Fig. 5).

Proterozoic T_{DM} model ages of 1.62 and 1.82 Ga and εNd_{445} of −8.8 and −11.2 for samples S73826 and S76514 respectively, support derivation from a mid-Proterozoic source. However, contribution from a partially volcanogenic terrain is indicated by 11% spilitic material in S76514. The effect of this on the isotopic composition of the sediment is minimal: a theoretical mixture of 10% basalt (Sm/Nd = 0.3, 8 ppm Nd, $\varepsilon Nd = +8$) and 90% material representing Proterozoic crust (Sm/Nd = 0.18, 30 ppm Nd, $\varepsilon Nd = -12$) produces a rock with Sm/Nd = 0.184, $\varepsilon Nd = -11.3$ which approximates to the Sm/Nd = 0.189, $\varepsilon Nd = -11.2$ results from S 76514. The εNd value of the continental crust end-member is only lowered by 0.8 epsilon units to compensate for 10% basalt/spilite 'contamination' and so that isotopic composition of the Kirkcolm greywackes is probably close to that of its continental provenance. This

similarity allows a comparison between the Kirkcolm Formation data and that from possible source regions.

The field of isotopic composition (as defined by currently available data) for Lewisian, Torridonian and Dalradian rocks north of the Iapetus Suture is plotted on Fig. 6, calculated at the time of deposition of the Kirkcolm Formation. One Torridonian sample has a similar isotopic composition to S76514 but all the others are too depleted in radiogenic Nd to represent the source of the Kirkcolm Formation detritus. Results from this study are therefore most compatible with a provenance in an 'along strike' mid-Proterozoic terrane and do not support derivation from the currently adjacent basement to the north and north-west. However, since the variable palaeocurrent pattern, involving opposing axial flows, is not supportive of a well-developed axial distribution system (Fig. 4a–c) the strike-slip separation of the Southern Uplands from its original provenance may be a preferable solution.

The Portpatrick and Galdenoch formations

The Portpatrick Formation greywacke sequence is dominated by southwesterly derived detrital andesite grains and andesitic pyroxenes eroded from a calc-alkaline ensialic arc (Styles *et al.* 1989). The isotopic composition of four samples cluster tightly giving εNd_{445} (time of deposition) between −2.9 and −2.3 and T_{DM} model ages between 1.28 and 1.35 Ga. Andesitic grains in total comprise 68% of S73805 (Fig. 5) and this sample thus gives the best available estimate of $\varepsilon Nd_{445} = -2.9$ for the isotopic composition of the source andesite.

The Galdenoch Formation is also a volcanic-rich greywacke sequence but it shows a wider

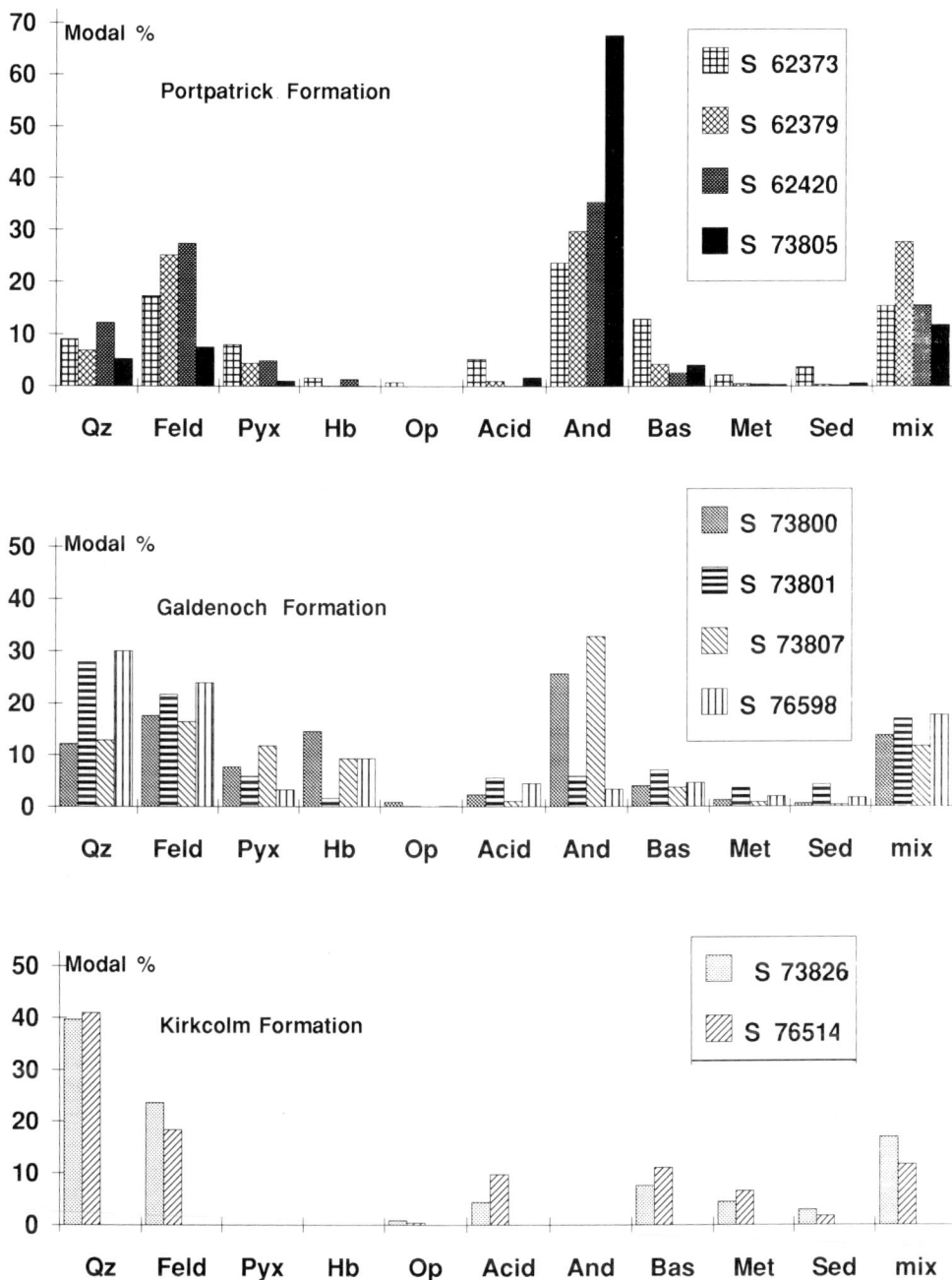

Fig. 5. Summary of point-count data for Southern Upland greywackes. All samples are medium grained sandstones collected from the Bouma 'a' division of turbidite units. 1000 points were counted in each thin section at 1 mm point spacing along traverse lines 1 mm apart. Qz, mono- and polycrystalline quartz grains excluding quartzite; Feld, K and Na/Ca feldspar grains; Pyx, pyroxene grains; Hb, hornblende grains; Op, all opaque mineral grains. Acid, acidic rock fragments; And, andesitic rock fragments; Bas, all other basic igneous fragments. Met, metamorphic rock fragments; Sed, sedimentary rock fragments; Mix, interstitial material and grains less than 0.01 mm diameter.

variation in sedimentary mineralogy than the Portpatrick Formation with quartz contents between 12 and 30% (Fig. 5). The 9–15% hornblende seen in the Galdenoch greywackes also suggest that its volcanic source region was mineralogically distinct from that of the Portpatrick Formation. Styles *et al.* (1989), noting the age difference between the Galdenoch and Portpatrick formations (Fig. 2), linked their provenances to two stages in the history of an evolving arc.

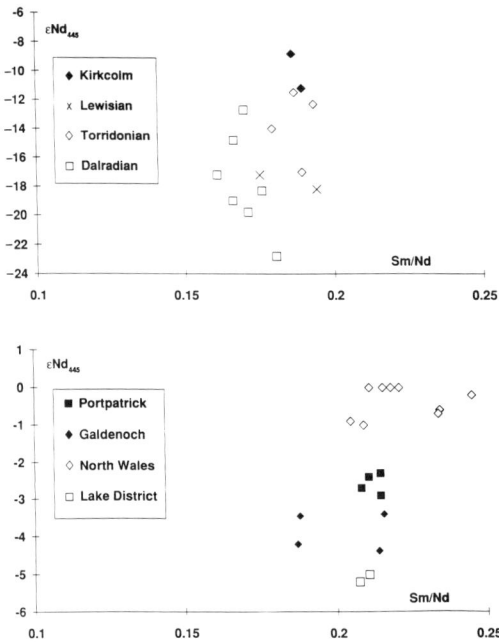

Fig. 6. Sm/Nd and εNd_{445} values of the Kirkcolm Formation compared with those of possible source rocks (upper diagram); and the Portpatrick and Galdenoch Formations with Ordovician andesitic and dioritic lithologies (lower diagram). Additional data from O'Nions *et al.* 1983, Evans 1990 and C. C. Rundle pers. comm.

On petrographic grounds the four Galdenoch Formation samples can be split into two pairs: S73800 and S73807 contain a higher proportion of andesitic detritus and pyroxene fragments and have Sm/Nd ratios of 0.216 and 0.214, whereas samples S73801 and 76598 contain more quartz and feldspar and have lower Sm/Nd ratios of 0.188 and 0.187. However, the mineralogical control over the Sm/Nd ratios is not mirrored by a correlation with Nd isotopic composition; the higher epsilon of −3.4 is shared by andesitic-rich S73800 and the more quartz-feldspar rich S73801. The other two samples have lower values of −4.2 and −4.4. From this is appears that the Galdenoch Formation is derived from

an intermediate–acid igneous source with a Nd-isotopic signature close to −4.

Comparison of the isotopic characteristics of the volcanogenic detritus with known Ordovician andesites can be made allowing for two assumptions: (1) that the age of deposition of the greywackes is close to the age of the parent andesite source; (2) that the composition of the greywackes approximates the composition of the parent igneous rock. The Nd isotopic compositions for the greywackes must clearly represent a minimum value for the source andesites at a given time because, if a small Proterozoic component is present in the detritus, the true value of the igneous rocks will be higher than the whole-rock value of the greywacke. Bearing these qualifications in mind, the Portpatrick and Galdenoch greywackes are compared with samples of approximately contemporaneous volcanic rocks from North Wales and the Lake district (Fig. 6), because no in situ Caradoc andesites are available from the Southern Uplands or elsewhere on the northern side of the Iapetus Suture, at least within Britain and Ireland. Note that these English and Welsh volcanic rocks cannot represent the source of the Southern Uplands volcanogenic detritus, but they can be used to constrain the isotopic composition of intermediate volcanic material generated in a late Ordovician ensialic marginal basin and arc environment respectively. Significantly, the Sm/Nd ratios and epsilon values at the time of deposition (445 Ma) for both the Portpatrick and Galdenoch Formations lie between the currently available Welsh and Lake District data and are thus consistent with contemporaneous arc and/or marginal ensialic arc provenances. The negative epsilon Nd values inferred for the volcanic detritus (accepting assumptions above) are consistent with incorporation of nonradiogenic, crustal Nd during original magma formation by either assimilation of continental crust or the introduction of subducted sedimentary material to the zone of magma generation. Mid-ocean ridge basalt (MORB) usually has $\varepsilon Nd > +7$ and so does not provide a suitable source rock.

Geochronology can test the contemporaneity of arc sources for volcanic detritus. Detrital volcanic clasts from the Portpatrick Formation have yielded $^{40}Ar–^{39}Ar$ cooling ages of 560–530 Ma which indicate that at least some of the volcanogenic detritus was not derived from a contemporaneous arc but from an older arc terrane (Kelley & Bluck 1989). However, the isotopic composition of the Portpatrick and Galdenoch formation detritus at 545 Ma (an average of the ages above) is between −1.5 and −2.1 and −2.4 and −3.5 respectively, and this is still entirely within the range of arc and marginal basin andesites and diorites represented by available British data. The Nd isotopic data therefore cannot be used to distinguish a contemporaneous from an older provenance for the volcanic components of the greywackes.

Marchburn, Shinnel and Ross formations

Three samples from O'Nions et al. 1983 can be assigned to specific greywacke formations, these are SU21, SU19a and SU47 which come from the Marchburn, Shinnel and Ross formations respectively. The εNd_{430} of -0.5 from the Marchburn Formation sample is one of the most radiogenic signatures so far recorded from Southern Uplands greywackes. The Marchburn Formation greywackes are heterogeneous in composition with significant components derived from ophiolitic ultramafic rocks and spilites. Their magnetic susceptibility is remarkably high (Floyd & Trench 1989) as a result of a high detrital magnetite content. This important basic detrital fraction is probably responsible for the high radiogenic signature and a comparison can be drawn with the elevated radiogenic signature of those Portpatrick Formation samples containing the most basaltic/spilitic grains. The Caradoc/Ashgill Shinnel Formation (Fig. 2) contains 57% detrital quartz accompanied by igneous acid clasts, quartzite, shale and spilite fragments making it petrographically rather similar to the Kirkcolm Formation. Shinnel Formation greywacke sample SU19a (O'Nions et al. 1983) has an Sm/Nd ratio of 0.1843, εNd 430 Ma of -9.8 and a T_{DM} of 1.65 Ga. This is within the range of isotopic data from the Kirkcolm Formation and demonstrates that within the limits of the current data set the Shinnel and Kirkcolm Formations are petrographically and isotopically similar.

The Ross Formation is significantly younger than the other formations discussed. It was deposited in the late Llandovery but has an Nd_{430} value of -5.9, between the Kirkcolm Formation values and those of the Portpatrick Formation. This is compatible with its intermediate composition; the detrital components include quartz, feldspar, acid igneous rocks and spilite and is consistent with the suggestion that Llandovery greywackes were formed by the erosion and reworking of the Ordovician formations (Leggett et al. 1979).

Concluding discussion

The Portpatrick and Kirkcolm formations represent the detritus of two end-member provenances for the greywackes of the Southern Uplands; volcanogenic andesitic detritus with εNd_{445} close to -2.5 ± 0.5 and a Proterozoic source region with εNd_{445} as low as -11.2. All the other greywackes of this study, with the possible exception of the Marchburn Formation, can be described in terms of variable proportions of these end-members.

During Llandeilo–Caradoc times greywackes were derived (and interbedded) from both a Proterozoic source (the Kirkcolm Formation) and a markedly different, ensialic volcanic terrain (the Galdenoch Formation). Later in the Caradoc greywackes derived from the S and SW were deposited (the Portpatrick Formation) containing abundant volcanic arc andesitic detritus. Since ophiolitic detritus is also present in some formations such as the Marchburn (Floyd 1982), models of the provenance of Southern Uplands sedimentary rocks must consider a range of volcanogenic end-members such as MORB and/ or arc/back-arc andesitic magmatic rocks. On petrographic criteria alone the greywackes could result from 20–30% dilution of Dalradian detritus with volcanogenic material but the Dalradian is too depleted in radiogenic Nd to be the source of even the Kirkcolm Formation detritus. Although the possibility can not be excluded entirely because of the limited nature of available data sets, the Proterozoic rocks currently exposed to the north of the Southern Uplands, such as the Dalradian assemblage, are unlikely to have been the provenance for the Southern Uplands greywackes. Thus, circumstantial support is given to current concepts of large scale sinistral strike-slip along the Southern Uplands Fault (e.g. Elders 1987).

Integration of palaeocurrent, petrographic and isotopic data sets suggest that, during the late Ordovician, a broadly similar Proterozoic basement existed on both sides of the Southern Uplands depositional basin. At the southern margin it formed the foundations of a calc-alkaline andesitic terrane (arc and/or back-arc) from which were derived the volcanic components of the Portpatrick and Galdenoch formations. Once delivered into the depositional basin the palaeocurrent evidence shows that the volcanic detritus was mainly carried axially towards the NE. The Nd isotopic evidence is not sufficient to establish whether or not the andesitic volcanicity preceded, or was contemporaneous with, greywacke deposition.

We thank J. S. Daly, P. Haughton and another for constructive reviews. J. Evans thanks C. C. Rundle for permission to quote unpublished data from Teighton

Howe andesite. This paper is published with the permission of the Director, British Geological Survey (NERC) and is part of the NERC Isotope Geosciences Laboratory publication series, No. 14.

References

BAMFORD, D., NUNN, K., PRODEHL, C. and JACOBS, B. 1977. LISPB-III. Upper crustal structure of Northern Britain. *Journal of the Geological Society, London*, **133**, 481–488.

BARNES, R. P., ANDERSON, T. B. & McCURRY, J. A. 1987. Along-strike variation in the stratigraphical and structural profile of the Southern Uplands Central Belt in Galloway and Down. *Journal of the Geological Society, London*, **144**, 807–816.

DAVIES, G. R. 1983. *The isotopic evolution of the British Lithosphere*. PhD thesis. Open University.

ELDERS, C. F. 1987. The provenance of granite boulders in conglomerates of the Northern and Central Belts of the Southern Uplands of Scotland. *Journal of the Geological Society, London*, **144**, 853–863.

EVANS, J. A. 1990. *Resetting of Rb-Sr whole-rock isotope systems during low-grade metamorphism, North Wales*. PhD thesis, London University.

FLOYD, J. D. 1982. Stratigraphy of a flysch succession: the Ordovician of W Nithsdale, SW Scotland. *Transactions of the Royal Society of Edinburgh: Earth Sciences*. **73**, 1–9.

—— & TRENCH, A. 1989. Magnetic susceptibility contrasts in Ordovician greywackes of the Southern Uplands of Scotland. *Journal of the Geological Society, London*, **146**, 77–83.

FROST, C. D. & WINSTON, D. 1987. Nd isotope systematics of coarse- and fine-grained sediments: examples from the Middle Proterozoic Belt Purcell Super Group. *Journal of Geology*, **95**, 309–327.

HAUGHTON, P. D. W. 1988. A cryptic Caledonian flysch terrane in Scotland. *Journal of the Geological Society, London*, **145**, 685–704.

KELLEY, S. and BLUCK, B. J. 1989. Detrital mineral ages from the Southern Uplands using ^{40}Ar–^{39}Ar laser probe. *Journal of the Geological Society, London*, **146**, 401–403.

KELLING, G. 1961. The stratigraphy and structure of the Ordovician rocks of the Rhinns of Galloway. *Quarterly Journal of the Geological Society of London*, **117**, 37–75.

—— 1962. The petrology and sedimentation of Upper Ordovician rocks in the Rhinns of Galloway, south-west Scotland. *Transactions of the Royal Society of Edinburgh*, **65**, 107–137.

——, DAVIES, P. & HOLROYD, J. 1987. Style, scale and significance of sand bodies in the Northern and Central belts, southwest Southern Uplands.

Journal of the Geological Society, London, **144**, 787–806.

LEGGETT, J. K. 1987. The Southern Uplands as an accretionary prism: the importance of analogues in reconstructing palaeogeography. *Journal of the Geological Society, London*, **144**, 737–752.

——, McKERROW, W. S. & EALES, M. H. 1979. The Southern Uplands of Scotland: a Lower Palaeozoic accretionary prism. *Journal of the Geological Society, London*, **136**, 755–770.

McCULLOCH, M. T. & WASSERBURG, G. J. 1978. Sm-Nd and Rb-Sr chronology of continental crust formation. *Science*, **200**, 1003–1011.

MICHARD, A., GURRIET, P., SOUDANT, M. & ALBEREDE, F. 1985. Nd isotopes in French Phanerozoic shales; external vs internal aspects of crustal evolution. *Geochimica et Cosmochimica Acta*, **49**, 601–610.

MILLER, R. G. & O'NIONS, R. K. 1984. The provenance and crustal residence ages of British sediments in relation to palaeogeographic reconstructions. *Earth and Planetary Science Letters*, **68**, 459–470.

MORRIS, J. H. 1987. The Northern Belt of the Longford-Down Inlier, Ireland and Southern Uplands, Scotland: an Ordovician back-arc basin. *Journal of the Geological Society, London*, **144**, 773–786.

NELSON, B. K. & DePAOLO, D. J. 1988. Comparison of isotopic and petrographic provenance indicators in sediments from Tertiary continental basins of New Mexico. *Journal of Sedimentary Petrology*, **58**, 348–357.

O'NIONS, R. K., HAMILTON, P. J. & HOOKER, P. J. (1983). A Nd isotope investigation of sediments related to crustal development in the British Isles. *Earth and Planetary Science Letters*, **63**, 229–240.

STONE, P., FLOYD, J. D., BARNES, R. P. & LINTERN, B. C. 1987. A sequential back-arc and foreland basin thrust duplex model for the Southern Uplands of Scotland. *Journal of the Geological Society, London*, **144**, 753–764.

STYLES, M. T., STONE, P. & FLOYD, J. D. 1989. Arc detritus in the Southern Uplands: mineralogical characterization of the 'missing' terrane. *Journal of the Geological Society, London*, **146**, 397–400.

THORPE, R. S., BECKINSDALE, R. D., PATCHETT, P. J., DAVIES, G. R. & EVANS, J. A. 1984. Crustal growth and late Precambrian-Palaeozoic plate tectonic evolution of England and Wales. *Journal of the Geological Society, London*, **41**, 521–536.

Geochemistry and provenance of Rhenohercynian synorogenic sandstones: implications for tectonic environment discrimination

P. A. FLOYD[1], R. SHAIL[1], B. E. LEVERIDGE[2] & W. FRANKE[3]

[1] *Department of Geology, University of Keele, Staffordshire ST5 5BG, UK*

[2] *British Geological Survey, St Just, 30 Pennsylvania Road, Exeter EX4 6BX, UK*

[3] *Institut für Geowissenenschaften und Lithospharenforschung, Justus-Liebig Universität, D-3600 Giessen, FRG*

Abstract: The provenance of synorogenic greywackes from Devonian flysch successions occupying structurally similar positions at opposite ends of the Rhenohercynian zone in SW England and Germany is evaluated. Greywackes from both regions are petrographically and chemically similar, although minor differences are seen in the relative proportions of lithics; the Gramscatho group being richer in volcanic and metavolcanic clasts, whereas the Giessen group is richer in metasedimentary clasts. Absolute abundances of Ni-Cr-V and Zr-Hf-Y vary to a limited extent in each group and reflect variable mafic detritus and heavy mineral inputs respectively. Framework mode parameters and chemical data indicate the Rhenohercynian greywackes were derived mainly from a calc-alkali, acidic, dissected continental arc source, with minor MORB-like and argillaceous metasedimentary components. Upper continental crust-normalized multi-element patterns for the greywackes are characteristic of the continental arc/active margin tectonic environment. However, 'Mid-Proterozoic' model Nd ages for the Gramscatho greywackes suggest that an active Devonian arc source is unlikely. Instead the range of chemical and isotopic composition displayed mainly reflects mixing between acidic arc terranes of Proterozoic age and Devonian (Lizard-type) oceanic crust. Petrographic and geochemical discrimination diagrams alone cannot resolve the temporal decoupling between source and basin and may lead to an erroneous interpretation of tectonic setting.

The sedimentary record in the Rhenohercynian zone of the northern European Variscides (Fig. 1) reflects the combination of an early to late Devonian rift event and a late Devonian to late Carboniferous crustal shortening event (Franke 1989).

The rift event was heterogeneous and resulted in a crude southwards increase in extension across the zone. A series of basins were developed on, for the large part, variably attenuated continental crust as indicated by the intraplate nature of contemporary volcanism (Floyd 1984). These basins were initially sourced by fluvial and neritic clastic sediments derived from Caledonian uplifts to the north. As rifting continued, this input was reduced and permitted the formation of reef and platform carbonates. In areas abandoned by neritic sedimentation, hemipelagic sediments were deposited (Engel *et al.* 1983*b*). However, in the more internal parts of the zone, rifting persisted until limited development of MOR-type oceanic crust was achieved (Bromley 1979; Kirby 1979; Floyd 1984; Grosser & Dorr 1986). The earliest sediments recorded here are Lower to Middle Devonian shales, radiolarian cherts and occasional thin greywackes (Engel *et al.* 1983*b*).

The crustal shortening event was brought about by the continued northwards migration of the collision processes already operating within the more internal sectors of the orogen (Franke 1989). Its effects are first realized in the Frasnian when, for the first time during the Devonian, a major uplifted source area (the Mid-German Crystalline Rise/Normannian High) became available at the southern margin of the Rhenohercynian zone (Engel & Franke 1983; Holder & Leveridge 1986*a*, *b*). During the Frasnian and Famennian, the most southerly parts of the zone were infilled by synorogenic deep-water clastic sediments derived from this source (Engel & Franke 1983). The transition from a rift basin to a foreland basin style of sedimentation was achieved by 'A-type' subduction of continental crust in the sense of Bally (1981) or Weber (1981) which may have succeeded to 'B-type' subduction in area where MOR-type crust had been developed.

As collision processes continued during the Tournaisian and Namurian, the locus of this clastic sedimentation migrated further northwards across the Rhenohercynian zone in a foreland basin style (Engel & Franke 1983; Seago & Chapman 1988), whilst the more

From Morton, A. C., Todd, S. P. & Haughton, P. D. W. (eds), 1991, *Developments in Sedimentary Provenance Studies.* Geological Society Special Publication No. 57, pp. 173–188.

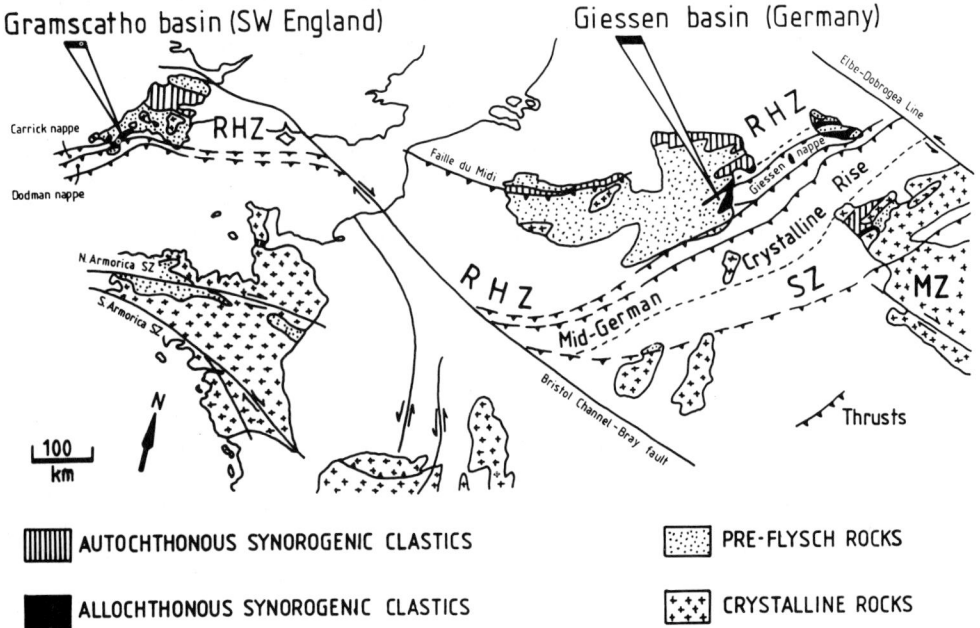

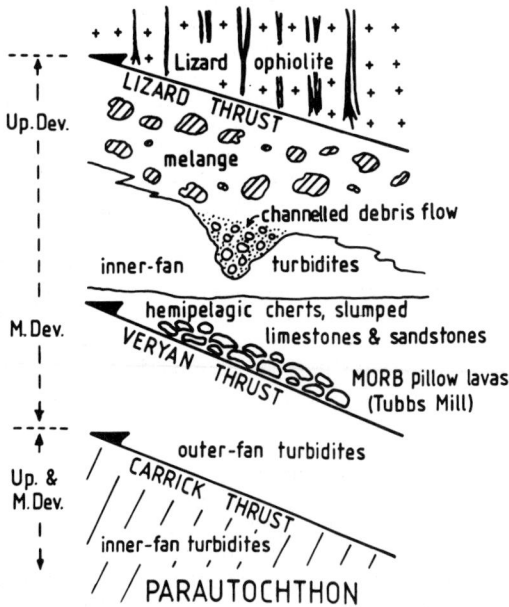

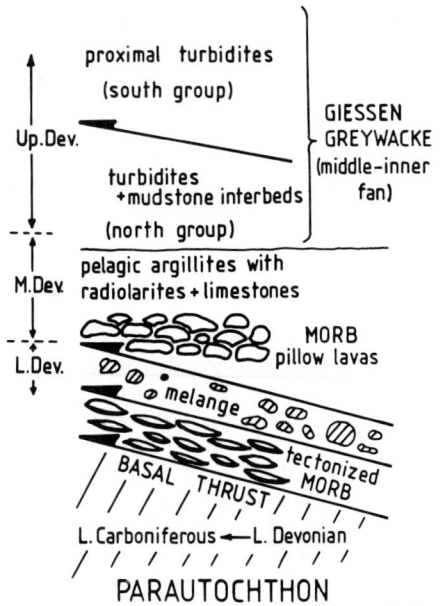

Fig. 1. Schematic cartoon showing the structural relationships between different lithological units within the Gramscatho (Holder & Leveridge 1986*b*) and Giessen (Birkelbach *et al.* 1988) greywacke-dominated sequences (not to scale). The Giessen proximal tubidites are not reliably dated and could be Lower Carboniferous. The map shows the relative positions of the Gramscatho and Giessen basinal sequences within the Rhenohercynian zone (RHZ) of northern Europe. SZ, Saxothuringian zone; MZ, Moldanubian zone.

internal zones underwent deeper levels of thrusting that resulted in the presently observed arrangement of lithostratigraphic units (Fig. 1).

Current plate tectonic models of the Rhenohercynian zone proposed to explain the initial rifting episode are: (a) a back-arc basin related to northerly directed subduction (Floyd 1982; Leeder 1982), (b) an intracratonic strike-slip system in which transtensional basins developed (Badham 1982; Barnes & Andrews 1986), or (c) a small ocean basin with southerly directed subduction below an active arc (Holder & Leveridge 1986a, b).

This paper chemically and petrographically compares temporally equivalent synorogenic clastic sediments from the most internal sectors of the Rhenohercynian zone preserved in SW England and W Germany. Although the outcrops are discontinuous along the strike of the Rhenohercynian zone, its continuity is emphasized by the similarity of the sedimentary successions, submarine bimodal volcanism and comparable metamorphic and deformational histories (Franke & Engel 1982; Holder & Leveridge 1986b). However at present it is equivocal whether the outcrops were part of a single narrow ocean basin or a series of smaller discontinuous basins. The object of this paper is to (a) evaluate the nature and composition of the greywacke source areas, and (b) to test the applicability of various techniques that may discriminate the tectonic environment.

The initial approach is based on recent petrographic and chemical studies which suggest that sandstone compositions may be used to determine their provenance and plate tectonic environment (e.g. Dickinson & Suczek 1979; Bhatia 1983; Bhatia & Crook 1986; Roser & Korsch 1988).

Rhenohercynian greywacke data base

Previous studies on Devono-Carboniferous greywackes in the German sector have demonstrated a gross decrease in total feldspar coupled with a concomitant increase in quartz and proportion of sedimentary and meta-sedimentary lithic fragments from the Devonian–Lower Carboniferous into the Upper Carboniferous (Huckenholz 1963; Henningsen 1978; Engel et al. 1983a). Chemical data on German Rhenohercynian sediments (including greywackes) show them to be markedly enriched in Cr and Ni relative to average abundances and according to Schulz-Dobrick & Wedepohl (1983) indicates the influence of ultramafic debris eroded from Caledonide ophiolite complexes.

The data presented here compare temporally equivalent greywacke successions in SW England (from the Gramscatho basin) and Germany (from the Giessen basin) of Middle/Upper Devonian age. The greywackes within these basins are mainly inner- to outer-fan turbidites that are now preserved within northward transported tectonic slices also containing MORB-like pillow lavas, pelagic sediments and sedimentary melanges (Fig. 1). Both the Gramscatho and Giessen sequences can be interpreted as representing dismembered fragments of Rhenohercynian ocean crust and turbidite deposited basin infill. Faunal control of the age of the greywacke successions, associated sediments and melange matrices tends to be limited, although the Gramscatho group probably spans the Middle to Upper Devonian (Sadler 1973; Barnes 1983; Le Gall et al. 1985; Cooper 1987; Wilkinson & Knight 1989), with a bias towards the younger age, whereas the Giessen Greywacke is largely Upper Devonian (Birkelbach et al. 1988).

Some 95 greywacke samples collected from the basal segment of turbidites (Ta unit) within the Gramscatho and Giessen sequences have been analysed for this comparative study. Detailed geology, basic petrography and related chemical data for the greywackes are given elsewhere (Holder & Leveridge 1986a; Floyd & Leveridge 1987; Birkelbach et al. 1988; Shail & Floyd 1988; Floyd et al. 1990). Apart from the bulk analysis of Gramscatho greywackes, pebble and cobble-sized acidic clasts from pre-melange channelized debris flows and from within the main melange of south Cornwall (Fig. 1), were also chemically analysed and represent new, additional data for this study. Various basic clasts and pillow lava sequences within the melange have enriched MORB-like compositions (Barnes 1984; Floyd 1984) similar to the Tubbs Mill lavas at the base of the Veryan nappe succession (Floyd 1984; Leveridge et al. 1990).

Petrographic features

During the Hercynian orogeny the original detrital mineralogy and textures of the greywackes were affected by microstructural deformation and metamorphism (as well as early diagenesis). These effects are mainly exhibited by the Gramscatho greywackes (rather than the Giessen samples) with the variable development of phyllosilicates, albitization of feldspars and matrix recrystallization. However, in terms of framework mode analysis and the relative proportions of individual clast types (as in the discriminant diagrams of Dickinson *et al.* 1983) the overall effect of alteration appears to be minimal with a small relative increase in quartz, Q in Q-F-L diagram and Qp in Qp-Lvm-Lsm diagram, due to degradation of feldspar and lithics (Shail & Floyd 1988).

However, even allowing for alteration, detailed petrography and framework mode analysis of the Gramscatho greywackes (Floyd & Leveridge 1987) and the Giessen greywackes (Floyd *et al.* 1990) demonstrate that they have many features in common which are probably characteristic for (Upper) Devonian Rhenohercynian synorogenic clastic sediments.

(a) Ta unit turbidites are of fine sand grain-size with mean and standard deviation of 0.15 ± 0.10 mm. There is no statistical significance in grain-size variations between the upper and lower stratigraphical units of the Gramscatho (Portscatho Formation), although the more proximal turbidites of the Giessen south group have twice the average grain-size of the northern group. Maximum grain-sizes are of an order of magnitude larger than the mean grain-size in both regions.

(b) Relative proportions of quartz, feldspar and lithic clasts in terms of the Q-F-L diagram are on average virtually the same (Fig. 2); error polygons, based on 1 standard deviation, are tight and overlapping for the two areas (Floyd & Leveridge 1987; Floyd *et al.* 1990). Average values for Gramscatho greywackes: $Q_{35}F_{31}L_{34}$ and Giessen greywackes: $Q_{33}F_{30}L_{37}$.

(c) The lithic components are dominated by acidic magmatic rocks, in particular devitrified and recrystallized rhyolites, quartz- and/or feldspar-phyric microcrystalline porphyries, medium-grained granites and some intermediate plutonics. When observed, basaltic components are generally highly degraded (to chlorite) in the Gramscatho greywackes, although smectite-replaced, plagioclase-phyric tachylites and basalts with quench textures (pillow lavas?) are relatively common in the Giessen sequence. Sediments are generally fine-grained argillites, together with metamorphosed (greenschist-facies) phyllosilicate-bearing equivalents.

(d) Zircon is the commonest recognizable detrital heavy mineral. It is not uniformly distributed throughout, neither is it specifically concentrated in the coarser grain-size fractions. Only the lower stratigraphic unit of the Gramscatho is characteristically high in zircon (Floyd & Leveridge 1987). Detrital tourmaline is less common, but again found in both successions.

The most significant difference between the Gramscatho and Giessen greywackes concerns the relative proportions of the various lithic clasts. In both cases the abundance of volcanic and metamorphic clasts is far greater than sedimentary clasts, with average proportions as follows, Gramscatho: $Lv_{54}Lm_{33}Ls_{13}$ and Giessen: $Lv_{29}Lm_{64}Ls_7$. The Gramscatho lithic clasts are predominantly volcanic and metavolcanic (invariably acidic), whereas the Giessen lithic clasts are mainly metasedimentary. There is also a

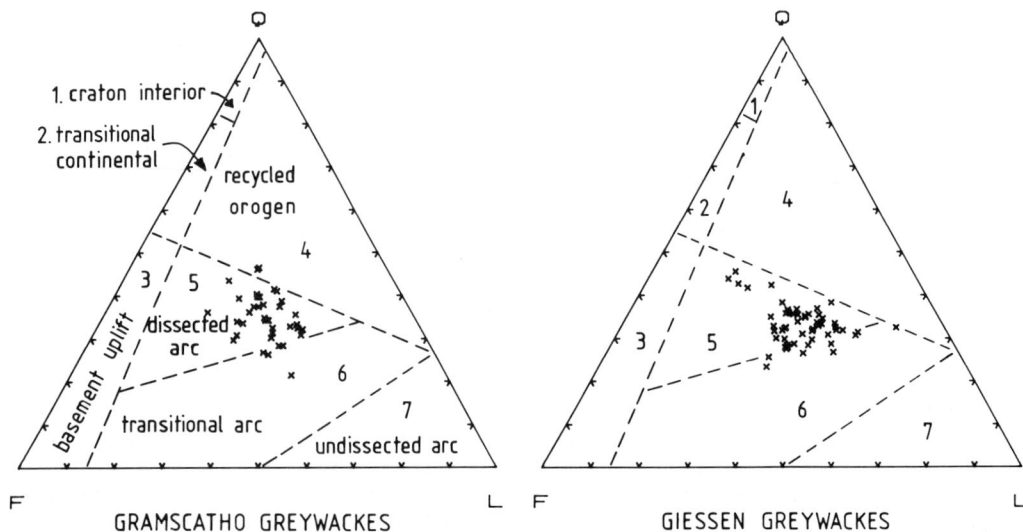

Fig. 2. Comparison of Rhenohercynian greywacke framework mode parameters, quartz (Q), feldspar (F) and lithic fragments (L) in the source discrimination diagram of Dickinson *et al.* (1983).

higher proportion of recognizable basaltic clasts within the Giessen greywackes. In terms of QFL framework modes (Fig. 2) a dissected arc environment (in part transitional to a recycled orogen) appears characteristic for the Hercynian greywackes studied here, although the lithic clast subpopulations indicate the availability of higher proportions of metasedimentary (basement?) material in the Giessen source area.

Chemical features of Rhenohercynian greywackes

Factors affecting sandstone chemistry

Before the Gramscatho and Giessen successions can be compared and evaluated in terms of chemical factors, it is necessary to consider briefly features which could affect their chemical composition, such as grain-size, degree of source weathering, diagenesis and metamorphism (Sawyer 1986; Wronkiewicz & Condie 1987).

As the greywackes studied here have a narrowly defined, relatively uniform grain-size and were collected from the same portion of each turbidite flow, variable hydraulic sorting by grain-size is not considered a major problem to direct chemical comparison. Also, irrespective of the portion of the turbidite unit sampled (of variable grain-size), ratios of many coherent elements remain constant throughout the unit, although absolute abundances may decrease in the coarser fractions due to dilution by quartz clasts (Spears & Amin 1981; Shail & Floyd 1988). For this reason element ratios make better chemical comparators than absolute abundances between sediment successions. Differential weathering at the source tends to mobilize and change the relative abundance of LIL elements (Nesbitt et al. 1980). A measure of the degree of weathering in the source is provided by the chemical index of alteration, CIA (Nesbitt & Young 1982), which suggests relatively low to moderately weathered sources for the Hercynian greywackes (CIA = 55–70; fresh grainte = 45–55) as might be expected. Diagenesis and subsequent metamorphism are two other factors which mainly affect the LIL elements (Hower et al. 1976), U (Colley et al. 1984) and possibly the light REE (Van Weering & Klaver 1985), although the general constancy of ratios (K/Rb, K/Ba, Ce/Sm, U/Th) within the Rhenohercynian greywackes suggests that they have not been systematically changed. However, Gramscatho samples collected from near the outer margins of granite aureoles clearly demonstrate the effect of contact metasomatism with lower than usual, but highly variable K/Rb ratios (200–30) and slightly enhanced light REE and U abundances.

Thus, to overcome many of the problems associated with the LIL elements, less mobile elements (REE, Th, Sc, Nb, Hf etc.) are generally used to characterize provenance and tectonic setting (Taylor & McLennan 1985; Bhatia & Crook 1986).

Chemical groups

Two chemostratigraphic units (Zr-rich lower unit, Cr-rich upper unit) have been recognized within the 3.5 km thick allochthonous Gramscatho (Portscatho Formation) succession (Floyd & Leveridge 1987). A similar distinction has not yet been recognized within the parautochthon which exhibits a range of Cr abundances apparently unrelated to stratigraphic level. The Giessen Greywacke is divided into two thrust slices (Fig. 1), the lower one located in the north and the upper one located in the south of the outcrop. Greywackes in both the north and south thrust 'groups' are broadly similar chemically (Floyd et al. 1990) and correspond chemically to the upper Cr-rich unit of the Gramscatho succession.

General composition

The greywackes from both areas have broadly similar chemical compositions (Table 1) and are derived from predominantly acidic precursors of magmatic origin (Fig. 3), as also demonstrated by the clast populations. The following chemical features are common to both successions and emphasize the predominance of rhyolitic and granitic rocks in the source area: (a) intercorrelations and high contents of K, Rb and Ba (K/Rb c. 230), (b) chondrite-normalized REE patterns with enriched light REE, small negative Eu anomalies (Eu/Eu* = 0.7–0.8) and flat heavy REE (cf. Floyd & Leveridge 1987), (c) generally high Zr and Hf contents reflect zircon released from acidic plutonic rocks. On the other hand, the high Cr, Ni, V and Ti contents (lower than those reported by Schulz-Dobrick & Wedepohl 1983) in both the Gramscatho and Giessen greywackes are indicative of the input of mafic detritus, becoming especially prominent in the upper Gramscatho unit. High Ca and Sr values would also reflect the presence of mafic material, although in this case their initial plagioclase host has been albitized with the subsequent redistribution of these elements.

Table 1. *Comparison of the average chemical composition of greywackes from the Giessen Greywacke Unit, Germany, and the Gramscatho Group, SW England (primary data by Floyd* et al. *1990 and Floyd & Leveridge 1987 respectively)*

| | Giessen | | | | Gramscatho | | | |
| | Northern group | | Southern group | | Lower group | | Upper group | |
	Av	SD	Av	SD	Av	SD	Av	SD
SiO_2	73.42	3.17	73.21	3.13	72.06	2.25	65.85	7.94
TiO_2	0.74	0.09	0.67	0.10	0.74	0.07	0.69	0.13
Al_2O_3	12.06	1.85	12.37	1.25	11.04	1.56	13.09	2.07
Fe_2O_3*	5.25	1.22	4.01	0.65	5.68	0.52	4.98	0.58
MnO	0.05	0.02	0.04	0.01	0.11	0.07	0.10	0.10
MgO	1.45	0.43	1.63	0.41	1.25	0.29	1.89	0.79
CaO	0.60	0.31	0.71	0.41	0.75	0.51	2.59	2.48
Na_2O	2.09	0.81	2.70	0.21	1.78	0.42	3.06	0.73
K_2O	1.67	0.47	2.00	0.35	1.37	0.35	1.68	0.71
P_2O_5	0.15	0.01	0.16	0.02	0.15	0.06	0.16	0.02
LOI	2.52	0.43	2.33	0.46	5.40	1.29	6.25	4.35
	N = 13		N = 45		N = 18		N = 26	
Total	100.00		99.83		100.33		100.34	
Ba	407	117	625	124	318	78	415	161
Ce	61	27	49	11	68	12	57	13
Cr	82	34	92	16	56	18	85	18
Cu	11	10	16	23	6	3	9	5
Ga	13	5	14	4	14	1	15	2
La	31	13	23	7	32	8	29	7
Nb	9	2	7	2	10	2	8	2
Nd	30	10	26	8	27	7	23	4
Ni	47	28	39	7	28	7	41	8
Pb	14	4	15	7	16	6	26	27
Rb	67	18	68	12	60	13	69	27
Sr	123	53	235	58	91	18	272	308
V	98	27	99	18	96	10	116	25
Y	27	7	20	3	29	5	24	3
Zn	96	75	64	34	71	15	63	10
Zr	318	190	206	37	325	84	202	43
Cs	3.6	1.2	4.0	1.0	3.4	0.8	4.2	2.4
Hf	7.8	3.8	5.8	0.8	9.1	3.0	6.2	1.3
Sc	11.1	2.4	11.0	1.8	8.3	1.3	9.0	2.2
Ta	0.90	0.11	0.87	0.06	1.14	0.12	0.95	0.15
Th	9.1	1.7	8.5	0.8	9.5	1.1	8.9	1.1
U	2.6	0.7	2.7	0.3	2.6	0.4	2.8	0.3
La	27.83	5.78	25.79	4.59	31.33	6.74	30.97	6.37
Ce	55.43	11.92	49.78	7.97	60.46	11.23	56.89	8.19
Pr	6.50	1.50	5.26	1.02	7.30	1.54	7.10	1.24
Nd	26.46	4.56	24.58	3.21	27.32	5.90	26.19	4.63
Sm	4.89	0.76	4.27	0.51	5.32	1.10	5.05	0.78
Eu	1.11	0.11	1.03	0.12	1.11	0.17	1.21	0.19
Gd	5.09	0.66	4.37	0.43	4.76	0.93	4.46	0.63
Dy	4.17	0.70	3.37	0.34	4.57	0.71	4.11	0.53
Ho	0.88	0.16	0.73	0.07	0.93	0.14	0.84	0.10
Er	2.45	0.56	1.91	0.33	2.71	0.39	2.38	0.31
Yb	2.33	0.47	1.94	0.19	2.56	0.37	2.14	0.31
Lu	0.38	0.07	0.33	0.03	0.42	0.06	0.35	0.05
	N = 7		N = 10		N = 8		N = 11	

Major oxides in wt %, trace elements in ppm. Major oxides and trace elements Ba to Zr by XRF Spectrometry (Geology Department, University of Keele); trace elements Cs to U by INAA (Universities Research Reactor, Risley); REE by ICP (RHBNC, Egham).
Av, average; SD, standard deviation; N, number of samples.

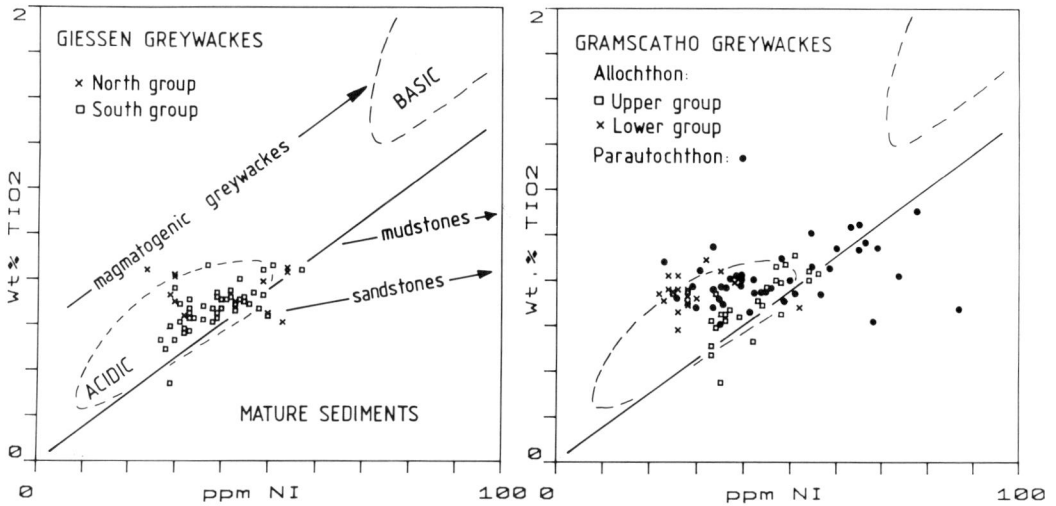

Fig. 3. TiO_2–Ni plot for Rhenohercynian greywackes showing their derivation from magmatic rocks of acidic composition. Acidic and basic fields, and trends for common mature sediments from Floyd *et al.* (1989).

Extracted clast compositions

Pebble- and cobble-sized clasts (rare examples up to 35 cm across) of cataclastic garnet-bearing granitic rocks and quartz-feldspar phyric rhyolites were extracted whole from the Gramscatho sequence (Fig. 1) and analysed. Both groups of rocks have suffered variable deformation prior to incorporation into the debris flows. In view of the preponderance of apparently similar clasts within the greywackes of both regions it was considered that their chemical composition might be representative of the acidic source component. The plutonic rocks are typical two-mica, megacrystic, calc-alkali granites and granodiorites which sometimes contain magmatic garnets and rare tourmaline. Chemically, the granitic and rhyolitic clasts (Fig. 4; Table 2) are dissimilar to the main granites of the Cornubian batholith (Exley *et al.* 1983) or tectonized basement xenoliths recovered from Cornish Upper Devonian volcanic rocks (Goode & Merriman 1987). Their restricted HFS element contents, low Nb (*c.* 5 ppm) and high Zr/Nb ratios (*c.* 16) suggest both groups were initially formed in a subduction-related, volcanic-arc environment (Fig. 4).

The garnets have features indicative of high-level in situ magmatic crystallization (such as, general euhedral form, lacking reaction rims and few inclusions), rather than those of high-pressure phenocrysts or xenocrysts (Leake 1967; Fitton 1972; Manning 1983). This is supported by their chemical composition; being Mn-rich almandines with 10–30 mol.% spessartine (Barnes 1982) which are typical of garnets in silicic rocks

crystallizing at depths < 12 km. (Green 1977). They are markedly different to the Mn-poor, restite-derived, alamandines of the Cornubian granites (Stone 1988) or the Ca-rich, high-pressure, almandine phenocrysts from the Permo-Carboniferous volcanic rocks of the Pyrenees (Gilbert & Rogers 1989). Garnets (and tourmalines), of unknown composition and parentage, are found as heavy mineral grains within both the Giessen and Gramscatho greywackes, although garnets are very rare in the latter succession.

Source characteristics

The above petrographic and chemical data indicate a common sorce area for Devonian greywacke sequences separated by *c.* 800 km along the Rhenohercynian zone. The source was composed of three main constituents: volcanic and plutonic acidic rocks, MORB-like mafic rocks (variably metamorphosed) and generally fine-grained metasediments (up to greenschist-facies of regional metamorphism). The plate tectonic environment of the source region can be chemically discriminated using the La-Sc-Th diagram (Bhatia & Crook 1986) which indicates that both greywacke groups were related to a continental island arc (cf. Floyd & Leveridge 1987; Floyd *et al.* 1990). This setting is supported by the analysed acidic clasts (see above) and framework mode parameters which indicate derivation from a dissected arc (Fig. 2).

The generation of the chemical range displayed by the greywackes and the model age of

Table 2. *New chemical data for granitic (samples 1–5) and rhyolitic (samples 6–9) pebbles and cobbles extracted from south Cornish, Devonian debris flows*

Sample no. Field no.	1 CF1	2 CF2	3 CF3	4 CF4	5 CP2	6 JP1	7 JP2	8 JP3	9 JP4
SiO_2	75.02	73.87	72.85	74.75	74.96	73.76	75.87	73.01	70.67
TiO_2	0.27	0.31	0.27	0.05	0.03	0.29	0.20	0.27	0.26
Al_2O_3	13.16	13.18	14.09	14.57	14.60	12.94	13.10	13.44	14.62
Fe_2O_3*	2.15	2.61	2.18	0.76	0.76	2.05	1.87	3.42	3.60
MnO	0.05	0.05	0.04	0.13	0.13	0.06	0.03	0.05	0.05
MgO	0.55	1.25	0.92	0.03	0.01	1.00	0.53	0.89	0.64
CaO	0.24	0.36	0.67	0.42	0.41	1.41	0.65	1.73	3.27
Na_2O	5.37	3.32	3.56	5.52	5.81	4.62	5.80	5.62	4.18
K_2O	1.75	3.31	3.60	2.40	2.49	0.82	0.66	0.23	0.95
P_2O_5	0.06	0.14	0.19	0.24	0.24	0.07	0.05	0.07	0.06
LOI	0.85	1.33	1.10	0.50	0.43	1.45	1.10	1.01	1.35
Total	99.49	99.73	99.47	99.37	99.88	99.49	99.87	99.74	99.63
Ba	405	664	919	299	277	728	505	190	726
Ce	31	34	34	20	15	34	28	16	17
Cr	15	24	16	12	14	30	16	9	15
Cu	9	4	7	1	1	10	40	18	12
Ga	13	15	13	12	14	11	10	10	13
La	20	15	19	10	10	11	11	10	14
Nb	10	8	8	7	6	7	6	4	5
Nd	20	26	14	12	7	15	13	8	6
Ni	9	9	7	2	2	11	9	6	6
Pb	1	10	11	11	10	8	2	2	8
Rb	25	79	81	74	72	20	13	4	21
Sr	83	131	164	61	60	123	69	126	207
V	25	34	26	3	1	56	42	35	27
Y	24	16	16	8	7	22	18	23	38
Zn	17	25	30	21	21	48	23	22	25
Zr	215	104	107	40	37	143	108	116	123

Major oxides in wt %, trace elements in ppm. LOI, Loss-on-ignition. Major oxide and trace element data determined by XRF Spectrometry (Geology Department, University of Keele).

their precursors can be determined using trace element mixing between possible end-member compositions and Nd-Sr systematics respectively. While it is realized that final sediment compositions involve many complex processes, we have initially invoked the simplest mixing scenario involving only two components. For example, binary plots of ratios of LIL and transition trace elements (Fig. 5) show linear correlations which suggest the chemical spectrum of the greywackes from both areas can be largely accounted for by mixing of two main components: acidic and basic. The Cr/Nb–Ni/Nb plot indicates that these components could be typical island arc intermediate rocks (or the Cornish acidic clasts) and a MORB-type rock similar to the Lizard ophiolitic dykes (Kirby 1984) in composition, rather than the more enriched Tubbs Mill and Harz MORBs. The La/Sc–Th/Sc plot, however, suggests that neither intermediate island arc rocks nor the Cornish acidic clasts are rich enough in LIL elements (like La and Th) to account for the data spread. In this case the

enriched 'acidic' end-member must be an admixture of a typical acidic island arc rock and a mudstone (roughly 60:40). Slight differences in the actual proportions mixed with the MORB end-member account for the scatter of data in this plot.

Nd and Sr isotope data for four Gramscatho greywackes (Table 3) are plotted as ε-values in Fig. 6, together with Hercynian granites from Cornubia and Armorica. While the upper unit Gramscatho samples (with high Cr etc.) have compositions and model ages (T_{DM}) similar to Hercynian granites and enriched continental margin volcanic rocks, the lower unit greywackes are distinct in having the oldest model ages of any Hercynian material. It is unlikely that the model age differences between the upper and lower Gramscatho greywacke units are due to unmixing of different source components during sedimentation (McLennan *et al.* 1989), as the grain-sizes and main framework mode parameters of the two groups are statistically similar.

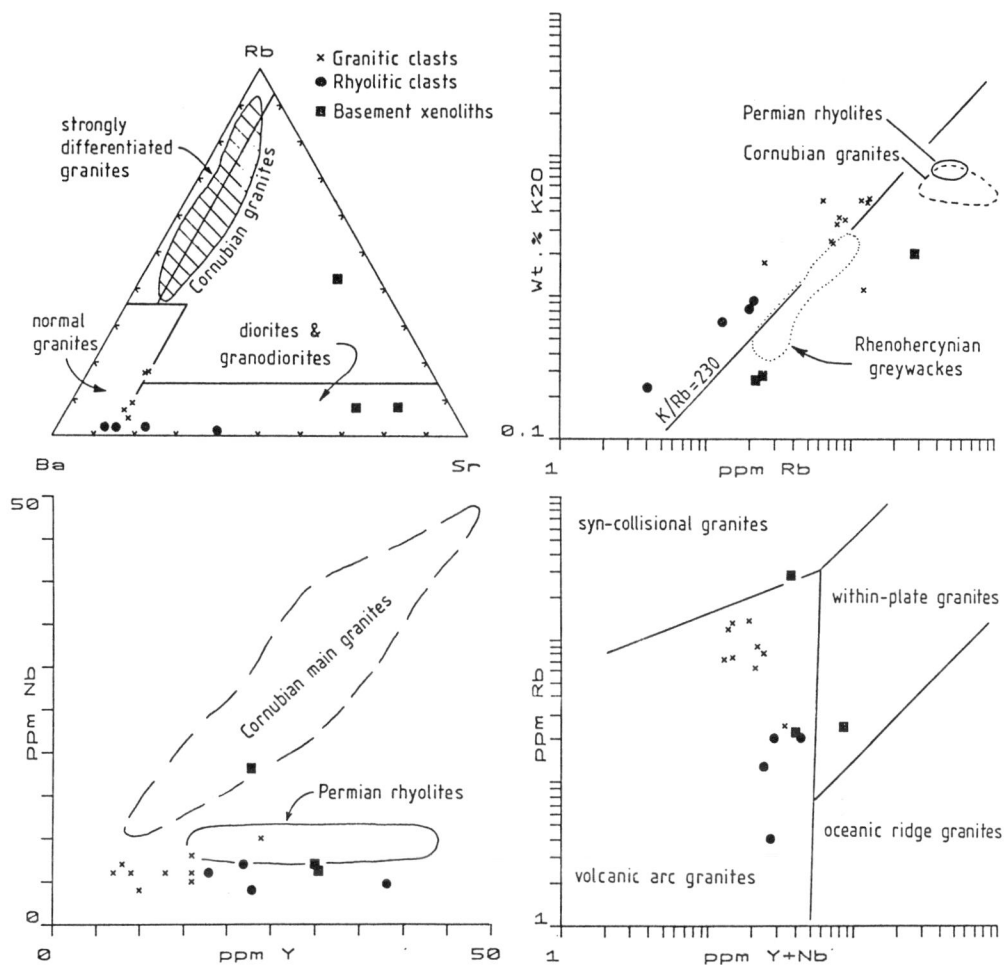

Fig. 4. Chemical features of extracted granitic and rhyolitic Gramscatho Group clasts compared with Cornubian granites (Exley *et al.* 1983), Permian rhyolites (Floyd, unpublished data) and basement xenoliths (Goode & Merriman 1987) from SW England. Rb–Sr–Ba plot from El Bouseily & El Sokkary (1975); Rb–Y + Nb plot from Pearce *et al.* (1984).

The significance of the trace element and isotope data is that the acidic component was not only derived from a dissected arc source, but that the arc was either (a) very old or (b) represented a series of arcs of different ages, that in either case were not contemporaneous with sediment generation or basin development. The basic component, on the other hand, was probably much younger and could have represented local Rhenohercynian oceanic crust that became exposed during thrusting and subsequently denuded. Based on the simple trace element mixing model the isotopic range of the four Gramscatho samples could be accommodated by mixing between two end-members, in this case, Lizard ophiolite and Rb-rich continental crust *c.* 2 Ga old (Fig. 6).

The source areas for the Rhenohercynian greywackes are generally considered to be the Normannian High and the Mid-German Crystalline Rise, both of which show a range of pre-Hercynian acidic continental rocks. For example, Armorica is composed of a collage of displaced terranes comprising late Proterozoic, Cadomian calc-alkali arcs (*c.* 700 Ma) and a Brioverian volcano-sedimentary succession, together with younger post-tectonic acid plutonics (*c.* 425 Ma) (Auvray & Maillet 1977; Denis &

Table 3. *Nd and Sr isotopic data for Gramscatho Group greywackes from lower (samples 1 & 2) and upper (samples 3 & 4) chemostratigraphic units*

Sample no. Field no.	1 GG-06	2 GG-15	3 GG-28	4 GG-44
Sm	7.03	5.55	6.11	5.93
Nd	36.60	28.86	31.16	32.73
$(^{147}Sm/^{144}Nd)$	0.11613	0.11633	0.11857	0.10956
$(^{143}Nd/^{144}Nd)_0$	0.511628	0.511629	0.511791	0.511939
$\varepsilon_{Nd}(t)$	−10.3	−10.3	−7.1	−4.3
T_{CHUR} (Ma) $\pm \sigma$	1374 ± 49	1373 ± 49	1088 ± 50	756 ± 46
T_{DM} (Ma) $\pm \sigma$	1658 ± 73	1658 ± 73	1456 ± 74	1177 ± 69
Rb	60.40	57.80	84.00	63.20
Sr	77.10	70.10	218.00	174.00
$(^{87}Sr/^{86}Sr)_0$	0.713187	0.712185	0.708077	0.707053
$\varepsilon_{Sr}(t)$	+129.8	+115.5	+57.2	+42.6
T_{Bulk} (Ma) $\pm \sigma$	667 ± 4	622 ± 4	647 ± 6	591 ± 6

Chemical and petrographic data given in Floyd & Leveridge (1987).
Trace elements in ppm. Bulk Earth values and decay constants from Hawkesworth & Norry (1983, p. 250).

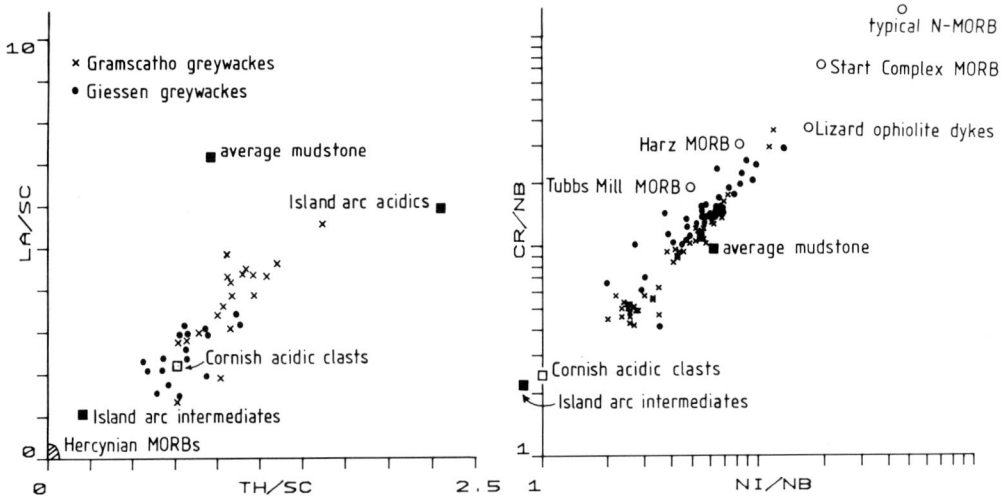

Fig. 5. Trace element ratio diagrams for Rhenohercynian greywackes showing derivation of compositional spectrum by mixing between a MORB-type, calc-alkali island arc magmatic rocks and an argillaceous component. Island arc data from Taylor *et al.* (1968), Taylor (1969) and Jakes & White (1972); average mudstone from Turekian & Wedephol (1961).

Dabard 1988; Strachan *et al.* 1989; Brown *et al.* 1990). Although these rocks might be a chemically suitable source they are generally younger than the model ages of the greywacke precursors implied by the isotope data and could only represent one possible, relatively young, acidic source input. To obtain the older average ages for the sediment precursors, much older material must also contribute to the model age. Ancient, 2 Ga old, pre-Cadomian relicts (Pentevrian) of limited extent or preserved as tectonic rafts within Cadomian plutons are present in Armorica (Calvez & Vidal 1978; Vidal *et al.* 1981; Roach 1988) and could therefore contribute to the older model ages. Suitable calc-alkali continental material was clearly available as a Gramscatho greywacke source in Armorica since c. 2 Ga ago, although the model ages probably reflect mixing between an ophiolitic end-member and different aged acidic end-members (mainly <700 Ma and 2 Ga), rather than a single very old (2 Ga) source. In this context mid-Proterozoic model Nd ages (not dissimilar to the Gramscatho ages) for S-type Hercynian leucogranites from Armorica are considered to be generated by mixing crustal source materials of widely different ages (Peucat *et al.* 1988).

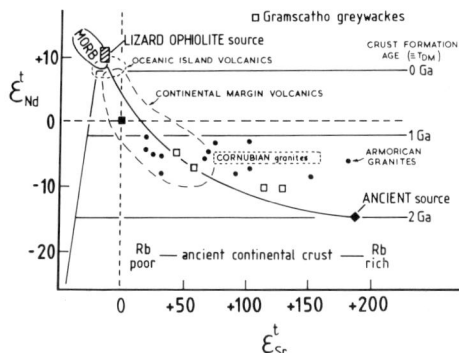

Fig. 6. Nd–Sr isotope diagram in terms of ε-values for four Gramscatho greywackes which approximately fit a mixing line between Lizard ophiolitic MORB (Davies 1981) and 2 Ga old Rb-rich continental crust. Data sources: oceanic island-arc volcanic rocks, continental margin volcanic rocks and crust formation age contours (DePaolo 1988); Hercynian granites (Bernard-Griffiths *et al.* 1985; Peucat *et al.* 1988; Davies 1981; Darbyshire & Shepherd 1985).

Chemical discrimination of tectonic environment

A number of major oxide and trace element-based diagrams reportedly determine the tectonic environment of sediments (Bhatia 1983, 1985; Roser & Korsch 1988), although the best discrimination is provided by ratios of stable trace elements in fine-grained greywackes (Bhatia & Crook 1986) that are quantitatively transfered from source to sink. Meaningful discrimination in simple binary plots may be adversely affected by sorting, heavy mineral content and

proportion of mafic input, such that some sedimentary series spread across a number of geologically unrelated tectonic fields. The La-Sc-Th and La-Zr-Th triangular diagrams of Bhatia & Crook (1986), while providing good discrimination, suffer from being based on a relatively small and geographically restricted data set of Australian greywackes. A brief literature survey of trace element data on Phanerozoic and Proterozoic greywackes worldwide indicates that there is, for example, considerable overlap between the continental arc and active continental margin environments, such that they cannot be adequately distinguished (cf. Van de Kamp & Leake 1985) and also, depending on the source composition the mafic input can often vary considerably in the same environment.

It is suggested that the full range of elemental composition for greywackes in different tectonic environments can be more adequately compared utilizing upper continental crust-normalized multi-element patterns. While not replacing specific discrimination diagrams employing diagnostic ratios, they have the advantage of showing the effect of variable mafic and heavy mineral input within a sedimentary suite, as well as providing a pointer to the tectonic environment. Figure 7 shows the normalized patterns for common tectonic environments based on averaged data (Table 4). The elements are arranged (from right to left) in order of increasing ocean residence times and comprise a relatively stable group (Th-Ta) and a more mobile group (Ni-K). A number of features are displayed (Fig. 7): (a) all patterns show negative Nb-Ta anomalies, the extent of whch can be measured by the Nb/Nb* ratio (actual normalized Nb abundance/calculated normalized Nb abundance extrapolated between Ni and Ti) and is typically low (*c.* 0.15–0.30) for sediment sources involving subduction-related magmatic rocks; the anomaly is often less for passive

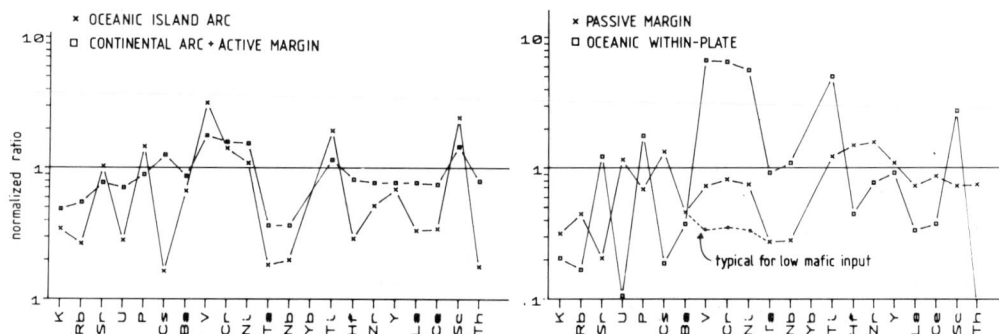

Fig. 7. Normalized multi-element patterns using averaged greywacke data (from Table 4) for different tectonic environments. Upper continental crust normalization values from Taylor & McLennan (1985).

Table 4. *Average composition of late Proterozoic and Phanerozoic greywackes associated with different tectonic environments*

	OIA	CAAM	PM	OWP
SiO_2	58.25	68.73	82.59	51.42
TiO_2	0.98	0.58	0.62	2.52
Al_2O_3	15.55	13.00	7.16	14.66
Fe_2O_3*	7.70	5.35	3.62	13.48
MnO	0.18	0.08	0.15	0.23
MgO	3.10	2.60	1.72	6.61
CaO	5.51	2.76	0.19	9.15
Na_2O	4.13	2.41	1.02	2.86
K_2O	1.17	1.65	1.09	0.70
P_2O_5	0.23	0.14	0.11	0.28
Ba	370	481	255	209
Ce	22	48	56	24
Cr	49	55	29	230
Cu	29	22	8	77
Ga	20	15	8	—
La	10	23	22	10
Nb	5	9	7	27
Nd	10	24	39	15
Ni	22	31	15	114
Pb	15	15	11	—
Rb	30	62	50	19
Sc	27	16	8	30
Sr	362	274	72	432
V	188	106	44	400
Y	15	17	24	20
Zn	88	73	49	122
Zr	99	146	302	146
Cs	0.6	4.7	4.9	0.7
Hf	1.7	4.7	8.8	2.6
Ta	0.4	0.8	0.6	2.0
Th	1.9	8.5	8.1	1.0
U	0.8	2.0	3.2	0.3
Max. no. samples*	57	139	125	27

* Number of samples averaged varies for some elements in each group, with the lowest number of samples represented by the Cs to U elements: OIA (11 samples), CAAM (42, but 93 for U & Th), PM (41) and OWP (9).

Data calculated from the literature. OIA, oceanic island arc; CAAM, continental arc + active margin; PM, passive margin; OWP, oceanic within-plate (ocean islands and seamounts). Major oxides in wt %, trace elements in ppm.

margins involving relatively old, reworked continental crust (c. 0.5), although is influenced by the proportion of mafic material (typically low) in this source; (b) positive V-Cr-Ni-Ti-Sc anomalies are indicative of variable mafic input, being only <1 for the passive margin environment; normalized values of c. 8 (V-Cr-Ni) for the oceanic intraplate environment represent about 90% reworked basaltic material in the greywackes; (c) only the passive margin environment shows a positive Ti-Hf-Zr-Y anomaly reflecting a heavy mineral input (mainly zircon); (d) soluble elements (Ba-K) with normalized values of <1 generally decrease with residence time, although Sr and P peaks are indicative of mafic source input for the oceanic intraplate and island arc environments.

Utilizing multi-element diagrams it can be immediately seen that the element patterns for the Gramscatho and Giessen greywacke groups are very similar (Fig. 8) with negative Nb-Ta anomalies (Nb/Nb* c. 0.35) and positive V-Cr-Ni anomalies (with clear differences between the upper and lower Gramscatho units). With the exception of a variable heavy mineral anomaly for the Gramscatho lower group and the Giessen north unit, stable elements have normalized value close to 1. The patterns exhibited indicate a source composed largely of subduction-generated rocks (continental arc + active margin), although the enhanced 'heavy mineral' anomaly is more typical of a passive margin.

Implications for tectonic environment discrimination

The use of petrographic and geochemical discrimination diagrams to distinguish plate-tectonic settings is now fairly well established after the early work of Dickinson & Suczek (1979) and Bhatia (1983). Such studies are based on the fundamental assumption that the nature and availability of lithologies within a source region are intimately related to the tectonic processes controlling the development of the adjacent depositional basin. If this relationship is valid, the sediment type(s) derived from this source may be used to classify the tectonic setting of the basin.

In this particular study, geochemical and petrographic discrimination techniques both suggest the source region(s) largely comprised a continental island arc with lesser amounts of variably metamorphosed MORB-like mafic rocks and fine-grained sediments. The logical outcome of the argument presented above would be that the Gramscatho and Giessen succession represent the infill of a continental fore-arc basin.

However, although we believe that in this particular case the discrimination techniques correctly identify the source type, it is largely unrelated to the true tectonic setting of the basin. The mid-Proterozoic Nd model ages of

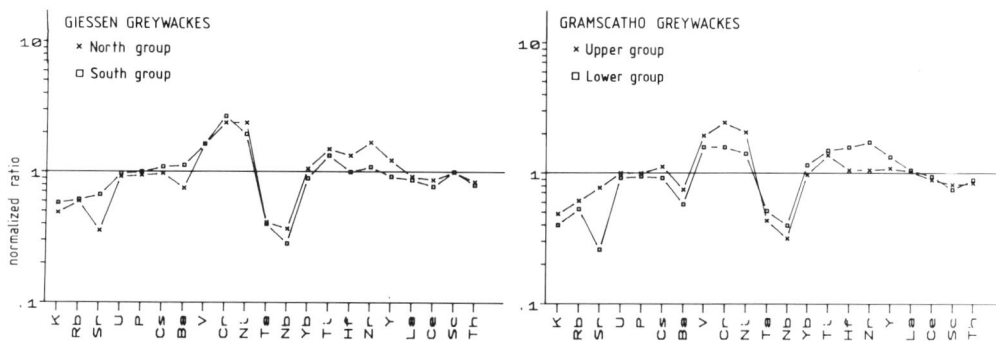

Fig. 8. Normalized multi-element patterns for Gramscatho and Giessen unit averages (from Table 1). All patterns show a general correspondence to the continental arc + active margin tectonic environment distribution illustrated in Fig. 7.

the sediments suggest that contemporaneous Devonian subduction processes could not have significantly contributed to this continental arc source. Although rifting in the Rhenohercynian zone had initiated in the Lower Devonian, it had not progressed much beyond the Red Sea stage by the time Upper Devonian convergence commenced (Davies 1984; Franke 1989). It therefore seems improbable that sufficient oceanic crust to generate a relatively mature arc, upon subduction, had ever been produced in the Rhenohercynian zone during the Devonian.

There is a clear discrepancy between the fore-arc setting implied by discriminatory techniques and the combined rift and foreland basin setting implied by non-discriminatory techniques. This is unlikely to be a local source effect due to the close similarity between the two separate areas. Mack (1984) has described situations in which the fundamental assumption regarding the relationship between source composition and basin tectonic setting is invalid.

We believe we have documented another such case. The main control on the compostion of sediments within the Giessen and Gramscatho successions was the nature of the pre-existing continental crust. This crust had been formed during several Precambrian orogenies and comprised a collage of arc-related terranes. Its tectonic setting was unrelated to that of the Rhenohercynian zone. Initial rifting of this crust, whether by the development of an intracratonic strike-slip system or back-arc basin would have resulted in the same erosional products.

Conclusions

(1) Devonian greywackes from the Gramscatho and Giessen successions at opposing ends of the Rhenohercynian zone have very similar petrographic and chemical features, and were probably derived from a common source type. Minor differences are seen in (a) the relative proportions of lithic clasts (the Gramscatho greywackes are dominated by volcanic and meta-volcanic fragments, whereas the Giessen greywackes are characterized by metasedimentary fragments), and (b) absolute abundances of Ni-Cr-V (representing variable mafic input) and Zr-Hf (representing variable zircon contents).

(2) Framework mode and chemical features indicate derivation from a calc-alkali, acidic, dissected continental arc with minor mafic and sedimentary components together with their metamorphosed equivalents. Upper continental crust-normalized multi-element patterns exhibit variable positive V-Cr-Ni-Ti-Sc anomalies, negative Nb-Ta anomalies with Nb/Nb* ratios of c. 0.35, stable element ratios either close to 1 or a positive Zr-Hf-Y anomaly, all features generally indicative of a continental arc/active margin tectonic environment.

(3) Trace element parameters suggest that the range of compositions displayed by Devonian Rhenohercynian greywackes can be largely accounted for by a simple two (or possibly three) component mixing model between (dominant) LIL-rich acidic arc material and MORB-type basalt like the Lizard ophiolite, with minor argillaceous sediment as an additional component.

(4) Granitic and rhyolitic pebbles and cobbles extracted from Devonian melanges in Cornwall have petrographic analogues within the clast population of the greywackes and were formed initially in a volcanic-arc environment. Some mafic clasts have MORB-like chemical features and quench textures similar to Rhenohercynian pillow lavas from the German sector Devonian succession.

(5) Limited Nd and Sr isotope data for the Gramscatho greywackes exhibit ε-values not dissimilar to Hercynian granites, and have variable mid-Proterozoic model Nd ages (T_{DM}). Isotopic relationships can be interpreted to suggest that the greywacke precursors were mixtures of ophiolitic MORB and Rb-rich continental arc material that was either early Proterozoic (*c.* 2 Ga) or (more likely) multiple arc terranes of variable age (*c.* 500–700 Ma and 2 Ga). Variation in greywacke model ages thus reflects weighted averages between variable proportions of different aged arc segments and MORB. The MORB component is considered to be penecontemporaneous Rhenohercynian oceanic crust.

(6) Petrographic and geochemical discrimination techniques alone cannot resolve the temporal decoupling between source and basin and may lead to an erroneous interpretation of tectonic setting. In this respect isotopic data on bulk sediments, extracted clasts and minerals may provide the necessary constraints on the relative age of the source and basin. In this study a combination of discrimination and isotope techniques have suggested that Rhenohercynian subduction with a contemporaneous Devonian arc is unlikely, but does not allow a distinction to be made between the strike-slip and back-arc models for the Rhenohercynian zone.

NATO grant 0011/87 and supplementary grant are gratefully acknowledged for providing financial support for field work in the UK and Germany. BEL publishes with the permission of the Director, British Geological Survey, Natural Environment Research Council. F. Darbyshire (Isotope Geology Unit, NERC) is thanked for making available the isotopic data on the Gramscatho greywackes.

References

AUVRAY, B. & MAILLET, P. 1977. Volcanisme et subductionau Proterozoique superieur dans le Massif Armoricain (France). *Bulletin Société Géologique de France*, **19**, 953–957.

BADHAM, J. P. N. 1982. Strike-slip orogens—an explanation for the Hercynides. *Journal of the Geological Society, London*, **139**, 495–506.

BALLY, A. W. 1981. Thoughts on the tectonics of folded belts. *In*: McCLAY, K. R. & PRICE, N. J. (eds). *Thrust and Nappe Tectonics*. Geological Society, London, Special Publication, **9**, 13–32.

BARNES, R. P. *The geology of south Cornish melanges*. PhD thesis, University of Southampton.

—— 1983. The stratigraphy of a sedimentary melange and associated deposits in south Cornwall. *Proceedings of the Geologists' Association*, **94**, 217–229.

—— 1984. Possible Lizard-derived material in the underlying Memneage Formation. *Journal of the Geological Society, London*, **141**, 79–85.

—— & Andrews, J. R. 1986. Upper palaeozoic ophiolitic generation and obduction in south Cornwall. *Journal of the Geological Society, London*, **143**, 117–124.

BERNARD-GRIFFITHS, J., PEUCAT, J. J., SHEPPARD, S. & VIDAL, Ph. 1985. Petrogenesis of Hercynian leucogranites from the southern Armorican massif: contribution of REE and isotopic (Sr, Nd, Pb and O) geochemical data to the study of source rock characteristics and ages. *Earth and Planetary Science Letters*, **74**, 235–250.

BHATIA, M. R. 1983. Plate tectonics and geochemical composition of sandstones. *Journal of Geology*, **91**, 611–627.

—— 1985. Composition and classification of Palaeozoic flysch mudrocks of eastern Australia: implications in provenance and tectonic setting interpretation. *Sedimentary Geology*, **41**, 249–268.

—— & CROOK, K. A. W. 1986. Trace element characteristics of greywackes and tectonic setting discrimination of sedimentary basins. *Contributions to Mineralogy and Petrology*, **92**, 181–193.

BIRKELBACH, M., DORR, W., FRANKE, W., MICHEL, H., STIBANE, F. & WECK, R. 1988. Die geologische Entwicklung der ostlichen Lahnmulde. *Jahresbericht und Mitteilungen des Oberrheinischen Geologischen Vereins*, **70**, 43–74.

BROMLEY, A. V. 1979. Ophiolitic origin of the Lizard Complex. *Journal of the Camborne School of Mines*, **79**, 25–38.

BROWN, M., POWER, G. M., TOPLEY, C. G. & D'LEMOS, R. S. 1990. Cadomian magmatism in the North Armorican Massif. *In*: D'LEMOS, R. S., STRACHAN, R. & TOPLEY, C. G. (eds) *The Cadomian Orogeny*. Geological Society, London, Special Publication, **51**, 181–213.

CALVEZ, J. Y. & VIDAL, Ph. 1978. Two billion years old relics in the Hercynian belt of western Europe. *Contributions to Mineralogy and Petrology*, **65**, 395–399.

COLLEY, S., THOMSON, J., WILSON, T. R. S. & HIGGS, N. C. 1984. Post-depositional migration of elements during diagenesis in brown clay and turbidite sequences in the North East Atlantic. *Geochimica et Cosmochimica Acta*, **48**, 1223–1236.

COOPER, J. A. G. 1987. A chert microfauna from the Gramscatho Group of the Lizard penisula, Cornwall. *Proceedings of the Geologists' Association*, **98**, 75–76.

DARBYSHIRE, D. P. F. & SHEPHERD, T. J. 1985. Chronology of granite magmatism and associated mineralization, S. W. England. *Journal of the Geological Society, London* **142**, 1159–1178.

DAVIES, G. 1981. Isotopic evolution of the Lizard Complex. *Abstracts, Proceedings of The Lizard Complex meeting*, Geological Society of London, 9.

—— 1984. Isotopic evolution of the Lizard Complex. *Journal of the Geological Society, London*, **141**, 3–14.

DENIS, E. & DABARD, M. P. 1988. Sandstone petrography and geochemistry of late Proterozoic sediments of the Armorican Massif (France)—a key to basin development during the Cadomian orogeny. *Precambrian Research*, **42**, 189–206.

DePAOLO, D. J. 1988. *Neodymian isotope geochemistry—an introduction.* Springer-Verlag, Berlin.

DICKINSON, W. R. & SUCZEK, C. A. 1979. Plate tectonics and sandstone compositions. *American Association of Petroleum Geologists Bulletin*, **6**, 2164–2182.

——, BEARD, L. S., BRAKENRIDGE, G. R., ERJAVEC, J. L., FERGUSON, R. C., INMAN, K. F., KNEPP, R. A., LINDBERG, F. A. & RYBERG, P. T. 1983. Provenance of North American Phanerozoic sandstones in relation to tectonic setting. *Geological Society of America Bulletin*, **94**, 222–235.

EL BOUSEILY, A. M. & EL SOKKARY, A. A. 1975. The relation between Rb, Ba and Sr in granitic rocks. *Chemical Geology*, **16**, 207–219.

ENGEL, W. & FRANKE, W. 1983. Flysch sedimentation—its relations to tectonism in the European Variscides. *In*: MARTIN, H. & ELDER, F. W. (eds) *Intracontinental fold belts.* Springer-Verlag, Berlin 289–321.

——, FLEHMIG, W. & FRANKE, W. 1983a. The mineral composition of Rhenohercynian flysch sediments and its tectonic significance. *In*: MARTIN, H. & EDER, F. W. (eds) *Intracontinental fold belts.* Springer-Verlag, Berlin, 171–184.

——, FRANKE, W. & LANGENSTRASSEN, F. 1983b. Palaeozoic sedimentation in the northern branch of the Mid-European Variscides—essay of interpretation. *In*: MARTIN, H. & EDER, F. W. (eds). *Intracontinental fold belts.* Springer-Verlag, Berlin, 267–288.

EXLEY, C. S., STONE, M. & FLOYD, P. A. 1983. Composition and petrogenesis of the Cornubian granite batholith and post-orogenic volcanic rocks in southwest England. *In*: HANCOCK, P. L. (ed.) *The Variscan foldbelt in the British Isles,* Adam Hilger, Bristol, 153–177.

FITTON, J. G. 1972. The genetic significance of almandine-pyrope phenocrysts in the calcalkaline Borrowdale volcanic group, northern England. *Contributions to Mineralogy and Petrology*, **36**, 231–248.

FLOYD, P. A. 1982. Chemical variation in Hercynian basalts relative to plate tectonics. *Journal of the Geological Society, London*, **139**, 505–520.

—— 1984. Geochemical characteristics and comparison of the basic rocks of the Lizard Complex and the basaltic lavas within the Hercynian troughs of S.W. England. *Journal of the Geological Society, London*, **141**, 61–70.

—— & LEVERIDGE, B. E. 1987. Tectonic environment of the Devonian Gramscatho basin, south Cornwall: framework mode and geochemical evidence from turbiditic sandstones. *Journal of the Geological Society, London*, **144**, 531–542.

——, FRANKE, W., SHAIL, R. & DORR, W. 1990. Provenance and depositional environment of Rhenohercynian synorogenic greywacke from the Giessen nappe, Germany. *Geologische Rundschau* (in press).

——, WINCHESTER, J. A. & PARK, R. G. 1989. Geochemistry and tectonic setting of Lewisian clastic metasediments from the early Proterozoic Loch Maree Group of Gairloch, N.W. Scotland. *Precambrian Research* **45**, 203–214.

FRANKE, W. 1989. Tectonostratigraphic units in the variscan belt of central Europe. *In*: DALLMEYER, R. D. (ed.) *Terranes in the Circum-Atlantic Palaeozoic orogens.* Geological Society of America, Special Paper, **230**, 67–90.

—— & ENGEL, W. 1982. Variscan sedimentary basins on the continent and relations with southwest England. *Proceedings of the Ussher Society*, **5**, 259–269.

GILBERT, J. S. & ROGERS, N. W. 1989. The significance of garnet in the Permo-Carboniferous volcanic rocks of the Pyrenees. *Journal of the Geological Society, London*, **146**, 477–490.

GOODE, A. J. J. & MERRIMAN, R. J. 1987. Evidence of crystalline basement west of the Land's End granite, Cornwall. *Proceedings of the Geologists' Association*, **98**, 39–43.

GREEN, T. H. 1977. Garnet in silicic liquids and its possible use as a P-T indicator. *Contributions to Mineralogy and Petrology*, **65**, 59–67.

GROSSER, J. & DORR, W. 1986. MOR-typ-basalte in Ostiichen Rheinischen Schiefergebirge. *Neues Jahrbuch für Geologie und Palaontologie, Monatshefte*, **12**, 705–722.

HAWKESWORTH, C. J. & NORRY, M. J. (eds) 1983. *Continental basalts and mantle xenoliths.* Shiva Publishing Ltd, Nantwich.

HENNINGSEN, D. 1978. Zusammensetzung und schuttung der Kulm-Grauwacken im Rheinischen Schiefergebirge-Ergebnisse und offene Fragen. *Zeitschrift der Deutschen Geologischen Gesellschaft*, **129**, 109–114.

HOLDER, M. T. & LEVERIDGE, B. E. 1986a. A model for the tectonic evolution of south Cornwall. *Journal of the Geological Society, London*, **143**, 125–134.

—— & —— 1986b. Correlation of the Rhenohercynian Variscides. *Journal of the Geological Society, London*, **143**, 141–147.

HOWER, J., ESLINGER, E. V., HOWER, M. E. & PERRY, E. A. 1976. Mechanism of burial metamorphism of argillaceous sediment. 1: Mineralogical and chemical evidence. *Geological Society of America Bulletin*, **87**, 725–737.

HUCKENHOLZ, H. G. 1963. Mineral composition and texture of greywackes from the Harz Mountains (Germany) and in arkoses from Auvergne (France). *Journal of Sedimentary Petrology*, **33**, 914–924.

JAKES, P. & WHITE, A. J. R. 1972. Major and trace element abundances in volcanic rocks of orogenic areas. *Geological Society of America Bulletin*, **83**, 29–40.

KIRBY, G. A. 1979. The Lizard Complex as an ophiolite. *Nature*, **282**, 58–60.

—— 1984. The petrology and geochemistry of dykes of the Lizard ophiolite complex, Cornwall. *Journal of the Geological Society, London*, **141**, 53–60.

LEAKE, B. E. 1967. Zoned garnets from the Galway granite and its aplite. *Earth and Planetary Science Letters*, **3**, 311–316.

LEEDER, M. R. 1982. Upper Palaeozoic basins of the British Isles—Caledonide inheritance versus Hercynian plate margin processes. *Journal of the Geological Society, London*, **139**, 479–491.

LE GALL, B., LE HERISSE, A. & DEUNFF, J. 1985. New palynological data from the Gramscatho Group at the Lizard front (Cornwall): palaeogeographical and geodynamical implications. *Proceedings of the Geologists' Association*, **96**, 237–253.

LEVERIDGE, B. E., HOLDER, M. T. & GOODE, A. J. J. 1990. *Geology of the country around Falmouth*. Memoir of the British Geological Survey, sheet 352.

MACK, G. H. 1984. Exceptions to the relationship between plate tectonics and sandstone composition. *Journal of Sedimentary Petrology*, **54**, 212–220.

MANNING, D. A. C. 1983. Chemical composition of garnets from aplites and pegmatites, penisular Thailand. *Mineralogical Magazine*, **47**, 353–358.

MCLENNAN, S. M., MCCULLOCH, M. T., TAYLOR, S. R. & MAYNARD, J. B. 1989. Effects of sedimentary sorting on neodynium isotopes in deep-sea turbidites. *Nature*, **337**, 547–549.

NESBITT, H. W., MARKOVICS, G. & PRICE, R. C. 1980. Chemical processes affecting alkalis and alkaline earths during continental weathering. *Geochimica et Cosmochimica Acta*, **44**, 1659–1666.

—— & YOUNG, G. M. 1982. Early Proterozoic climates and plate motions inferred from major element chemistry of lutites. *Nature*, **299**, 715–717.

PEARCE, J. A., HARRIS, N. B. W. & TINDLE, A. G. 1984. Trace element discrimination diagrams for the tectonic interpretation of granitic rocks. *Journal of Petrology*, **25**, 956–983.

PEUCAT, J. J., JEGOUZO, P., VIDAL, Ph. & BERNARD-GRIFFITHS, J. 1988. Continental crust formation seen through the Sr and Nd isotope systematics of S-type granites in the Hercynian belt of western Europe. *Earth and Planetary Science Letters*, **88**, 60–68.

ROACH, R. A. 1988. Pentevrian basement fragments within the Cadomian Perros granitoid complex at Port Beni, the Trogor, N. Brittany. *Proceedings of the Ussher Society*, **7**, 106.

ROSER, B. P. & KORSCH, R. J. 1988. Provenance signatures of sandstone-mudstone suites determined using discriminant function analysis of major element data. *Chemical Geology*, **67**, 119–139.

SADLER, P. M. 1973. An interpretation of new stratigraphic evidence from south Cornwall. *Proceedings of the Ussher Society*, **2**, 535–550.

SAWYER, E. W. 1986. The influence of source rock type, chemical weathering and sorting on the geochemistry of clastic sediments from the Quetico Metasedimentary Belt, Superior Province, Canada. *Chemical Geology*, **55**, 7–95.

SCHULZ-DOBRICK, B. & WEDEPOHL, K. H. 1983. The chemical composition of sedimentary deposits in the Rhenohercynian belt of central Europe. *In*:

MARTIN, H. & EDER, F. W. (eds) *Intracontinental fold belts*, Springer-Verlag, Berlin, 211–229.

SEAGO, R. D. & CHAPMAN, T. J. 1988. The confrontation of structural styles and the evolution of a foreland basin in central SW England. *Journal of the Geological Society, London*, **145**, 789–800.

SHAIL, R. & FLOYD, P. A. 1988. An evaluation of flysch provenance—example from the Gramscatho Group of southern Cornwall. *Proceedings of the Ussher Society*, **7**, 62–66.

SPEARS, D. A. & AMIN, M. A. 1981. A mineralogical and geochemical study of turbidite sandstones and interbedded shales, Mam Tor, Derbyshire, U.K. *Clay Minerals*, **16**, 333–345.

STONE, M. 1988. The significance of almandine garnet in the Lundy and Dartmoor granites. *Mineralogical Magazine*, **52**, 651–658.

STRACHAN, R. A., TRELOAR, P. J., BROWN, M. & D'LEMOS, R. S. 1989. Cadomian terrane tectonics and magmatism in the Armorican Massif. *Journal of the Geological Society, London*, **146**, 423–426.

TAYLOR, S. R. 1969. Trace element chemistry of andesites and associated calc-alkaline rocks. *Bulletin of the Oregon Department of Geology & Mining Industries*, **65**, 43–63.

—— & MCLENNAN, S. M. 1985. *The continental crust; its composition and evolution*. Blackwell Scientific Publications, Oxford.

——, EWART, A. & CAPP, A. C. 1968. Leucogranites and rhyolites: trace element evidence for fractional crystallization and partial melting. *Lithos*, **1**, 179–186.

TUREKIAN, K. K. & WEDEPOHL, K. H. 1961. Distribution of the elements in some major units of the Earth's crust. *Geological Society of America, Bulletin*, **72**, 175–192.

VAN DE KAMP, P. C. & LEAKE, B. E. 1985. Petrography and geochemistry of feldspathic and mafic sediments of the northeastern Pacific margin. *Transactions of the Royal Society of Edinburgh: Earth Sciences*, **76**, 411–449.

VAN WEERING, T. C. E. & KLAVER, G. Th. 1985. Trace element fractionation and distribution in turbidites, homogeneous and pelagic deposits: the Zaire Fan, southeast Atlantic Ocean, *Geomarine Letters*, **5**, 165–170.

VIDAL, Ph., AUVRAY, B., CHARLOT, R. & COGNE, J. 1981. Precadomian relicts in the Armorican Massif: their age and role in the western and central European Cadomian-Hercynian belt. *Precambrian Research*, **14**, 1–20.

WEBER, K. 1981. The structural development of the Rheinische Schiefergebirge. *In*: ZWART, H. J. & DORNSIEPEN, U. (eds) *The Variscan orogen in Europe*. Geologie en Mijnbouw, **60**, 149–159.

WILKINSON, J. J. & KNIGHT, R. R. W. 1989. Palynological evidence from the Porthleven area, south Cornwall: implications for Devonian stratigraphy and Hercynian structural evolution. *Journal of the Geological Society, London*, **146**, 739–742.

WRONKIEWICZ, D. J. & CONDIE, K. C. 1987. Geochemistry of Archaean shales from the Witwatersrand Supergroup, South Africa: source-area weathering and provenance. *Geochimica et Cosmochimica Acta*, **51**, 2401–2416.

Sedimentary petrology and the archaeologist: the study of ancient ceramics

CHRISTOPHER M. GERRARD

Cotswold Archaeological Trust, Corinium Museum, Park Street, Cirencester, Glos. GL7 2BX, UK

Abstract: Ceramics are essentially modified sediments. This paper describes the objectives of the scientific study of ceramics and the range of appropriate methodologies which have been developed in recent years with special reference to ceramic petrology. The paper is illustrated by a number of examples taken from the author's own work on prehistoric and medieval ceramics from Britain and Spain which demonstrate how the techniques of sedimentology have been modified to provide greater insight into ancient technologies and patterns of social and economic data.

Coarse and decorated pottery, bricks, tiles and clay pipes are all ceramic artefacts, and taken together they are the single most common find on archaeological sites from the Neolithic period to the present day. On some sites the material collection may number tens of thousands of sherds and weigh several tons. Because ceramics survive so well and reflect changing aspects of culture and society they can help us to locate, date and understand changing patterns of ancient occupation. In the study of ceramics many of the techniques which archaeologists use are common to, or developed from, the earth sciences because ceramics are effectively modified sediments. This paper provides a brief introduction to the world of the ceramic petrologist.

The basic archive

Archaeologists are interested to know three key things about the ceramics. How old is it? How was it made? Where did it come from? Essentially, these are questions which address dating, ancient technology and economy. Research normally begins with a basic description, an archive (accompanied by drawings) of the diagnostic sherds in a collection: a 'type series'. For pottery this will involve attention to the description and measurement of bases (ring base, disc base, etc), body (conical, spherical, etc.), neck (convex, concave), rims (rim angle, rim edge treatments, etc), handles (number, attachments), spouts (position), and surface treatments (glazing, burnishing, etc.). Only with this information is it possible to go on and look for typological parallels with other sites for relative dating or to concentrate on well dated ceramics from primary contexts which can be dated using absolute dating techniques such as radiocarbon dating or dendrochronology.

It is increasingly common to add to this basic archive some description of 'fabrics' or 'pastes'. This will involve some kind of assessment as to 'colour' (using a standard Munsell chart or a chromameter, though some countries prefer their own schemes e.g. Llanos & Vegas 1974), 'hardness' (using Moh's scale with modifications where no appropriate minerals are available; ceramics are mostly in the 2–5 range), 'feel' and 'fracture' (both subjective but useful nonetheless if the terms are standardized). Keys are now available which help with the identification of principal inclusions, based on the observation of simple traits such as colour, shape and reactions with HCl. These keys were designed to tighten up on unsound nomenclature based on uncertain identification of principal inclusions in the sherd (Peacock 1977; Davies & Hawkes undated). Some attempts have been made to go further and describe the frequency, sorting (size-distribution) and rounding of inclusions but, in practice, the size of inclusions is often all that can be recorded in Roman, medieval and later pottery because the fabric is so fine.

Armed with a basic archive of information about his ceramics the archaeologist must now decide which questions he wishes to ask before he can proceed. In some cases, it may simply not be worthwhile academically or financially to continue the study, where material is unprovenanced, unstratified or duplicates work undertaken elsewhere. In every case the questions asked must be appropriate to the research aims of the project, which will depend on the quality, quantity and date range of the collection, the level and orientation of previous research and the facilities, expertise and finances available.

From Morton, A. C., Todd, S. P. & Haughton, P. D. W. (eds), 1991,
Developments in Sedimentary Provenance Studies.
Geological Society Special Publication No. 57, pp. 189–197.

At this stage research can go beyond simple description of the physical characteristics of the pottery and begin to focus upon questions of ancient technology and provenance. The former requires attention to fabric composition, the production method (wheel-turned or hand-made) and the firing conditions (e.g. temperature). The latter concentrates upon attributing pottery to a particular region so that trade patterns and economy can be reconstructed.

A range of characterization techniques are available which aim to define more closely the mineralogical or chemical characteristics of the pottery. Most modern ceramic studies aim to cover a range of compositional, technological and stylistic analyses in order to best approach specific problems unique to a particular project. Peacock (1977) recommends that systematic scientific investigation should proceed logically from petrological analysis to elemental analysis.

Mineralogical characterization

Mineralogical characterization, including petrological work, is to be regarded as a necessary first step beyond the construction of a basic archive. The available techniques are still relatively cheap and the results can add significant detail to the basic archive by identifying principal inclusions in the pattery fabric. Chemical analyses, on the other hand, tend to produce lists of trace and major elements in various concentrations which are difficult to relate to a piece of pottery in the hand.

The same techniques of preparation are used for most ceramics as for hard rocks to produce 30 µm thin sections which can be examined in the usual way in plain or polarized light (see Middleton et al. 1985 for a review of grain sampling procedures). Friable pottery can be impregnated with resin.

From the thin-sections it is possible to identify minerals and voids, their abundance, shape and orientation, as well as alterations in the fabric due to firing or after deposition/burial in the ground. Several sections in different orientations may be useful to illustrate manufacturing techniques, for example, to show joins between applied decorations and body sherds.

Some of the inclusions in a typical ceramic thin section may be rather unfamiliar to geologists. When examined under a petrological microscope, pottery is seen to consist of fired clay (usually 600–1000°C), added rock fragments and other 'temper'. This temper is added to give greater manipulability to a 'sticky' clay and can help the piece to survive both the firing

process where conditions are not well controlled (for example, in a bonfire) and repeated thermal shocks when in use later on (for example, for a cooking pot or crucible). These added fragments may be distinctive by their sorting, frequency or angularity and can include rock fragments of all kinds such as gabbro (Peacock 1988) or basalt (Shepard 1966) and other more unusual additions such as blood, straw, crushed pottery, dung, and slag. There may be unintentional additions too which are 'natural' to the clay, such as metal-working residues and shells, some of which can be difficult to identify.

Some of the inclusions deliberately added to ceramic clays may be appropriate to different firing technologies and functional requirements. A recent study of Anglo-Saxon ceramics from Canterbury demonstrated the inadequacies of a calcium carbonate-rich clay heated to over 900°C. A switch to texturally standardized non-calcareous clays appears to have occurred when pots needed to be fired to higher temperatures as production became more commercialized (Mainman 1982). Elsewhere work on thermal shock resistance in ceramic bodies has been useful in expanding our understanding of the properties of minerals under different firing regimes (Woods 1986).

Any kind of petrological work on ceramics should be accompanied by a series of short background studies, especially if the intention is to attribute a distinctive suite of inclusions to a particular source. These may include some acquaintance with the local solid, drift geology, and geomorphology; examination of available thin-sections of bedrock facies; the collection of traditional local ceramics, bricks and tiles from known centres of production, usually medieval or later but possibly stamped Roman; drawing and assessment of exposed lithostratigraphic profiles from sites of known ancient clay extraction; interviews with local potters; experimental firings of ceramic bricks with accompanying detail on weight loss, shrinkage, weight, etc. The aim of these studies should be to provide a ceramic reference collection with a broad geographical coverage which will provide comparative baseline data for the thin-section work. The analysis of raw materials used in pottery making is sometimes referred to as 'ceramic ecology' (Matson 1965). The need for raw material studies to be undertaken alongside petrological analysis has been stressed because many suitable potting clays are not mapped adequately or explained in standard geological documentation, and the solid geology may be insufficient evidence on which to base interpretations (Howard 1982).

For prehistoric pottery it is usually the identification of minerals and rock fragments which is of primary interest, clay minerals being too fine grained. In theory, once identified, the source of the inclusions can be pinpointed and something said about trade and changing patterns of economy assuming the chronological sequence is long enough. In practice, the successful identification of clay or tempering sources will depend on the distinctiveness or uniqueness of the mineral suites and rock types and the degree of variation in the surrounding geology. Usually some initial study is advisable to assess compositional variation in both clays and ceramics, and important results can be achieved when the right conditions are fulfilled (Bishop *et al.* 1988). Even where this is not possible, petrology can still be usefully employed to discriminate scientifically between similar fabrics and construct some general models of production and supply (Vince 1981, for southern England 10th to 13th centuries; Vince 1985, for Saxon and medieval London).

The techniques employed at the microscope will depend on the range of material analysed and the time available for study. Much medieval and later pottery is tempered with quartz which is less than 100 µm and little else may be visible or identifiable. In this case, some kind of textural analysis may be appropriate (Streeten 1982 for medieval ceramics from Kent) and it may be possible to distinguish between production centres and even phases of the production process on this basis. Textural analysis is one technique which was directly borrowed from the field of sedimentology (Darvill & Timby 1982). The recent analysis of some late medieval Spanish ceramics has shown at least three phases in clay preparation visible in thin section with observable differences in the size-frequency of added quartz found in kiln furniture (cockspurs), coarse wares and fine wares as increasing attention is paid to refinement and textural standardization (Gerrard & Gutierrez 1990a).

From this type of study it is clear that multiple compositional groupings at a single site can be due to different choices made in the production process as well as to trade in pottery. Ethnographical investigations, though not necessarily providing direct analogues for ancient production cycles, are often helpful in the interpretation of archaeological phenomena (Peacock & Williams 1986). Record forms have been published to regulate the collection and analysis of clay samples and the study of fired ware-fabrics at modern traditional potters' workshops. For the study of medieval Spanish ceramics for example it was useful to 'model' the different ways in which clay had been refined and mixed by traditional modern potters so as to demonstrate ways in which textural variability might occur in the same pot, at the same kiln site and between kilns (Gerrard & Gutierrez 1990b).

Other methods of mineralogical characterization are open to the archaeologist. X-ray diffraction (XRD) is strictly limited in use in ceramic study because most clay minerals are altered or destroyed by heating to above 500°C. XRD, like thermal analysis, is unsuitable for the study of the complex mineralogy of vitrified high-fired clays like earthenwares, stonewares, china or porcelain but both techniques can be helpful in determining the firing temperatures of low-temperature prehistoric pottery which retains its original crystalline structure. Mineralogical characterization and grain identification is usually restricted to petrological techniques and the analysis of heavy minerals from sandy fabrics whose minerals may be characteristic of the parent geological deposit.

Chemical analysis

In chemical analysis the chemical constituents of a fabric are analysed as present in major, minor and trace amounts. Techniques applied to the study of ceramics include: optical emission spectroscopy, X-ray fluorescence (XRF), electron microprobe, neutron activation analysis (NAA) and X-ray photoelectron spectroscopy.

Ideally the observed chemical compositions and concentrations will reflect the chemical patterns of the natural clay resource but, in practice, pottery and clay compositions, however carefully determined, may never match because of the addition of temper, the blending of different clays, the preparation of the clay before throwing and the chemical alterations or contamination which can take place during firing and after burial (Kilikoglou *et al.* 1988). It has been suggested that pottery composition may be influenced even by the presence of impurities in the water used to wet the clay (Rye & Duerden 1982). These complications can mean that pottery produced in different places shows overlapping ranges of elements, although similarity measures have been developed to account for some of these effects (Manonsen *et al.* 1988).

Two popular techniques of recent years have been XRF and NAA. Both these are useful for characterization, especially when used in conjunction with petrological evidence, and they have been used mainly on medieval and classical pottery with no identifiable inclusions except commonplace minerals (see Boardman &

Schweizer 1973 for archaic Greek pottery; Wilson 1978 for a general review). In one recent study of Spanish lustreware ceramics of similar decoration, texture and mineral composition it has proved possible to distinguish imported lustrewares from different sources (Hughes & Vince 1986).

In the future, the use of surface analysis (Baillie & Stern 1984) and electron probe microanalysis (Freestone 1982) is expected to become more widespread. the microprobe, for example, when attached to an SEM is useful for spot size chemical information on the main constituents, individual minerals, and for defining internal morphologies, although it is less useful for average information or bulk characterization.

The use of mineralogical and chemical analyses has brought about a fundamental change in the way that ceramics are studied. Attention has now shifted away from art-historical studies of decoration toward the description and characterization of fabrics. Roles within archaeology have now become more specialized and the study of artefacts is to be found increasingly in the hands of the specialist analyst rather than the excavation director. With this separation of roles has come the clearer definition of descriptive vocabularies and a swift advance in technological mastery. Attention to the relationship between technical and cultural phenomena is

now developing. The case study which follows is an illustration of what can be achieved.

A ceramic case study from Bronze Age Spain

A study which illustrates the typical pattern of archaeological investigation and the uses of ceramic petrology in an area of varied geology is that of Moncin in Aragon, northeastern Spain (Fig. 1). The Eneolithic and Bronze Age ceramics (dated 2650–1300 BC in 8 cultural phases) examined were excavated from a deep 3.5 m thick stratigraphy. Two series were selected for mineralogical study; decorated ceramics including Maritime Bell Beakers and various Bronze Age types; and plain pottery.

The research strategy was designed to test out a new subsistence model for the 3rd and 2nd millennium BC in Spain and Portugal called the 'Policultivo Ganadero' (Harrison 1985). This model suggests that between 3000 BC and 2000 BC cereals dominated the productive economy, then livestock rearing emphasized a pastoral element, while secondary products (e.g. cheeses, yoghurts, etc.) became increasingly important.

From this model one can make certain predictions, which should be observable in patterns of material culture surviving in the archaeological

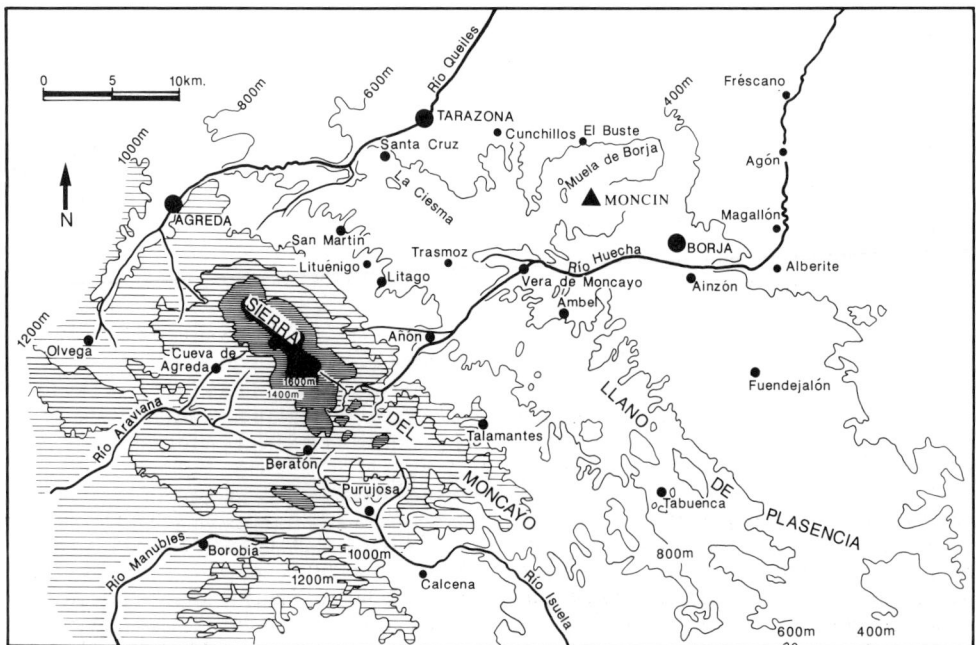

Fig. 1. Location map for Moncin with population centres indicated. The site is marked by the solid triangle on the Muela de Borja with the Sierra de Moncayo to the southwest.

record: the Eneolithic should show a range of imported materials and a few of them would be expected to be brought from great distances: in Bronze Age systems the number and volume of imported goods should decline, relative to their abundance in the Eneolithic: these changes may be gradual or abrupt, and need not follow the same pattern from one site or region to another.

The emphasis in the research was placed upon defining pottery sources, and therby directions of movement and their changing patterns, over time. With a view to future identification of fabrics and inclusions in hand specimen and in the field, mineralogical analysis through ceramic petrology was selected as the most effective method of evaluation.

The ceramic complex

Background

Several types of evidence were considered as supporting the interpretation of the petrography of the ceramics from the site. They were: geological mapping reports and geomorphological reviews; thin sections of all the major bedrock facies; thin sections of a selection of local medieval and early modern pottery from known centres; thin sections of the Moncin ceramics themselves and comparative material from a number of other contemporary sites in the vicinity.

Geology

The geology, geomorphology and vegetation of the Muela de Borja, on which Moncin is located, has been geologically mapped and published (Pellicer Correllano 1984). The relevant geological and geomorphological information is summarized on Fig. 1. There are three terrains.

(1) The tabular Tertiary terrain which covers most of the study area and consists mainly of marls and Neogene limestones. There are three prominent Tertiary limestone formations on the right bank of the river Ebro: La Muela, La Plana and the Borja formation, separated from each other by valleys draining into the Ebro depression. They are composed of Neogene calcareous limestone bodies with occasional marls, siltstones and gypsum covering Jurassic and Triassic conglomerate bodies and basement rocks, mainly quartzites and slates of Palaeozoic age. The two important rivers, the Queiles and Huecha, have up to five alluvial terraces and consist of clays, sands, detritic gravels, limestones and siliceous rocks depending on the hard rock geology in the immediate locality.

(2) The Somontano de Moncayo is the complex range of foothills between the Sierra de Moncayo and the Tertiary terrain below. It lies parallel to the NW–SE fold axis of the Sierra and the Ebro valley stretching from the villages of Anon to San Martin de Moncayo. The Somontano is a mantle of mixed eroded material with active erosion surfaces exposing Triassic Bunter sandstones, Keuper Marls, Jurassic Limestones, and Miocene conglomerates.

(3) The Sierra de Moncayo is the southernmost of the northern group of Iberian Sierras. It is a huge anticlinal complex reaching 2313 m altitude and separates Castile from Aragon. On its highest slopes it has the classic features of a glaciated landscape: lower down, faulting has exposed Cambrian, Ordivician and Silurian bedrock.

The geological background was supplemented by an examination of hard rock samples in thin section. At the Department of Earth Sciences at the University of Zaragoza there is an extensive reference collection from identified geological exposures to which the inclusions found in the ceramics can be referred.

Petrographical analysis

The 85 Moncin pottery thin-sectioned could be divided into ten main pottery wares on the basis of the inclusions present and then subdivided into individual fabrics on the basis of the size-frequency of inclusions. Two distinctive groups contained inclusions that were consistent with local production at Moncin because they contained rounded micrite from the clay around the site and variable quantities and sizes of angular crushed pottery temper (Fig. 2), angular crushed veined calcite (Fig. 3) and rare zoned rhombic dolomite (Fig. 4). Differences between these fabrics and those from an early Iron Age site 15 km away reinforces the interpretation that these fabrics are locally made.

A third fabric contained elongated rounded siltstone and may be derived from a local river source (Fig. 5). A further six fabrics contained greywacke fragments (Fig. 6), and fibrous radial chert (Fig. 7) but lacked the micrite which appears to be distinctive of local fabrics. These inclusions did not match with the local geology at Moncin but compared with the known geology and petrographical evidence from thin sections of hard rock exposures from the Somontano de Moncayo about 15 km from the site.

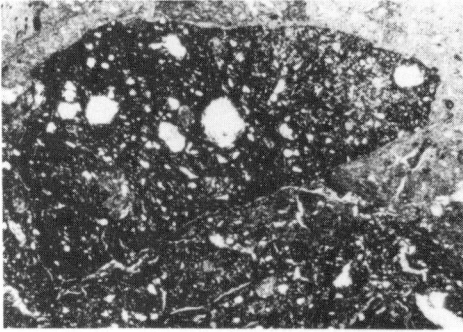

Fig. 2. Moncin fabric 1.1. Ceramic grog added as temper.

Fig. 3. Moncin fabric 2.1. Abundant, crushed coarse fragments of micritic limestone.

Fig. 4. Moncin fabric 2.2. Dolomitised limestone showing distinctly zoned rhombic shape.

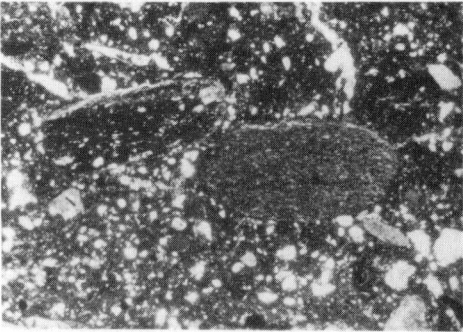

Fig. 5. Moncin fabric 5.1. Rounded, elongate silt-stone fragments.

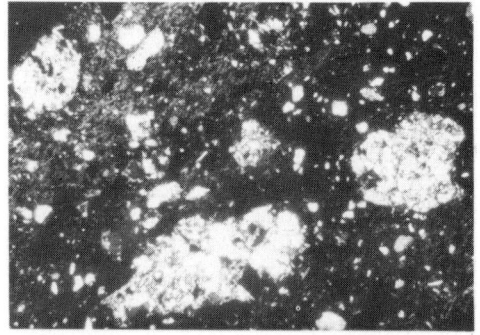

Fig. 6. Moncin fabric 6.1. Greywacke fragments.

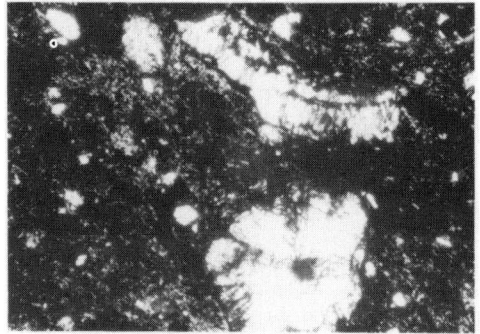

Fig. 7. Moncin fabric 6.2. Radial, fibrous chert.

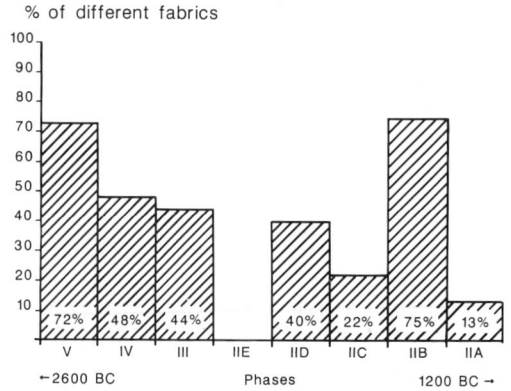

Fig. 8. Graph showing the percentages of different imported pottery fabrics (shaded) for each cultural phase at Moncin. Phases IIe and IIb had only five samples between them.

One fabric was truly exotic and contained metamorphic rock fragments whose nearest source would appear to be the Pyrennean fringe alluvial fans which contain up to 15% reworked metamorphic lithologies from the Pyrenean axial zone. The nearest source would be about 90 km away.

Fabrics and typology

Some fabrics among the assemblage at Moncin are represented solely in the decorated ceramics, while others occur only in the coarser wares. There is a tendency for the better clays of finer texture to prevail among the decorated pottery. Where these vessels can be reconstructed, they are the smaller and more delicate types.

Parallel technologies were used in the manufacture of both decorated and coarse wares. The use of grog (crushed pottery) and calcite as tempers throughout the Eneolithic and Bronze Age at Moncin may have been intended to neutralise the effects of thermal shock in bonfire firing, particularly in the case of cooking pottery. It is true the calcite has a pleasing aesthetic quality on the surface of burnished pottery but it is difficult to discount entirely the influence of firing technology as being the primary reason for its inclusion as a temper.

Why two parallel technologies for ceramic production (local grog and calcite gritted ceramics) were functioning on the same site for at least a thousand years is hard to say. One suggestion is that the availability of temper changed according to post depositional site formation processes or perhaps these two traditions were thought of as interchangeable or even independent.

Not all the larger storage vessels were made at Moncin, as might have been expected given their great size. Some were brought in from the Moncayo area 15–20 km. distant, even though identically decorated ones were being made at Moncin itself. This suggests exchange in a commodity kept in a jar, rather than the jar being traded as an object in its own right. The exotic fabric with the metamorphic grains, probably made on the left bank of the Ebro, is a plain coarse pot.

Chronological patterns

When the number of local wares is expressed as a percentage of the total sample (decorated and coarse wares) for each phase (Fig. 8), it is apparent that the percentage of local pottery increases after the Full Bronze Age of phase IID, while the proportion of imported fabrics declines throughout the second millennium. The general trend on Fig. 8 is interrupted by periods IIE and IIB; this is likely to reflect sampling variation, since only one sherd was obtained from period IIE, and four from IIB, on account of the extensively reworked strata which comprise their deposits.

This pattern can be seen best amongst the decorated pottery where there is a steady decrease in the percentage of decorated pottery imports in the classes represented. In the late Eneolithic, Maritime Bell Beakers are 71% imported fabrics (in three wares), and Ciempozuelos Bell Beakers 57% imported (in four wares); but in the later Bronze Age, only 25% of the Boquique wares are imported (one ware), and only 12.5% of the incised ware (one ware). Through succeeding cultural phases, more decorated pottery is made at Moncin, and the social intercourse, so much in evidence in Eneolithic material culture, is apparently reduced or switched to another cultural medium which has left fewer traces in the archaeological record.

Marked patterning can also be picked out in the diversity of fabrics. Amongst the nine fabrics whose petrographic characteristics are consistent with local production a total of three fabrics are represented in phase V, five in phase IV, increasing to a total of six in phase IIa. In the same three phases, those which contain more than 15 examples, the number of exotic or regional fabrics decreases from eight in phase V, to seven in phase IV, and finally to only two in phase IIa. Further examples from other phases are needed to reinforce any preliminary conclusion but these data would be consistent with the notion of greater variation in the local Moncin fabrics at the same time as the variety of exotic/regional fabrics is declining.

Not only is the range of exotic fabrics greater in the Eneolithic phases V and IV, but some eight fabrics are represented for the last time. After cultural phase III there are only two new fabrics. The majority of the fabrics recognized are therefore present throughout the cultural sequence and of the twenty-three possible fabrics fifteen are present both before and after cultural phase III. The patterns of future trade were encoded in the earliest phases of the site's occupation in the late third millennium, and thereafter few new links were opened up. There is no evidence for technological innovation: grog, calcite and the same range of rock fragments remain as the ceramic tempers.

Conclusions

In this case study a direct correspondence can be established between raw clay, petrological inclusions and finished pottery. This allows the ceramic petrology to be used as an appropriate technique for pinpointing possible production sources and, given the long chronology for this site, used to provide insights into redistribution mechanisms. In particular, these conclusions bear out the predictions derived from the Policultivo Ganadero model. The results constitute an empirical test which could now be repeated on other sites.

This study and its conclusions have a clear corollary in the work of sedimentary geologists and for the archaeologist there are issues raised by the study of the ceramic petrology which might otherwise remain beyond grasp. In this case the petrology is only one part of the archaeological investigation and the ideas presented here await supportive evidence from faunal studies which might also indicate trade and movement. Here and elsewhere petrography will continue to have an important role in archaeological research design.

I would like to thank A. Gutierrez and C. Johnson, J. Turner for their advice in preparing this paper and for making unpublished data available to me. The Moncin information forms part of a report to be included in the excavation monograph currently being prepared by G. Moreno Lopez and R. Harrison. The three figures are reproduced with their kind permission and were drawn by S. Grice.

References

BAILLIE, P. J. & STERN, W. B. 1984 Non-destructive surface analysis of Roman terra sigillata: a possible tool in provenance studies. *Archaeometry*, **26**, 62–68.

BISHOP, R. I., CANOUTS, V, DE ATLEY, S. P., QOYAWAYMA, A. & AIKINS, C. W. 1988. The formation of ceramic analytical groups: Hopi pottery production and exchange, A. C. 1300–1600. *Journal of Field Archaeology*, **15**, 317–337.

BOARDMAN, J. & SCHWEIZER, F. 1973. Clay analysis of archaic Greek pottery. *Annual of the British School at Athens*, **68**, 267–283.

DARVILL, T. C. & TIMBY, J. 1982. Textural analysis: a review of potentials and limitations. *In*: FREESTONE, I. C., JOHNS, C. & POTTER, T. (eds) *Current research in ceramics: thin section studies*. British Museum Occasional Paper, **32**, 73–87.

DAVIES, S. M. & HAWKES, J. W. undated. *Pottery Recording System*, Trust for Wessex Archaeology.

FREESTONE, I. C. 1982. Applications and potential of electron probe micro-analysis in technology and provenance investigations of ancient ceramics. *Archaeometry*, **24**, 99–116.

GERRARD, C. M. & GUTIERREZ, A. 1990a. El analisis de secciones delgadas y la caracterizacion macroscopica de algunas ceramicas medievales. *Boletin del Museo de Zaragoza*, (in press).

—— & —— 1990b. The thin section analysis and macroscopic characterisation of some medieval and post-medieval pottery from Spain. *In*: FREESTONE, I. C. & MIDDLETON, A. (eds) *Recent developments in ceramic petrology*, British Museum Occasional Paper, (in press).

HARRISON, R. J. 1985. The 'Policultivo Ganadero', or the Secondary Products Revolution in Spanish Agriculture, 5000–1000bc. *Proceedings of the Prehistoric Society*, **51**, 75–102.

HOWARD, H. 1982. Clay and the archaeologist. *In*: FREESTONE, I. C., JOHNS, C. & POTTER, T. (eds) *Current research in ceramics: thin section studies*. British Museum Occasional Paper, **32**, 145–157.

HUGHES, M. J. & VINCE, A. G. 1986. Neutron activation analysis and petrology of Hispano-Moresque pottery. *In*: OLIN, J. S. & BLACKMAN, M. J. (eds) *Proceedings of the 24th International Archaeometry Symposium*. Smithsonian Institution Press, 353–368.

KILIKOGLOU, V., MANIATIS, Y. & GRIMANOS, A. P. 1988. The effect of purification and firing of clays on trace element analysis, *Archaeometry*, **30**, 37–46.

LLANOS, A. & VEGAS, J. I. 1974. *Ensayo de un metodo para el estudio y clasificacion tipologica de la ceramica*. Estudios de Arqueologia Alavesa.

MAINMAN, A. 1982. Studies of Anglo-Saxon pottery from Cornwall. *In*: FREESTONE, I. C., JOHNS, C. & POTTER, T. (eds) *Current research in ceramics: thin section studies*. British Museum Occasional Paper, **32**, 93–100.

MATSON, F. R. 1965. Ceramic ecology: An approach to the study of the early cultures of the Near East. *In*: MATSON, F. R. (ed.) *Ceramics and Man*. Methuen & Co, 202–17.

MANONSEN, H., KREUSER, A. & WEBER, J. 1988. A method for grouping pottery by chemical composition, *Archaeometry*, **30**, 47–57.

MIDDLETON, A., FREESTONE, I. F. and LEASE, M. N. 1985. Textural analysis of ceramic thin sections: evaluation of grain sampling procedures, *Archaeometry*, **27** (1), 64–74.

PEACOCK, D. P. S. 1977. Ceramics in Roman and medieval archaeology. *In*: PEACOCK, D. P. S. (ed.) *Pottery and early commerce*. Academic Press, 21–32.

—— 1988. The gabbroic pottery of Cornwall, *Antiquity*, **62**, 302–304.

—— & WILLIAMS, D. L. 1986. *Amphorae and the Roman economy*. Longman.

PELLICER CORELLANO, F. 1984. Geomorfologia de las Cadenas Ibericas entre el Jalon y el Moncayo. *Cuadernos de Estudios Borjanos*, t. 11–14.

RYE, O. S. & DUERDEN, P. 1982 Papuan pottery sourcing by PIXE: preliminary studies. *Archaeometry*, **24**, 59–64.

SHEPARD, A. O. 1966. Rio Grande glaze-paint pottery: a test of petrographic analysis. *In*: MATSON, F. T. (ed.) *Ceramics and man*, 62–87.

STREETEN, A. D. F. 1982. Potters, kilns, and markets in medieval Kent: a preliminary study. *In*: LEACH, P. E. (ed.) *Archaeology in Kent to AD1500*. CBA

Research Report, **40**, 87–94.

VINCE, A. G. 1981. The medieval pottery industry in southern England 10th to 13th centuries. *In:* HOWARD, H. & MORRIS, E. C. (eds) *Production and distribution: a ceramic standpoint*, 309–321.

—— 1985. The Saxon and medieval pottery of London: a review. *Medieval Archaeology*, **29**, 25–94.

WILSON, A. N. 1978. Elemental analysis of pottery and in the study of provenance: a review, *Journal of Archaeological Science*, **5**, 219–236.

WOODS, A. J. 1986. Form, fabric and function: some observations on the cooking pot in antiquity. *In:* KINGERY, W. (ed.) *Ceramics and Civilization vol 2. Technology and Style.* American Ceramics Society, 157–172.

A local source for the Ordovician Derryveeny Formation, western Ireland: implications for the Connemara Dalradian

JOHN R. GRAHAM[1], JOHN P. WRAFTER[2], J. STEPHEN DALY[2]
& JULIAN F. MENUGE[2]

[1] *Department of Geology, Trinity College, Dublin 2, Ireland*
[2] *Department of Geology, University College, Dublin 4, Ireland*

Abstract: Facies and palaeocurrent data indicate that the conglomeratic Ordovician Derry-veeny Formation is the deposit of an alluvial fan whose apex was located a few kilometres east of the present outcrop. Clasts in the conglomerates include migmatite, schist, gneiss, granite, acid porphyry, spilite and vein quartz. Sillimanite bearing migmatite clasts closely resemble rocks of the neighbouring Lough Kilbride Formation, a unit of the Connemara Dalradian (Lower Proterozoic) basement from which they are presumed to be derived. This local source for the conglomerates is also suggested by the similarity of the Nd-model ages (*c.* 2.15 Ga) from the metamorphic clasts and from the Lough Kilbride Formation. Moreover peraluminous granite clasts give the same model ages, suggesting the granites are intracrustal melts from the same source. Interpretation of the Connemara Dalradian as a suspect terrane requires that it docked before Upper Ordovician times. Differences between the clast assemblage in the Derryveeny Formation and the presently exposed Lough Kilbride Formation are related to stripping of the original cover, emphasizing the value of clastic detritus in the study of the uplift history of metamorphic basement.

Conglomerates are strongly favoured in provenance studies, because they are generally the least travelled clastic detritus and they provide the largest samples of the ambient source rocks. Yet, because of the very existence of these conglomerates, the source area of the clasts must have changed in character during uplift and erosion such that exact matches between clasts and source are rarely achieved. With care, differences between clasts and what is presently exposed in the inferred source area can be reasonably interpreted and linkages established. Conversely, clastic detritus is an important source of evidence in studying the uplift history of metamorphic basements for which the present exposures yield only an incomplete history.

This study focuses on part of the northwest (Laurentian) margin of the Caledonides which is generally agreed to have been an active, destructive continental margin during Cambrian and Ordovician times. In Britain and Ireland the late Precambrian basement rocks of this margin are termed the Dalradian Supergroup (Harris & Pitcher 1975). These always lie to the northwest of demonstrable Lower Palaeozoic rocks except in South Mayo, western Ireland, where the Dalradian rocks of Connemara lie to the south of extensive Lower Palaeozoic outcrop (Fig. 1). The Connemara Dalradian rocks are unusual both in their anomalous position, significantly east of where they might be expected from along-strike interpolation (Leake *et al.* 1984),

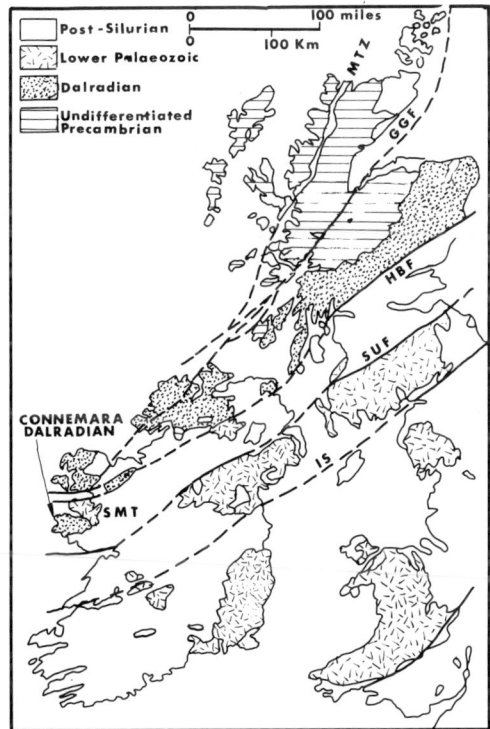

Fig. 1. General location and outline geology of the British Isles Caledonides. GGF, Great Glen Fault; HBF, Highland Boundary Fault; IS, Iapetus Suture; MTZ, Moine Thrust Zone; SMT, South Mayo Trough; SUF, Southern Uplands Fault.

From Morton, A. C., Todd, S. P. & Haughton, P. D. W. (eds), 1991,
Developments in Sedimentary Provenance Studies.
Geological Society Special Publication No. 57, pp. 199–213.

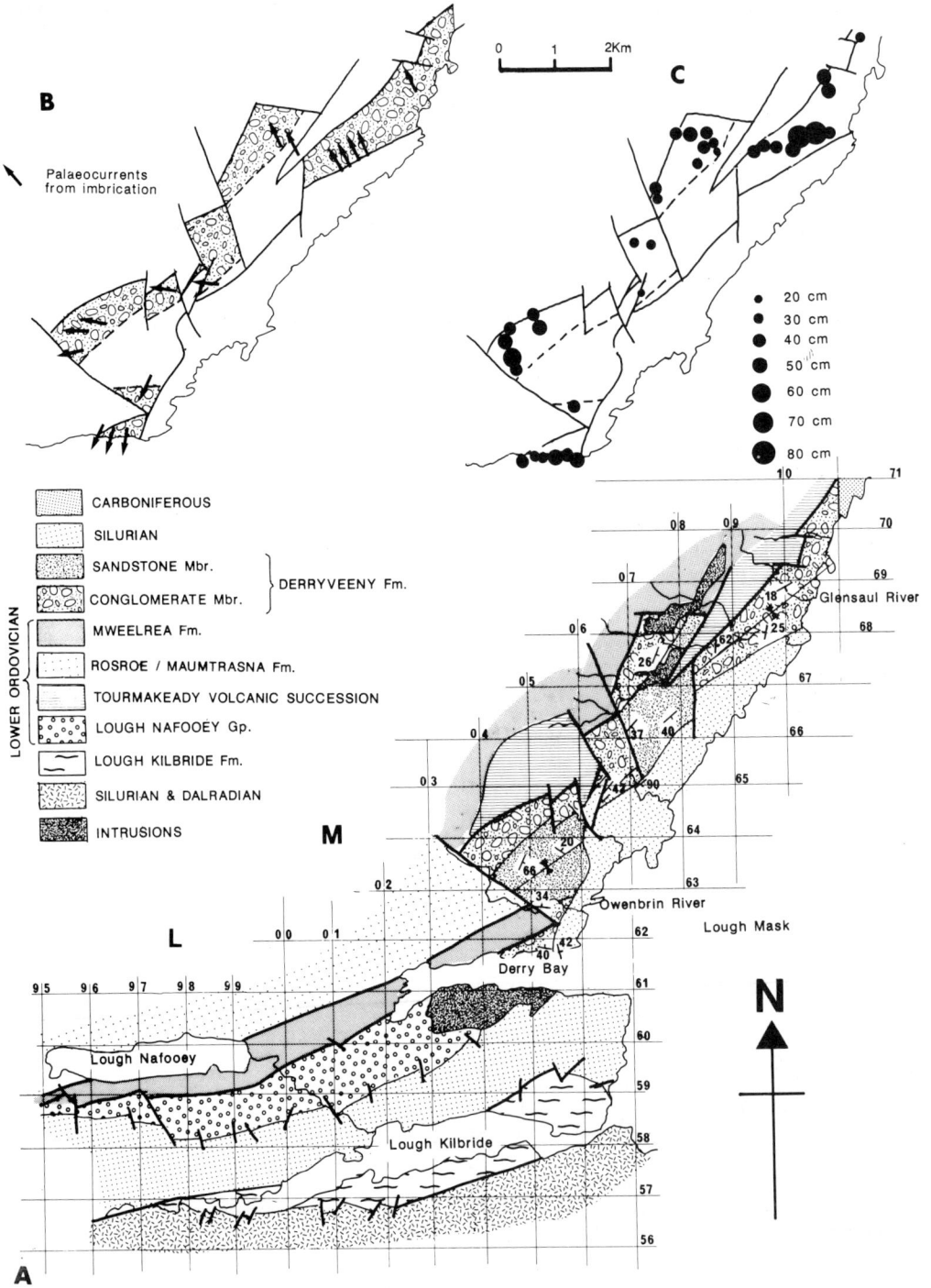

Fig. 2. (**A**) Simplified geological map of the area west of Lough Mask. Irish national Grid, subzones L and M, is superimposed in 1 km squares. (**B**) Palaeocurrent directions derived from imbrication for the Conglomerate Member of the Derryveeny Fm. (**C**) Maximum particle size data from the Conglomerate Member of the Derryveeny Fm.

and also in their late high temperature–low pressure metamorphism associated with major gabbroic intrusions (Yardley *et al.* 1987). Recent analogies of the Caledonides with the western margins of the Americas have alerted geologists to the possibilities of major strike-slip movements, particularly on the Laurentian margin (Hutton 1987). This has led to the suggestion that Connemara is an exotic terrane which arrived in its present position at some stage during the Ordovician (Dewey 1982; Bluck & Leake 1986; Hutton & Dewey 1986; Hutton 1987).

The Lower Palaeozoic rocks of South Mayo contain numerous conglomeratic formations which are important sources of provenance information (see Graham *et al.* 1989 for summary). These can be used to test the various, generally poorly constrained, hypotheses. This paper documents information from a unit of coarse-grained, locally derived conglomerates

and associated sandstones, the Derryveeny Formation, which we believe provides a clear linkage to the late Proterozoic Dalradian metasediments which presently crop out to the south. Such linkages are most convincingly established by using a combination of field, petrographic and geochemical techniques.

Stratigraphy and field relations

The Derryveeny Formation consists of boulder and cobble conglomerates and sandstones. It crops out only in an area, 10×3 km in dimensions, west of Lough Mask (Figs 2 & 3). It lies west of Upper Llandovery sediments of the Lough Mask and Kilbride Formations, and east of the Arenig to Llanvirn rocks of the Tourmakeady Volcanic Succession and the conglomeratic Maumtrasna Formation. Early workers considered the rocks of the Derryveeny Formation

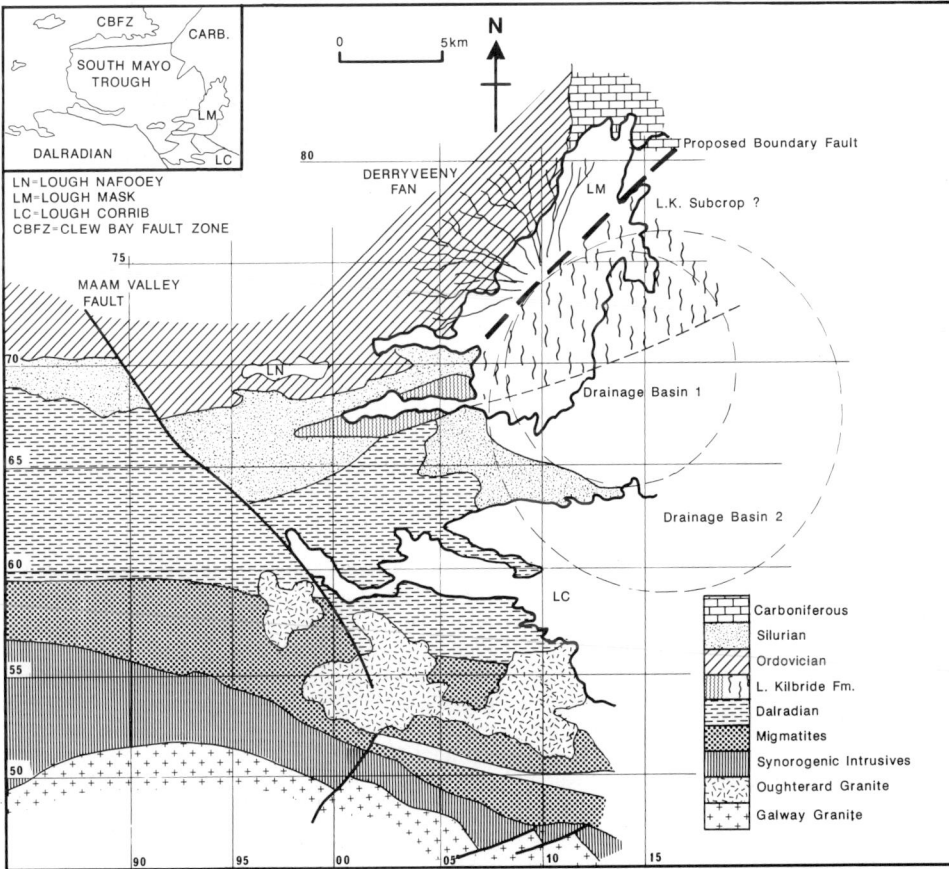

Fig. 3. Outline geology of eastern Connemara showing the disposition of the main migmatite belt and the likely minimum and maximum drainage areas for the Derryveeny fan as discussed in the text (geology after Graham *et al.* 1985; Yardley *et al.* 1987).

to be either part of the Tourmakeady Volcanic Succession (Geikie & Kilroe 1897; Gardiner & Reynolds 1909, 1910) and by inference of Arenig age, or to be lateral equivalents of the Maumtrasna conglomerates (Kilroe 1907) then thought to be Upper Ordovician but since shown to be Llanvirn in age (Graham 1987; Harper *et al.* 1988). The separate and distinctive nature of the Derryveeny Formation and its outcrop pattern were first presented by Graham *et al.* (1985) and a brief description is given in Graham *et al.* (1989).

In the east, the strike of the Derryveeny Formation is parallel with the overlying, apparently conformable Silurian (Fig. 2). However on the shores of Derry Bay (M005 617), the strikes are perpendicular and the red sandstones of the Upper Llandovery Lough Mask Fm. rest unconformably on boulder conglomerates of the Derryveeny Fm.

The western contacts of the Derryveeny Formation are with both the Arenig rocks of the Tourmakeady Volcanic Succession and the Llanvirn rocks of the Maumtrasna Formation. Both these lower Ordovician units, which are separated by an unconformity, dip and young to the northwest, whereas the Derryveeny Formation dips and youngs to the southeast. All the exposed contacts (M099 691, M090 682, M063 649) are faults. However the general parallelism of strike with this western contact, which transgresses the lower Ordovician formations, may suggest that these faulted contacts modify an original unconformity. The most likely age of the Derryveeny Formation based on field relationships is post-Llanvirn but pre-Llandovery.

The Derryveeny Formation comprises two main lithologies; conglomerates, dominantly of boulder grade, and medium- to coarse-grained sandstones. The conglomerates consistently underlie the sandstones such that an informal division into a lower conglomerate and an upper sandstone member is possible (Fig. 2). Most of the outcrop is poorly exposed with reasonable quality of exposure confined to the major stream sections and the shores of Derry Bay. Even here exposures are of very limited size and quality and little can be said about bed geometries. Minimum thicknesses of 220 m for the conglomerates and 400 m for the sandstones are probable. The conglomerates are clast supported and well bedded where seen on favourable exposures with the layering picked out by sandstone interbeds and lenses. These sandstone interbeds demonstrate that the prominent alignment of elongate clasts commonly represents clast imbrication. The sandstones are much less well exposed than the conglomerates and usually appear massive. However there are some localities displaying decimetre scale cross beds and flat beds although exposures are too small to allow palaeocurrent measurements. It is likely that the massive aspect of most exposures is apparent rather than real. The upward transition from the conglomerate to the sandstone member is relatively abrupt. The sequence is interpreted as being of alluvial fan origin on the basis of the very coarse grain size showing rapid lateral changes downcurrent and the radial palaeocurrent pattern described below.

Palaeocurrent data

Although imbrication is well developed in the Derryveeny conglomerates, the exposure quality is such that accuracy is limited to an octant of the compass (see Graham 1988 for technique). Better quality exposure on the shores of Derry Bay locally allows the measurement of *a*-axis orientation following the technique of Nilsen (1968). The mean orientation of 141° ($n = 30$), which is expected to be transverse to flow, agrees well with the northeast palaeocurrent determined from imbrication. The results are summarized in Fig. 2B which demonstrates that the direction is constant for any given area independent of stratigraphic position within the conglomerate member, but that directions vary from area to area. This variation is systematic and gives a radial fan pattern suggesting overall derivation from the east. A simple upstream projection of the palaeocurrents (Fig. 2B), whilst not allowing for the limited tectonic shortening, strongly suggests a point source situated some 2–3 km east of the present outcrop and a minimum fan radius of *c*. 6–7 km. The suggestion of an alluvial fan from the palaeocurrent pattern is consistent with the sedimentary facies. The lack of debris flows may be due to the preserved deposits being too far down the fan surface to be reached by such flows or source rock and/or climatic conditions which did not favour debris flows. However, it is possible that some debris flow deposits may have gone unrecognized due to the poor quality of the exposure. Maximum particle size (MPS) was measured as the mean of the ten largest clasts in a conglomerate bed within 5 m laterally of the relevant traverse. A summary of these data is given in Fig. 2C. Variation in MPS is a function both of geographical and stratigraphical position. This is most easily seen in the Glensaul River section where the conglomerate member is exposed both east and west of an inlier of the Tourmakeady Volcanic Succession. Both eastern and western sections show a coarsening up–fining up sequence of conglomerates but those in the eastern section are persistently coarser than those in the western section. Thus the pattern of MPS is consistent with and supportive of the imbrication data. Average MPS of > 0.75 m also suggests that these clasts have travelled only a short distance. Three individual clasts exposed in different beds in the Glensaul River at Tourmakeady have cross-sectional dimensions of 1.75 m × 1.5 m (migmatite), 1.6 m × 1.2 m (porphyry), and 1.4 m × 1.35 m (migmatite) strongly supporting a very limited transport distance. Whilst correlation of similar stratigraphic level in this area can only be approximate, downstream decrease in MPS is of the order of 10–15 cm km^{-1}. This is consistent with many of the smaller alluvial fans described in the literature, e.g. Denny (1965, 1967); Hooke (1968); Gloppen & Steel (1981) and with our estimation of a fan radius of c. 8 km.

Empirical equations between fan area and drainage basin area have been established (Bull 1964, 1977; Hooke 1968; Denny 1965). These equations are generally given in the form where

$$A_f = c.A_d n$$

where

A_f = area of fan
A_d = area of drainage basin
n generally varies from 0.8 to 1.0 (Bull 1977)
c is dependent on factors such as lithology, climate, slope and tectonic setting and shows considerable variation.

For the Derryveeny fan we have taken a fan radius of 8 km and assumed a simple semi-circular plan form to give a value for A_f of 100.5 km^2 and we have assumed a mean value for n of 0.9. The value of c is poorly constrained and we have substituted the extreme values given by Hooke (1986) for 0.4 and 1.3. These give values for A_d of 280 km^2 and 86 km^2 respectively and we suggest that the drainage area supplying the Derryveeny fan most likely lay within this range. It is not possible to fix the shape of the drainage basin so that on Fig. 3 we show the simplest pattern of circular areas with radii of 9.94 and 5.23 km. Elongation of the drainage basin perpendicular to the proposed boundary fault could extend drainage further south-east than shown on Fig. 3.

Conglomerate clast composition

The conglomerates of the Derryveeny Formation are polymict, consisting mainly of metamorphic, plutonic and volcanic clasts. At reasonable quality exposures most of the clasts can be assigned to one of the following lithological groups: silicic volcanic and hypabyssal rocks, intermediate/basic volcanic rocks, leucogranites, biotite granites, basic intrusive rocks, metamorphic rocks, vein quartz, chert and jasper. The abundance of these clast groups was recorded at several localities throughout the formation (Table 1). Marked stratigraphical variation in clast composition is observed in the western outcrops, particularly well seen in the most continuous stream section at Derryveeny (M072 680 to M080 678). In the lower parts of the succession silicic volcanic and hypabyssal clasts and leucogranites predominate. Metamorphic clasts, mainly schists and gneisses, are present in only small amounts in the lower parts but increase in

Table 1. *Abundance of various clast types at selected localities in Conglomerate Mbr of the Derryveeny Fm*

Clast Type	Locality								
	Derryveeny–Tourmakeady area						Derry Bay–Owenbrin valley		
	a	b	c	d	e	f	x	y	z
Silicic volcanic & hypabyssal	3	3	2	1		1	1	2	2
Basic/Intermediate volcanic			2	1	1		2	2	1
Leucogranite	3	3	2	1	1	1			
Biotite granite			2	3	3	3	2	2	2
Basic intrusive			1					1	
Metamorphic		1	2	4	4	4	4	4	4
Vein quartz		1	1			1		1	1
Chert								1	
Jasper			1					1	

Key: 1, <5%; 2, 5–25%; 3, 25–50%; 4, 50–75%.

importance upwards such that they comprise the bulk of the clasts in the upper half of the conglomerate member. A distinctive biotite granite, often xenolithic, is also common at these upper levels, while the silicic volcanic and leucogranite clasts are most abundant in the middle levels of the conglomerate member but are uncommon elsewhere. The abundance of vein quartz in the finer grain sizes increases stratigraphically upwards, its occurrence being closely related to that of the metamorphic clasts suggesting derivation from the same source. More detailed clast counting was carried out at four localities where exposures were sufficiently clean to identify most clasts without constant hammering (Fig. 4). The latter are all interpreted as being in the middle to upper part of the conglomerate member but are from different geographical locations. They are essentially similar in composition but demonstrate a relationship between size and composition which has been noted elsewhere (e.g. Haughton 1989), i.e. that granites are more common in the larger sizes and that vein quartz, chert and jasper are generally confined to the smaller sizes.

Further subdivision could be made of the compositional groups recognized in the field on investigation of a representative suite of clasts in the laboratory.

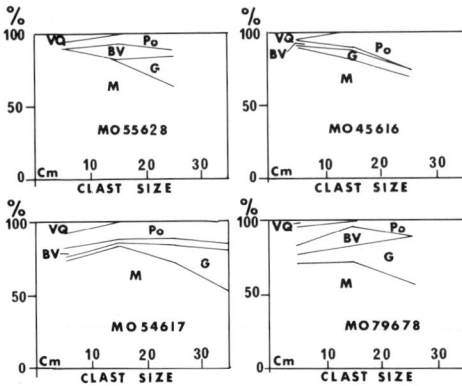

Fig. 4. Composition of Derryveeny conglomerates for different size windows; sample localities given by grid references. M, metamorphic; G, granite; BV, intermediate/basic volcanic; Po, silicic volcanic and hypabyssal; VQ, vein quartz, chert and jasper.

Silicic volcanic and hypabyssal clasts

This group is dominated by quartz–feldspar porphyries. They consist of 20–30% phenocrysts, mainly feldspar, quartz and biotite, set in a fine

grained matrix which, in hand specimen, may be red, green or purple. The phenocrysts range from 1–5 mm in size. Plagioclase is the dominant phenocryst phase comprising 40–60% of the total, quartz (25–40%) and biotite (5–15%) are less abundant, while K-feldspar occurs only rarely. The plagioclase is subhedral, commonly zoned, and sericitized and saussaritized. Where fresh it is determined to have a composition varying from calcic oligoclase at the crystal margins to calcic andesine, and in some cases sodic labradorite, in the core. Normal zoning is also suggested by the pattern of alteration, i.e. more calcic cores more intensely altered than less calcic rims. Quartz phenocrysts are often blue in colour, are anhedral to subhedral and often exhibit glomerophyric texture. Biotite is usually partly or wholly replaced by chlorite. The groundmass, which is predominantly felsic in composition, in places containing small amounts of biotite, may be either microcrystalline or consist of a fine-grained polygranular mosaic of quartz and feldspar. Weak flow texture was observed within some of the clasts examined. A well developed tectonic fabric, defined by flattening of the phenocrysts, is present in some samples and occasionally this fabric is mylonitic. A dacitic composition is suggested by the occurrence of plagioclase as the dominant feldspar phenocryst, while the high amount of quartz phenocrysts, the generally low colour index and the SiO_2 content (74.9%) suggests a rhyolitic composition. Rb/Sr ratios are <0.25.

Intermediate–basic volcanic clasts

This group is represented by spilitized intermediate and basic volcanic rocks, the original compositions (confirmed by geochemistry) of which were andesite and basalt respectively. The more abundant basic clasts are dark green to black, aphanitic rocks. They consist of 40–50% plagioclase laths pervasively sericitized with interstitial chlorite and opaques. Minor amounts (up to 5%) of quartz also occur. Green porphyritic andesites contain up to 15% subhedral plagioclase (andesine where fresh) phenocrysts, often altered to sericite and epidote in a fine-grained groundmass of quartz, plagioclase, chlorite and opaque minerals.

Basic intrusive clasts

Rare clasts of altered basic intrusive rocks comprise dolerites and subordinate gabbros. Both types consist of subhedral to euhedral sericitized plagioclase (*c.* 40%), anhedral quartz (*c.* 5%) and chlorite (*c.* 55%). The chlorite contains opaque inclusions and encloses the plagioclase representing a relict ophitic texture. In the dolerites the plagioclase is lath shaped and exhibits variolitic texture.

Leucogranites

These are medium to coarse grained leucocratic, pink coloured rocks, characterized by high quartz contents (40–50%), plagioclase as the dominant feldspar and low mafic mineral content (*c.* 5%). Plagioclase is subhedral, weakly zoned, calcic oligoclase. to sodic andesine in composition and exhibits variable alteration to sericite, epidote, and calcite, the cores being particularly affected. K-feldspar is present as anhedral orthoclase, occasionally microperthitic, clouded by fine opaque aggregates. Biotite has been almost completely replaced by chlorite and opaque aggregates. The leucogranite clasts plot in the granite field of Streckeisen (1976), and compare well with the average granite composition of Le Maitre (1976), except for their lower K_2O contents. They also have low Rb and quite high Sr concentrations, with an average Rb/Sr ratio of 0.16 and are moderately peraluminous with Mol $(Al_2O_3/CaO + Na_2O + K_2O)$ values of 1.03 to 1.20.

Biotite granites

These S-type granite clasts are distinctive due to the presence of abundant biotite and also differ from the leucogranites in containing muscovite, having orthoclase as the dominant feldspar, and in commonly containing metasedimentary inclusions.

The biotite granites are medium grained, nonporphyritic, and have a high colour index. Quartz, anhedral in shape, is abundant (*c.* 35–40%). Orthoclase (*c.* 20–25%) is also anhedral while plagioclase (*c.* 15–20%) is subhedral. Perthite is absent. Feldspar alteration is variably developed with orthoclase going to sericite and plagioclase (sodic andesine when fresh) going to sericite and calcite. The most striking feature is the high (*c.* 15–20%) content of anhedral to subhedral, partly chloritized biotite. Muscovite (2–5%) is usually intergrown with biotite. Anhedral cordierite (altered to pinite) may be accompanied by biotite (with rounded zircon inclusions) and/or sillimanite. Occasionally a weak foliation is defined by a preferred alignment of micas. Despite having a high mafic mineral content, the granites still plot in the granite (*sensu stricto*) field on the Streckeisen diagram. However, on the total alkalis (TAS) diagram of Middlemost (1985) they fall within the fields of either granodiorite or tonalite, although they differ markedly from mean compositions of these types having anomalously high TiO_2, Fe_2O, MgO and low CaO and NaO contents. They also have anomalously high concentrations of Cr and Ni, quite high Zr contents and mean Rb/Sr ratios of 0.73. Furthermore these granites are strongly peraluminous (Mol $(Al_2O_3/CaO + Na_2O + K_2O) = 1.7$ to 2.3) in contrast to the leucogranites.

Xenolithic biotite granites and inclusion-free varieties are equally common. Apart from the inclusions they are identical in all respects. The xenolithic varieties contain between 10 and 25% of metasediment and occasional inclusions of quartz up to 20 mm in length. Importantly the presence of variable amounts of xenoliths does not produce any appreciable chemical affect. The inclusions occur either as pods, up to 50 mm in diameter, or as continuous or semi-continuous bands 10–20 mm thick. Their contacts with the enclosing granite are irregular. A fabric is sometimes imparted to the rock by the presence of bands or the alignment of inclusions. The inclusions consist of abundant biotite, in addition to feldspar, quartz, muscovite $\pm$ sillimanite, $\pm$ garnet, $\pm$ cordierite, $\pm$ hercynite. Sillimanite is common and replaced biotite as it grew. Typically a well developed foliation defined by biotite and sillimanite and overgrown by late muscovite is present. The mineral assemblages indicate high temperature–low pressure metamorphic conditions.

There appears to be a gradation between the biotite-rich inclusions and single biotite crystals occurring as part of the enclosing granite. Biotite clots containing only a few crystals of biotite are common, and appear to represent an intermediate stage between the inclusions and single crystals. This may suggest that at least some of the biotite within the granite proper may have been provided by the breakdown of the inclusions into their constituent crystals. Biotite crystals within inclusions are sub-idioblastic with straight grain boundaries, whereas biotite at the margins of an inclusion have irregular contacts with the enclosing granite suggesting some reaction between the inclusions and the granite magma. Isolated biotite grains show similar anhedral shape and ragged grain boundaries and are identical to the biotite in the non-xenolithic biotite granites.

Migmatites

Two varieties of migmatite clasts are present which differ in texture but not in mineralogy. Stromatic migmatites are regularly banded on a scale of 50–300 mm, more quartzo-feldspathic leucosome being interlayered with finer grained quartz–feldspar–poor melanosome. Schollen migmatites are heterogeneous rocks consisting of irregularly oriented blocks and pods of restite with a mainly quartzo-feldspathic leucosome matrix.

In the schollen migmatites granitoid leucosome and restite comprise roughly equal proportions of the rock. The leucosomes contain quartz, plagioclase and potassium feldspar, as well as biotite, muscovite, and occasionally cordierite commonly altered to pinite. The restite is rich in biotite and sillimanite and also contains muscovite, quartz, plagioclase, orthoclase and opaques. Sillimanite occurs as fibrolite growing over biotite parallel to cleavage planes. Biotite is sub-idioblastic in shape. Muscovite is seen to replace other minerals such as orthoclase, biotite and sillimanite and is therefore considered not to be part of the peak metamorphic assemblage. Within the restite a foliation is developed by biotite and sillimanite but because of the irregular orientation of the restite bodies no single fabric direction is dominant. However, a crude banding on a scale of 5–30 mm is imparted to the rock by the interbanding of material dominated by leucosome with material dominated by restite. The irregular orientation of the restite within the leucosomes indicates formation of the migmatite in the presence of a melt. The occurrence of K-feldspar + sillimanite suggests that temperatures exceeded the muscovite + quartz out melting reaction. Furthermore an anatectic origin for the leucosome material is considered likely because of the occurrence of biotite and sillimanite within the leucosomes, the probable result of incomplete separation of melt from restite. The restites within the migmatites are similar to the metasedimentary inclusions present in the xenolithic biotite granites.

Gneiss

Quartzo-feldspathic and pelitic (up to 25% biotite) gneisses contain highly altered plagioclase and orthoclase (often microperthitic), variably chloritised biotite and poikiloblastic garnet with quartz inclusions. Occasional anhedral grains of chromite are probably detrital. The possibility that some of the quartz-feldspar rich layers crystallized, at least partly, from a melt cannot be excluded. Banding, possibly anatectic in origin, is defined by 2–6 mm thick layers alternately quartz–feldspar rich and quartz–feldspar poor with a parallel biotite–muscovite foliation. The quartz–feldspar rich layers are typified by granoblastic polygonal textures, sometimes destroyed by mylonitization.

Schists

Metasedimentary schists vary from psammitic to semi-pelitic and are mineralogically similar to the gneisses. There is compositional variation from psammitic to semi-pelitic schists. A strong foliation is defined by the alignment of biotite and muscovite as well as the flattening of quartz. A crenulation cleavage is sometimes developed.

Mylonites

Some schists and gneisses, that are otherwise similar to those described above, possess a strong mylonitic fabric. The mylonitization is expressed by elongation of quartz grains, grain reduction especially of quartz, highly undulose extinction of many mineral phases, and crystallization of small, new grains at the boundary between larger quartz grains.

Summary of the metamorphic clast assemblage and comparison with the biotite granites.

It is important to note that there are no sharp boundaries between the metamorphic clast types described above, rather a complete gradation exists between schist, gneiss, stromatic migmatite and schollen migmatite. This sequence is accompanied by an increase in grain size, a greater degree of differentiation between quartz–feldspar rich material and mafic material, and higher grade metamorphic assemblages.

In the metamorphic clasts, garnet is not observed to coexist with cordierite, a probable reflection of its progressive replacement by biotite and sillimanite as higher metamorphic grades are reached. Evidence for biotite growth continuing after the development of garnet comes from the latter being frequently replaced by and wrapped around by the former. Therefore the garnet predates at least the latter stages of the development of the main biotite ± sillimanite fabric. In addition cordierite may occur within the migmatite leucosomes.

The source of the metamorphic clasts consisted of psammitic and semi-pelitic rocks showing a variety of mineral asemblages and textures. High temperature–low pressure metamorphic conditions are indicated for the sillimanite–cordierite bearing assemblages.

An anatectic origin for the biotite granites is consistent with both textural and geochemical data. Inclusions within the biotite granites have identical mineralogy to the restite material in the migmatites. Furthermore both the biotite granites and the migmatitic leucosomes are granitic (*sensu stricto*) in composition and both contain cordierite and ragged biotite grains. Only high grade, never low grade, metamorphic assemblages are present in the metasedimentary inclusions.

The chemical composition implies a liquidus for the granites at unreasonably high temperatures of $> 1000°C$ (Chappell & White 1988) and suggests the presence of a substantial restite component. The abundance of peraluminous minerals, the high quartz content, together with the highly mafic character indicates a very low feldspar content and may be a reflection of low alkali and calcium contents in the magma source rocks, a feature typical of sedimentary material that has undergone a cycle of weathering. Their chemical composition satisfies most of the criteria listed by Miller (1985) for origin of magmas from pelitic material, except for Rb and Rb/Ba values that are too low. This can simply be explained by a significant greywacke component in the source rocks.

Composition of Derryveeny Formation sandstones

A reconnaissance examination of ten thin sections of sandstones from the sandstone member of the Derryveeny Formation shows a consistent composition independent of geographical or stratigraphical position. Lithic fragments within the sandstones are comparable with clast types in the conglomerates with a dominance of metamorphic grains, particularly schists. Chlorite–biotite grains are common and the generally micaceous appearance in the field is a locally distinctive criterion. The compositions suggest derivation from a similar source area to that which supplied the conglomerate clasts.

Summary of provenance data

The source of clasts for the Derryveeny Formation conglomerates was situated only a few km east of the present outcrop. The initial supply was from a source consisting predominantly of silicic volcanic and hypabyssal rocks, and leucogranites, presumably of high level origin, with subordinate metasedimentary clasts. This gradually changed to one containing more intermediate/basic volcanic rocks and, more importantly, anatectic biotite granites and a wide range of metamorphic lithologies, including migmatites, indicative of high temperature–low pressure metamorphism. This sequence of clast arrivals could represent unroofing of a metamorphic terrain from silicic volcanic and high level intrusive rocks downwards to deeper level anatectic S-type granites and their enveloping high-grade metamorphic rocks. However, in view of the clear petrographical and geochemical distinction between the granite types, an alternative scenario to simple unroofing is that the leucogranites, porphyries and volcanic rocks on the one hand and the biotite granites, metasediments and migmatites on the other may be from adjacent areas of differing geology with the metamorphic area contributing an increasing proportion of the sediment as the drainage basin evolved by nick point retreat.

Possible source of the clasts

Basic and intermediate volcanic rocks, jasper and chert are the main lithologies comprising the Lough Nafooey Group which crops out immediately to the south of, and is probably in fault contact with, the Derryveeny Formation. Thus it represents the most obvious source for these lithologies. The basic intrusive clasts could have a source in the deeper intrusive equivalents of the Lough Nafooey Group.

The mainly underformed silicic volcanic and granitoid clasts have no obvious source lying to the southeast. Silicic volcanic rocks of Lower Ordovician age do crop out immediately to the east and north (Tourakeady Valcanic Succession), the opposite direction to that of their derivation. However, these rocks may originally have had a much greater areal extent to the south and east and may underlie much of the present Derryveeny outcrop. The similarity between the silicic volcanic and leucogranite clasts suggests that both may be derived from this terrain although leucogranites are no longer represented in the local geology.

Similarly, biotite granites are no longer represented in the local geology. A clue to the source of the biotite granites is provided by the nature of their metasedimentary inclusions, which are mineralogically and texturally similar to migmatite restite within the migmatites of the Connemara Dalradian rocks to the south, and those of the Lough Kilbride Formation in particular.

The Connemara Dalradian metasedimentary rocks and their associated intrusions are a potential source for all of the metamorphic clast types recognized in the Derryveeny Formation. They lie just to the south of the Derryveeny Formation across some major faults and consist of quartzites, psammites, pelites, marbles, gneisses and amphibolites. Psammitic and semi-pelitic schists and gneisses are especially common in the upper parts of the stratigraphy which crop out in northern Connemara, closest to the Derryveeny Formation. In the southern part of Connemara there is an extensive belt of syn-metamorphic intrusive rocks ranging in com-position from ultrabasic bodies through to intrusive acidic gneisses.

The dominant regional metamorphism is a low pressure (Buchan) type and the meta-morphic grade increases southwards across the dominant strike from garnet zone to migmatites immediately north of the intrusive gneisses (Yardley *et al.* 1987; Fig. 3). However restricted areas of high grade rocks occur in both NW Connemara, adjacent to ultrabasic intrusive rocks, and in the lough Kilbride Formation of NE Connemara just south of the Lough Nafooey basic rocks and only a few km south of the Derryveeny Formation.

The migmatites associated with the intrusive gneisses in the southern part of Connemara have metamorphic assemblages (abbreviations after Kretz 1983) in the pelitic restites of:

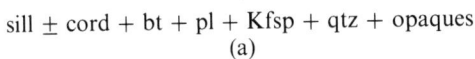

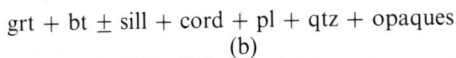

$$\text{sill} \pm \text{cord} + \text{bt} + \text{pl} + \text{Kfsp} + \text{qtz} + \text{opaques}$$
$$(a)$$

$$\text{grt} + \text{bt} \pm \text{sill} + \text{cord} + \text{pl} + \text{qtz} + \text{opaques}$$
$$(b)$$

Assemblage (a) is associated with granitic leuco-somes with occasional cordierite while (b) is associated with trondhjemitic leucosomes con-taining andalusite and cordierite (Barber & Yardley 1985; Treloar 1985). Muscovite, where it occurs, is not considered to be primary.

The migmatites of the Lough Kilbride Forma-tion in NE Connemara are characterized by the assemblage in pelitic restite of:

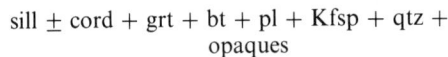

$$\text{sill} \pm \text{cord} + \text{grt} + \text{bt} + \text{pl} + \text{Kfsp} + \text{qtz} + \text{opaques}$$

In all the migmatites described above biotite-sillimanite (fibrolite variety) defines the main foliation, which in Connemara is considered to be S3. It is difficult to discriminate between these two migmatite areas as potential sources on mineralogical evidence alone. However two fac-tors argue against the main migmatite belt and in favour of the Lough Kilbride migmatites. A petrographic argument is the absence of mig-matite clasts with trondhjemitic leucosomes and also of many typical Dalradian lithologies such as quartzites, marbles and amphibolites which currently crop out between the Derryveeny For-mation and the main migmatite belt. A sedimen-tological argument can also be made by consi-dering the estimates of drainage basin size presented above (Fig. 3). It would seem unlikely that any significant contribution to the detrita would be made by the main belt of Connemara migmatites even taking the largest likely drain-age basin, unless their along-strike continuation is radically offset by faults concealed beneath the later cover. In contrast a source from the Lough Kilbride Formation, probably with a greater outcrop width, provides a much simpler expla-nation.

Comparison with the Lough Kilbride Formation

In the only published summary of the Lough Kilbride Formation it was described as a series of crushed and retrogressed impure psammites and semipelites with occasional lenses of ultrabasic rock (Graham *et al.* 1989). Our observations indicate an abundance of migmatites showing nebulitic and schollen structure. Blocks with a metasedimentary appearance, which are fre-quently graphitic, are dispersed irregularly within quartz–feldspar rich granitic material. At some localities the amount of leucosome is much reduced so that it forms only a minor part of the rock. Well developed stromatic migmatites have not been observed. The contacts between restite and leucosome are diffuse and irregular, as in the migmatite clasts of the Derryveeny Formation. The leucosomes are granitic consisting of quartz, plagioclase, orthoclase, biotite and muscovite, and cordierite.

There are also some differences between the migmatite clasts and the in situ migmatites:

(a) stromatic migmatites appear to be more commonly represented in the clasts;
(b) leucosome : restite ratio is generally higher in the clasts;
(c) the transition to true granites is not observed within the Lough Kilbride Forma-tion as it appears to be within the clast assemblage.

Despite these differences, the Lough Kilbride Formation represents a plausible candidate for the source of most of the Derryveeny clasts. The differences can be explained if the rocks of the Lough Kilbride Formation are considered to be the deeper-level equivalents of the eroded material now represented as clasts in the Derryveeny Formation. The migmatites of the Lough Kilbride Fm. may have experienced a higher degree of melt extraction and more disruption of the original metasedimentary layering because of this magma movement, which may be largely a function of the degree of partial melting. The upward movement of melt of granitic composition may explain the absence of true granite at the present exposure level of the Lough Kilbride Formation, and its occurrence as clasts in the Derryveeny Formation which represent material from a higher structural level.

Thus to a large extent the source of much of the conglomerates is presently represented, albeit at a more deeply eroded level, immediately to the south of the Derryveeny outcrop, within what are now represented by the Lough Nafooey Group and the Lough Kilbride Formation.

Isotopic data

Rb, Sr, Sm and Nd were separated from mineral and whole-rock samples by conventional ion exchange techniques. $^{87}Sr/^{86}Sr$ and $^{143}Nd/^{144}Nd$ ratios, and isotope dilution concentration determinations of Rb, Sr, Sm and Nd, were performed using a semi-automated VG Micromass 30 mass spectrometer in the Geology Department of University College, Dublin. Maximum measured blanks were as follows: Sm, 0.25 ng; Nd, 1 ng; Rb, 1 ng; Sr, 2 ng. These have negligible effects on calculated ages and initial ratios and no corrections have been applied to the data quoted. Rb-Sr ages were calculated using a decay constant for ^{87}Rb or 1.42×10^{-11} a^{-1}. Sm–Nd ages were calculated using a decay constant for ^{147}Sm of $6.54 \times 10^{-12} a^{-1}$. Further analytical details are given in Table 2.

Muscovite–whole-rock pairs from two schist clasts (Table 2) yield Rb-Sr ages of 471 ± 8 Ma and 462 ± 7 Ma. In view of the very low metamorphic grade of the Derryveeny Formation, these ages are likely to represent cooling events in the source. If it is assumed that these ages reflect the time that the rocks last passed through about 500°C, a reasonable blocking temperature for the Rb-Sr system in muscovite (Jager 1979), then the younger date (462 Ma) provides a maximum age for the deposition of the Derryveeny Formation i.e. Upper Llanvirn

using the McKerrow et al. (1985) timescale, and Llandeilo on most others. This is consistent with the field evidence that the Derryveeny Formation is a post-Llanvirn and pre-Llandovery deposit and thus of Upper Ordovician age.

These ages are similar to the mean value (455 ± 10 Ma) of muscovite ages from the Dalradian of Connemara (Elias et al. 1988) and Scotland (Dempster 1985) and suggest a comparable early Caledonian cooling history in the Derryveeny source region. The slightly older ages of the clasts compared to currently exposed outcrop may be explained by an origin higher in the structural pile. Such rocks would have cooled through the closing temperature of muscovite somewhat earlier.

Whole rock Sm-Nd and Rb-Sr isotopic data (Table 2) from a variety of clasts provide some constraints on the age and nature of the source. Sm-Nd model ages (T_{DM}, De Paolo 1981) range from 1680 Ma for a leucogranite to between 1916 and to 2319 Ma for a migmatite, S-type granite and schists. A single sandstone sample has a model age of 2091 Ma. These ages overlap with those of Dalradian Supergroup metasediments including the available data from Connemara (Daly & Menuge 1989). The mean model age of the migmatite and schist clasts (2158 Ma) is identical to that of two migmatite samples from the Lough Kilbride Formation supporting their petrographic similarities. In addition, one S-type (biotite) granite clast has a model age that falls within the range of the migmatite and schist clasts and is identical to one of the Lough Kilbride migmatites. These results suggest that the S-type granite may be derived by partial melting of parents similar to the schist and migmatite clasts and could represent a high-level equivalent of the presently exposed Lough Kilbride Formation migmatites. This is supported by Rb-Sr data (Table 2). Assuming an intrusion age of 475 Ma, its initial $^{87}Sr/^{86}Sr$ ratio (0.722, Table 2) is slightly less than, but comparable with that of the schist clasts (c. 0.726) and with the range from the Connemara Dalradian at the same time (0.7142 to 0.7419, Elias et al. 1988).

The lower T_{DM} (1680 Ma) and $^{87}Sr/^{86}Sr_{475}$ values of the leucogranite suggest an origin separate from the biotite granite in agreement with the chemical differences noted above. However even this sample falls within the range of the Dalradian data and a Dalradian protolith cannot be ruled out.

Discussion and implications

Derivation of much of the Derryveeny Formation from stripping of the higher levels

Table 2. Sm-Nd data on model ages from clasts in the Derryveeny Fm and from the Lough Kilbride Fm

Lab. no.	Sm	Nd	$^{147}Sm/^{144}Nd$	$^{143}Nd/^{144}Nd$	T_{DM}	Rb	Sr	$^{87}Rb/^{86}Sr$	$^{86}Sr/^{87}Sr$	$^{87}Sr/^{86}Sr_{475}$	Ms-WR age
Derryveeny Formation sandstone											
80701	1.84	9.79	0.1136	0.511665 (14)	2091						
Derryveeny Formation clasts											
50220[1]	6.87	38.79	0.1071	0.511560 (26)	2113	89.92	163.8	1.592	0.73267 (5)	0.7219 (2)	
50213[2]	5.21	30.43	0.1035	0.511822 (14)	1680	49.00	208.8	0.680	0.71962 (6)	0.7150 (1)	
81107[3]	2.94	16.18	0.1099	0.511460 (22)	2319	35.53	121.0	0.852	0.73199 (4)	0.7262 (1)	
50219[4]	6.34	34.67	0.1106	0.511741 (12)	1916						
80301[5]	4.32	24.92	0.1048	0.511440 (14)	2239	68.72	148.3	1.344	0.73435 (5)	0.7253 (2)	471 (8)
80301 ms						223.5	67.5	9.660	0.79014 (6)		
M8[6]						34.69	137.0	0.734	0.73147 (6)	0.7265 (1)	462 (7)
M8 ms						242.4	24.7	28.92	0.91678 (4)		
Migmatized pelite, L. Kilbride Formation											
3/88-12	8.46	52.90	0.0967	0.511422 (12)	2104						
3/88-13	6.39	38.40	0.1006	0.511394 (12)	2216						
Standards											
La Jolla (6 runs)				0.511866							
BCR-1 (4 runs)				0.512620							
NBS K-feldspar								8.057			
SRM987									0.71031 (7)		

Analytical techniques were those described by Menuge (1988) as modified by Menuge & Daly (1990). $^{143}Nd/^{144}Nd$ ratios were determined from spiked samples and are normalized to $^{146}Nd/^{144}Nd = 0.7219$. $^{87}Sr/^{86}Sr$ ratios are corrected for mass fractionation by normalizing to an $^{86}Sr/^{88}Sr$ value of 0.1194. Quoted errors (in parentheses) in $^{143}Nd/^{144}Nd$ and $^{87}Sr/^{86}Sr$ ratios are within-run precisions; reproducibility of $^{147}Sm/^{144}Nd$ and $^{87}Rb/^{86}Sr$ ratios are 0.1% and 1% respectively. All errors are quoted at the $2\sigma_m$ level.
Derryveeny Formation clasts: 1, biotite granite (M079 678); 2, Leucogranite (M079 678); 3, pelitic schist (M054 617); 4, migmatite (M079 678); 5, semi-pelitic schists (M041 633); 6, pelitic schist (M054 617); ms, muscovite; WR, whole rock.

of the Lough Kilbride Formation carries several important implications which help to constrain hypotheses of the regional geological development.

(1) Little or no lateral movement is likely between the Derryveeny Formation and the Lough Kilbride Formation in post-sedimentation times. Thus, whatever its earlier history, Connemara appears to be docked relative to South Mayo during the Upper Ordovician.

(2) Clasts in the Derryveeny Formation contain mineral assemblages and fabrics similar to those associated with D3 deformation and metamorphism in the Connemara Dalradian (Yardley *et al.* 1987) and therefore indicate a post-D3 age for the Derryveeny Formation. D3 is considered to have occurred at around 480 ± 10 Ma (Elias *et al.* 1988; Yardley *et al.* 1987). This is consistent with age constraints imposed by the Rb-Sr muscovite–whole-rock dating of schist clasts and conclusions from the field data.

(3) High grade metamorphism which led to the formation of migmatites may have been responsible for the production of the granites. These granites rose to higher levels in the crust and were subsequently removed by erosional unroofing. Such granites may be a common product of migmatites associated with high temperature metamorphism, but because of erosion this association may only rarely be preserved. Inclusions of metasedimentary fragments with fabrics similar to S3 in Connemara implies a maximum age of 480 ± 10 Ma for the granites.

(4) The Derryveeny Formation, whilst containing volcanic and high level intrusive clasts, shows no sign of low grade metasedimentary clasts. It would thus seem that these rocks must have been stripped during an earlier period of uplift or alternatively have bypassed the present Derryveeny outcrop because the onset of subsidence allowing sediment accumulation occurred only part way through this uplift event. In view of the general linking of uplift and subsidence and the known pulsatory uplift history of Connemara, our preference is for the former. Widespread pulses of uplift are suggested at 455 ± 10 and at 440 ± 5 Ma by Elias *et al.* (1988) and the latter of these was correlated with the major Taconic event of North America. Recent well constrained data on major southward thrusting of Connemara gives ages of 447 ± 4 ma (Tanner *et al.* 1989) which are basically ascribed to the same Taconic event. Despite gross imprecision in relating the numerical and biostratigraphical time scales, these ages are all within the Upper Ordovician. Thus a relationship between this late uplift documented in Connemara and the deposition of the Derryvenny Formation seems

reasonable. It is possible that this late Taconic event also caused the pre-Silurian deformation of the Derryveeny Formation and that its deposition is related to the earlier (455 ± 10 Ma) uplift pulse recorded in Connemara.

(5) Both Elias *et al.* (1988) and Tanner *et al.* (1989) ascribe these upper Ordovician uplift and deformational events to the docking of Connemara with South Mayo and, on a wider scale, of the Midland Valley block with more northerly terranes. Whilst the evidence for docking having taken place by this time is strongly supported by the information presented here, justification of their hypotheses also requires proof of irreconcilable source-clast mismatches in lower Ordovician rocks.

(6) A strong case has been made for linking the Derryveeny clasts with the Lough Kilbride Formation which crops out to the south. However it has also been shown that derivation was from a point source to the east and not the south. This can be explained by examining the regional geology on a larger scale. Throughout western Ireland there is a major strike swing from the typical NE–SW Caledonoid trend of Scotland and northern Ireland to an E–W strike (Fig. 1). This major bend is well seen some 15 km to the north of the area discussed here at the eastern end of Clew Bay, but a similar swing must occur near the present eastern limit of exposure of the Connemara Dalradian. The structure in the Lower Ordovician rocks west of Lough Mask (Fig. 2; Graham *et al.* 1985, 1989) suggests that such a change also occurs in this area. Thus we suggest that the Derryveeny fan is located along the fault bounded northern margin of Connemara which has swung to a more northeasterly trend. Such a predicted subcrop pattern for the Connemara Dalradian may be testable geophysically. It is unlikely to be a coincidence that a small upper Ordovician basin should locate in this area. Where tectonic shortening is accompanied by strike slip displacements, as is generally suggested for the Caledonides, such small basins commonly locate at the bends in major fault systems. Such an origin has been suggested for Silurian and Devonian basins in South Mayo (Graham 1981; Graham *et al.* 1989).

(7) The cause of the high grade metamorphism in the Lough Kilbride Formation is not obvious. Elsewhere in Connemara such metamorphism is associated with ultrabasic-basic to acidic intrusions. Could this also explain the metamorphism here? The occurrence of basic volcanic and intrusive clasts in the Derryveeny Formation as well as the close proximity of the Lough Nafooey Group is certainly consistent with such

an idea but would have major implications for the lower Ordovician history of the area.

J. P. W. acknowledges financial support from University College Dublin and Department of Education, Ireland. J. S. D. and J. F. M. acknowledge support from EOLAS grant SC052/87. The work was instigated when J. R. G. was in receipt of NBST grant SRG/85/82.

References

BARBER, J. P. & YARDLEY, B. W. D. 1985. Conditions of high grade metamorphism in the Dalradian of Connemara, Ireland. *Journal of the Geological Society, London*, **142**, 87–96.

BLUCK, B. J. & LEAKE, B. E. 1986. Late Ordovician to Early Silurian amalgamation of the Dalradian and the adjacent Ordovician rocks in the British Isles. *Geology*, **14**, 917–919.

BULL, W. B. 1964. Geomorphology of segmented alluvial fans in western Fresno County, California. *United States Geological Survey Professional Paper* **352-E**, 89–129.

—— 1977. The alluvial fan environment. *Progress in Physical Geography*, **1**, 222–270.

CHAPPELL, B. W. & WHITE, A. J. R. 1988. Some supracrustal (S-type) granites of the Lachlan Fold Belt. *Transactions of the Royal Society of Edinburgh: Earth Sciences*, **79**, 169–181.

DALY, J. S. & MENUGE, J. F. 1989. Nd isotopic evidence for the provenance of Dalradian Supergroup sediments in Ireland. *Terra Abstracts*, **1**, 12.

DEMPSTER, T. J. 1985. Uplift patterns and orogenic evolution in the Scottish Dalradian. *Journal of the Geological Society, London*, **142**, 111–128.

DENNY, C. S. 1965. Alluvial fans in the Death Valley region California and Nevada. *United States Geological Survey Professional Paper*, **466**.

—— 1967. Fans and pediments. *American Journal of Science*, **265**, 81–105.

DE PAOLO, D. J. 1981. Neodymium isotopes in the Colorado Front Range and crust-mantle evolution in the Proterozoic. *Nature*, **291**, 193–196.

DEWEY, J. F. 1982. Plate tectonics and the evolution of the British Isles. *Journal of the Geological Society, London*, **139**, 371–412.

ELIAS, E. M., MACINTYRE, R. M. & LEAKE, B. E. 1988. The cooling history of Connemara, western Ireland, from K-Ar and Rb-Sr age studies. *Journal of the Geological Society, London*, **145**, 649–660.

GARDINER, C. I. & REYNOLDS, S. H. 1909. On the igneous and associated sedimentary rocks of the Tourmakeady district (County Mayo). *Quarterly Journal of the Geological Society of London*, **65**, 104–154.

—— & —— 1910. The igneous and associated sedimentary rocks of the Glensaul district (County Galway). *Quarterly Journal of the Geological Society of London*, **66**, 253–280.

GEIKIE, A. & KILROE, J. R. 1897. *Annual report of the Geological Survey of the United Kingdom for 1896*, 49–50.

GLOPPEN, T. G. & STEEL, R. J. 1981. The deposits, internal structure and geometry in six alluvial fan—fan delta bodies (Devonian-Norway)—a study in the significance of bedding sequences in conglomerates. *In*: ETHRIDGE, F. G. & FLORES, R. M. (eds) *Recent and ancient non-marine depositional environments: models for exploration.* Society of Economic Palaeontologists & Mineralogists Special Publication, **31**, 49–70.

GRAHAM, J. R. 1981. The 'Old Red Sandstone' of County Mayo, northwest Ireland. *Geological Journal*, **16**, 157–173.

—— 1987. The nature and field relations of the Ordovician Maumtrasna Formation, County Mayo, Ireland. *Geological Journal*, **22**, 347–369.

—— 1988. Collection and analysis of field data. *In*: TUCKER, M. E. (ed.) *Techniques in Sedimentology*, 5–62.

——, LEAKE, B. E. & RYAN, P. D. 1985. *The geology of South Mayo (map)* 1:63,360. University of Glasgow.

——, —— & —— 1989. *The geology of South Mayo, western Ireland.* University of Glasgow.

HARPER, D. A. T., GRAHAM, J. R., OWEN, A. W. & DONOVAN, S. K. 1988. An Ordovician fauna from Lough Shee, Partry Mountains, Co. Mayo, Ireland. *Geological Journal*, **23**, 293–310.

HARRIS, A. L. & PITCHER, W. S. 1975. The Dalradian Supergroup. *In*: HARRIS, A. L. *et al.* (eds) *A correlation of the Precambrian rocks of the British Isles.* Geological Society, London, Special Report, **6**, 52–75.

HOOKE, R. LEB. 1968. Steady-state relationships on arid-region alluvial fans in closed basins. *American Journal of Science*, **266**, 609–629.

HAUGHTON, P. D. W. 1989. Structure of some Lower Old Red Sandstone conglomerates, Kincardineshire, Scotland: deposition from late-orogenic antecedent streams? *Journal of the Geological Society, London*, **146**, 509–525.

HUTTON, D. H. W. 1987. Strike-slip terranes and a model for the evolution of the British and Irish Caledonides. *Geological Magazine*, **124**, 405–425.

—— & DEWEY, J. F. 1986. Palaeozoic terrane accretion in the western Irish Caledonides. *Tectonics*, **5**, 1115–1124.

JAGER, E. 1979. Introduction to geochronology and the Rb-Sr method. *In*: JAGER, E. & HUNZIKER, J. C. (eds) *Lectures in isotope geology*, Springer-Verlag, Berlin, 1–26.

KILROE, J. R. 1907. The Silurian and metamorphic rocks of Mayo and north Galway. *Proceedings of the Royal Irish Academy*, Section B, **26**, 129–160.

KRETZ, R. 1983. Symbols for rock-forming minerals. *American Mineralogist*, **68**, 277–279.

LEAKE, B. E., TANNER, P. W. G., MACINTYRE, R. M. & ELIAS, E. 1984. Tectonic position of the Dalradian rocks of Connemara and its bearing on the evolution of the Midland Valley of Scotland. *Transactions of the Royal Society of Edinburgh: Earth Sciences*, **75**, 165–171.

LE MAITRE, R. W. 1976. The chemical variability of some common igneous rocks. *Journal of Petrology*, **17**, 589–637.

MCKERROW, W. S., LAMBERT, R. St. J. & COCKS, L. R. M. 1985. The Ordovician, Silurian and Devonian periods. *In*: SNELLING, N. J. (ed.) *The Chronology of the Geological Record*. Geological Society, London, Memoir, **10**, 73–80.

MENUGE, J. F. 1988. The petrogenesis of massif anorthosites: a Nd and Sr isotopic investigation of the Proterozoic of Rogaland/Vest-agder, SW Norway. *Contributions to Mineralogy and Petrology*, **98**, 363–373.

—— & DALY, J. S. 1990. Proterozoic evolution of the Erris Complex, NW Mayo, Ireland: Neodymium isotope evidence. *In*: GOWER, C. F., RIVERS, T. & RYAN, B. (eds) *Mid-Proterozoic geology of the southern margin of proto Laurentia-Baltica*. Geological Association of Canada Special Paper (in press).

MIDDLEMOST, E. A. K. 1985. *Magmas and magmatic rocks: an introduction to igneous petrology*. Longmans.

MILLER, C. F. 1985. Are strongly peraluminous magmas derived from pelitic sedimentary sources? *Journal of Geology*, **93**, 673–89.

NILSEN, T. H. 1968. *The relationship of sedimentation to tectonics in the Solund Devonian district of south-western Norway*. Norges geologiske Undersokelse, **259**.

STRECKEISEN, A. L. 1976. To each plutonic rock its proper name. *Earth Science Reviews*, **12**, 1–34.

TANNER, P. W. G., DEMPSTER, T. J. & DICKIN, A. P. 1989. Time of docking of the Connemara terrane with the Delaney Dome Formation, western Ireland. *Journal of the Geological Society, London*, **146**, 389–392.

TRELOAR, P. J. 1985. Metamorphic conditions in Central Connemara, Ireland. *Journal of the Geological Society, London*, **142**, 77–86.

YARDLEY, B. W. D., BARBER, J. P. & GRAY, J. R. 1987. The metamorphism of the Dalradian rocks of western Ireland and its relation to tectonic setting. *Philosophical Transactions of the Royal Society of London*, **321A**, 243–270.

Petrological and geochemical determination of provenance in the southern Welsh Basin

T. McCANN

Department of Geology, University of Leicester, University Road, Leicester LE1 7RH, UK
Present address: Department of Geology, University College, Belfield, Dublin 4, Ireland

Abstract: The provenance of Llandeilo to upper Llandovery sediments from the Welsh Basin has been investigated by petrographic and geochemical methods. Sandstone-dominated formations were deposited from high-concentration turbidity currents. Petrographic data suggest derivation from a non-collisional recycled or cratonic setting. This is broadly confirmed by analysis of microconglomerate lithic fragments and the major element geochemistry. Mudstone-dominated formations were deposited from dilute turbidity currents and as hemipelagic sediments. The mudstone geochemistry indicates a tectonic setting transitional between an active continental margin and a passive margin. This study demonstrates that neither set of analytical methods are individually adequate for provenance reconstruction and it is advisable to use a variety of techniques for greater confidence in interpretation.

In recent years a number of detrital modal discriminants aimed at the determination of tectonic setting of ancient basins have been developed (e.g. Crook 1974; Dickinson & Suczek 1979). These have been complemented by similar work on modern sediments of known plate tectonic setting (e.g. Valloni & Maynard 1981; Potter 1986). Both are normally restricted to sandstones, although, mudstone geochemistry is increasingly seen as an important area especially by the petroleum industry. The Welsh Basin provides an opportunity to test the use of petrographical and geochemical discriminators against presently accepted tectonic models for the area, which are derived from geochemical studies of volcanic rocks (e.g. Bevins *et al.* 1984; Kokelaar *et al.* 1984*a*).

Tectonic history of the Welsh Basin

The Lower Palaeozoic sediments of the Welsh Basin form part of the Eastern Avalonia terrane and overlie a late Precambrian age lower continental crust comprising calc-alkaline rocks (Watson & Dunning 1979; Thorpe 1979). The early Palaeozoic Welsh Basin was a relatively rapidly subsiding area of continental crust separated from the more stable Midland Platform by an arcuate array of steep (at the surface) faults (Woodcock 1984). The area was the site of a NE–SW oriented, elongate, fault bounded, backarc or marginal basin located along a destructive plate margin on the southern side of the Iapetus Ocean.

Geochemical evidence from the volcanic centres in Wales record a period of late Tremadoc arc volcanism followed by Arenig–Caradoc backarc extension (Kokelaar *et al.* 1984*a*) with the main arc being situated to the northwest in the Leinster–Lake District Zone. The recorded changes, both in the geochemistry and the style of volcanism, between the Tremadoc and mid-Ordovician represent a transition from a volcanic arc to a more marginal basin setting (Bevins *et al.* 1984; Kokelaar 1988). The detailed Ashgill and Silurian setting of the basin is uncertain, although a number of models have been proposed, including a passive continental margin (Davies & Cave 1976), a forearc (Okada & Smith 1980) and a strike-slip continental borderland, similar to present day California (Woodcock 1984).

Age, previous work and depositional setting

The study area is located on the west coast of Wales (Fig. 1) and comprises a succession of 14 formations (Fig. 2) which range in age from Llandeilo (*Nemagraptus gracilis* Zone) to upper Llandovery (*Monograptus turriculatus* Zone) spanning a cumulative period of 30–35 Ma (Harland *et al.* 1982). The formations are identified on the basis of the dominant lithology. There are few studies of the petrographical and geochemical composition of the sediments from this part of the Welsh Basin, apart from work by Bjørlykke (1971), James (1971, 1981), Keeping (1881), Smith (1956) and Wood & Smith (1959).

From Morton, A. C., Todd, S. P. & Haughton, P. D. W. (eds), 1991,
Developments in Sedimentary Provenance Studies.
Geological Society Special Publication No. 57, pp. 215–230.

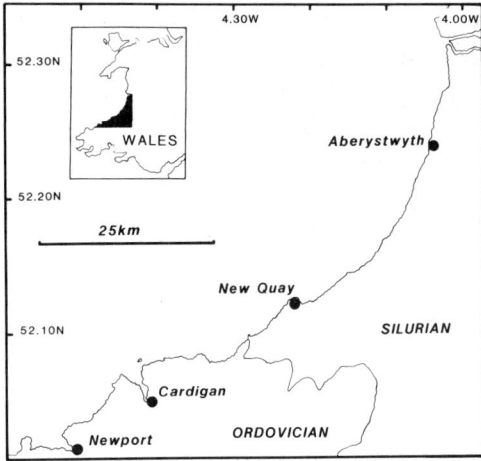

Fig. 1. Location and simplified geological map of the study area.

Sandstone-dominated formations (e.g. Newport Sands Formation, Aberystwyth Grits Formation) were deposited mainly from high-density turbidity currents and, locally, slide deposits whereas mudstone-dominated formations (e.g. Tresaith Formation, Gwbert Formation) were deposited from low-density turbidity currents (Fig. 3). Hemipelagites are locally abundant in

some of the mudstone-dominated formations (e.g. Tresaith Formation) and predominate in others (e.g. Gaerglwyd Formation, Parrog Formation) (McCann 1990). The controls on sedimentation appear to have been dominantly eustatic although tectonic influence was also important.

Palaeocurrent evidence reveals that the source area was to the S/SW and sediments were transported parallel to the NW–SE-oriented basin axis. Much of this southerly area was subjected to volcanic activity, some of it quite extensive, in the period immediately preceding the onset of deposition in the study area (Allen 1982; Bevins *et al.* 1984, 1989; Kokelaar *et al.* 1984*a,b*, 1985; Thorpe *et al.* 1989).

The sediments of the area represent a series of 'Type 1' turbidite systems (*sensu* Mutti & Normark 1987) which are composed dominantly of unchannelled sandstone lobes. They are particularly characteristic of basins where tectonic activity produces and maintains narrow depositional basins (Mutti & Normark 1987). Type 1 systems commonly have only one major site of sediment input, for example a major river or delta complex, and it is probable that such a sediment source was active on the southern margin of the Welsh Basin from Llandeilo–Llandovery times.

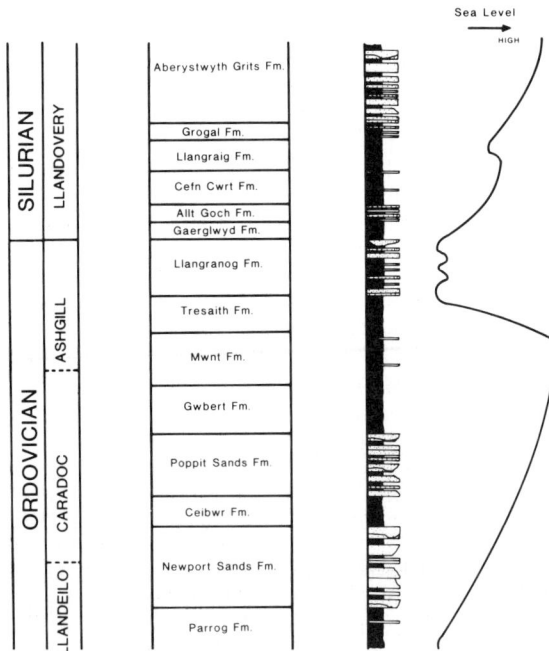

Fig. 2. Generalized stratigraphic column of the study area relating the lithological divisions to eustatic sea-level changes. Sea level curve after Leggett *et al.* (1981).

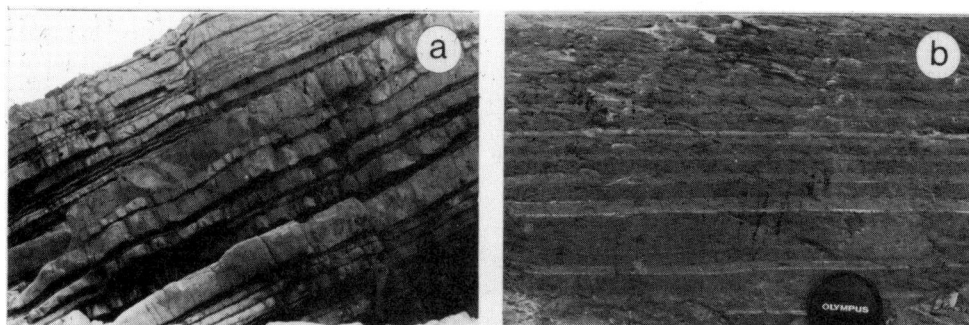

Fig. 3. Representative outcrop of lithological units in the study area. (**a**) Aberystwyth Grits Formation (sandstone dominated) at New Quay. Outcrop youngs to the left. Person for scale. (**b**) Gwbert Formation (mudstone dominated) at Gwbert, just north of Cardigan. Lens cap (5.0 cm diameter) for scale.

Sandstone petrography and framework modes

The petrography of the Newport Sands, Poppit Sands, Llangranog and Aberystwyth Grits formations was examined in detail using the Gazzi-Dickinson point-counting technique (see Ingersoll *et al.* 1984; Table 1). 146 samples were analysed, counting 200–500 points per thin section. All thin sections are stored in the National Museum of Wales, Cardiff (NMW89.13G.T1–161).

Excluding the matrix, which varies between the formations from Newport Sands (15.83%), Poppit Sands (19.33%), Llangranog (16.89%) and Aberystwyth Grits (19.71%), and the patchy calcite cements (1–4%) found in the Llangranog and Aberystwyth Grits formations, the rocks are made up of three main constituents.

Quartz (Q). Monocrystalline quartz grains (Qm) are commonly clear although they may be embayed or contain vacuole rims indicative of a volcanic origin (e.g. Newport Sands Formation, Poppit Sands Formation, Aberystwyth Grits Formation). Strained quartz crystals are most common; the lack of any common orientation to the strain shadows suggests that they were strained in the source area. Inclusions within the monocrystalline quartz grains are common particularly muscovite (sericite), zircon and tourmaline. Polycrystalline quartz (Qp) is either chert or strained quartz with sub-grains bounded by crystal faces. Contacts between the sub-grains are straight to crenulate. Zircon inclusions have been noted.

Feldspar (F). Plagioclase is the dominant feldspar in the succession. It has an albite–oligoclase composition (confirmed by electron microprobe analysis) which shows no stratigraphic variation. Post-depositional albitisation of feldspars, however, is common and thus the values obtained may not reflect the original composition. K-feldspar is much less common. Crystals of perthite are ubiquitous, though rare. Feldspars are frequently altered to sericite and other clay minerals and may be replaced by calcite.

Lithic grains (L). The most common lithic fragments are intraformational sedimentary rocks such as siltstone, mudstone and fine-grained sandstone. Volcanic rock fragments are commonly intermediate or acid in composition. Probable precursors of the felsic rock fragments were andesites, trachytes, granites and rhyolites. The rare metamorphic rock fragments are predominantly quartz and mica schists.

Accessory minerals (present in all formations) include muscovite, pyrite, chlorite-mica stacks, chlorite and zircon in decreasing order of abundance. Granophyre is present in all formations except for the Llangranog Formation. Other accessory minerals and their occurrences include tourmaline (Aberystwyth Grits Formation), biotite (Newport Sands Formation, Poppit Sands Formation, Aberystwyth Grits Formation), epidote (Poppit Sands Formation) and titanite (Newport Sands Formation, Aberystwyth Grits Formation).

Provenance and tectonic discrimination

The data are divided into two groups, the older formations which were deposited prior to the late Ashgill glacio-eustatic fall in sea level (Brenchley & Newall 1984) and the younger formations which were deposited during the low sea-level stand and the subsequent transgression. On the QFL diagram the older formations plot at the boundary of the continental block and

Table 1. Framework grain mode parameters of sandstones from the southern Welsh Basin

Formation	N		Q	F	L	Qm	F	Lt	Qp	Lvm	Lsm
Parrog	1	Mean	7.75	86.82	5.43	7.75	82.17	10.08	30.77	46.15	23.08
		Range	—	—	—	—	—	—	—	—	—
Newport Sands I	25	Mean	6.76	86.4	6.84	6.76	71.96	21.28	17.56	66.83	15.61
		Range	(2.2–11.46)	(69.5–93.68)	(2.81–19.39)	(2.81–19.39)	(52.38–85)	(3.34–30.0)		(29.91–86.66)	(2.81–19.39)
Mean	85.1	Mean	7.67		7.11		20.47	13.89		22.52	
		Range	(73.2–93.3)		(1.4–20.39)		(60.77–91.33)		(6.4–48.15)		(0.0–48.21)
Llangranog	10	Mean	9.33	87.07	3.59	9.33	77.33	13.33	15.48	73.1	11.42
		Range	(5.26–16.6)	(79.9–91.21)	(1.52–6.62)	(5.26–16.58)	(68.45–83.33)	(9.4–19.23)	(3.86–35.44)	(54.16–86.95)	(0.0–22.64)
Allt Goch	1	Mean	9.83	87.28	2.89	9.83	78.03	12.14	19.05	76.19	4.76
		Range	—	—	—	—	—	—	—	—	—
Aberystwyth Grits	58	Mean	11.39	82.7	5.9	10.85	72.55	16.6	18.64	65.9	15.46
		Range	(3.29–19.62)	(68.84–94.74)	(0.62–15.1)	(3.29–19.63)	(54.35–89.44)	(4.84–31.43)	(2.66–43.75)	(6.66–90.9)	(0.0–60.0)

Q, total quartzose grains; F, total feldspar grains; L, total lithic fragments; Qm, monocrystalline quartz grains; Lt, total lithic fragments including polycrystalline quartzose grains; Qp, total polycrystalline quartzose grains; Lvm, total volcanic and metavolcanic lithic fragments; Lsm, total sedimentary and metasedimentary lithic fragments.

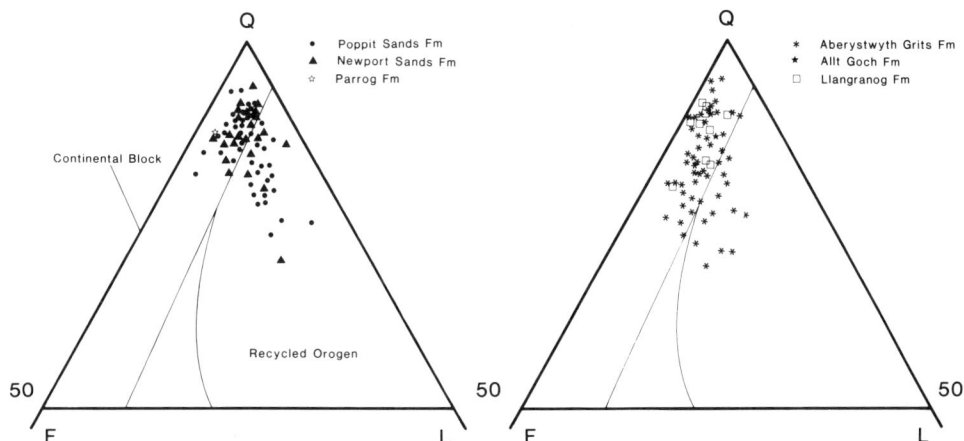

Fig. 4. Triangular QFL plot showing mean framework modes for sandstones from the southern Welsh Basin: Q, total quartzose grains; F, total feldspar grains; L, total lithic fragments (after Dickinson & Suczek 1979).

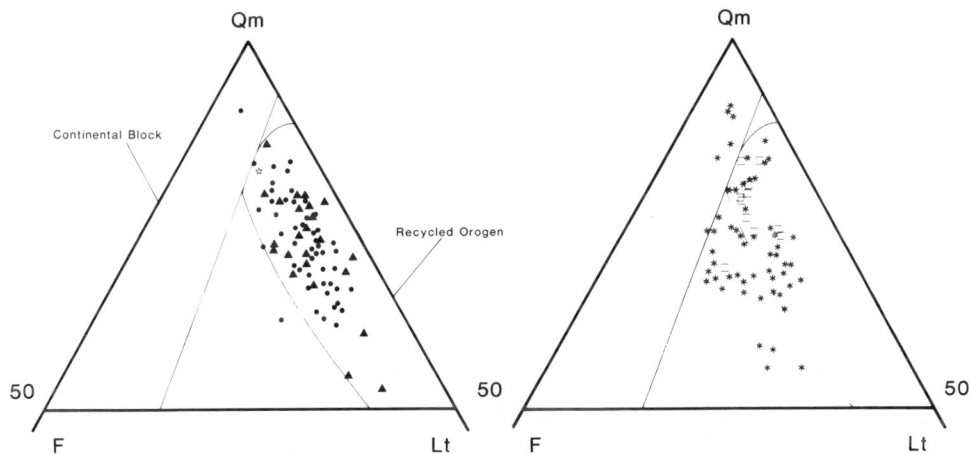

Fig. 5. Triangular Qm–F–Lt plot showing mean framework modes for sandstones from the southern Welsh Basin: Qm, monocrystalline quartz grains; F, total feldspar grains; Lt, total lithic fragments including polycrystalline quartzose grains (after Dickinson & Suczek 1979). Legend as for Fig. 4.

recycled orogen provinces (Fig. 4). The younger formations also plot on this boundary but are skewed more towards the continental block province. According to Dickinson & Suczek (1979) sediments plotting within the continental block provenance are derived either from stable shields and platforms or from areas of uplift. Within recycled orogens, sediments are derived mainly from sedimentary strata and subordinate volcanics.

It is possible to increase the discrimination of the source area by assigning polycrystalline quartz (Qp) to the total lithics mode (Lt) in the Qm–F–Lt diagram (Fig. 5). The older forma-

tions plot in the recycled orogen province while the younger formations plot on the boundary of the recycled orogen and continental block provinces. All of the data points plot in the quartzose recycled area and were probably derived from sediments whose ultimate source was cratonic. Sedimentological factors, however, may have locally enhanced quartz content and caution should therefore be exercised when interpreting the provenance of quartz-rich sediments (cf. Mack 1984). Further discrimination of source is afforded by the Qp–Lvm–Lsm diagram showing the polycrystalline quartz (Qp) sub-population of the framework grain modes (Fig.

6). Almost all of the data points plot outside the two delimited fields indicating that the source area was neither a collisional nor an arc orogen.

Microconglomerate petrography

Twenty thin sections were examined by transmitted light microscopy and the electron microprobe, to determine the composition of selected lithic fragments. The majority of these thin sections were microconglomerates from either the Poppit Sands or Aberystwyth Grits formations. Intraformational sedimentary lithic clasts predominate although felsic fragments (typically rhyolites or andesites) and tuffs, trachytes and granitic clasts are also common. Both formations show similar assemblages of lithic fragments confirming the similar nature of the volcanic rocks in the source area and also its longevity.

An interesting feature is the occurrence in both rock suites of multicyclic sedimentary lithic fragments (cf. Zuffa 1987). This is in agreement with the evidence from analysis of the framework grains. The lack of mafic fragments is more problematic. Some of the volcanic centres in the source area contained extensive basalt sheets and pillow lavas (e.g. Fishguard Volcanic Complex) and yet there is very little evidence of this in the sandstone mineralogy of the area. This is either a function of the lower preservation potential of basaltic fragments, as opposed to that of fine-grained felsic fragments, or else that the basalts did not, at the time of deposition, form part of the source area.

Three polished sections (one from the Poppit Sands Formation and two from the Aberystwyth Grits Formation) were examined using the electron microprobe to determine the chemical composition of some of the volcanic lithic clasts. A total of seven clasts were analysed and their total alkali and SiO_2 contents determined. Based on these values, the lithic fragments may be classified as dacites, basalts and mugearites (cf. Cox et al. 1979). The samples suggesting a mugearite composition were both from the Aberystwyth Grits Formation and this is entirely in keeping with a possible derivation from the Skomer Volcanic Group (Lower Llandovery) (Stillman & Francis 1979; Thorpe et al. 1989). Basalts and dacites are recorded from the majority of the Tremadoc–Llandeilo volcanic episodes (Allen 1982).

Chemical classification of rocks

The major and trace element geochemistries for 29 mudstones and 32 sandstones were determined by means of X-ray fluorescence using a Philips PW1400 X-ray generator following the procedure of Marsh et al. (1983) and Weaver et al. (1983). The sandstones chosen were all medium grained. These results were then used to provide the basis for geochemical analysis of the sediments.

The majority of the sandstones have a SiO_2 range of 70.7–85.67 wt%, low Fe_2O_3 (total Fe as Fe_2O_3) and MgO contents of between 1.9 and 3.01 (Table 2). Chemical classification of the sandstones indicates that they are quartz-rich to

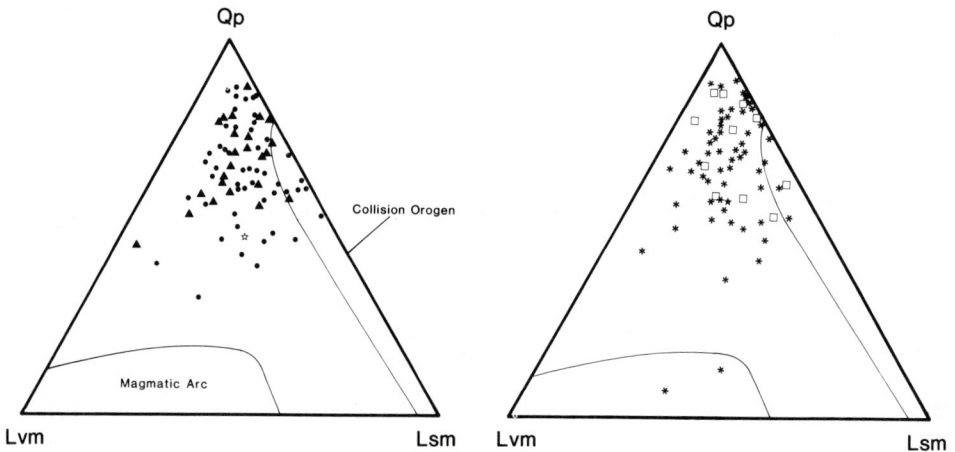

Fig. 6. Triangular Qp–Lvm–Lsm plot showing mean framework modes for sandstones from the southern Welsh Basin: Qp, polycrystalline quartzose grains; Lvm, total volcanic and metavolcanic lithic fragments; Lsm, total sedimentary and metasedimentary lithic fragments (after Dickinson & Suczek 1979). Legend as for Fig. 4.

Table 2. *Representative chemical analyses of sandstones and mudstones from the southern Welsh Basin*

Pellet No. / P.S.	SANDSTONES					MUDSTONES				
	L11584 Llan.	L11009 Formation	L11593 N.S.	L11581 P.S.	L11587 Llan.	L11002	L10995 Cwrt	L10997	L11 Formation	N.S.
SiO_2	78.0	73.0	86.7	62.4	70.3	58.1	62.0	61.2	60.8	62.0
TiO_2	0.34	0.51	0.33	0.91	0.73	1.09	1.11	1.2	0.95	0.98
Al_2O_3	12.2	13.0	9.4	20.4	16.0	22.0	23.4	22.5	17.9	20.6
Fe_2O_3	0.8	0.7	0.6	1.2	1.0	7.7	7.3	9.5	9.0	10.7
MnO	0.04	0.04	0.1	0.08	0.15	0.07	0.09	0.06	0.2	0.18
MgO	1.9	2.2	1.1	2.7	2.4	2.5	2.0	1.8	2.7	2.4
CaO	0.1	0.3	0.2	0.4	1.2	0.2	0.1	0.2	0.1	0.2
Na_2O	0.3	1.6	2.3	1.3	1.5	3.1	1.6	1.0	1.6	1.5
K_2O	0.39	1.35	0.7	3.18	2.11	4.38	3.75	3.93	2.47	3.4
P_2O_5	0.13	0.23	0.1	0.28	0.16	0.12	0.07	0.24	0.18	0.17
Total	94.25	92.92	101.57	92.73	95.48	99.35	101.61	101.63	97.62	100.31
V	98	66	39	130	91	170	178	164	113	135
Cr	59	73	18	85	68	115	122	122	77	82
Ba	406	225	104	434	269	575	558	655	291	437
La	21	15	bd	bd	bd	36	38	47	23	36
Ce	47	32	bd	bd	bd	68	80	100	49	78
Nd	22	16	bd	bd	bd	29	29	43	21	33
Nb	8	8	5	15	12	19	19	23	16	17
Zr	140	145	110	219	205	237	183	217	242	182
Y	28	23	16	40	28	36	30	39	24	33
Sr	63	51	37	87	70	117	119	116	74	89
Rb	72	46	29	132	92	166	192	158	92	137
Th	10	8	4	14	10	16	12	17	10	14
Ga	14	13	10	24	20	28	32	37	24	21
Zn	47	74	52	114	83	113	105	126	127	115
Ni	16	19	11	39	36	42	41	44	48	37

N.S., Newport Sands; P.S., Poppit Sands; Llan, Llangranog; A. Goch, Allt Goch; A.G., Aberystwyth Grits; bd, below detection limits.
Total Fe as Fe_2O_3.
Major oxide in wt%. trace elements in ppm.
Full chemical analyses can be obtained as Supplementary Publication No. SUP 18068 (5pp) from the Society Library or from the British Library Document Supply Centre, Boston Spa, West Yorkshire, UK.

quartz-intermediate (Fig. 7). Large-ion-lithophile (LIL) elements such as K, Rb, Sr and Th show a range of abundances.

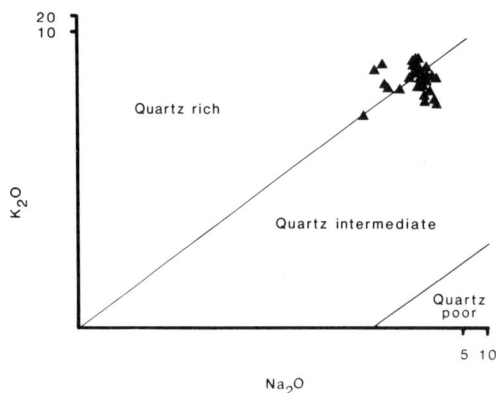

Fig. 7. Analysis of quartz-richness of southern Welsh Basin sandstones based on major element geochemistry (after Crook 1974).

Geochemical analysis of provenance

Geochemical analysis of sediments may indicate the plate tectonic setting. Three main tectonic provenances are defined by Roser & Korsch (1986): (a) *Passive Continental Margin (PM)*—mineralogically mature (quartz-rich) sediments deposited in plate interiors at stable continental margins or intracratonic basins (equivalent to the 'trailing-edge tectonic setting' of Maynard *et al.* 1982); (b) *Active Continental Margin (ACM)*—quartz-intermediate sediments derived from tectonically active continental margins on or adjacent to active plate boundaries (e.g. trench, forearc and backarc settings); (c) *Oceanic Island Arc (OIA)*—quartz-poor volcanogenic sediments derived from oceanic island arcs (i.e. sediments derived from an island arc source and deposited in a variety of settings including forearc, intra-arc and backarc basins and trenches).

The following sections consider the geochemistry of the Welsh Basin sediments in terms of provenance utilizing a variety of approaches.

Quartz richness

Crook (1974) subdivided sandstones on the basis of SiO_2 content and the relative K_2O/Na_2O ratio into three classes and assigned each to a plate tectonic environment. All of the sandstone samples from the study area may be classified as

either quartz-rich (average 89% SiO_2, $K_2O/Na_2O > 1$) or quartz-intermediate (average 68–74% SiO_2, $K_2O/Na_2O < 1$) (Table 3). Based on the K_2O/Na_2O ratio, four of the sandstone formations may be classified as quartz-rich. The SiO_2 wt% of these samples are, however, lower than the average value of 89% suggested by Crook (1974) and are closer to the values for quartz-intermediate sandstones. It is best, therefore, to consider the samples as falling on the quartz-rich/quartz-intermediate boundary. Modern equivalents of quartz-rich sediments adjoin Atlantic-type continental margins on the trailing edge (PM) of continents, whereas quartz-intermediate sediments are more indicative of Andean-type (ACM) margins.

Table 3. *Mean SiO_2 wt% and K_2O/Na_2O values of sandstone-dominated formations from the southern Welsh Basin*

Formation	N	SiO_2 wt%	K_2O/Na_2O
Newport Sands	7	81.21	2.19
Poppit Sands	9	74.9	1.19
Llangranog	3	85.67	0.43
Allt Goch	2	75.35	1.03
Aberystwyth Grits	12	70.7	1.26

Major element analysis of sandstones (Fig. 8)

Bhatia (1983), in a study of the geochemistry of sandstones from Australia, devised a series of plots to differentiate four main tectonic settings. The most discriminating parameters are TiO_2 wt%, Al_2O_3/SiO_2, K_2O/Na_2O and $Al_2O_3/(CaO + Na_2O)$ ratios all plotted against $Fe_2O_3 + MgO$ wt%. In the TiO_2 wt% and Al_2O_3/SiO_2 plots (Fig. 8a,b) the majority of the points fall within the passive margin (PM) field although some also plot within the active continental margin field (ACM). The observed vertical distribution of the points is interpreted as a function of interelement variations.

Plotting the ratio of K_2O/Na_2O produces a distribution where most of the formations fall within the PM field (Fig. 8c). Displacement is here affected by the degree of maturity of the sediments, maturity being directly reflected in the relative feldspar ratios. In Fig. 8d the majority of the formations plot within the PM field. Again, displacement of data points is a function of sediment maturity.

The distribution of most of the points in or around the PM field of Bhatia (1983) suggests

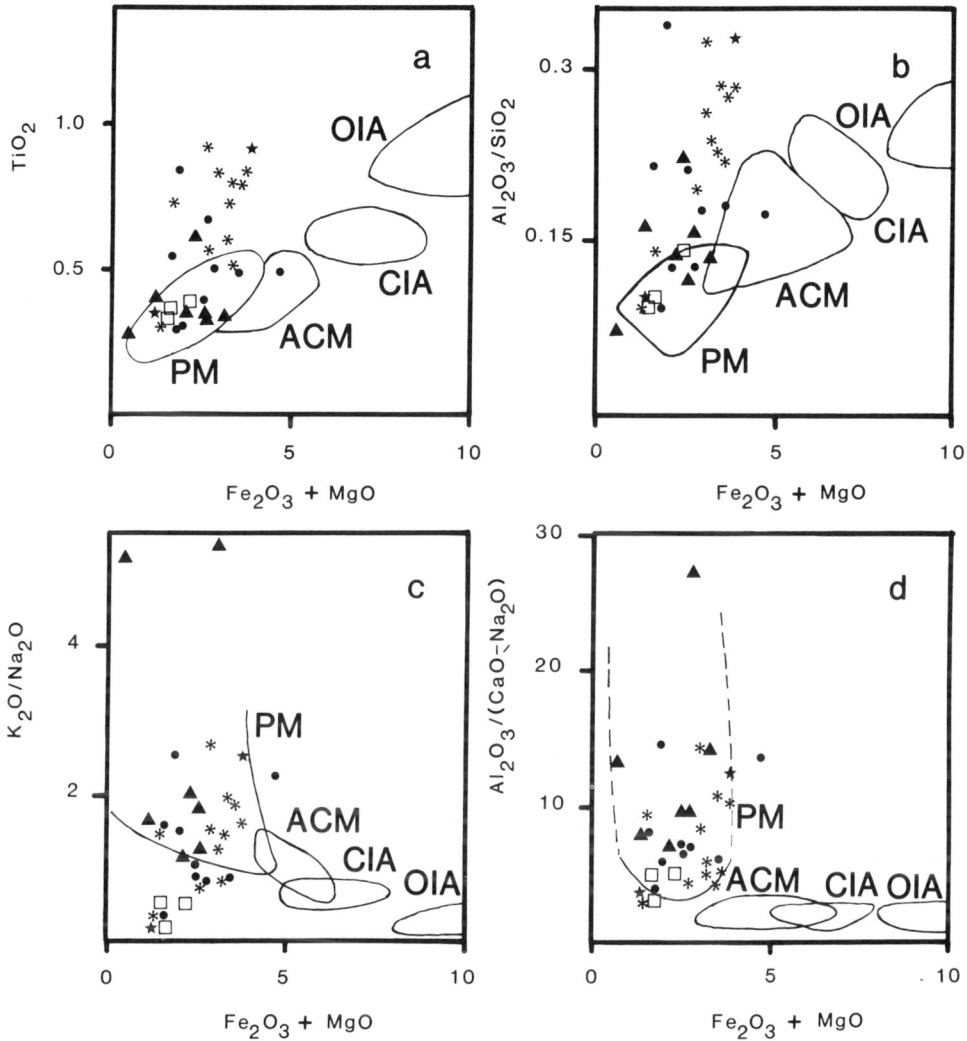

Fig. 8 Bivariate plots for the discrimination of plate tectonic setting of sandstones from the southern Welsh Basin: PM, passive margin; ACM, active continental margin; CIA, continental island arc; OIA, oceanic island arc (after Bhatia 1983). Legend as for Fig. 4.

that they may be derived from Atlantic-type rifted continental margins, remnant ocean basins adjacent to collision orogens and inactive or extinct convergent margins. Within the PM field sediments are generally highly matured, being derived from the recycling of other sedimentary and metamorphic rocks on platforms or recycled orogens.

K/Rb diagram (Fig. 9)

This plot may be used to distinguish those sediments derived from rocks of acid and intermedi-ate compositions from those derived from rocks of basic composition. The relatively high K/Rb of the sediments from the Welsh Basin is indicative of derivation from acid and intermediate source rocks with some input from basic sources.

K₂O/Na₂O v. SiO₂ diagram (Fig. 10)

The mudstone samples from the Welsh Basin plot astride the Active Continental Margin (ACM) and the Passive Margin (PM) border of Roser & Korsch (1986). This position suggests that the presence of arc-derived material may be

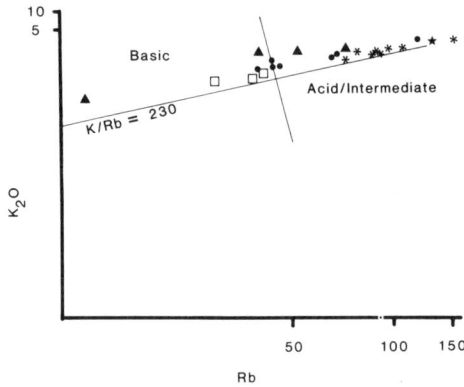

Fig. 9. Distribution of K (log wt%) and Rb (ppm) in the southern Welsh Basin sandstones relative to a K/Rb ratio of 230 (= Main Trend of Shaw 1968). Boundary line between acid/intermediate and basic compositions after Floyd & Leveridge (1987). Legend as for Fig. 4.

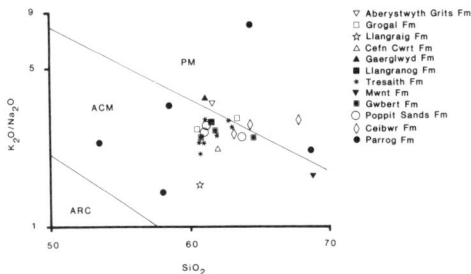

Fig. 10. Tectonic discrimination diagram for mudstones from the southern Welsh Basin: PM, passive margin; ACM, active continental margin; ARC, arc (after Roser & Korsch 1986).

discounted. It should be noted, however, that the distribution of the data points does not show any stratigraphic trend. According to Roser & Korsch (1986) the location of data points on the diagram is primarily controlled by the nature of volcanism, the extent of plutonism and related erosional levels. The effect of mineralogical maturation through sediment recycling is a secondary consequence. The data points from the Welsh Basin correspond with the Greenland and Torlesse terranes (New Zealand) of Roser & Korsch (1986). The Greenland Terrane is a recycled quartzose sandstone (PM) with most of the data points lying within the PM field. The Torlesse Terrane, however, was derived from an ACM tectonic setting compatible with the quartz-intermediate nature of the sandstones.

Trace element concentrations

Very high levels of Cr (e.g. 100–1500 ppm) and Ni (e.g. 50–600 ppm) have been used by a variety of authors (e.g. Hiscott 1984; Haughton 1988; Wrafter & Graham 1989) to indicate an ultramafic provenance for the sediments. The low levels of Cr (17–125 ppm) and Ni (3–47 ppm) recorded in the Welsh sandstone-dominated formations suggests either some basic input into the system or else that the trace elements could have travelled into the depositional basin as adsorbed ions on clays. Vanadium levels are relatively high in both the sandstones (31–159 ppm) and mudstones (102–193 ppm). These levels are higher than the levels commonly recorded in sandstones (20 ppm) and given that V is concentrated in basic rocks they suggest some basic input into the depositional system.

Bhatia (1985) used trace elements to geochemically determine the tectonic setting of mudstones. He distinguished four main tectonic provenances using concentrations and inter-element ratios of Nb, Zr, Y, Rb, Sr, Th, Ba, Cr and Ni. The results obtained herein are not definitive with all four tectonic settings being indicated. The high Ni and Cr contents suggest a passive margin origin while the Rb/Sr value is closest to that of a continental island arc (Table 4). The Th and Zr/Th values fall between those suggesting a continental island arc and an oceanic island arc, but are closer to the former. The Ba/Sr ratio suggests a continental island arc setting. The remaining three factors all compare most favourably with the active continental margin tectonic setting (Table 4).

Multi-element diagram normalized to Post-Archaean Shale (Fig. 11)

Multi-element diagrams may be used to examine the distribution of trace elements in mudstones. The plot compares a range of elements from seven of the mudstone-dominated formations against a normalized post-Archaean average shale (PAAS; Taylor & McClennan 1985). The elements are arranged such that those elements mainly derived from acidic source rocks plot on the left-hand side of the diagram while those derived from basic and ultrabasic source rocks are plotted on the right-hand side. The main feature to note is the degree of conformity of the values with those of the PAAS. Most values are evenly spread out around the PAAS although certain elements (e.g. V, Cr, Fe) do suggest a greater amount of basic input.

Table 4. *Trace element geochemical parameters for mudstone-dominated formations from the southern Welsh Basin*

Element	Mean (N = 29)	OIA	CIA	ACM	PM
Nb	17.27	3.7	9.0	16.5	15.8
Zr/Nb	12.2	38.0	21.0	11.0	10.0
Nb/Y	0.51	0.17	0.35	0.5	0.54
Rb/Sr	1.47	0.29	1.31	2.9	5.8
Th	13.34	5.5	16.2	28.0	22.0
Zr/Th	15.82	28.0	12.0	7.0	7.0
Ba/Sr	6.33	2.5	6.3	8.7	17.6
Cr	111.38	39.0	55.0	58.0	100.0
Ni	42.24	15.0	18.0	26.0	36.0

Values for the tectonic settings OIA (oceanic island arc), CIA (continental island arc), ACM (active continental margin) and PM (passive margin) after Bhatia (1985).

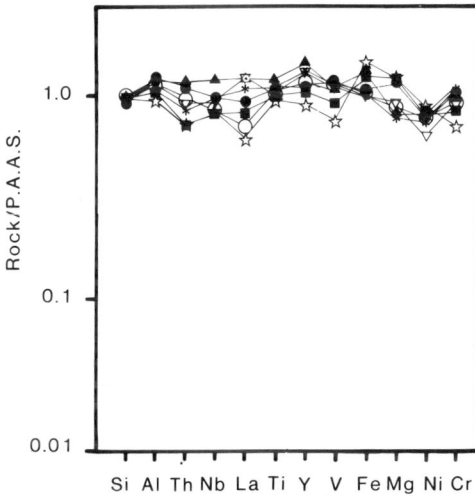

Fig. 11. Multi-element diagram normalised to Post-Archean average shale (after Taylor & McClennan 1985). Legend as for Fig. 10.

Discussion

It has been suggested that the tectonic setting of the Welsh Basin was a passive margin prior to the Tremadoc, an active margin back-arc basin from the early Ordovician to the Caradoc and a non-volcanic active margin or collision zone basin from the Ashgill to the Devonian (Pickering *et al.* 1988; Woodcock 1990). Thus the tectonic setting of the basin for the entire period of deposition of the sediments was that of an active margin. Volcanic activity was particularly pronounced in the early Ordovician but had ceased, in the southern Welsh Basin, by the Llanvirn. The eruption of the Skomer Volcanic Group (early Llandovery) was the only volcanic episode

coeval with deposition within the southern Welsh Basin. Caradoc volcanism was confined to North Wales and was not a detrital source for the sediments of the depositional area.

The signatures from the sedimentary provenance indicators, however, do not agree with the active margin setting. Both the framework grains and the major element geochemistry of the sandstones suggest a passive margin tectonic setting for the area. The mudstone geochemistry, particularly that of the trace elements, is more variable, suggesting a variety of tectonic settings ranging from continental island arc to passive margin. What is remarkable about both sets of data, however, is the lack of stratigraphic variation between the data sets. Indeed, the provenance signature of the sediments shows a remarkable degree of uniformity over the entire period of deposition within the depositional basin (c. 35 Ma).

As mentioned earlier, palaeocurrent evidence within the region shows that sediment was chiefly derived from the south and southeast. Transport directions are similar for the majority of the formations; the only exceptions being some south-directed current directions to the north of Aberystwyth (McCann & Pickering 1989 and *in* Loydell *et al.* 1990) and westerly-directed currents in some of the mudstone-dominated formations, these latter produced as a result of lateral transport of sediment at lobe margins (McCann 1990).

The passive margin signature was derived, therefore, from this southern landmass. There are two ways in which this could have occurred. Firstly, detritus may have been derived from the calc-alkaline Precambrian basement and associated Lower Palaeozoic rocks and therefore records the original tectonic signature (PM) of this basement and not the setting of the active basin.

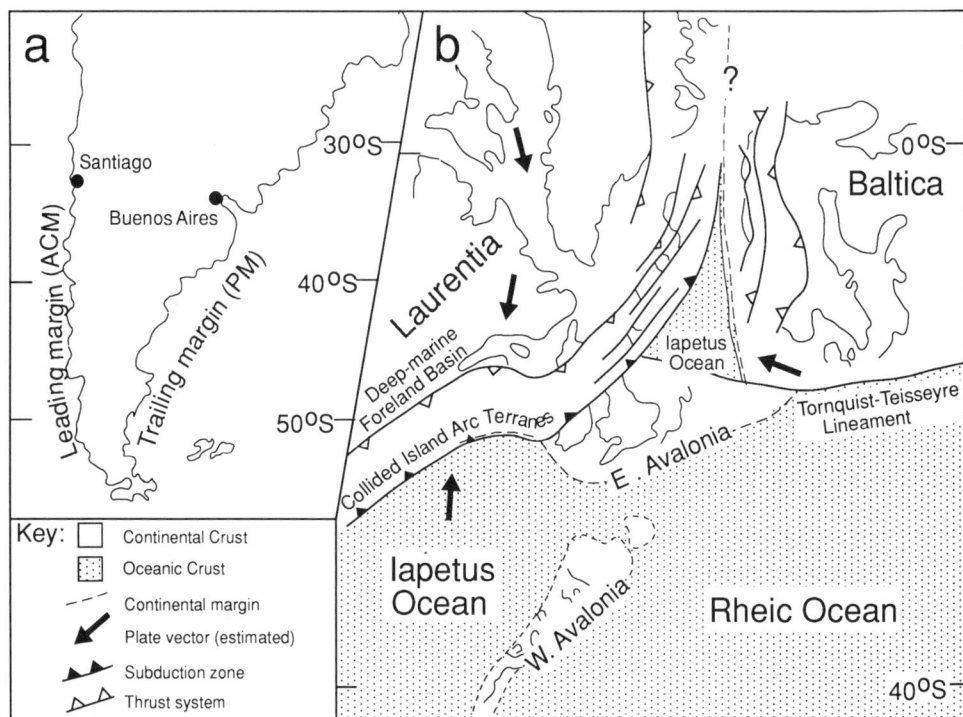

Fig. 12. Comparative diagram showing (**a**) present day tectonic setting of South America, and (**b**) Ashgill–Llandovery palaeogeographic reconstruction of the North Atlantic region ((**b**) modified from Pickering 1989 and Vannier *et al.* 1989). The key refers to (**b**) only.

Alternatively, the sediment could have been derived distally from the trailing passive margin of the southern landmass. The landmass is generally considered to have been narrow, separating the Welsh Basin from the Rheic Ocean to the south. During the period of deposition there was an active continental margin to the north of the Welsh Basin (with the closing Iapetus Ocean) while to the south of the microcontinent of Eastern Avalonia there was a passive margin to the Rheic Ocean (Fig. 12). This situation is somewhat analogous to the southern end of present day South America where the Chilean side is a leading margin (ACM) while the Argentinian side is a trailing margin (PM) (Fig. 12). It is also interesting to note that both the southern tip of South America and the microcontinent of Eastern Avalonia (which contains the Welsh Basin) are of similar size (Fig. 12). In such situations, where two disparate settings exist adjacent to one another, then the tectonic setting, as deduced from beach sands, may be an amalgamation of both (Potter 1984, 1986). The analogy is supported by evidence suggesting that sediment was carried across the Tornquist–

Teisseyre Lineament from the area of Baltica into the Welsh Basin (Pickering 1989). It is, however, difficult to determine with any degree of certainty which of the two models are the most likely. It does seem unlikely that sediment from the passive margin could find its way into the back-arc Welsh Basin and thus it is more probable that the reflected signature is a relict one.

Sandstone provenance

Much of the sand-grade sediment was deposited as part of elongate turbidite systems developed from point sources in the south/southeast. As noted earlier such elongate turbiditic bodies tend to derive their sediment from point sources, for example, large rivers or delta systems. Turbidite deposition is very much controlled by eustatic changes in sea level (Mutti & Normark 1987). Maximum input of detritus is during low sea-level stands when sediment sources (e.g. rivers) can prograde over the shelf area and directly funnel sediment loads into the deeper marine basins (Stow *et al.* 1985; Vail *et al.* 1977). This sediment discharge, therefore, would be a more accurate reflection of the tectonic setting suggested by the hinterland rather than that of the depositional basin.

A qualitative approach, based on individual lithic fragments, does provide more information about the volcanic successions. Unfortunately, given that many of the volcanic successions are lithologically similar, it is not possible to trace the erosion of particular volcanic centres. Furthermore, while Zuffa (1987) has subdivided volcanic fragments into coeval and ancient types, this classification has only been used for modern successions. It is doubtful, given the significant degree of alteration which can occur in volcanic fragments, whether this classification could be applied to ancient sequences. Certainly in this situation, with the majority of the volcanic fragments being classified as 'felsic', it is probable that they could have been derived from Precambrian as well as Ordovician sources.

The limited sample size of Bhatia's (1983) sandstone provenance model (69 samples used) may explain the relatively poor correlation obtained when using the model to examine the sediments of the southern Welsh Basin. While all 32 Welsh samples plotted in or around the passive margin field there tended to be a vertical spread for which there was no explanation. Although this was primarily a function of inter-element variability and/or maturity neither reason was taken into consideration by the fields of Bhatia (1983), nor was there any explanation in the text that such a spread might occur. It is suggested that caution be exercised when applying this model to ancient sequences, about which little is known in terms of provenance, as it may be misleading.

Mudstone provenance

The depositional environments of the mudstones are more variable than those of the sandstones occurring both as turbidite-related deposits and also as general background sedimentation (e.g hemipelagites) within the basin. Some of the sediment may also be redeposited sediment derived from shelf areas and possibly input as a result of storm activity on the shelf. The more disparate tectonic settings, therefore, may be related to the fact that the mudstones are derived, not only from point sources (in the case of muds which formed part of the turbidite systems) but also from the more general area of the shelf.

None of this, however, explains the diverse provenance attributions derived from the use of the trace element geochemistry parameters of Bhatia (1985). The spread of the results suggests that there are some fundamental flaws in Bhatia's (1985) arguments.

According to Bhatia (1985), the active continental margin (ACM) and passive margin (PM) mudstones are similar in most immobile trace elements and may be distinguished from the mudstones of other tectonic settings by their significantly higher Th and Nb percentages and Nb/Y ratio and lower Zr/Th ratio. They may be distinguished from each other by the higher Rb/Sr and Ba/Sr ratios and Cr and Ni percentages of the passive margin setting. The increased Cr and Ni in passive margin tectonic settings is a result of enrichment and adsorption of these elements with the increased phyllosilicate content. The decrease in Rb/Sr and Ba/Sr is due to the loss of Sr and feldspar with the increased weathering and recycling of passive margin type sediments.

The geochemical behaviour, however, of some of these elements is extremely variable. For example, Dimberline (1987), notes that Rb/Sr ratios may be reduced due to the presence of additional Sr incorporated in diagenetic carbonate. The presence of diagenetic carbonates in some of the mudstone-dominated formations may, therefore, have depressed the values. As noted earlier, the distribution of Sr is affected to an extent by the presence of Ca. Fairbridge (1972) also notes that the Sr content of sedimentary rocks is variable because of the many influences on Sr in low temperature deposition. Mudstones seem to have an ability to concentrate Sr due to ion exchange properties of the clay minerals and concentrations of up to 298 ppm have been reported (Fairbridge 1972). This would cast considerable doubt on the advisability of using Sr, either alone or in combination with other elements, for provenance determination in mudstones. The geochemistry of Ba is very close to that of Sr and, therefore, it too would appear to be an unreliable provenance indicator.

The behaviour of Rb is largely controlled, in sedimentary processes, by its adsorption on clay minerals (Fairbridge 1972). While both the degree of adsorption and the presence or absence of certain clay minerals may be related to provenance, the link is not a firm one and, therefore, it does not appear to be a particularly reliable element to use. Dimberline (1987) noted that high Ni contents could be produced by having Ni concentrated preferentially in organic matter. The presence of organic matter in laminated hemipelagites in the Gaerglwyd Formation and parts of the Parrog and Tresaith formations could, therefore, have resulted in a high Ni value.

Based on geochemical criteria alone, the model of Bhatia (1985) is questionable. In the present study area 29 samples, were used while Bhatia (1985) used a total of 23 samples to define all of his tectonic provinces. Furthermore his sample numbers for the various settings were as follows: oceanic island arc (9); continental island arc (9); active continental margin (2), and passive margin (3). As noted earlier all four of Bhatia's (1985) tectonic provinces are represented in the Welsh succession, an extremely unlikely occurrence given the history of the source area. It would, therefore, appear that there are a number of problems with his model, namely (a) the relatively small sample size, particularly of some of the tectonic settings (e.g. passive margin); (b) the lack of recognition of the complexity of the geochemical history of the sediments used in creating his model; and (c) the use of certain elements, primarily Rb, Sr and Ba, which as noted above, show very poor correlation in terms of provenance as their distribution is affected by many variables.

Conclusions

The geotectonic setting for the southern part of the Welsh Basin has been discussed by a variety of authors (Bevins 1982; Bevins et al. 1984; Kokelaar et al. 1984a; Siveter et al. 1989, Woodcock 1990). The current consensus, based largely on geochemical evidence, is that the Welsh Basin formed in an ensialic marginal basin on a continental margin which developed on the southern side of the Iapetus Ocean, subduction being initiated during the Tremadoc (Kokelaar 1979). The two dominant signatures, namely that of the

pre-Tremadoc passive margin and the post-Tremadoc active continental margin, are supported by both the petrographic and geochemical analysis of the sediments deposited in the basin over a period of 30–35 Ma. The sediments, particularly the sandstones, reflect the longevity of the relict passive margin signature. This signature may be enhanced by the palaeogeographic position of the source area (southern Eastern Avalonia) for the sandstone detritus. The mudstones are apparently more sensitive and thus reflect the active continental margin signature. There are serious flaws in certain geochemical provenance models for both sandstones and mudstones, as outlined above, and extreme caution should be exercised when applying them. Certainly they should not be used as the sole indicators of sedimentary provenance in areas which are poorly understood geologically.

The bimodal characteristics of the volcanic rocks in the source area are not truly reflected in many of the techniques. While this may be accounted for in framework grains by the easier weathering/alteration of basic fragments and their lower transport stability, it is of note that only a few of the geochemical techniques suggest any form of basic input. The most useful tools are the spider plot of trace elements and the K/Rb plot. In summary, given that the geology of provenance areas is commonly complex, it is best to use a variety of techniques, both petrographic and geochemical, in conjunction with other geological parameters to provide as complete a reconstruction as possible.

I would like to thank N. G. Marsh, M. J. Norry, K. T. Pickering and two anonymous referees for helpful comments and discussions. This work was partly funded by monies from the Whittaker Fund (Department of Geology, Leicester University) and the Tarquin Teale Memorial Fund.

References

ALLEN, P. M. 1982. Lower Palaeozoic volcanism in Wales, the Welsh Borderland, Avon and Somerset. In: SUTHERLAND, D. S. (ed.) Igneous Rocks of the British Isles, J. Wiley and Sons, Chichester, 65–91.

BEVINS, R. E. 1982. Petrology and geochemistry of the Fishguard Volcanic Complex, Wales. Geological Journal, 17, 1–21.

——, KOKELAAR, B. P. & DUNKLEY, P. N. 1984. Petrology and geochemistry of lower to middle Ordovician igneous rocks in Wales: a volcanic arc to marginal basin transition. Proceedings of the Geologist's Association, 95, 337–347.

——, LEES, G. J. & ROACH, R. A. 1989. Ordovician intrusions of the Strumble Head-Mynydd Preseli region, Wales: lateral extensions of the Fishguard Volcanic Complex. Journal of the Geological Society, London, 146, 113–123.

BHATIA, M. R. 1983. Plate tectonics and geochemical composition of sandstones. Journal of Geology, 91, 611–27.

—— 1985. Composition and classification of Paleozoic Flysch mudrocks of eastern Australia: implications in provenance and tectonic setting interpretation. Sedimentary Geology, 41, 249–268.

BJØRLYKKE, K. 1971. Petrology of Ordovician sediments from Wales. *Norsk Geologisk Tidsskrift*, **51**, 123–139.

BRENCHLEY, P. J. & NEWALL, G.1984. Late Ordovician environmental changes and their effect on faunas. *In*: BRUNTON, D. L. (ed.) *Aspects of the Ordovician System.* Palaeontological Contributions from the University of Oslo, **295**, Universitetsforlaget, 65–79.

COX, K. G., BELL, J. D. & PANKHURST, R. J. 1979. *The interpretation of igneous rocks.* George Allen & Unwin, London.

CROOK, K. A. W. 1974. Lithogenesis and geotectonics: the significance of compositional variation in flysch arenites (greywackes). *Society of Economic Paleontologists and Mineralogists, Special Publication*, **19**, 304–310.

DAVIES, W. & CAVE, R. 1976. Folding and cleavage determined during sedimentation. *Sedimentary Geology*, **15**, 89–133.

DICKINSON, W. R. & SUCZEK, C. A. 1979. Plate tectonics and sandstone compositions. *American Association of Petroleum Geologists, Bulletin*, **63**, 2–31.

DIMBERLINE, A. J. 1987. *Geology of the Wenlock turbidite system, Wales.* PhD thesis, University of Cambridge.

FAIRBRIDGE, R. W. 1972 (ed.) *The encyclopedia of geochemistry and environmental sciences.* Van Nostrand Reinhold Company, New York.

FLOYD, P. A. & LEVERIDGE, B. E. 1987. Tectonic environment of the Devonian Gramscatho basin, south Cornwall: framework mode and geochemical evidence from turbiditic sandstones. *Journal of the Geological Society, London*, **144**, 531–542.

HARLAND, W. B., COX, A. V., LLEWELLYN, P. G., PICKTON, C. A. G., SMITH, A. G. & WALTERS, R. 1982. *A geologic time scale.* Cambridge University Press.

HAUGHTON, P. 1988. A cryptic Caledonian flysch terrane in Scotland. *Journal of the Geological Society, London*, **145**, 685–704.

HISCOTT, R. N. 1984. Ophiolitic source rocks for Taconic-age flysch: Trace-element evidence. *Geological Society of America Bulletin*, **95**, 1261–1267.

INGERSOLL, R., BULLARD, T. F., FORD, R. L., GRIMM, J. P., PICKLE, J. D. & SARES, S. W. 1984. The effect of grain size on detrital modes: a test of the Gazzi-Dickinson point counting method. *Journal of Sedimentary Petrology*, **54**, 103–116.

JAMES, D. M. D. 1971. Petrography of the Plynlimon Group, west central Wales. *Sedimentary Geology*, **6**, 255–270.

—— 1981. Petrographic evidence bearing on plate tectonics of the Upper Ordovician Welsh Basin. *Geological Magazine*, **118**, 95–96.

KEEPING, W. 1881. The geology of central Wales. *Quarterly Journal of the Geological Society of London*, **37**, 141–177.

KOKELAAR, B. P. 1979. Tremadoc to Llanvirn volcanism on the southeast side of the Harlech Dome (Rhobell Fawr), N Wales. *In*: HARRIS, A. L., HOLLAND, C. H. & LEAKE, B. E. (eds) *The Caledonides of the British Isles—reviewed.* Geological Society, London, Special Publication, **8**, 591–596.

—— 1988. Tectonic controls of Ordovician arc and marginal basin volcanism in Wales. *Journal of the Geological Society, London*, **145**, 759–775.

——, BEVINS, R. E. & ROACH, R. A. 1985. Submarine silicic volcanism and associated sedimentary and tectonic processes, Ramsey Island, SW Wales. *Journal of the Geological Society, London*, **142**, 591–613.

——, HOWELLS, M. F., BEVINS, R. E., ROACH, R. A. & DUNKLEY, P. N. 1984a. The Ordovician marginal basin of Wales. *In*: KOKELAAR, B. P. & HOWELLS, M. F. (eds) *Marginal Basin Geology: Volcanic and associated sedimentary tectonic processes in modern and ancient marginal basins.* Geological Society, London, Special Publication, **16**, 245–269.

——, ——, & —— 1984b. Volcanic and associated sedimentary and tectonic processes in the Ordovician marginal basin of Wales. A field guide. *In*: KOKELAAR, B. P. & HOWELLS, M. F. (eds) *Marginal Basin Geology: Volcanic and associated sedimentary tectonic processes in modern and ancient marginal basins.* Geological Society, London, Special Publication, **16**, 291–322.

LEGGETT, J. K., MCKERROW, W. S., COCKS, L. R. M. & RICKARDS, R. B. 1981. Periodicity in the early Palaeozoic marine realm. *Journal of the Geological Society, London*, **138**, 167–176.

LOYDELL, D. K. ET AL. 1990. Discussion on the palaeocurrent evidence of a northern structural high to the Welsh Basin during the late Llandovery. *Journal of the Geological Society, London*, **147**, 885–891.

MACK, G. H. 1984. Exceptions to the relationship between plate tectonics and sandstone composition. *Journal of Sedimentary Petrology*, **54**, 212–220.

MARSH, N. G., TARNEY, J. & HENDRY, G. L. 1983. Trace element geochemistry of basalts from Hole 504B, Panama Basin, Deep Sea Drilling Project Legs 69 and 70. *Initial Reports of the Deep Sea Drilling Project*, **69**, 747–763.

MAYNARD, J. B., VALLONI, R. & YU, H. 1982. Composition of modern deep sea sands from arc-related basins. *In*: LEGGETT, J. K. (ed.) *Trench-Forearc Geology.* Geological Society, London, Special Publication, **10**, 551–561.

MCCANN, T. 1990. *Palaeoenviromental evolution of an Ordovician–Silurian deep marine sedimentary succession in the Welsh Basin.* PhD thesis, University of Leicester.

—— & PICKERING, K. T. 1989. Palaeocurrent evidence of a northern structural high to the Welsh Basin during the late Llandovery. *Journal of the Geological Society, London*, **146**, 211–212.

MUTTI, E. & NORMARK, W. R. 1987. Comparing examples of modern and ancient turbidite systems: problems and concepts. *In*: LEGGETT, J. K. & ZUFFA, G. G. (eds) *Marine Clastic Sedimentology: Concepts and Case Studies.* Graham & Trotman, London, 1–38.

OKADA, H. & SMITH, A. J. 1980. The Welsh 'geosyncline' of the Silurian was a fore-arc basin. *Nature*, **288**, 352–354.

PICKERING, K. T. 1989. The destruction of Iapetus and Tornquist's Oceans. *Geology Today*, 160–166.

PICKERING, K. T., BASSETT, M. G. & SIVETER, D. J. 1988. Late Ordovician-early Silurian destruction of the Iapetus Ocean: Newfoundland, British Isles and Scandinavia—a discussion. *Transactions of the Royal Society of Edinburgh: Earth Sciences*, **79**, 361–382.

POTTER, P. E. 1984. South American beach sands and plate tectonics. *Nature*, **311**, 645–648.

—— 1986. South America and a few grains of sand: Part 1—Beach Sands. *Journal of Geology*, **94**, 301–319.

ROSER, B. P. & KORSCH, R. J. 1986. Determination of tectonic setting of sandstone-mudstone suites using SiO_2 content and K_2O/Na_2O ratio. *Journal of Geology*, **94**, 635–50.

SHAW, D. M. 1968. A review of K-Rb fractionation trends by covariance analysis. *Geochimica et Cosmochimica Acta*, **32**, 573–602.

SIVETER, D. J., OWENS, R. M. & THOMAS, A. T. 1989. *Silurian Field Excursions: a geotraverse across Wales and the Welsh Borderland*. National Museum of Wales, Geology Series, Cardiff, 10.

SMITH, A. J. 1956. *Investigations of the sedimentation and the sedimentary history of the Aberystwyth Grits series and its lateral equivalents*. PhD thesis, University College of Wales, Aberystwyth.

STILLMAN, C. J. & FRANCIS, E. H. 1979. Caledonide volcanism in Britain and Ireland. *In*: HARRIS, A. L., HOLLAND, C. H. & LEAKE, B. E. (eds) *The Caledonides of the British Isles—reviewed*. Geological Society, London, Special Publication, **8**, 557–77.

STOW, D. A. V., HOWELL, D. G. & NELSON, C. H. 1985. Sedimentary, tectonic, and sea-level controls. *In*: BOUMA, A. H., NORMARK, W. R. & BARNES, N. E. (eds) *Submarine Fans and Related Turbidite Systems*. Springer-Verlag, New York, 15–22.

TAYLOR, S. R. & McCLENNAN, S. M. 1985. *The Continental Crust: its Composition and Evolution*. Blackwell Scientific Publications, Oxford.

THORPE, R. S. 1979. Late Precambrian igneous activity in Southern Britain. *In*: HARRIS, A. L., HOLLAND, C. H. & LEAKE, B. E. (eds) *The Caledonides of the British Isles—reviewed*. Geological Society, London, Special Publication, **8**, 579–584.

THORPE, R. S., LEAT, P. T., BEVINS, R. E. & HUGHES, D. J. 1989. Late-orogenic alkaline/subalkaline

Silurian volcanism of the Skomer Volcanic Group in the Caledonides of south Wales. *Journal of the Geological Society, London*, **146**, 125–132.

VAIL, P. R., MICHUM JR., R. M. & THOMPSON, S. 1977. Seismic stratigraphy and global changes of sea level, part 3: Relative changes of sea level from coastal onlap. *In*: PAYTON, C. E. (ed.) *Seismic Stratigraphy—applications to hydrocarbon exploration*. American Association of Petroleum Geologists, Memoir **26**, 63–81.

VALLONI, R. & MAYNARD, J. B. 1981. Detrital modes of recent deep-sea sands and their relation to tectonic setting: a first approximation. *Sedimentology*, **28**, 75–83.

VANNIER, J. M. C., SIVETER, D. J. & SCHALLREUTER, R. E. L. 1989. The composition and palaeogeographical significance of the Ordovician ostracode faunas of southern Britain, Baltoscandia, and Ibero-Amorica. *Palaeontology*, **32**, 163–222.

WATSON, J. & DUNNING, F. W. 1979. Basement-cover relations in the British Caledonides. *In*: HARRIS, A. L., HOLLAND, C. H. & LEAKE, B. E. (eds) *Caledonides of the British Isles—reviewed*. Geological Society, London, Special Publication, **8**, 67–91.

WEAVER, B. L., MARSH, N. G. & TARNEY, J. 1983. Trace element geochemistry of basaltic rocks recovered at Site 516, Rio Grande Rise, by DSDP Leg 72. *Initial Report of the Deep Sea Drilling Project*, **72**, 451–456.

WOOD, A. & SMITH, A. J. 1959. The sedimentation and sedimentary history of the Aberystwyth Grits (Upper Llandoverian). *Quarterly Journal of the Geological Society of London*, **114** [for 1958], 163–195.

WOODCOCK, N. H. 1984. Early Palaeozoic sedimentation and tectonics in Wales. *Proceedings of the Geologists' Association*, **95**, 323–335.

—— 1990. Sequence stratigraphy of the Palaeozoic Welsh Basin. *Journal of the Geological Society, London*, **147**, 537–547.

WRAFTER, J. P. & GRAHAM, J. R. 1989. Ophiolitic detritus in the Ordovician sediments of South Mayo, Ireland. *Journal of the Geological Society, London*, **146**, 213–215.

ZUFFA, G. G. 1987. Unravelling hinterland and offshore paleogeography from deep-water arenites. *In*: LEGGETT, J. K. & ZUFFA, G. G. (eds) *Marine Clastic Sedimentology: Concepts and Case Studies*. Graham & Trotman, London, 39–61.

Controls on the petrographic evolution of an active margin sedimentary sequence: the Larsen Basin, Antarctica

DUNCAN PIRRIE

British Antarctic Survey, Natural Environment Research Council, High Cross,
Madingley Rd, Cambridge CB3 0ET, UK
Present address: Camborne School of Mines, Redruth, Cornwall TR15 3SE, UK

Abstract: The sedimentary fill of the Cretaceous–Tertiary Larsen Basin, located at the northern tip of the Antarctic Peninsula, records the evolution of the Antarctic Peninsula source terrain. Uplift, and the possible renewal of arc volcanism within this area, is documented by the wide petrographic spread in the lower Gustav Group (Barremian–Coniacian). The Hidden Lake Formation (Coniacian–Santonian) records a major pulse in proximal calc-alkaline arc volcanism. The overlying Marambio Group shows a change from lithic-volcanic sandstones to quartz–feldspar-rich sandstones, reflecting a change in source terrain. An abrupt change in sandstone composition in the upper Santa Marta Formation (Campanian) may reflect both a switch in volcanism and plutonism from the east to the west coast of the Antarctic Peninsula (i.e. away from the basin margin), and a change to more mafic volcanism. The Lopez de Bertodano Formation (Campanian–Palaeocene) dominantly reflects a change to a quartzo-feldspathic source terrain (the Trinity Peninsula Group). This is also related to both a decrease in the intensity of arc volcanism and an increase in distance from the location of concurrent volcanism in relation to the site of deposition. Controls on petrography other than simple arc-unroofing (e.g. location and nature of arc volcanism and depositional setting) play an important role in sandstone petrography. These controls on sandstone composition should always be considered in interpreting sequences from active plate margin settings.

The relationship between sandstone petrography and plate tectonic setting has been studied by many authors (e.g. Dickinson & Suczek 1979; Ingersoll & Suczek 1979). In particular, studies of active margin sedimentary sequences have been used to reconstruct the evolution of the associated magmatic arc (e.g Garzanti 1985; Dorsey 1988). However, the relationship between source area and sediment petrography is complex, and many factors may affect the final sediment composition (e.g. climate, depositional setting) (Mack 1978, 1984), although source terrain is usually considered to be the dominant control. The aim of this paper is to document the petrographic evolution of the Larsen Basin, Antarctica and to discuss the controls on the petrographic evolution of active margin basins. The Larsen Basin developed in a back-arc setting, in relation to a magmatic arc, represented today by the Antarctic Peninsula. The observed petrographic evolution will be compared with the geological evolution of the source terrain and the sedimentary basin.

Regional setting and stratigraphy

The Antarctic Peninsula represents the eroded roots of a Mesozoic to Tertiary volcanic arc that formed during the northeastward subduction of proto-Pacific oceanic crust beneath the margin of Gondwana (Storey & Garrett 1985). Arc volcanism and magmatism was predominantly calc-alkaline in nature, with the major locus of volcanism and magmatism migrating north-westwards with time (Pankhurst 1982). In the Late Cretaceous, magmatism switched from the east to the west coast of the Antarctic Peninsula, with a possible change to more mafic volcanism (Storey & Garret 1985; Pankhurst *et al.* 1988). This was due to a change in the angle of subduction along the arc, with the subduction of younger oceanic crust. The Larsen Basin, located on the eastern side of the Antarctic Peninsula (Fig. 1), represents a back-arc basin developed in relation to the magmatic arc (Macdonald *et al.* 1988). The tectonic evolution of the region is, however, complex, related to the Early Cretaceous opening of the Weddell Sea (Macdonald *et al.* 1988), and strike-slip or oblique extension along the eastern margin of the Antarctic Peninsula, which may have controlled the development of sedimentary basins along this margin (Storey & Nell 1988). Nevertheless the evolution of the Larsen Basin is closely related to the magmatic arc, with coarse-clastic

From Morton, A. C., Todd, S. P. & Haughton, P. D. W. (eds), 1991,
Developments in Sedimentary Provenance Studies.
Geological Society Special Publication No. 57, pp. 231–249.

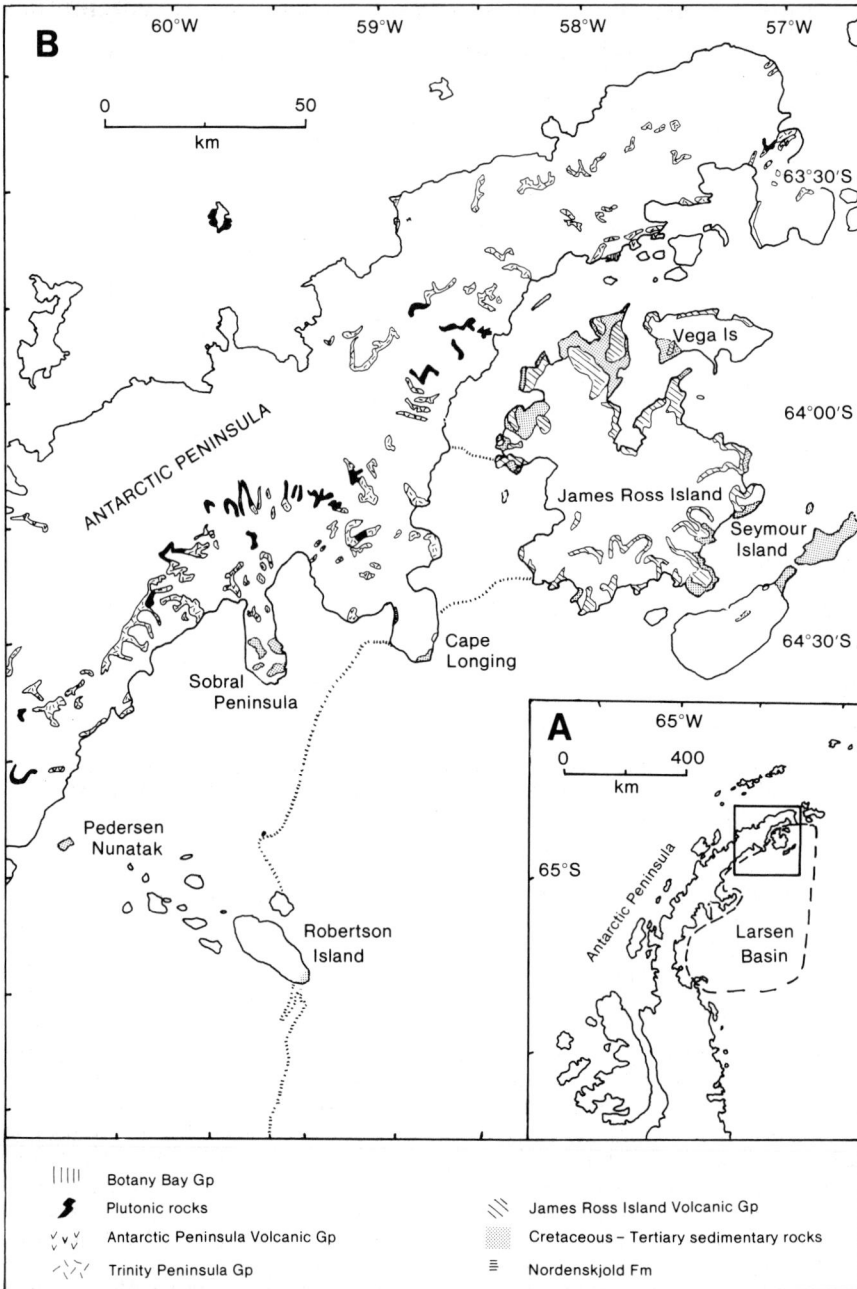

Fig. 1. (A) Map showing the location of the Larsen Basin, east of the northern Antarctic Peninsula. **(B)** Sketch geological map of the northern Antarctic Peninsula.

sedimentation beginning within the basin contemporaneously with a major interval of arc volcanism and plutonism dated at about 130 Ma (Hauterivian–Barremian, Late Jurassic–Early Cretaceous) (Pankhurst 1982).

The Larsen Basin sedimentary succession is best exposed in the James Ross Island area (Fig. 2). A sedimentary sequence approximately 5–6 km thick and Barremian to Oligocene in age, is thought to overlie radiolarian-rich mudstones

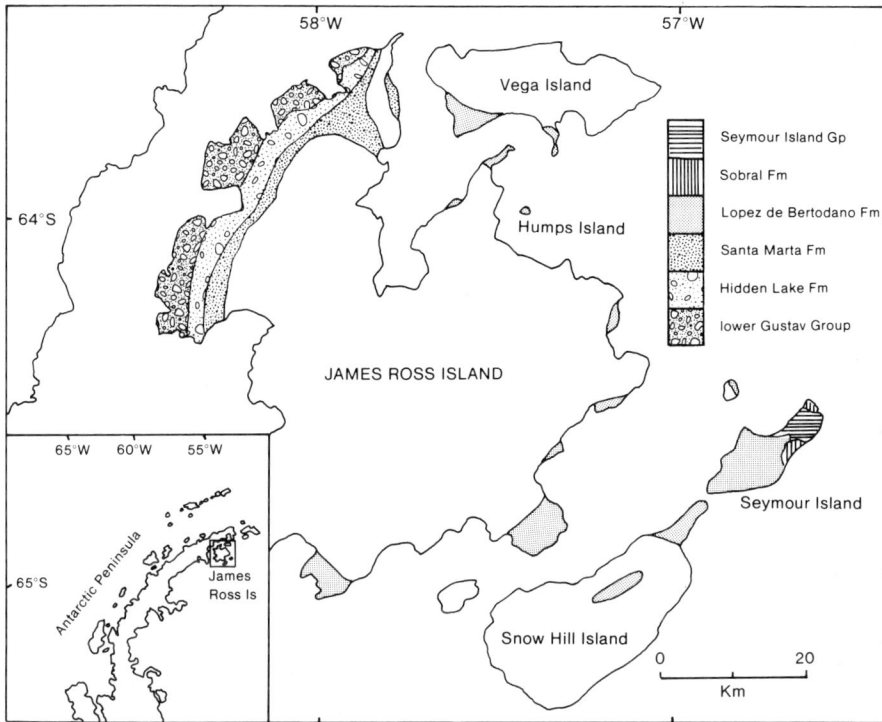

Fig. 2. Geological sketch map of the James Ross Island area showing the exposures of the lower Gustav Group (Lagrelius Point, Kotick Point and Whisky Bay Formations), Hidden Lake Formation, Santa Marta Formation, Lopez de Bertodano Formation, Sobral Formation and the Seymour Island Group.

and tuffs of the Late Jurassic–Early Cretaceous Nordenskjöld Formation (Macdonald *et al.* 1988). The basin fill forms a regressive megasequence, and has been divided into three groups (Fig. 3) (Elliot & Trautman 1982; Ineson *et al.* 1986, Olivero *et al.* 1986). The Gustav Group of Barremian-Santonian age is about 2.1 km thick and is exposed on northern James Ross Island (Fig. 2) (Ineson *et al.* 1986). The lower three formations of the group (the Lagrelius Point, Kotick Point and Whisky Bay Formations) represent deposition within a deep-marine slope-apron to submarine fan complex (Ineson 1989). The western basin margin was a tectonically active fault zone flanked by a slope apron, interrupted at intervals by coarse-grained gravelly submarine fan sedimentation. Sedimentation was influenced by both intra- and extra-basinal controls (Ineson 1985, 1989). Widespread shallowing, related to partial basin inversion, led to the deposition of the Hidden Lake Formation (the uppermost unit of the Gustav Group) during the Coniacian–Santonian (Macdonald *et al.* 1988). The Hidden Lake Formation represents deposition within fan-delta, shelf and slope settings (Ineson 1989; A. G. Whitham pers. comm. 1989). Deposition was closely related to a major pulse of proximal arc volcanism.

The Gustav Group is conformably overlain by the Marambio Group, which represents the onset of shallow-marine, typically fine-grained sedimentation. The Marambio Group spans the Santonian to Palaeocene, and has been subdivided into three formations: the Santa Marta, Lopez de Bertodano and Sobral Formations (Fig. 3) (Olivero *et al.* 1986). The Santa Marta Formation of Santonian–Campanian age is approximately 1 km thick (Olivero *et al.* 1986; Pirrie 1989). The formation is exposed on northern James Ross Island (Fig. 2) and can be divided into two facies associations representing deposition within mid-outer shelf (Association 1) and inner shelf (Association 2) environments (Pirrie 1989, 1990). Concurrent arc-volcanism is reflected by the presence of accretionary lapilli and airfall ash beds within Association 1. The Lopez de Bertodano Formation is exposed throughout the James Ross Island group, with the most continuous outcrops on Vega and Seymour islands (Fig. 2) (Zinsmeister 1982; Macel-

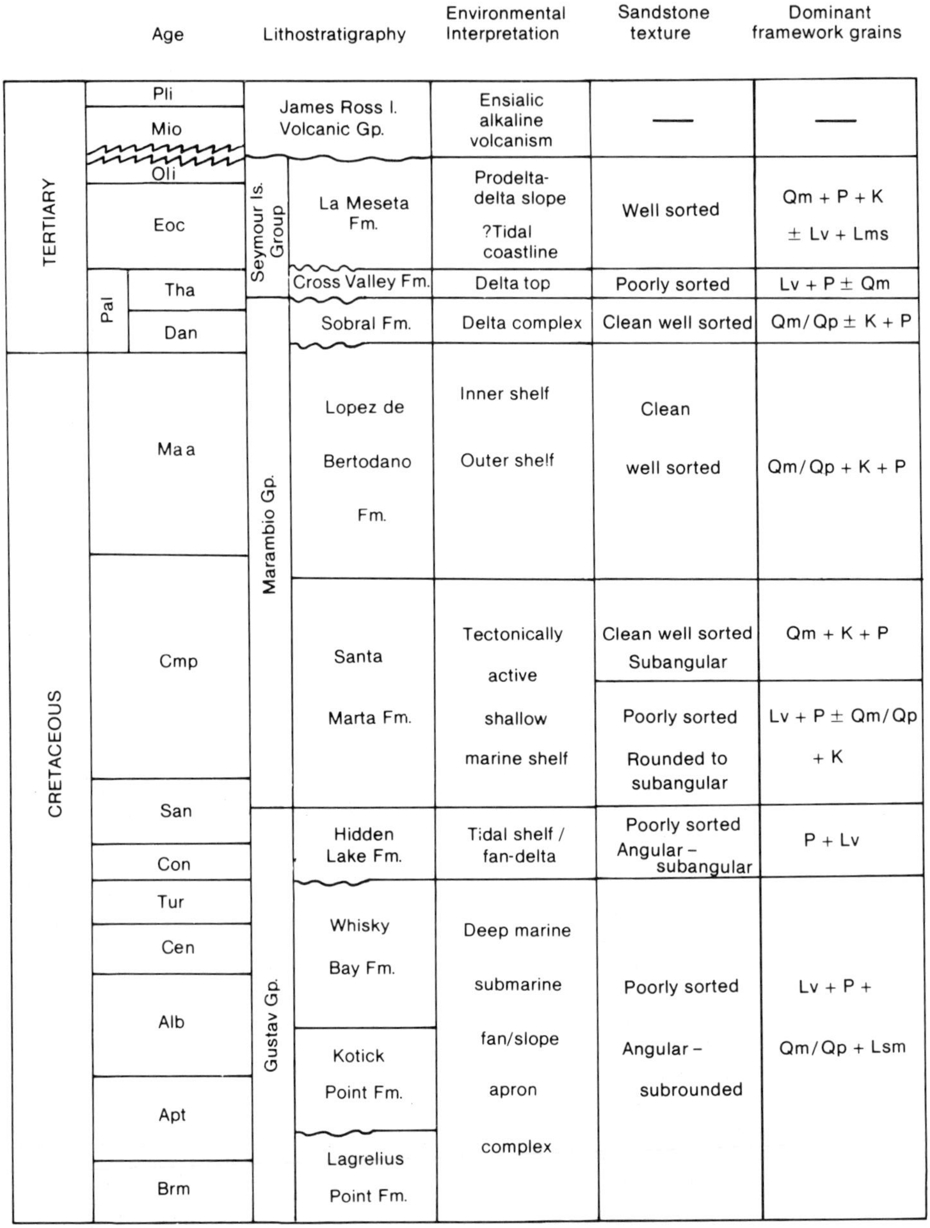

Age			Lithostratigraphy	Environmental Interpretation	Sandstone texture	Dominant framework grains
TERTIARY	Pli		James Ross I. Volcanic Gp.	Ensialic alkaline volcanism	—	—
	Mio					
	Oli	Seymour Is. Group	La Meseta Fm.	Prodelta-delta slope ?Tidal coastline	Well sorted	Qm + P + K ± Lv + Lms
	Eoc					
	Pal — Tha		Cross Valley Fm.	Delta top	Poorly sorted	Lv + P ± Qm
	Pal — Dan		Sobral Fm.	Delta complex	Clean well sorted	Qm/Qp ± K + P
CRETACEOUS	Maa	Marambio Gp.	Lopez de Bertodano Fm.	Inner shelf / Outer shelf	Clean well sorted	Qm/Qp + K + P
	Cmp		Santa Marta Fm.	Tectonically active shallow marine shelf	Clean well sorted / Subangular / Poorly sorted / Rounded to subangular	Qm + K + P / Lv + P ± Qm/Qp + K
	San / Con	Gustav Gp.	Hidden Lake Fm.	Tidal shelf / fan-delta	Poorly sorted / Angular – subangular	P + Lv
	Tur / Cen / Alb / Apt		Whisky Bay Fm. / Kotick Point Fm.	Deep marine submarine fan/slope apron complex	Poorly sorted / Angular – subrounded	Lv + P + Qm/Qp + Lsm
	Brm		Lagrelius Point Fm.			

Fig. 3. Age, lithostratigraphy, inferred environmental interpretation, sediment petrography and texture for the formations within the Larsen Basin (after Ineson *et al.* 1986; Olivero *et al.* 1986; Macdonald *et al.* 1988).

lari 1988). The formation spans the Campanian to Palaeocene interval (Macellari 1988), with deposition in a range of shelf settings (Macellari 1988; Pirrie & Riding 1988). The uppermost formation of the Marambio Group (the Sobral Formation of Palaeocene age) unconformably overlies the Lopez de Bertodano Formation and was deposited within shelf to deltaic settings (Sadler 1988).

The overlying Seymour Island Group, exposed only on Seymour Island, comprises the Cross Valley and La Meseta Formations (Elliot & Trautman 1982), of Palaeocene to ?Oligocene age (Zinsmeister 1982), separated by a major unconformity (Sadler 1988). Deposition was within a tidally influenced deltaic setting (Elliot & Trautman 1982).

Previous work

Previous studies on the petrography of the Larsen Basin are limited. Prior to this study, no modal sandstone petrography or clay mineralogy data were available for the sedimentary succession exposed on James Ross Island. A number of authors have examined the Larsen Basin sediments exposed on Seymour and Robertson islands, Cape Longing, Sobral Peninsula and Pedersen Nunatak (see Fig. 1) (Elliot & Trautman 1982; Farquharson 1983; del Valle & Medina 1985: summarized in Macdonald *et al.* 1988). Elliot & Trautman (1982) examined the Cross Valley and La Meseta Formations of the Seymour Island Group, and showed a trend with time from litharenites to sublitharenites, which was interpreted as representing a change from a predominantly volcanic source area to a metamorphic/plutonic source region. Farquharson (1983) suggested that the basin fill sequence represented a change from an undissected to a dissected source terrain. Other more limited studies have described the presence of Late Cretaceous quartzose sediments at Robertson Island (del Valle & Medina 1985), and Pezzetti & Krissek (1986) described litharenite to lith-arkose compositions for the La Meseta Formation on Seymour Island.

Previous work has shown that the sandstone petrography of the Larsen Basin sediments can be matched with the major geological units which form the northern Antarctic Peninsula (see Fig. 1). The peninsula adjacent to James Ross Island is composed predominantly of calc-alkaline plutons and volcanic rocks (the Antarctic Peninsula Volcanic Group) (Hamer 1983), which intrude and unconformably overlie an accretionary complex (the Trinity Peninsula Group), dominantly composed of siliciclastic sedimentary and metasedimentary rocks (Smellie 1987). Derivation from this region is also supported by palaeocurrent data which show a dominant southeasterly palaeoflow (Pirrie 1987, 1989; Ineson 1989).

Sandstone petrography and clay mineralogy

Methodology

In order to elucidate the petrographic evolution of the basin a suite of 60 sandstone samples were analysed using standard point-count techniques and 40 mudrocks were analysed using X-ray diffraction. In addition, selected mineral phases were analysed by microprobe. Representative samples were chosen from both the Gustav and Marambio Groups. The sandstones were point-counted for all grains using the 'QFL' method (Dickinson 1970), with a minimum of 500 points counted per slide. Using this method, all sand-sized monomineralic fragments within a section are counted as single grains, even if they occur within polymineralic lithic grains. Thirty eight of these samples were then counted for lithic grains only, with 200 grains counted per section. Although the interdependence of grain size and mineralogy is reduced by using the QFL methodology (Ingersoll *et al.* 1984), medium grained sandstones were selected where possible. All sandstones point-counted were stained for plagioclase and K-feldspar, using the method of Houghton (1980). The clay mineralogy of 40 samples was examined by XRD analysis of the <2 μm fraction. All analyses were carried out by the author on a Philips (PW 1730) diffractometer, housed in the Department of Earth Sciences, University of Cambridge, using Cu kα radiation. Samples were prepared as smear slides, and were analysed between 2–36° 2τ for untreated, glycolated, heated at 400°C and heated to 550°C samples. Sample preparation methodology followed Jeans (1978). Microprobe analyses were carried out by T. Alabaster, using a wavelength dispersive, Cambridge Instruments microscan 9 microprobe, at the Open University. A summary of the results is given in Table 1: full tabulated data have been deposited with The Society Library and British Library at Boston Spa, W. Yorkshire, UK, as Supplementary Publication No. SUP 18066 (11 pages).

Vitrinite reflectivity and diagenetic studies imply limited post-depositional burial and diagenetic alteration of the sequences studied (Whitham & Marshall 1988; D. Pirrie unpublished data). Sandstones within the Gustav Group are typically cemented by chlorite and calcite, with zeolite cements (heulandite–clinoptilolite) predating calcite in the Hidden Lake Formation. The dominant cements within sandstones in the Marambio Group are non-ferroan calcites, which in some samples partially replace labile grains, although relict framework grains are still recognized. The absence of the diagenetic transformations of smectite to illite (or interlayered illite–smectite) or the alteration of clinoptilolite–heulandite to analcime suggests that none of the sequence studied has been heated to >80°C (Lee & Klein 1986). Thus the original framework grain mineralogy and petrography has undergone only limited diagenetic modification.

Sandstone petrography

The sandstones studied are arkosic and lithic

Table 1. Summary of sandstone modal data for each of the stratigraphic units studied, and abbreviations used

All clasts point-count data range and averages (in brackets)

Stratigraphic unit	No.	Qm	Qp	Qt	P	K	F	Lv	Lsm	Lt	Musc.	Biotite	Opaques	Hbl.	Glauc.	Px.
lower Gustav Gp. (Kotick Pt./Whisky Bay fms.)	9	0–33.2 (12.8)	0–14.8 (3.6)	0–33.6 (16.5)	8.5–97.0 (26.8)	0–4.1 (0.8)	9–99.0 (29.1)	0–70.0 (35.8)	0–57.7 (18.0)	0–78.4 (54.3)	0	0–0.7 (0.1)	0–3.2 (0.8)	0	0–0.5 (0.05)	0
Hidden Lake Formation	13 (3.1)	0.3–10.6 (2.9)	0–6.8 (6.5)	0.3–21.3 (35.2)	23.7–50.3 (0.03)	0–0.3 (37.3)	24.3–52.8 (49.6)	37.4–62.4 (4.2)	0.3–9.3 (56.2)	42.2–65.3 (0.02)	0–0.3 (0.07)	0–0.5 (2.1)	0.6–4.8 (0.01)	0–0.2 (0.01)	0–0.2 (1.8)	0.3–4.9
Santa Marta Formation Assoc. 1	23	1.5–31.8 (10.9)	0.4–5.8 (3.1)	2.6–38.2 (14.5)	5.8–31.0 (17.8)	0–18.0 (2.8)	6.1–47.3 (21.4)	6.7–89.2 (57.6)	0.7–9.5 (1.4)	14.4–90.1 (64.2)	0–0.4 (0.06)	0–2.9 (0.7)	0–3.6 (1.2)	0–1.7 (0.4)	0–2.2 (0.2)	0–2.1 (0.5)
Santa Marta Formation Assoc. 2	3	14.1–52.1 (37.0)	0.6–2.3 (1.5)	15.1–58.5 (47.1)	9.4–17.0 (14.0)	1.4–12.6 (7.4)	17.5–27.0 (22.9)	6.5–62.5 (25.7)	3.2–15.5 (8.0)	14.5–67.4 (35.4)	0–0.8 (0.3)	0.6–3.7 (1.9)	1.4–2.4 (2.0)	0.6–2.0 (1.1)	0–2.3 (0.9)	0
Lopez de Bertodano Formation	12	32.3–67.9 (42.2)	0–6.5 (3.3)	38.2–74.3 (49.1)	4.8–21.6 (16.1)	2.0–19.2 (13.0)	16.1–47.7 (31.6)	1.5–16.2 (8.5)	2.0–17.6 (9.6)	3.9–33.4 (19.3)	0–0.8 (0.3)	0–2.4 (1.2)	0–2.4 (1.4)	0–4.9 (0.7)	0.3–9.1 (3.6)	0–1.8 (0.2)

Lithic clasts only point-count data, range and averages (in brackets)

Stratigraphic unit	No.	Lv(and)	Sn	Glass	Lvt	Lp	Qpt	Qp 2–3	Qp > 3	Ls	Lsm
lower Gustav Gp. (Kotick Pt./Whisky Bay fms.)	5	5.2–94.1 (45.2)	0–85.9 (17.2)	0	6.5–94.1 (62.4)	0	0.9–19.9 (6.3)	0–0.9 (0.8)	0.9–18.5 (5.5)	0–0.5 (0.1)	4.4–86.5 (31.1)
Hidden Lake Formation	12	41.9–97.6 (85.6)	0–30.0 (3.7)	0–40.2 (4.5)	72.9–98.1 (94.8)	0–0.4 (0.07)	0–8.8 (3.8)	0–2.7 (0.8)	0–7.7 (3.0)	0–4.7 (0.4)	1.9–20.0 (9.3)
Santa Marta Formation Assoc. 1	17	70.3–95.0 (86.6)	0–1.2 (0.2)	0–28.2 (3.7)	78.4–98.5 (90.5)	0–0.5 (0.03)	0–7.2 (3.1)	0–2.3 (0.8)	0–5.6 (2.3)	0–0.8 (0.08)	0.3–15.5 (6.2)
Lopez de Bertodano Formation	3	36.2–70.0 (50.4)	0–6.8 (2.3)	0–5.5 (1.8)	36.2–70.0 (54.6)	0	2–23.2 (11.6)	0–18.8 (11.6)	2.0–6.8 (7.2)	0	28.0–40.6 (33.8)

Explanation of petrographical parameters and abbreviations employed in this study (after Ingersoll & Suczek 1979)

$Qt = Qm + Qp$

where Qt = Total quartzose grains
Qm = Monocrystalline quartz grains
Qp = Polycrystalline quartz grains

$F + P + K$

where F = Total feldspar grains
P = Plagioclase feldspar grains
K = Potassium feldspar grains

$Lt + L + Qp$

where Lt = Total lithic grains
L = Total lithic grains − Qp

$L = Lp + Lv + Ls(m) + Lm$

where Lp = Lithic plutonic grains
Lv = Lithic volcanic grains
Ls(m) = Lithic sedimentary and metasedimentary grains
Lm = Lithic metamorphic grains

$Lvt = Lv(and) + sh + glass$

where Lv(and) = Lithic volcanic andesitic grains
sh = Glass shards
glass = pumice + volcanic glass fragments

arenites (*sensu* Pettijohn *et al.* 1972). The framework grains and major accessory minerals are
briefly described below.

Quartz. Mono-crystalline quartz (Qm) and polycrystalline quartz (Qp) occur throughout the
sequence (Fig. 4a). Undulose (> 5°) and non-
undulose Qm are recognized. Embayed quartz
and inclusions are rarely observed. Qp with > 3
constituent crystals are more abundant than Qp
grains with 2–3 constituent crystals, Qp grains
mainly show sutured grain contacts and marked
undulosity.

Feldspar. Plagioclase (P) and K-feldspar (K)
occur throughout the sequence studied, with a
marked change in the ratio of plagioclase to
total feldspar from about 1 in the Hidden Lake
Formation to about 0.5 in the Lopez de Bertodano Formation.

Plagioclase is very abundant within the Gustav Group and the Santa Marta Formation.
Grains are typically fresh and unaltered, and
range from large, euhedral, compositionally
zoned crystals (Fig. 4b) to subangular broken
crystals commonly showing twinning. The grains
range from oligoclase to labradorite in composition, with most grains falling in the andesine
field.

K-feldspar is less abundant than plagioclase
within the Gustav Group and Santa Marta Formation. Microline and orthoclase are predominant, rare sanidine, perthitic and microperthitic
feldspars also occur. K-feldspar grains are
typically small and fractured, showing variable
rounding, but otherwise relatively unaltered.

Lithic fragments. Lithic-volcanic (Lv), lithic-
sedimentary (Ls), lithic-metasedimentary (Lsm)
and lithic-plutonic (Lp) grains occur in varying
proportions throughout the sequence. Lv grains
predominate, and include glass shards, pumice,
accretionary lapilli, and andesitic and rhyolitic
fragments. The glass shards (commonly cuspate)
and pumice are locally very abundant. Some
shards appear relatively unaltered whilst others
are altered to clay minerals including chlorite.
Pumice grains are frequently larger than other
detrital grains. Extensive replacement of glass by
calcite and chlorite is common.

Andesitic grains composed of plagioclase
phenocrysts within a fine-grained or aphanitic
groundmass (Fig. 4c) are the main Lv grain type
present. The groundmass is commonly altered to
calcite, chlorite and other clay minerals, and
opaque minerals are common. Relatively unaltered and altered grains may occur together
within the same thin section. Extensive replacement by calcite also occurs, with only ghost

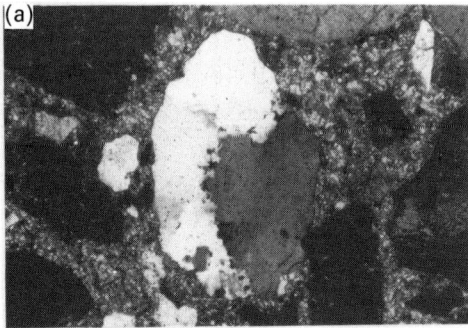

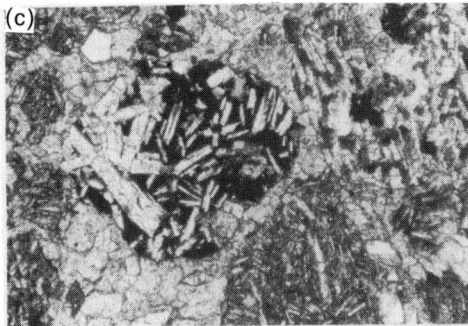

Fig. 4. (**a**) Photomicrograph (crossed polars) of a
polycrystalline quartz grain with >3 sub-grains,
which show undulose extinction and sutured grain
contacts. Sample number D.8223.2. Marambio
Group. Field of view 1.4 mm (**b**) Thin-section photomicrograph (crossed polars) showing a fractured
compositionally zoned euhedral plagioclase grain.
Sample number D.8651.14. Hidden Lake Formation.
Maximum width of field 1.4 mm (**c**) Photomicrograph (crossed polars) showing a lithic-rich sandstone. Lithic fragments are andesitic, composed of
plagioclase phenocrysts within a fine-grained groundmass. Sample number D.8664.63. Santa Marta Formation (Association 1). Field of view 1.4 mm.

outlines of the original grains retained. Rare possible rhyolitic grains also occur.

Sedimentary and metasedimentary grains are a less common component than Lv. Ls grains are dominantly radiolarian-rich mudstones, thought to represent clasts of the Late Jurassic–Early Cretaceous Nordenskjöld Formation. Contemporaneous sedimentary intraclasts are difficult to recognize in thin section. Lsm grains are more abundant, including quartz–mica aggregates. These grains typically have >3 quartz sub-grains which have undulose extinction and sutured grain contacts. Finer grained Lsm grains include deformed phyllosilicate aggregates; Lp clasts are very rare with mainly granophyric clasts recognized.

The dominant accessory minerals present within the sandstones examined include biotite, muscovite, hornblende, pyroxene, opaque minerals and glauconite, along with less common titanite, zircon, garnet, ilmenite, ?apatite and ?rutile. The biotite and hornblende are dominant within the Marambio Group. Biotite grains are typically splayed by the growth of calcite along the cleavage. Clinopyroxene is typically very fresh and locally very abundant within the lower part of the sequence. Electron-microprobe analysis indicates an augite composition for the clinopyroxene, with high Na_2O and MnO contents and low CrO contents (Fig. 5). Opaque grains, which include pyrite, occur throughout, although some pyrite grains are diagenetic in origin.

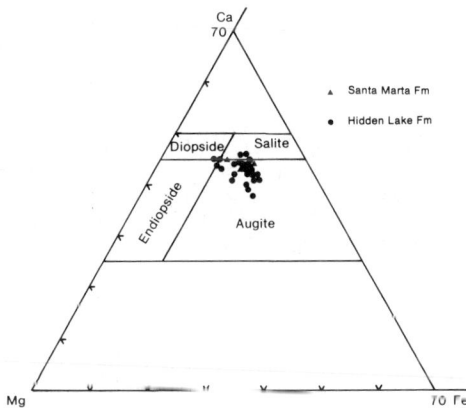

Fig. 5. Ternary plot (Ca:Mg:Fe) showing the predominantly augite composition of clinopyroxenes from the Hidden Lake and Santa Marta Formations.

Point count data from 60 counts for all clasts and 38 for lithic-clasts only are shown in Figs 6 and 7 and are summarized in Table 1. Data for all clast types were plotted on QtFL, QmFLt and QmPK ternary diagrams (Fig. 6). The QtFL plot emphasizes sediment maturity by recasting the Qp grains within the Lt field. The lower Gustav Group (Kotick Point and Whisky Bay Formations) typically shows a wide scatter on the QtFL, QmPK, QpLvmLsm and LmLvLs ternary diagrams, with a more tightly constrained field on the QmFLt plot. This reflects the wide compositional range of samples from this unit, although it may also be partially a function of the small data set. In contrast, the Hidden Lake Formation plots as a coherent petrographic suite on all of the ternary diagrams (Figs 6 & 7). The formation is composed of feldspathic litharenites as shown by the QtFL/QmFLt and QmPK diagrams (Fig. 6), with the lithic framework grains dominated by Lvm (see Fig. 7).

Association 1 of the Santa Marta Formation shows a compositional trend from litharenites to arkosic arenites. This is shown on the QtFL and QmFLt ternary diagrams as a diagonal field extending from a composition of $Qt_5F_5L_{90}$ to $Qt_{40}F_{45}L_{15}$. Compositions are typically enriched in Qm and K but plot in the same field on the QpLvmLsm and LmLvLs ternary diagrams as the Hidden Lake Formation. Due to limited availability of samples, only three sandstones from Association 2 of the Santa Marta Formation were counted. One of three plots within the same field as Association 1, whilst the other two are enriched in Qm and K relative to Association 1.

The Lopez de Bertodano Formation shows the most quartzose field for any of the samples examined, with an average modal composition of $Qt_{55}F_{30}L_{15}$ on the QtFL ternary plot, and a less quartzose more lithic-rich field on the QmFLt plot. The formation also shows enrichment in K-feldspar relative to P and decreased lithic content. On the basis of the ternary diagrams the four major stratigraphic units studied can be distinguished petrographically, although individual fields do show some overlap.

In summary, the sandstone modal data show that the four stratigraphic units studied are petrographically distinct. The lower Gustav Group shows a wide compositional range, whilst the Hidden Lake Formation is composed of feldspathic litharenites, representing a coherent petrographic suite. Association 1 of the Santa Marta Formation shows a compositional trend from litharenites to arkosic arenites, whilst Association 2 is typified by more quartzose sandstones with sub-equal feldspar and lithics. The Lopez de Bertodano Formation is categorized by a more quartzose suite of sandstones with sub-equal feldspar and lithic clasts.

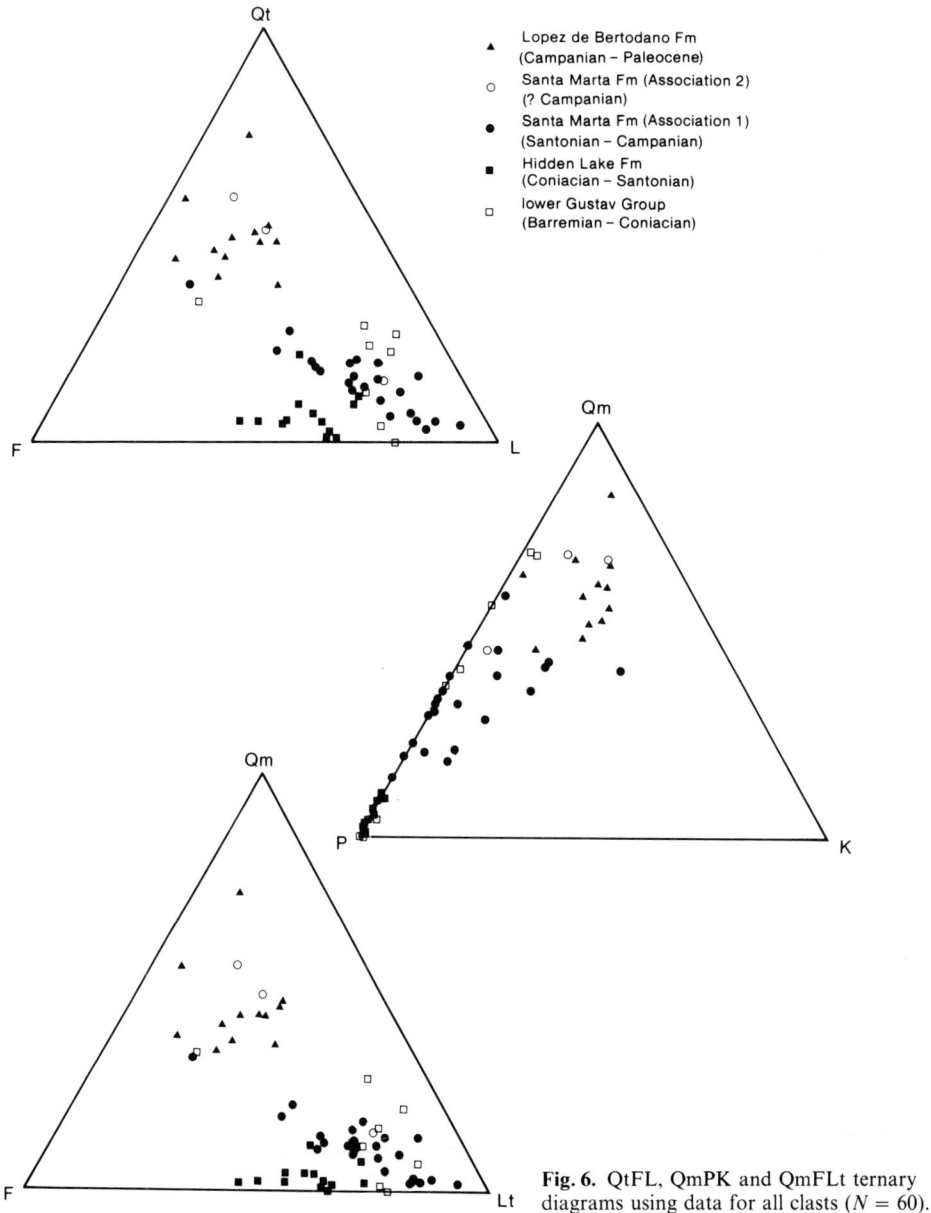

Fig. 6. QtFL, QmPK and QmFLt ternary diagrams using data for all clasts ($N = 60$).

Clay mineralogy

The clay mineralogy ($< 2\,\mu m$ fraction) of 40 samples of fine-grained facies including silty sandstones, mudstones, and beds of possible airfall ash origin were examined by XRD. Samples are only available from the Hidden Lake, Santa Marta and Lopez de Bertodano Formations.

All of the samples analysed are dominated by a smectitic clay mineral. 30% of samples analysed only have smectite present, whilst the other samples include variable quantities of smectite–chlorite, mica, chlorite and/or kaolinite and zeolite minerals of the clinoptilolite–heulandite and laumontite groups. Samples with varied clay compositions are shown plotted against stratigraphic height in Fig. 8. Zeolite minerals are more abundant in the Hidden Lake Formation, and beds attributed to be of direct volcaniclastic

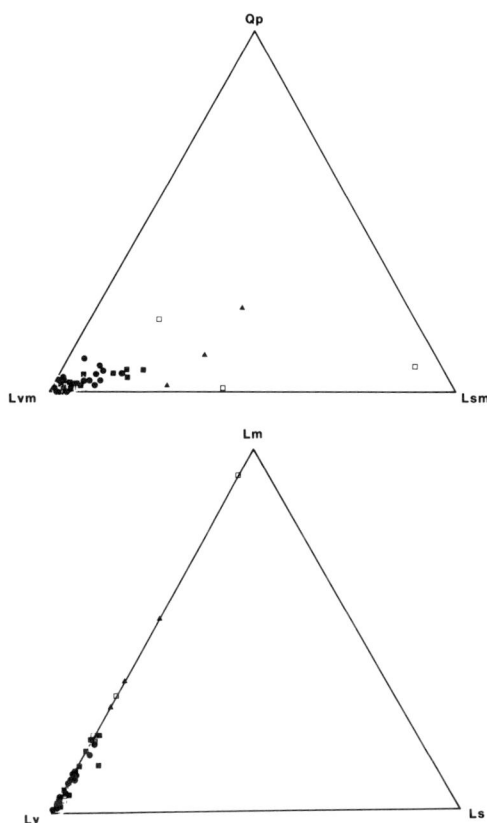

Fig. 7. Sandstone modal data for the lithic clasts only (N = 38) plotted on QpLvmLsm and LmLvLs ternary diagrams. Symbols as in Fig. 6.

air-fall origin. Mica and smectite occur throughout. Chlorite/kaolinite are only common within the upper part of the Santa Marta and Lopez de Bertodano Formations. Beds composed solely of smectite occur within mudrocks from the Hidden Lake and basal Santa Marta Formations, and also include beds of probable airfall origin within the Santa Marta and Lopez de Bertodano Formations.

Source region

In combination with palaeocurrent data (Ineson 1985, 1989; Pirrie 1987, 1989), the petrographical and mineralogical data presented here support a northwesterly derivation from the northern Antarctic Peninsula. In addition, all of the clasts can be matched with known geological units or units inferred to be present within this region. The arc-derived framework grains (including plagioclase, Lv (andesitic), Lv (rhyolitic), glass shards, pumice and lapilli) can all be matched with the predominantly calc-alkaline andesitic suites of the Antarctic Peninsula Volcanic Group (Hamer 1983); the accessory minerals pyroxene, hornblende and ilmenite were also derived from this source. The smectite clay which predominates within the fine-grained facies, was also probably derived from this source, either as reworked detrital clay or as authigenic clays formed during the break-down of volcanic ashes deposited within the basin. Comparable smectite-dominated clay mineral assemblages, with subordinate mica, attributed to the alteration of volcanic detritus, have been

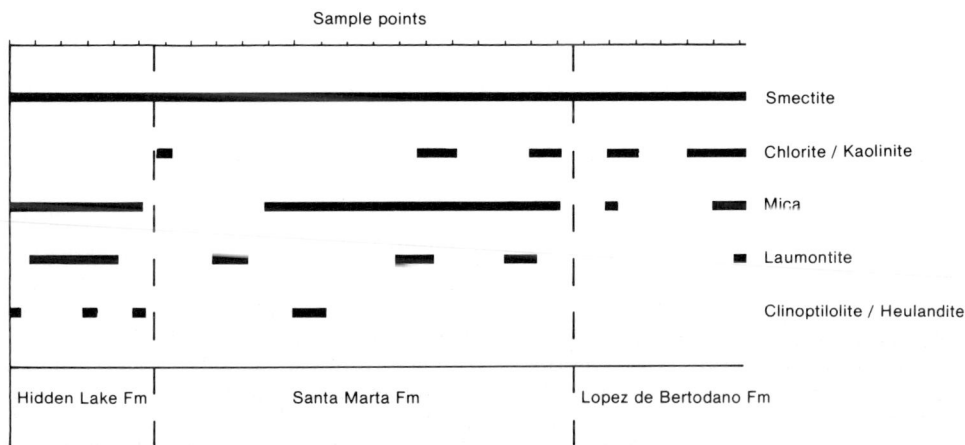

Fig. 8. Diagrams showing clay mineral compositions of fine-grained beds from the Hidden Lake, Santa Marta and Lopez de Bertodano Formations plotted against stratigraphic height. No vertical scale intended.

reported (e.g. Jeans *et al.* 1982; Andrews 1987). Some of the fine-grained beds represent primary air-fall ashes (Facies D of Pirrie & Riding 1988), whilst the remaining fine-grained sediment was derived from either reworked volcanic ashes, or altered fine-grained volcaniclastic detritus derived from the flanks of the arc.

The rare lithic plutonic (Lp) grains were largely derived from granophyric plutons; similar plutonic rocks are known to occur in the arc. In addition some of the mono- and poly-crystalline quartz and K-feldspar may have been derived from a plutonic source, associated with the Antarctic Peninsula Volcanic Group. However, the majority of the quartz and lithic metasedimentary clasts were probably derived from the Trinity Peninsula Group. The Trinity Peninsula Group rocks exposed on the northern Antarctic Peninsula adjacent to James Ross Island are composed of low-grade quartzose metasediments (Smellie 1987), petrographically similar to the Lsm grains seen within the sequence. In addition the predominance of undulosity of $> 5°$ within Qm grains, and > 3 component grains in Qp clasts, supports derivation from a low-rank metamorphic terrain rather than from volcanic rocks of the Antarctic Peninsula Volcanic Group (Basu *et al.* 1975). The K-feldspar seen within the sequence may also have been derived from the Trinity Peninsula Group although Smellie (1987) has recognized considerable alteration of K-feldspar within the Trinity Peninsula Group. A second possible source for the K-feldspar could be rhyolitic rocks within the Antarctic Peninsula Volcanic Group. These would also provide a source for the sanidine observed. Rare lithic sedimentary (Ls) clasts of the Late Jurassic–Early Cretaceous Nordenskjöld Formation also occur; fault-related uplift and re-exposure of these sediments within fault scarps along the western basin margin has previously been invoked to explain the presence of large glide blocks of Nordenskjöld Formation within the Gustav Group (Ineson 1985), and the presence of reworked Late Jurassic–Early Cretaceous palynomorphs within the basin (J. B. Riding pers. comm. 1989).

Petrographic evolution of the basin

The sandstone detrital modes for the stratigraphic units studied are summarized in Fig. 9. On the QmFLt ternary diagram the lower Gustav Group plots within a range of fields from dissected arc to lithic recycled. The Hidden Lake Formation plots within the undissected arc field whilst Association 1 of the Santa Marta Formation plots dominantly within the undissected to transitional arc fields. Two sandstones from Association 2 of the Santa Marta Formation plot within the mixed field. The Lopez de Bertodano Formation plots mainly within the dissected arc to mixed provenance fields. The QtFL diagram shows a similar pattern, with a wide spread of points for the lower Gustav Group. The Hidden Lake and Santa Marta Formations show a temporal trend from undissected to transitional arc fields, and the Lopez de Bertodano Formation plots within the dissected arc to recycled orogen fields. On both the QpLvmLsm and LmLvLs discrimination diagrams the data broadly plot within the magmatic arc fields. In the QpLvmLsm plot the data for the Hidden Lake and Santa Marta Formations cluster at the Lvm pole of the ternary diagram. On the LmLvLs diagram the data are more comparable with the field for fore-arc magmatic terrains, rather than the back-arc terrains which typically have both a magmatic arc component and a rifted continental margin component.

The data of Elliot & Trautman (1982), Farquharson (1983) and del Valle & Medina (1985) are plotted on the QtFl ternary plot in Fig. 10. These data suggest a trend in time from a 'basement uplift' source for the ?Barremian Sobral Peninsula sediments towards a recycled orogen field for the La Meseta Formation of Eocene–Oligocene age. The data of del Valle & Medina (1985) from Robertson Island for sediments of ?Campanian age plot within the craton interior to recycled orogen fields. The Palaeocene Cross Valley Formation records a pulse in arc volcanism (Elliot & Trautman 1982).

Discussion

The petrographical data for the basin suggest a temporal evolution from a complex source terrain in the Barremian (reflected by the wide spread of data) through undissected, transitional and dissected arc terrains and finally to a recycled-orogen source terrain on the basis of standard ternary diagrams. This assumes that the dominant control on sediment petrography within the basin was the changing composition of the source area (the Antarctic Peninsula). Although this probably does represent the background control on the petrography, a complex interplay of other controls may also be superimposed upon this petrographic trend, which should not therefore be considered as simple arc unroofing (Farquharson 1983). The observed petrographic shifts are thought to be due to the interplay of (a) arc tectonics and the location

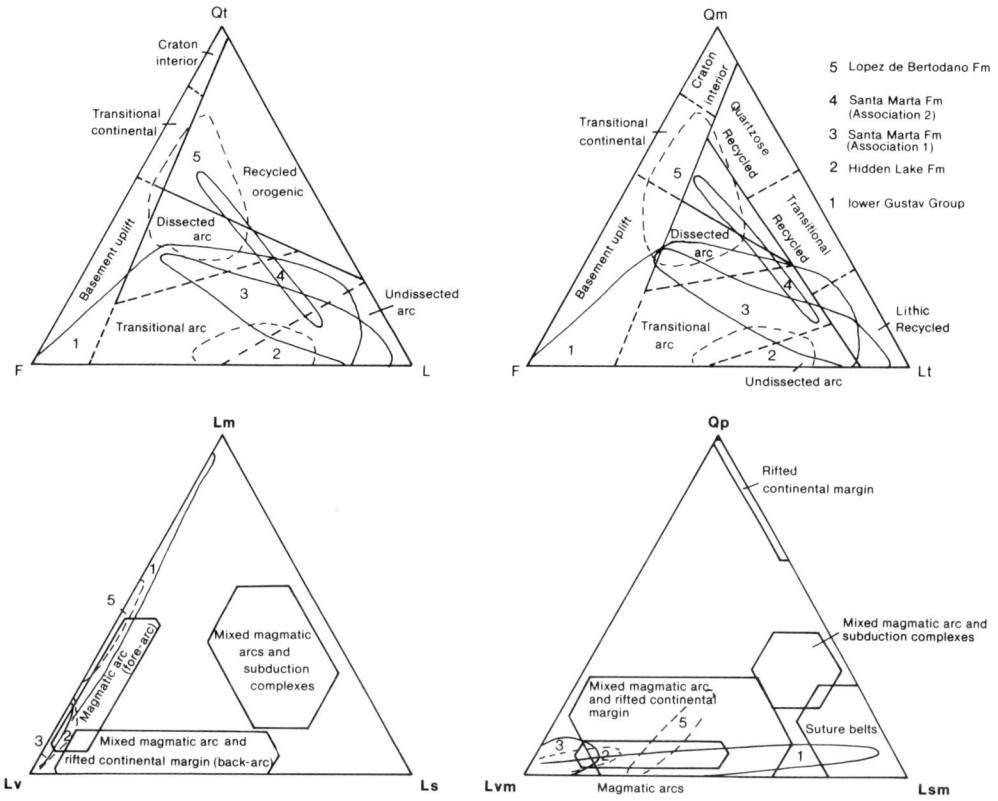

Fig. 9. Sandstone modal data plotted on QtFL and QmFLt tectonic discrimination diagrams (defined by Dickinson *et al.* 1983). Modal data for lithic clasts only plotted on QpLvmLsm and LvLmLs ternary diagrams, with the tectonic discrimination fields defined by Ingersoll & Suczek (1979).

and nature of arc volcanism, (b) source area relief and climate, (c) depositional environment, process and sedimentation rate and (d) the changing source area composition (Mack 1978, 1984; Suttner *et al.* 1981; Basu 1985; Morris & Busby-Spera 1987). It should be noted that all of these controls inter-relate. This is summarized in Fig. 11, and discussed further below.

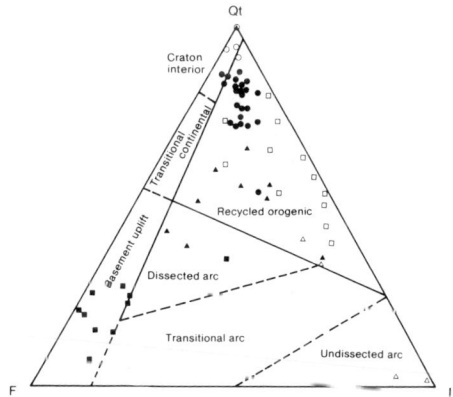

Fig. 10. Previous sandstone modal data for the Larsen Basin plotted on a QtFL ternary diagram, showing the tectonic discrimination fields defined by Dickinson *et al.* (1983). Data for Sobral Peninsula, Cape Longing and Pedersen Nunatak from Farquharson (1983), for Robertson Island from del Valle and Medina (1985) and for the Cross Valley and La Meseta Formations from Elliot & Trautman (1982).

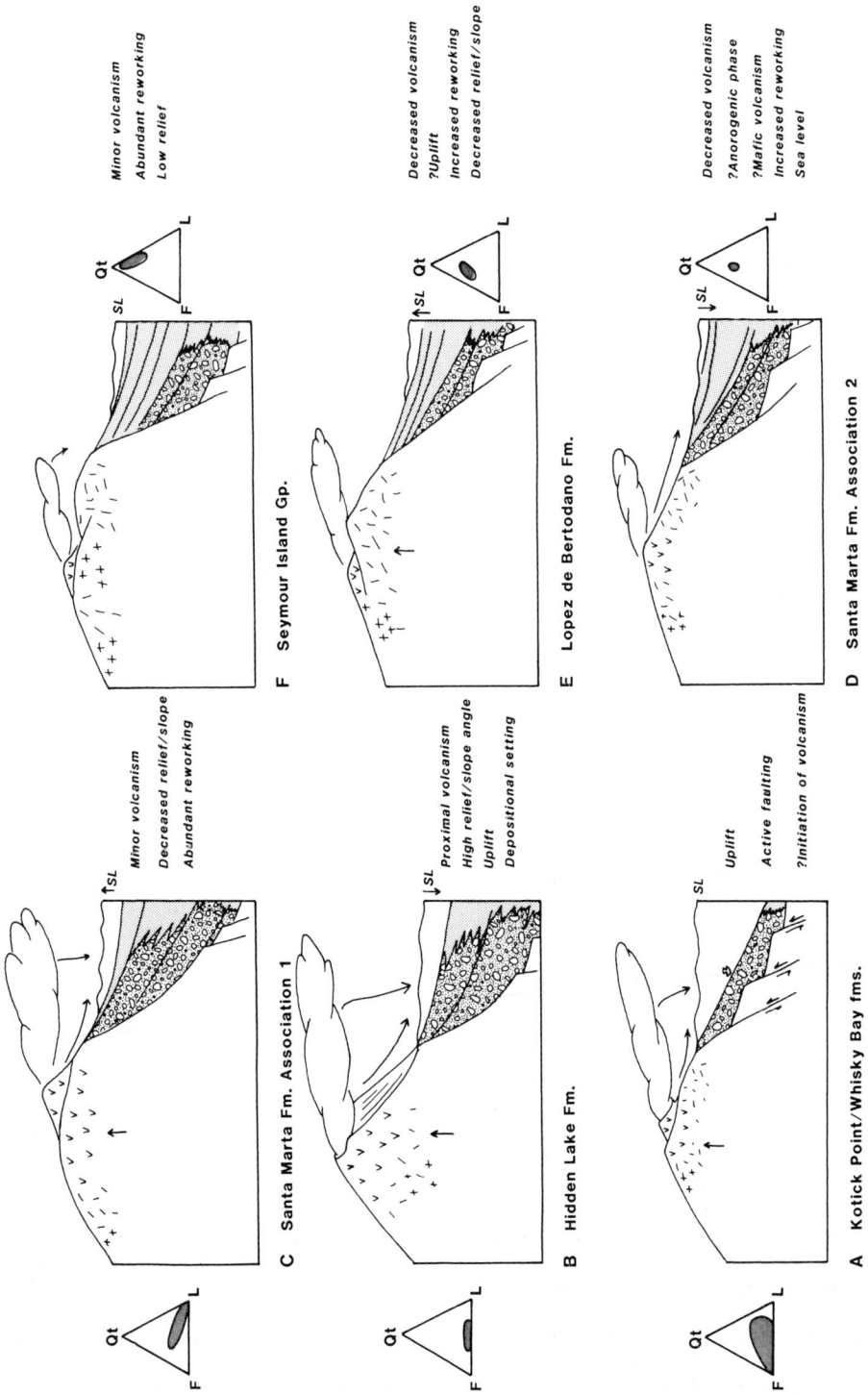

Fig. 11. Summary diagram showing the sandstone modal data and likely controls on sandstone composition for each stratigraphic unit studied.

Arc volcanism and tectonics

Arc tectonics and the location and nature of arc volcanism are major controls on the petrographic evolution of active margin sequences. Volcanism was probably coeval with sedimentation throughout the Larsen Basins history, yet had a varying influence on the petrographic evolution. Within the lower Gustav Group pulses of volcanic derived detritus indicate the widespread onset or renewal of volcanism within this region. This is consistent with a documented phase of arc magmatism in the Early Cretaceous (Pankhurst 1982). Active faulting along the basin margin led to the deposition of Late Jurassic–Early Cretaceous Nordenskjöld Formation clasts within the basin (Ineson 1985). Further evidence for this reworking comes from recycled Late Jurassic–Early Cretaceous palynomorphs within this part of the sequence (J. B. Riding pers. comm. 1989).

The Hidden Lake Formation records a major pulse in arc volcanism, or more proximal arc volcanism, probably associated with arc uplift. The formation is predominantly composed of zoned, euhedral plagioclase crystals (with a plagioclase to total feldspar ratio of about 1), andesitic lithic-volcanics and abundant pumice. The fine-grained facies from this unit are typified by an abundance of ?authigenic zeolite group minerals, associated with the abundance of volcaniclastic detritus. The presence of very fresh clinopyroxene crystals (augite) with high Na_2O and MnO contents and the low CrO contents supports an andesitic volcanic source. In addition, fission track age dating of zircons from this formation supports sediment derivation from coeval volcanism (I. Evans pers. comm. 1989).

Association 1 of the Santa Marta formation records minor coeval arc volcanism with sediment deposition from air-fall processes (Pirrie 1989). However, although the sandstones are dominated by lithic-volcanic clasts, it is likely that much of the sediment was derived from the erosion of older volcanic suites, rather than from coeval volcanism (Zuffa 1985). This is implied by the presence of both relatively unaltered and altered Lv grains within the same samples, and the decreased abundance of zoned, euhedral plagioclase grains. Fission track dating of zircons in sandstones from this sequence has given ages of 109 ± 15 Ma, supporting reworking of earlier volcanics (I. Evans pers. comm. 1989).

The marked petrographic shift seen in Association 2 of the Santa Marta Formation probably reflects a lull in calc-alkaline arc volcanism. However, it may also be related to a change to more mafic volcanism within the arc. Glauconite grains, some of which show relict igneous textures, are abundant within this unit, and may reflect the diagenetic alteration of more mafic volcanic grains (Jeans et al. 1982). Several authors have documented a change to more mafic magmatism in the Late Cretaceous, related to a switch of magmatism from the east to the west coast of the Antarctic Peninsula (Moyes & Storey 1986; Pankhurst et al. 1988). This change in magmatism has been interpreted as a response to an increased angle of subduction (due to the subduction of younger oceanic crust) along the Antarctic Peninsula margin (Pankhurst et al. 1988). The apparent petrographic shift seen in Association 2 of the Santa Marta Formation is probably related to this major tectonic event.

The Lopez de Bertodano Formation was sourced predominantly from the Trinity Peninsula Group, although minor or distal coeval arc volcanism is suggested by the presence of air-fall ashes within the formation (Pirrie & Riding 1988). However, Macellari (1986) recognized a pulse in calc-alkaline arc volcanism in the Late Maastrichtian on Seymour Island, which has also been recognized on Vega Island. A further proximal calc-alkaline volcanic pulse is also recognized within the Cross Valley Formation (Elliot & Trautman 1982).

In summary, although the volcanic arc was active throughout the basin history, the sandstone petrography can be related to the location and nature of arc volcanism. Proximal coeval volcanism deposited fresh plagioclase crystals and lithic-fragments into the basin, whilst distal coeval volcanism sourced air-fall detritus (glass shards, pumice, lapilli). Non-coeval volcanics are dominated by lithic fragments which were reworked into the basin. The volcanic grains were dominantly a product of calc-alkaline volcanism, although a change to more mafic volcanism is possible in the Late Cretaceous.

Source area relief and climate

Source area relief and climate can influence sandstone petrography. Source area relief partially controls the residence time of sediment within the regolith prior to deposition (Mack 1978, 1984). Increased residence time may lead to increased diagenetic alteration of labile grains causing a shift in the final petrography towards more stable components (e.g. quartz). Climate may also influence sandstone petrography (Suttner et al. 1981). Recent work by Mack & Jerzykiewicz (1989) showed that climate (mainly precipitation level) influenced the ratio of plagioclase to lithic rock fragments in andesitic terrains.

Within the source area for the Larsen Basin relief was directly related to tectonic events within the arc. For example, coeval arc-volcanism during the deposition of the Hidden Lake Formation was likely to have been linked to arc uplift. Consequently, the arc would have had high relief, with related decreased sediment residence times. This would have restricted the degree of alteration of labile grains prior to deposition, enhancing the volcaniclastic signature. In contrast, volcanism was less active (or less proximal) during the deposition of the Santa Marta Formation. Consequently relief along the arc was probably lower, increasing the degree of weathering of labile grains prior to their deposition within the basin, and enhancing the more quartz–K-feldspar rich petrographic signal.

Decreased arc relief is likely throughout much of the deposition of the Lopez de Bertodano Formation, allowing extensive weathering and alteration of labile volcaniclastic detritus within the source area. As a result of this, the coarser-grained sandstones are petrographically enriched in quartz and K-feldspar, yet the finer-grained beds retain a volcanic-related signal being composed predominantly of smectite (the weathering product of the andesitic volcanics) along with an increased abundance of chlorite/kaolinite up section.

The possible influence of climatic change during basin history on the petrographic signature, is more difficult to constrain. Palaeoclimate studies of Late Cretaceous–Tertiary fossil wood derived from the arc terrain imply a warm/cool temperate climate, with a possible mean annual precipitation of 1000–2000 mm (Francis 1986). Mack & Jerzykiewicz (1989) showed that sandstones derived from andesitic source terrains in humid climates were enriched in plagioclase in relation to lithic fragments, whilst the opposite is seen in sandstones from arid climates. This simple parameter may be used to infer decreased precipitation during the deposition of Association 1 of the Santa Marta Formation relative to the Hidden Lake Formation, based on the ratio of lithic grains to plagioclase plus lithic grains (see Mack & Jerzykiewicz 1989). However, in this example it is more likely that this petrographic shift is related to waning proximal arc volcanism, rather than changing climate.

Depositional environment, process and rate

Increased sediment reworking and sediment transport will effectively decrease the labile composition of a sandstone in favour of more stable grains (Mack 1978). Consequently the environment, process and rate of sediment deposition will modify the sediment composition. For example, the highly volcaniclastic sandstones of the Hidden Lake Formation on northern James Ross Island were deposited within fan-delta to tidally influenced marine shelf settings (Ineson 1989; A. G. Whitham pers. comm. 1989). Sedimentation was controlled directly by arc volcanism, generating rapid episodic deposition. In addition, the distance between the sediment source and the site of deposition is likely to have been small, with rapid sedimentation rates. Both of these processes would increase the likelihood of the preservation of abundant labile grains. In contrast, Association 2 of the Santa Marta Formation was deposited within an inner shelf setting (Pirrie 1989, 1990). The sandstones are clean and well-sorted, implying considerable reworking within this high energy setting prior to deposition. This would enhance the quartz-rich signal within this part of the sequence by preferentially removing the labile grains by mechanical fragmentation (Mack 1978). In addition, this sequence reflects a basin-wide regression (Pirrie 1989) and a decreased sedimentation rate, increasing the time available for sediment breakdown. The inferred shift of the site of coeval volcanism from the east to the west coast of the Antarctic Peninsula (i.e. away from the basin margin) increases the distance of sediment transport prior to deposition, increasing the loss of labile grains.

Changing source area composition

Elliot & Trautman (1982) and Farquharson (1983) have interpreted the changing petrography within the basin in terms of an arc unroofing trend. However, the sequence typically shows derivation from either the calc-alkaline Antarctic Peninsula Volcanic Group or from the quartzose low-grade meta-sediments of the Trinity Peninsula Group. During phases of proximal coeval arc volcanism the sediment supply is dominated by volcanic material (e.g the Hidden Lake and Cross Valley Formations), whilst during periods of volcanic quiescence, the sediment petrography switches back to the low-grade metasediments of the Trinity Peninsula Group. The true end-member of an arc-unroofing trend (i.e. plutonic-derived grains) are scarce within the basal Marambio Group, only becoming more apparent within the Tertiary La Meseta Formation (Elliot & Trautman 1982). Consequently, rather than representing an arc-unroofing event, the petrographic record shows temporal changes in the dominant sediment source. Similar changes along the entire length of the arc/basin margin are likely. As volcanism migrated north with time in the Antarctic Peninsula region, quartzose sediments may be predicted to occur in the south of the basin prior to quartzose sediments in the north.

The data presented by Farquharson (1983) for more southerly outcrops within the Larsen Basin, can be compared with the petrographic evolution in the James Ross Island area proposed here. The highly feldspathic sandstones of ?Barremian age from Sobral Peninsula probably represent proximal concurrent arc volcanism, with compositional controls comparable to those for the Hidden Lake Formation. The Cape Longing sandstones may be comparable in age to the upper Whisky Bar or lower Hidden Lake Formations, but are petrographically distinct, being considerably more quartzose than the Hidden Lake Formation, although concurrent arc volcanism is shown by the presence of shard-rich air-fall beds. The more quartzose nature of the data may reflect a differing source terrain (with an increased supply from the Trinity Peninsula Group) as predicted. However, differences in point counting procedure, in particular the use of staining to differentiate feldspar from quartz used in this study, but not by Farquharson (1983), may also have affected the modal data. The data from the Late Cretaceous sediments at Pedersen Nunatak (Farquharson 1983) are again more quartzose than similar age sediments from the James Ross Island area. As already discussed this may be related to the earlier northwestward migration of concurrent arc volcanism in this area relative to the James Ross Island area (Pankhurst 1982). The sediments studied by Farquharson (1983) are exposed up to 180 km further south than northern James Ross Island, hence diachronous changes in the sandstone composition along the basin may be expected. The highly quartzose sediments reported by del Valle & Medina (1985) from Robertson Island may also be related to this change.

Conclusions

The sedimentary rocks which infill the Larsen Basin in the James Ross Island area, were de-

rived from the Antarctic Peninsula. The lower Gustav Group (Barremian–Coniacian) shows a wide petrographic scatter; this reflects a period of arc uplift, and possible initiation or renewal of arc-volcanism. The Hidden Lake Formation records a major pulse of proximal calc-alkaline andesitic volcanism in the ?Coniacian–Santonian. Subsequently the Santa Marta Formation (Santonian–Campanian) was sourced from both older and coeval volcanic rocks of the Antarctic Peninsula Volcanic Group, with a switch to more quartzose sedimentation coinciding with a change in depositional setting. The more quartzose sequence reflects increased sediment supply from the Trinity Peninsula Group, probably documenting a marked lull in arc volcanism. This coincides with the shift of arc volcanism away from the basin margin, and a change to more mafic volcanism in the Late Cretaceous. The overlying Lopez de Bertodano Formation is enriched in quartz and K-feldspar relative to the underlying formations. This represents the change from a dominantly active to predominantly quiescent basin margin, with increased sediment supply from the Trinity Peninsula Group. Concurrent arc volcanism was either of decreased intensity or at a greater distance from the site of deposition.

Although the sequence broadly shows a trend, similar to that expected for arc unroofing, changes in the depositional environment, nature and location of arc-volcanism, source area relief, sedimentation rate and arc tectonics all contributed to the final sediment composition. All of these processes are inter-related. In sequences which lack good regional geological control, change in sediment composition from feldspathic to lithic-volcanic to quartzose, should not necessarily be attributed solely to arc unroofing; other possible controls should be considered.

The author is grateful to C. V. Jeans, D. Long and T. Abrahams for assistance with XRD analysis, and to T. Alabaster for microprobe analyses. I. Goddard prepared excellent thin-sections from variable rocks. A. G. Whitham and J. R. Ineson provided rock samples. Comments on earlier drafts of this manuscript by G. Kelling, G. Nichols, S. P. Todd, D. I. M. Macdonald, J. D. Marshall, J. L. Smellie, D. A. V. Stow, T. H. Tranter and A. G. Whitham are all gratefully acknowledged. G. Clarke kindly typed the manuscript.

References

ANDREWS, J. E. 1987. Jurassic clay mineral assemblages and their post-depositional alteration: Upper Great Estuarine Group, Scotland. *Geological Magazine*, **124**, 261–271.

BASU, A. 1985. Reading provenance from detrital quartz. *In*: ZUFFA, G. G. (ed.) *Provenance of arenites*. Reidel Publishing Co, 231–247.

——, YOUNG, S. W., SUTTNER, L. J., JAMES, W. C.

& MACK, G. H. 1975. Re-evaluation of the use of undulatory extinction and polycrystallinity in detrital quartz for provenance interpretation. *Journal of Sedimentary Petrology*, **45**, 873–882.

DEL VALLE, R. A. & MEDINA, F. A. 1985. *Cape Marsh Geology, Robertson Island, Antarctica*. Contribucion del Instituto Antartico Argentino, No. **309**.

DICKINSON, W. R. 1970. Interpreting detrital modes of greywacke and arkose. *Journal of Sedimentary Petrology*, **40**, 695–707.

—— & SUCZEK, C. A. 1979. Plate tectonics and sandstone compositions. *American Association of Petroleum Geologists Bulletin*, **63**, 2164–2184.

——, BEACH, L. S., BRACKENRIDGE, G. R., ERJAVEC, J. L., FERGUSON, R. C., KNEPP, R. A., LINDBERG, F. A. & RYBERG, P. T. 1983. Provenance of North American Phanerozoic sandstones in relation to tectonic setting. *Geological Society of America Bulletin*, **94**, 222–235.

DORSEY, R. J. 1988. Provenance evolution and unroofing history of a modern arc–continental collision: evidence from petrography of Plio-Pleistocene sandstones, eastern Taiwan. *Journal of Sedimentary Petrology*, **58**, 208–218.

ELLIOT, D. H. & TRAUTMAN, T. A. 1982. Lower Tertiary strata on Seymour Island, Antarctic Peninsula. *In*: CRADDOCK, C. (ed.) *Antarctic Geoscience*. University of Wisconsin Press, Madison, Wisconsin, 287–297.

FARQUHARSON, G. W. 1983. Evolution of Late Mesozoic sedimentary basins in the northern Antarctic Peninsula. *In*: OLIVER, R. L., JAMES, P. R. & JAGO, J. B. (eds) *Antarctic Earth Science*. Australian Academy of Science, Canberra and Cambridge University Press, Cambridge, 323–327.

FRANCIS, J. E. 1986. Growth rings in Cretaceous and Tertiary wood from Antarctica and their palaeoclimatic implications. *Palaeontology*, **29**, 655–684.

GARZANTI, E. 1985. The sandstone memory of the evolution of a Triassic volcanic arc in the southern Alps, Italy. *Sedimentology*, **32**, 423–433.

HAMER, R. D. 1983. Petrogenetic aspects of the Jurassic-Early Cretaceous volcanism, northernmost Antarctic Peninsula. *In*: OLIVER, R. L., JAMES, P. R. & JAGO, J. B. (eds) *Antarctic Earth Science*. Australian Academy of Science, Canberra and Cambridge University Press, Cambridge, 338–342.

HOUGHTON, H. F. 1980. Refined technique for staining plagioclase and alkali feldspars in thin section. *Journal of Sedimentary Petrology*, **50**, 629–631.

INESON, J. R. 1985. Submarine glide blocks from the Lower Cretaceous of the Antarctic Peninsula. *Sedimentology*, **32**, 659–670.

—— 1989. Coarse-grained submarine fan and slope apron deposits in a Cretaceous back-arc basin, Antarctica. *Sedimentology*, **36**, 793–819.

——, CRAME, J. A. & THOMSON, M. R. A. 1986. Lithostratigraphy of the Cretaceous strata of west James Ross Island, Antarctica. *Cretaceous Research*, 7, 141–159.

INGERSOLL, R. V. & SUCZEK, C. A. 1979. Petrology and provenance of Neogene sand from Nicobar and Bengal Fans, DSDP sites 211 and 218. *Jour-*

nal of Sedimentary Petrology, **49**, 1217–1228.

——, BULLARD, T. F., FORD, R. L., GRIMM, J. P., PICKLE, J. D., & SARES, S. W. 1984. The effect of grain size on detrital modes: A test of the Gazzi-Dickinson point-counting method. *Journal of Sedimentary Petrology*, **54**, 105–117.

JEANS, C. V. 1978. The origin of the Triassic clay assemblages of Europe with special reference to the Keuper Marl and Rhaetic of parts of England. *Philosophical Transactions of the Royal Society of London*, **A289**, 549–639.

——, MERRIMAN, R. J., MITCHELL, J. G. & BLAND, D. J. 1982. Volcanic clays in the Cretaceous of southern England and northern Ireland. *Clay Minerals*, **17**, 105–156.

LEE, Y. I. & KLEIN, G. DE V. 1986. Diagenesis of sandstones in the back-arc basins of the western Pacific Ocean. *Sedimentology*, **33**, 651–675.

MACDONALD, D. I. M., BARKER, P. F., GARRETT, S. W., INESON, J. R., PIRRIE, D., STOREY, B. C., WHITHAM, A. G., KINGHORN, R. R. F. & MARSHALL, J. E. A. 1988. A preliminary assessment of the hydrocarbon potential of the Larsen Basin, Antarctica. *Marine and Petroleum Geology*, **5**, 34–53.

MACELLARI, C. E. 1986. *Late Campanian–Maastrichtian ammonite fauna from Seymour Island (Antarctic Peninsula)*. Paleontological Society Memoir No. **18**.

—— 1988. Stratigraphy, sedimentology and paleontology of late Cretaceous/Paleocene shelf deltaic sediments of Seymour Island. *In*: FELDMANN, R. M. & WOODBURNE, M. O. (eds) *Geology and Paleontology of Seymour Island*. Geological Society of America Memoir, 169, 25–54.

MACK, G. H. 1978. The survivability of labile light-mineral grains in fluvial, aeolian and littoral marine environments: the Permian Cutler and Cedar Mesa formations, Moab Utah, *Sedimentology*, **25**, 587–604.

—— 1984. Exceptions to the relationship between plate tectonics and sandstone composition. *Journal of Sedimentary Petrology*, **54**, 212–220.

—— & JERZYKIEWICZ, T. 1989. Detrital modes of sand and sandstone derived from andesitic rocks as a palaeoclimatic indicator. *Sedimentary Geology*, **65**, 35–44.

MORRIS, W. R. & BUSBY-SPERA, C. 1987. Relationship between sea level fluctuations and petrologic variation in forearc basin sedimentary rocks of Upper Cretaceous Rosario Formation at San Carlos, Baja California del Norte, Mexico. *American Association of Petroleum Geologists Annual Convention 1987, Abstracts volume.*

MOYES, A. B. & STOREY, B. C. 1986. The geochemistry and tectonic setting of gabbroic rocks in the Scotia Arc region. *British Antarctic Survey Bulletin*, **73**, 51–69.

OLIVERO, E. B., SCASSO, R. A. & RINALDI, C. A. 1986. *Revision of the Marambio Group, James Ross Island, Antarctica*. Instituto Antartico Argentino Contribucion, No. **331**.

PANKHURST, R. J. 1982. Rb-Sr geochronology of Graham Land, Antarctica. *Journal of the Geo-*

logical Society, London, **139**, 701–711.

——, HOLE, M. J. & BROOK, M. 1988. Isotope evidence for the origin of Andean granites. *Transactions of the Royal Society of Edinburgh: Earth Sciences*, **79**, 123–133.

PEZZETTI, T. G. & KRISSEK, L. A. 1986. Re-evaluation of the Eocene La Meseta Formation of Seymour Island, Antarctic Peninsula. *Antarctic Journal of the United States*, **11**, 75.

PETTIJOHN, F. J., POTTER, P. E. & SIEVER, R. 1972. *Sand and sandstone*. Springer-Verlag, New York.

PIRRIE, D. 1987. Orientated calcareous concretions from James Ross Island, Antarctica. *British Antarctic Survey Bulletin*, **75**, 41–50.

—— 1989. Shallow marine sedimentation within an active margin basin, James Ross Island, Antarctica. *Sedimentary Geology*, **63**, 61–82.

—— 1990. A new sedimentological interpretation for part of the Santa Marta Formation, James Ross Island. *Antarctic Science*, **2**, 77–78.

—— & RIDING, J. B. 1988. Sedimentology, palynology and structure of Humps Island, Northern Antarctic Peninsula. *British Antarctic Survey Bulletin*, **80**, 1–19.

SADLER, P. M. 1988. Geometry and stratification of uppermost Cretaceous and Paleogene units on Seymour Island, northern Antarctica Peninsula. *In*: FELDMANN, R. M. & WOODBURNE, M. O. (eds) *Geology and Paleontology of Seymour Island, Antarctic Peninsula*. Geological Society of America Memoir, **169**, 303–320.

SMELLIE, J. L. 1987. Sandstone detrital modes and basinal setting of the TPG, northern Graham Land, Antarctic Peninsula: A preliminary review. *In*: MCKENZIE, G. D. (ed.) *Gondwana Six: structure, tectonics, and geophysics*. Geophysical Monograph 40, American Geophysical Union, Washington, 199–207.

STOREY, B. C. & GARRETT, S. W. 1985. Crustal growth of the Antarctic Peninsula by accretion, magmatism and extension. *Geological Magazine*, **122**, 5–14.

—— & NELL, P. A. R. 1988. Role of strike slip faulting in the tectonic evolution of the Antarctic Peninsula. *Journal of the Geological Society, London*, **145**, 333–337.

SUTTNER, L. J., BASU, A. & MACK, G. H. 1981. Climate and the origin of quartz arenites. *Journal of Sedimentary Petrology*, **51**, 1235–1246.

WHITHAM, A. G. & MARSHALL, J. E. A. 1988. Syndepositional deformation in a Cretaceous succession, James Ross Island, Antarctica. Evidence from vitrinite reflectivity. *Geological Magazine*, **125**, 583–591.

ZINSMEISTER, W. J. 1982. Review of the Upper Cretaceous–Lower Tertiary sequence on Seymour Island, Antarctica. *Journal of the Geological Society, London*, **139**, 779–785.

ZUFFA, G. G. 1985. Optical analyses of arenites: influence of methodology on compositional results. *In*: ZUFFA, G. G. (ed.) *Provenance of arenites*, Reidel Publishing Co, 165–189.

An integrated approach to provenance studies: a case example from the Upper Jurassic of the Central Graben, North Sea

BERNARD HUMPHREYS[1,3] ANDREW C. MORTON[1],
CLAIRE R. HALLSWORTH[1], ROBERT W. GATLIFF[2] & JAMES B. RIDING[1]

[1] *British Geological Survey, Keyworth, Nottingham NG12 5GG, UK*
[2] *British Geological Survey, 19 Grange Terrace, Edinburgh EH9 2LF, UK*
[3] *Present address: Lemigas, P. O. Box 1089/Jkt., Cipulir-Kebayoran Lama, Jakarta 12230, Indonesia*

Abstract: The provenance of Upper Jurassic sandstones in a small half graben in the southern Central Graben of the North Sea has been investigated by studying a range of parameters: the composition of feldspars, the nature of rock fragments and their cathodoluminescence properties, heavy mineral assemblages, clay mineral assemblages, whole rock geochemistry and the occurrence of reworked palynomorphs. The combined result of these investigations is that rocks of Triassic, Permian and Carboniferous age are believed to have supplied detritus to the Upper Jurassic basin and to have diluted the contemporaneous bioclastic deposits. However, each individual provenance indicator tended to emphasize the importance of one particular source. Only an integrated study provided a complete picture of the interplay of different sources during deposition. Such integrated studies also identify the limitations of individual provenance indicators, in particular the adverse role of burial diagenesis in the removal of diagnostic grains such as volcanic feldspars and certain heavy mineral grains.

In the past decade, much of the progress made in provenance research has concentrated on detailed consideration of parameters such as framework grain types (e.g. Dickinson 1985), heavy mineral assemblages (Morton 1985), the compositional variations within one mineral species such as detrital feldspar (e.g. Trevena & Nash 1981) and certain heavy minerals such as garnet (Haughton & Farrow 1989; Morton, this volume), sediment geochemistry (e.g. Floyd & Leveridge 1987), and isotopic dating of detrital clasts and individual grains (e.g. Elders 1987; Haughton 1988). However, most studies have stressed the importance of one provenance indicator in isolation which, while often giving strong indications of source, may not provide the entire picture. To maximize the amount of provenance data from a given basin, and to minimize the adverse effects of diagenesis and regional variations in lithology and grain size, an approach which integrates various methods should be more useful. To this end, we present a case study from the Argyll Embayment (Central Graben, North Sea), in which provenance is considered from study of six variables: feldspar chemistry, the nature and cathodoluminescene properties of rock fragments, heavy mineral assemblages, clay mineral assemblages, whole rock and trace element geochemistry and reworked palynomorphs. The example has been chosen

because the potential source rocks appear to be well constrained, both in terms of location and possible lithologies, and because the potential source rocks could actually be sampled in borehole cores in the study area. The purpose of this paper is to compare the results from the various provenance indicators and to account for any discrepancies that occur. By this approach we hope to draw attention to the limitations and errors of adopting a single parameter approach to provenance studies.

Geological setting

The study area lies within UK Block 31/26 at the southern end of the Central Graben, southeast of the Argyll, Innes and Duncan Fields (Fig. 1). A seismic section across the Block shows a basin, the Argyll Embayment, with a sediment fill that thickens eastwards towards a fault-bounded high, informally called here the Quadrant 31 Ridge (Fig. 2). The Argyll Embayment is connected at its northern end to the main Central Graben, and is bordered by relatively gentle slopes on its southern and western margins.

This study is largely based on the study of core from Well 31/26-4 which was drilled in the deeper part of the Argyll Embayment and proved *c.* 100 m (330 ft.) of Upper Jurassic sedi-

From Morton, A. C., Todd, S. P. & Haughton, P. D. W. (eds), 1991, *Developments in Sedimentary Provenance Studies.* Geological Society Special Publication No. 57, pp. 251–262.

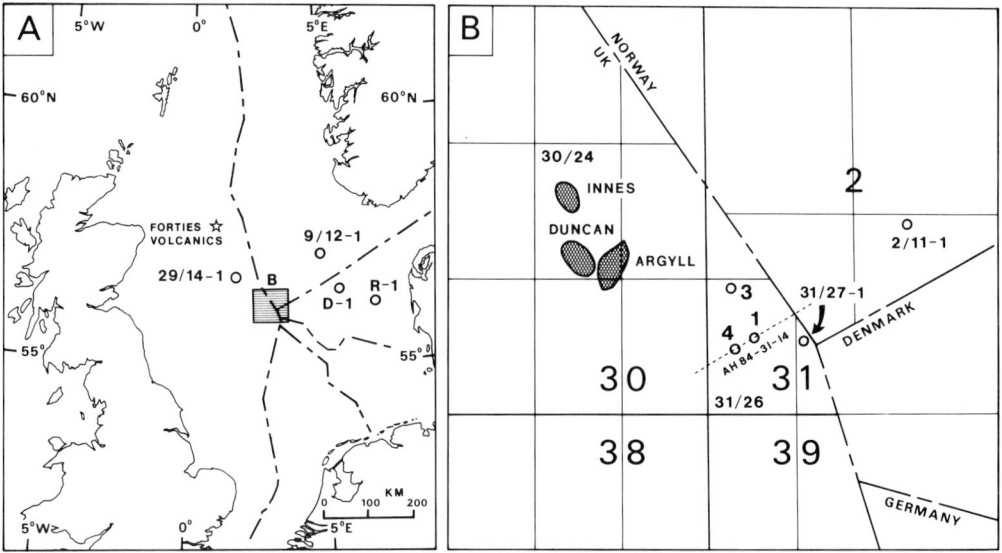

Fig. 1. Map of North Sea (**A**) showing position of the study area (shaded) enlarged in (**B**). The position of Danish Wells D-1 & R-1, Norwegian Wells 2/11-1 & 9/12-1, seismic line AH 84-31-14 and possible volcanic sources (UK Well 29/14-1 and the Forties igneous province) that are mentioned in the text are also shown.

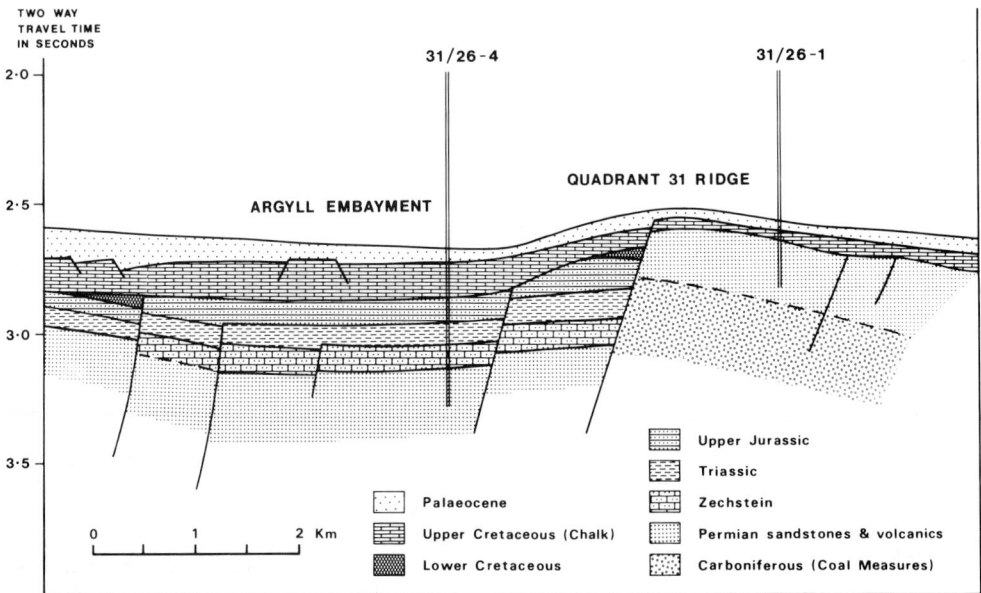

Fig. 2. Geological sketch across the Argyll Embayment and Quadrant 31 Ridge, based on interpretation of seismic line AH 84-31-14, showing the ridge and basin topography and the relative position of Wells 31/26-1 and 31/26-4.

ments overlying Triassic mudstones and silt-stones. A rich assemblage of palynomorphs from the Jurassic interval indicates an early to mid-Volgian age for the sediments. The sediments consist of bioturbated muddly siltstones and very fine sandstones passing up into nodular spiculite beds. They contain abundant cal-careous microfossils and belemnite guards, the latter being conspicuous in slabbed cores. The sediments are clearly marine, and judging by the assemblage of trace fossils which include *Helminthoidea, Chondrites, Teichichnus* and large crustacean-like burrows, were deposited on an inner shelf above storm wave base. They re-semble the bioturbated Upper Jurassic Fulmar Formation sediments described by, amongst others, Johnson *et al.* (1986). Reference will also be made to a thin sequence (*c.* 5 m) of sandstones of indeterminate age, sandwiched between the Triassic and Cretaceous chalk, that was cored in Well 31/26-3 on the flanks of the basin. It is less easy to determine the environment of deposition of these medium- to coarse-grained sediments. They are coarser grained than those from Well 31/26-4, but include sporadic shell debris, phos-phatic fragments and glauconite indicating shal-low marine deposition. Well 31/26-1 drilled on the Quadrant 31 Ridge proved Permian sand-stones (Rotliegendes Group) above intercalated sedimentary and volcanic rocks of early Permian age. These may overlie Carboniferous sediments (Westphalian Coal Measures) at depth; the latter have been drilled on the eastern side of the Quadrant 31 Ridge. As far as can be ascertained from available data, Jurassic and Lower Creta-ceous strata are absent across much of the Quad-rant 31 Ridge. This suggests that, during the Mesozoic, the Quadrant 31 Ridge may have been a site of sediment production and that this sediment may have been locally shed into the Argyll Embayment. Apart from the poten-tial source rocks cored in Well 31/26-1 and the Triassic sediments from Well 31/26-4, the following locally occurring strata have also been sampled for provenance investigation: Rotliegendes Group and Devonian clastics from the vicinity of the Argyll Field (Block 30/24), Mid-Jurassic basalts from Quadrant 29, and Westphalian Coal Measures (sidewall cores) from Well 31/27-1.

Methods

Polished thin sections were prepared for 52 core samples from Wells 31/26-1, -3, and -4. Cathodo-luminescence observations were made with a Techno-syn cold cathode luminescence model 8200 mark 2. The chemistry of the detrital feldspars was determined on selected carbon-coated thin sections using a Cam-bridge Instruments Microscan 5 electron microprobe with a Link Systems model 290 energy-dispersive X-ray analyser.

For clay mineral analysis, samples of both sand-stone and mudstone were first disaggregated by gentle rolling using a mortar and pestle. They were then immersed in distilled water and subjected to intense ultrasonic agitation to ensure that all the constituent grains were separated. Clays in suspension were decanted through a <32 μm wet sieve and then centri-fuged to collect the <2 μm fraction. Clays were pre-pared as oriented smears on glass slides and were run in air-dried, glycerolated and heated states using a Philips PW1130 X-ray diffractometer.

Heavy minerals were separated from ultrasonically cleaned 63–125 μm fractions of disaggregated sand-stones by gravity settling in bromoform stabilised with ethanol (specific gravity 2.80). The resulting separ-ations were mounted in Canada Balsam and 200 non-opaque grains were identified and counted using a petrographic microscope to establish mineral proportions.

Major and trace element analyses were made by X-ray fluorescence techniques at the University of Nottingham, while rare earth element determinations were obtained by neutron activation at the Imperial College Reactor Centre.

Provenance indicators

Feldspar compositions

Compared with the feldspar assemblages of most other North Sea Jurassic sandstones (Humph-reys & Lott 1990), the assemblage from the Upper Jurassic sandstones of Well 31/26-4 is unusually diverse. Calcic plagioclases with com-positions up to An_{54} (labradorite) are common and, together with alkali feldspars with only moderately high potassium contents, fall within the known compositional range of volcanic feldspars (Trevena & Nash 1981; Helmold 1985). The remainder of the assemblage comprises Na-plagioclases with less than *c.* An_{15} (albites) and potassium-rich alkali feldspars, principally orthoclase and microcline, of either plutonic, metamorphic or recycled sedimentary origin (Fig. 3). Of the 107 feldspar grains analysed from this well, 32 were considered to be of volcanic origin. Core samples from Well 31/26-1 on the adjacent Quadrant 31 Ridge revealed that Permian volcanic rocks including both tholeiites and more evolved lithologies with euhedral lath-like crystals of Ca-plagioclase are interbedded with sandstones containing a simple K-feldspar and Na-plagioclase assemblage. The combined Permian feldspar assemblage is virtually ident-ical to that from the Upper Jurassic of Well

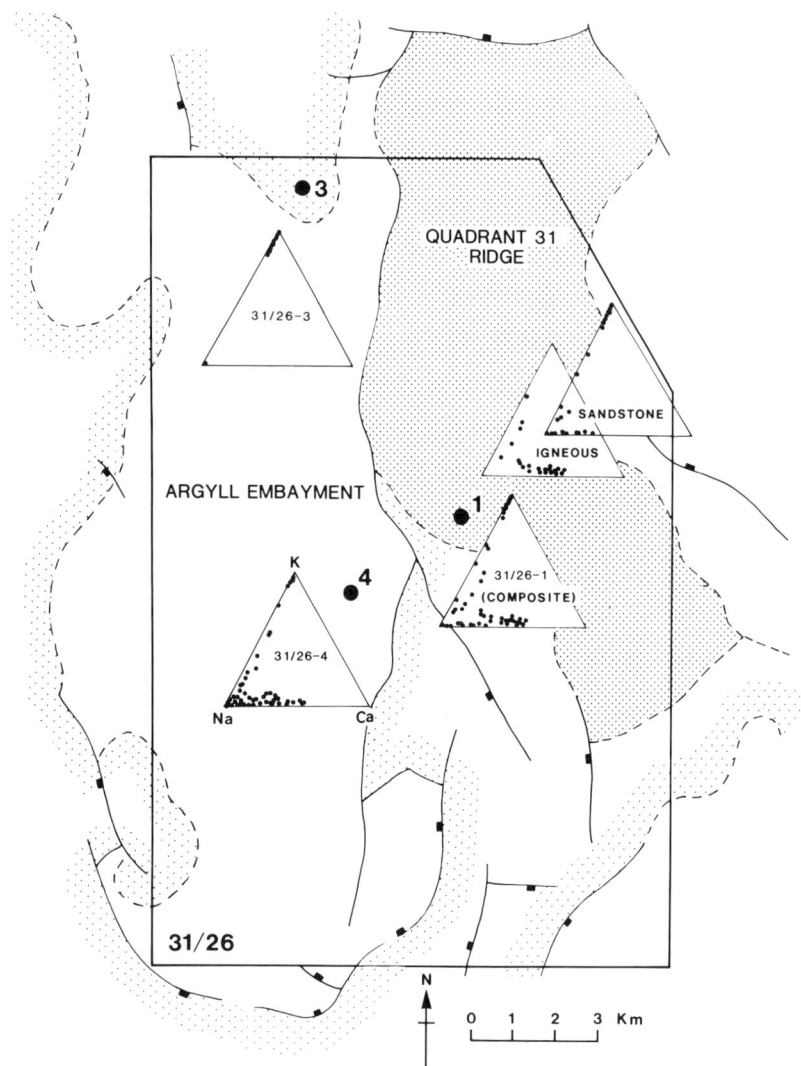

Fig. 3. K–Na–Ca triangular plots of feldspar compositions in possible reservoir sandstones (Wells 3 & 4) and in Permian rocks (Well 1) as determined by electron microprobe analyses. The composition diagram for Well 1 was compiled from data from two different lithologies, Permian basalts and Permian sandstones, separate triangular plots for which are also shown. Light stipple denotes slope or terrace features around the margins of the basin, dense stipple indicates approximate area of the high. Faults are indicated by continuous black lines showing direction of downthrow.

31/26-4 (Fig. 3) suggesting a clear Permian provenance for a component of the Upper Jurassic succession.

By contrast, the sandstones from Well 31/26-3 have a simple K-feldspar plus albite assemblage. These feldspars could have been derived solely from Palaeozoic sandstones on the Quadrant 31 Ridge, but other sources are equally possible. However, the recorded assemblage in this well has been modified by diagenetic processes; severely etched feldspars and oversized pores are seen in thin section, and calcic plagioclases,

which are generally most susceptible to dissolution during burial diagenesis, may have been lost from the assemblage. Provenance interpretations from feldspar compositions are clearly ambiguous for Well 31/26-3.

Rock fragments and cathodoluminescence

Little has been published on the possible use of cathodoluminescence (CL) in provenance studies save for one article by Matter & Ramseyer

(1985). The rocks cored on the Quadrant 31 Ridge show distinctive CL characteristics. The Permian sandstones are notable for their green as well as the usual bright blue luminescent feldspars, whereas the igneous rocks show many distinctive features: fine-grained igneous lavas have a distinct mauve luminescence with vesicles and pink euhedral laths; other volcanic rocks show random meshworks of bright blue luminescent subhedral feldspar laths, and irregularly zoned calcite cements occluding voids left after the dissolution of phenocrysts. There is little evidence that any of this distinctive material has been recycled into the medium- to coarse-grained sandstones of Well 31/26-3. Several volcanic rock fragments with a dark, non-luminescent groundmass and pink luminescent crystal laths do occur and may relate to the volcanic rocks on the Quadrant 31 Ridge. Reworked carbonate clasts with CL zoning patterns and textures atypical of sedimentary calcite (in one case with lath-shaped pseudomorphs) may also have a volcanic affinity. These grains are, however, a minor component compared with other rock fragments in these sandstones. In particular, pink luminescent phosphatic clasts are very common, sometimes with glauconite in borings and/or fractures, together with polycrystalline quartz, cherts and dolostone clasts. The majority of rock fragments appear to be of recycled sedimentary origin.

On the basis of rock fragments and cathodoluminescence a volcanic input to Well 31/26-3 from the Quadrant 31 Ridge is at best tenuous. Unfortunately, in Well 31/26-4, where there is clear evidence from the feldspars of a volcanic input, the detritus is generally too fine grained for a study of rock fragments to be informative. The rock fragments present are mostly chert. However, at the base of the sequence a pebble lag comprising phosphatic pebbles (containing calcitic sponge spicules and sporadic glauconite grains), echinoderm plates and belemnite guards indicates local reworking of sedimentary detritus.

Heavy minerals

A rather restricted heavy mineral assemblage was obtained from Wells 31/26-3 and 31/26-4 consisting of zircon, rutile, tourmaline and apatite, which are stable under burial conditions, together with less stable minerals such as garnet, staurolite and very rare epidote. Despite the evidence of volcanic feldspars in Well 31/26-4, no heavy minerals diagnostic of basaltic volcanism such as clinopyroxene were observed.

Although heavy minerals resulting from basaltic volcanism are notoriously unstable during burial diagenesis, some might have survived along with the Ca-plagioclase feldspars of volcanic derivation; clearly this has not happened. Despite this lack of source-specific heavy minerals, the data are nevertheless useful in that they indicate that the sandstones from the two wells were derived from different sources. This is based on ratios of stable heavy mineral grains with similar densities, rutile : zircon and apatite : tourmaline, which are different between the two wells (Fig. 4). Different proportions of garnet in the two wells may also reflect their different provenance, but could also reflect different intensities of intrastratal solution. The ratios of zircon : rutile and apatite : tourmaline in Well 31/26-3 are similar to those observed in Jurassic sandstones elsewhere in the Central Graben, but those in Well 31/26-4 are unusual for the area. The low proportion of apatite is particularly distinctive, because first cycle detritus from the North Sea borderlands generally has a high apatite : tourmaline ratio. The low apatite : tourmaline ratio suggests a phase (or phases) of recycling, as the ratio can only be lowered by subjecting the suite to subaerial weathering (following Morton 1986). Reworking from a sequence deposited in a terrestrial environment under warm humid conditions promotes apatite dissolution (Mackie 1927). The low apatite : tourmaline ratio in Well 31/26-4 is therefore consistent with reworking of Carboniferous sandstones.

Clay minerals

The detrital clay component of sediments still receives little attention in most provenance studies of clastic sequences, despite a number of interesting case studies in which recycled clays have been successfully used as indicators of source area (e.g. Karlin 1980; Hurst 1982, 1985; Wilmot 1985; Sawyer et al. 1988). However, great care is needed when interpreting clay assemblages because the mineralogy may often have been modified during burial diagenesis. In the mud-rich Upper Jurassic succession from Well 31/26-4 it seems imperative to consider the source of the clays. A vertical profile of clay mineral assemblages over the Jurassic interval shows some significant changes from the stable illite + chlorite assemblage in the underlying Triassic (assemblage A, Fig. 5). The lowermost Upper Jurassic mudstones (assemblage B, Fig. 5) comprise chlorite, discrete illite and ordered mixed-layer illite–smectites, all of which could

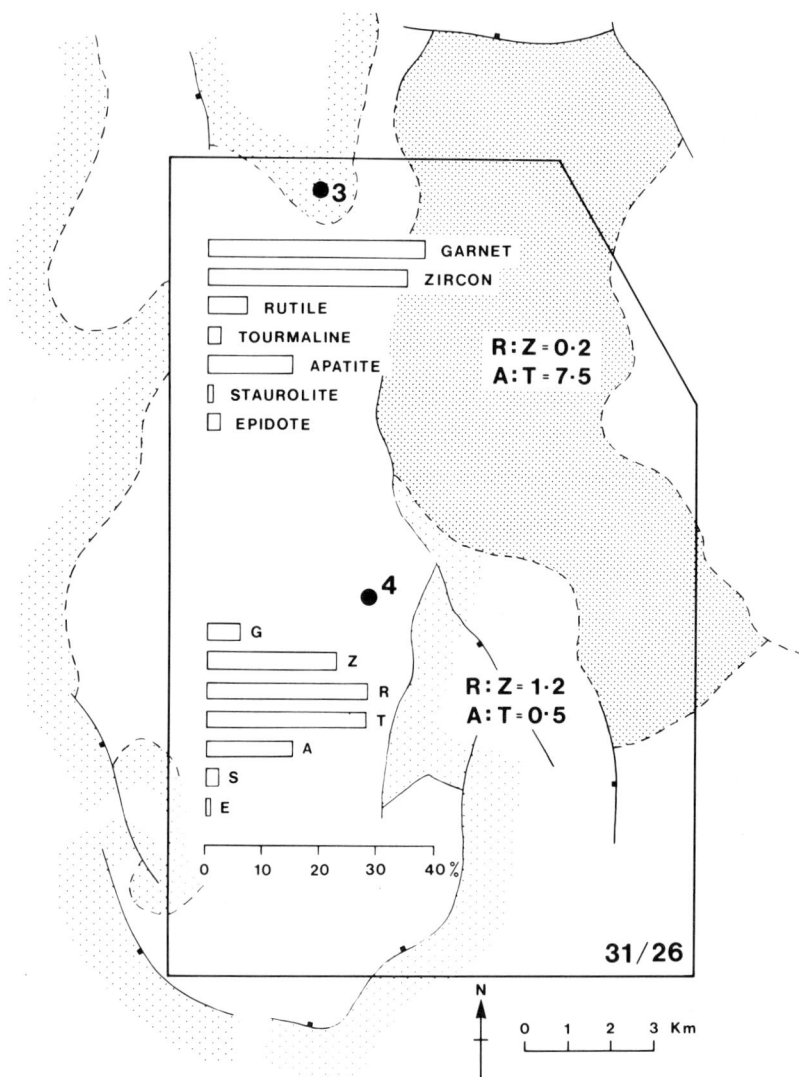

Fig. 4. A summary of heavy mineral data for sandstones from Wells 3 & 4 presented as bar scales showing mean proportions of heavy mineral species. Ratios of heavy mineral species with similar densities, rutile : zircon (R : Z) and apatite : tourmaline (A : T), illustrate clear differences between assemblages in different parts of the Argyll Embayment.

have been locally derived from the underlying Triassic mudstones. There is, however, an upward decrease in the amount of chlorite reflecting perhaps dilution of Triassic detritus by an illite-rich source. Such a common clay mineral is not source specific. The decrease in chlorite, however, also precludes a low-grade metamorphic source for the illite.

Higher in the sequence, assemblage C includes significant proportions of a mixed-layer clay showing the characteristics of randomly interlayered chlorite–smectite (Fisher 1988), namely slight expansion from a basal spacing of 14.4 Å

to 15.7 Å with glycerolation, and collapse to *c.* 12 Å with heating to 600°C. Mixed-layer chlorite–smectites are reported from many different geological environments (Reynolds 1988) including saline lacustrine deposits, from the weathering of minerals such as chlorite (e.g. Jackson 1963; Churchman 1980), as an authigenic clay during diagenesis (Helmold & Van de Kamp 1984; Chang *et al.* 1986), and in association with basalts and volcanogenic material (e.g. Bain & Russell 1981). Assemblage C is considered essentially detrital on thin section evidence because the clay is either grain-support-

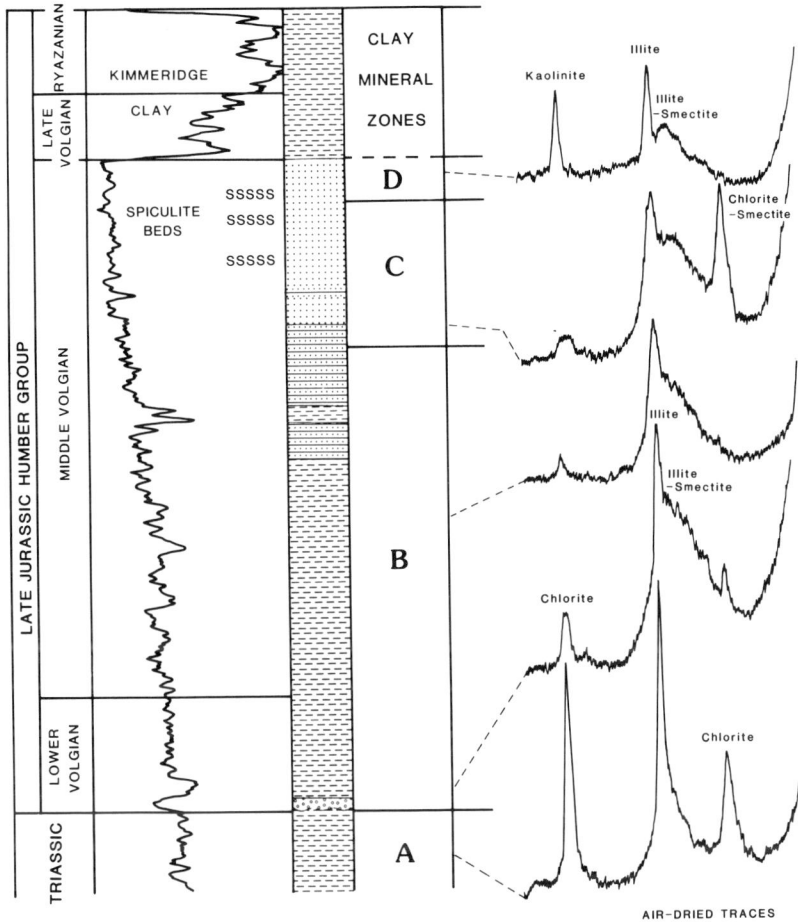

Fig. 5. Clay mineral zones and the occurrence of spiculite beds (SSSSS) in Upper Jurassic siltstones and fine-grained sandstones of Well 31/26-4.

ing or confined to laminae. In view of the geological setting of the Argyll Embayment, the chlorite–smectite was probably sourced either from the Permian basalts during erosion of the Quadrant 31 Ridge, or more likely, from prolonged weathering of chlorite following a waning in the rate of erosion on a Triassic land surface. The gradual flooding of exposed highs by the encroaching sea may in itself have been the main cause of this reduced erosion. The inference of a waning clastic supply from highs is supported here by the marked increase in shell debris moving up the Jurassic sequence and more especially sponge spicules; very fine grained sediments are inimical to sponges.

Towards the top of the arenaceous interval the appearance of detrital kaolinite (assemblage D) may signify the onset of erosion of Carboniferous bedrocks (Hurst 1985), a suggestion supported by the occurrence of reworked Carboniferous spores over the same interval. Kaolinite is recorded in Westphalian mudstones from Wells 31/27-1 and 39/2-1 and kaolinite is a common cement in Westphalian sandstones (e.g. Cowan 1989). An alternative cause of a kaolinite influx could be a climatic change to conditions favouring intense sub-tropical leaching on exposed highs, although persuasive arguments against this idea were forwarded by Hurst (1985). Because intense kaolinization of the source area is likely to deplete heavy mineral assemblages of apatite, the presence of abundant apatite in the heavy mineral separations of Wells 31/26-3 and 31/26-4 also argues against acidic leaching (Morton 1986). Kaolinite is an important detrital clay in the overlying mudstones of the Kimmeridge Clay Formation which, though not cored in 31/26-4, have been studied in the

nearby Norwegian Well 2/11-1 (Bjørlykke *et al.*
1975). Kaolinite appears to be present right
across the Kimmeridgian basin to the onshore
UK succession (Scotchman 1987). Whatever its
precise provenance, the appearance of kaolinite
marks a major change in the lithology of the
source area. This factor does not appear to have
been detected by the other provenance indicators.

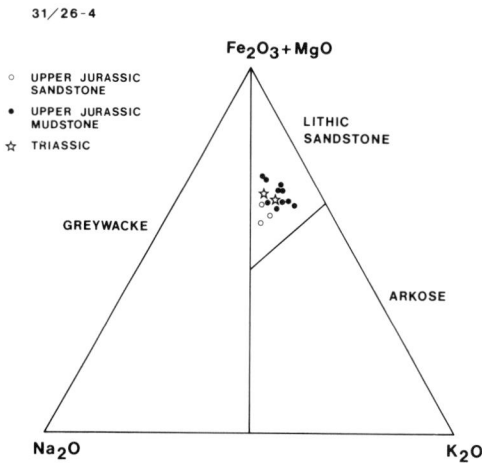

Fig. 6. Major element chemical classification of
Upper Jurassic and Triassic sediments from 31/26-4,
all of which fall within the lithic sandstone field.
Diagram from Blatt *et al.* (1980).

Geochemistry

The advantage of using a geochemical approach
to provenance is that data from sandstones and
shales can be compared, although interpretations must account for lithological variations.
The technique is based on whole rock analysis
and there are inevitable problems in establishing
whether trace elements are concentrated in the
fine fraction or as inclusions in some of the
framework grains. The overall range of compositions from the sandstones and mudstones is
relatively narrow indicating a relatively constant
source or sources. Blatt *et al.* (1980) devised a
scheme for chemical classification of sandstones
using a ternary plot with $Fe_2O_3 + MgO$, Na_2O
and K_2O as poles. All the data from Well
31/26-4, from both Jurassic and Triassic sediments and including both sandstones and mudstones, plot within a tightly constrained area
within the lithic sandstone field (Fig. 6). On the
basis of this major element plot, therefore, there
is no reason to doubt that the Jurassic sediments

could have been derived from underlying Triassic beds. Consideration of trace and rare earth
element data, however, shows that the Upper
Jurassic has on average lower Nb, Zr and Hf and
a higher La/Th ratio than the local Triassic
suggesting incorporation of other source materials (Table 1). The petrographical evidence
favouring a basaltic input is considered by comparing Upper Jurassic compositions with those
of Permian volcanic rocks from Well 31/26-1
and the Danish sector (data from Dixon &
Fitton 1981) and Jurassic volcanic rocks from
two volcanic centres, one sampled in Well 29/14-
1 and the other from the Forties area (Ritchie *et
al.* 1988). The Jurassic material has alkalic affinities and is too enriched in Nb, Zr and Hf to be a
possible contaminant, but the Permian basalts
appear to be tholeiitic with low Nb, Zr and Hf,
and provide suitable source material (Table 1).
This is easier to demonstrate using various plots
which aid discrimination between all potential
sources (Figs 7 & 8). The Nb-Zr plot rules out
the possibility of involvement of mid-Jurassic
volcanics, but pre-Triassic components such as
the Permian Rotliegendes Group, Permian volcanic rocks and Devonian sediments could all be
regarded as possible sources (Fig. 7). A triangular
plot of Hf/3–Th–Ta (Fig. 8) shows data for the
Triassic Smith Bank Formation and some data
points from the Permian Rotliegendes Group
overlapping that from the Upper Jurassic. It is
evident from this plot that dilution of volcanic
detritus has occurred by Triassic Smith Bank
Formation and also possibly by Rotliegendes
Group and Devonian sediments.

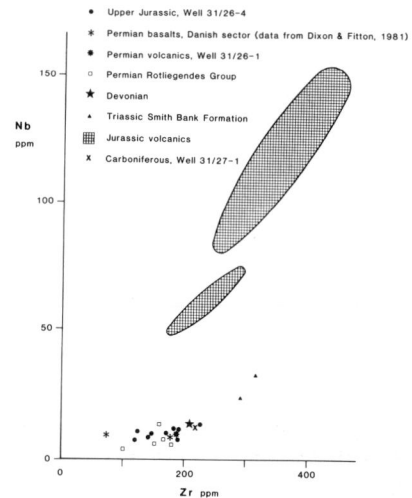

Fig. 7. A Zr–Nb plot (ppm) comparing analysed
sandstones and mudstones from Well 31/26-4
with possible source lithologies.

Table 1. *Major (wt %) trace and rare earth element (ppm) chemical data for representative samples of Upper Jurassic sediments and possible local provenances*

	31/26-4 U. Jurassic Sandstone	31/26-4 U. Jurassic Mudstone	31/26-4 Smith Bank Fm.	Quad. 30/24 Smith Bank Fm.	31/26-1 Permian volcanics	Quad. 30/24 Rotliegendes	Quad. 30/24 Devonian	31/27-1 (side-wall cores) Carboniferous
SiO_2	74.10	59.64	64.86	51.04	47.35	78.22	57.06	72.83
Al_2O_3	4.61	9.50	13.18	13.29	9.52	7.48	22.81	8.08
Fe_2O_3	2.05	4.06	4.97	4.82	5.06	3.60	3.93	4.79
MgO	0.81	1.43	3.75	5.59	3.87	1.54	2.25	1.05
CaO	8.20	8.96	2.02	8.17	13.04	1.49	0.34	0.34
Na_2O	0.75	0.93	1.95	0.93	0.86	0.39	0.62	0.28
K_2O	1.00	1.98	2.84	3.56	3.72	2.60	6.73	0.35
MnO	0.04	0.06	0.05	0.08	0.35	0.01	0.01	0.11
P_2O_5	0.14	0.22	0.17	0.17	0.40	0.08	0.19	0.02
TiO_2	0.46	0.80	0.87	0.79	1.50	0.43	1.07	0.84
LOI	8.06	12.21	5.39	11.65	14.40	3.76	5.03	6.78
Total	100.22	100.14	100.04	100.08	100.07	99.60	100.02	96.87
Ba	397	203	1586	683	974	944	560	5210
Cr	351	225	130	105	271	435	na	434
Cu	12	30	17	14	14	11	4	20
Ga	7	10	15	20	16	9	37	na
Nb	7	12	44	28	43	19	21	15
Ni	31	74	49	65	98	24	46	29
Pb	11	13	8	10	15	21	11	145
Rb	29	67	91	107	71	71	204	17
S	16782	31847	1277	989	636	549	216	3099
Sc	10	14	12	14	20	12	20	19
Sr	175	195	224	187	310	193	326	142
V	45	98	90	105	180	76	132	133
Y	17	24	33	38	33	13	48	23
Zn	12	23	50	na	na	na	na	na
Zr	177	145	292	219	240	172	218	218
La	28.44	38.50	39.44	49.90	33.10	28.10	na	12
Ce	43.44	53.56	98.67	162.60	121.60	104.30	189	20
Nd	24.50	26.04	45.00	34.20	31.80	25.20	na	25
Sm	3.63	4.32	7.39	6.93	7.56	4.83	na	12
Eu	0.76	0.98	1.62	1.57	2.42	1.11	na	na
Tb	0.49	0.50	0.98	0.82	0.91	0.47	na	na
Yb	1.46	1.60	1.60	2.16	1.63	1.26	na	na
Lu	0.21	0.30	0.44	0.37	0.30	0.25	na	na
Ta	0.36	0.78	1.67	1.55	2.62	1.09	na	<2
Hf	4.76	3.68	8.32	6.16	6.55	7.50	na	<3
Th	3.87	6.08	13.49	11.20	5.56	6.27	na	1
U	2.12	2.93	2.95	3.11	1.93	1.62	na	na

Total iron presented as Fe_2O_3. LOI is loss on ignition. Major and trace elements determined by XRF, rare earth elements, Hf, Ta, Th and U by INAA. na, not analysed.

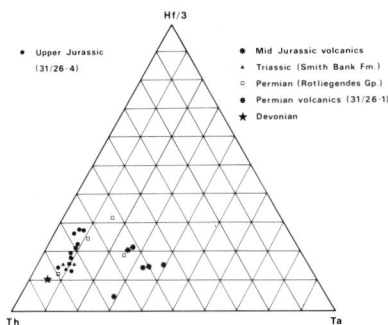

Fig. 8. A triangular plot using hafnium (Hf), thorium (Th) and tantalum (Ta) to discriminate chemically between different provenances.

These plots show that the Upper Jurassic of Well 31/26-4 is geochemically unrelated to Jurassic volcanics, but that Permian volcanic rocks are likely to have contributed to the Upper Jurassic sediment supply. Contamination by Rotliegendes and possibly Devonian rocks cannot be ruled out on geochemical grounds.

Reworked palynomorphs

Together with other clastic detritus, biogenic debris can sometimes be reworked from older sediments. Reworked palynomorphs are particularly useful indicators of sedimentary provenances (e.g. Muir 1967; Windle 1979; Streel & Bless 1980; Guy-Ohlson *et al.* 1987; Eshet *et al.* 1988; Batten, this volume). They can be distinguished from contemporaneous palynomorphs on the basis of factors such as different optical properties, particularly colour (related to their thermal maturation history), and their state of preservation; reworked palynomorphs have often been compressed during a previous burial history and are more likely to show damage due to transport during reworking. A restricted geographical distribution is another feature of reworked assemblages, contrasting with the wide distribution of indigenous species.

The Upper Jurassic of Well 31/26-4 has yielded reworked Carboniferous miospores. These are persistently recorded in the upper *c.* 30.5 m (100 ft.) of spicule-rich Upper Jurassic sands, where the reworked assemblage is dominated by *Densosporites* spp. Reworked Carboniferous spores have not been recorded from the Upper Jurassic succession nearby in the southern Central Graben, although they have been reported from northeast Scotland (Windle 1979), which is suggestive of a relatively localised

source of Carboniferous material. Significant numbers of reworked Carboniferous miospores occur through the cored interval where detrital kaolinite of possible Carboniferous origin was detected.

The underlying Triassic of Well 31/26-4 is barren of palynomorphs which is the case throughout much of the southern Central Graben. Palynology is thus unable to assist in confirming a Triassic provenance.

Intrabasinal components

A discussion of provenance would be incomplete without giving due consideration to reworked biogenic components of the sediments in Well 31/26-4, which consist of siliceous spicules, belemnite guards and a variety of carbonate microfossils. Spiculites are particularly numerous towards the top of the Jurassic succession (Fig. 5) and biogenic grains form between 2% and 41% of the total framework grains in the top *c.* 26 m (150 ft.) interval of Upper Jurassic sediments. This biogenic component formed within the basin and is best considered in terms of sites of sediment production and mechanisms of dispersal rather than source areas. The spicule debris comprises both ovoid or bean-shaped microscleres (rhaxes) with subordinate single axis megascleres (monaxons) which have been liberated following death and disintegration of the skeletal sponge structure (Rützer & MacIntyre 1978). The good preservation of delicate, lanceolate-shaped monaxon spicules indicates that they have suffered little transport and thus originated not far from their present location. Judging by studies of Recent and ancient sponge communities, sponge growth at the time was likely to have been favoured by relatively low rates of sedimentation, the presence of pelagic swells and quiet well aerated waters (Hartman *et al.* 1980). Local sites of biogenic sediment production therefore probably existed on submarine highs that protruded above the basin fill of muddy sediments. Storm and bottom currents would have assisted mechanical disintegration of the sponge structure and local transport of the spicules.

Discussion

The results of this brief study demonstrate the limitations of relying on individual provenance techniques and reveal how much information can be lost by not assimilating data from several different techniques.

In the southern part of the Argyll Embayment, classic heavy mineral analysis failed to detect a volcanic input into the basin in the vicinity of Well 31/26-4, although the feldspar data strongly suggested such an origin. Geochemical evidence allowed assessment of possible different volcanic sources and indicated that other sedimentary sources were also involved. In particular, the geochemistry provided support for a Permian volcanic source on the Quadrant 31 Ridge, but precluded a role for Mid-Jurassic volcanic rocks located to the north and west of the study area. Unlike the other techniques, and possibly contradicting the geochemical evidence of relatively uniform sources with time, the clay mineralogy provided clear evidence of changes in source areas. Late in the Jurassic the Quadrant 31 Ridge may no longer have been a major site of sediment production as evidenced by a gradual increase in the proportion of biogenic components before the area was covered, at least in part, by a very thin cover of mud deposition that is ascribed to the Kimmeridge Clay Formation. The influx of detrital kaolinite in the area may indicate erosion of Carboniferous sediments. Heavy mineral and palynological evidence also implies a significant input from the Carboniferous.

We have fewer data on the Mesozoic sandstones on the northern flank of the Argyll Embayment. The different ratios of stable heavy mineral grains in the sandstones of Well 31/26-3 compared with that of the Upper Jurassic in Well 31/26-4 may reflect a major new source or sources of arenaceous detritus, the provenance of which has yet to be determined. The simple feldspar assemblage in Well 31/26-3 compared with 31/26-4, though modified by diagenesis, can also be interpreted as indicating a different source. The only local source of coarse detritus known to the authors in the study area is Rotliegendes Group sandstones, a contribution from which is not improbable on geochemical grounds. Sedimentary inputs from other sources also occurred supplying phosphatic rock fragments and shell debris.

This study has demonstrated how different techniques may identify different sources and provide contradictory evidence on provenance. It is therefore hoped this study has illustrated some of the complexities of provenance analysis that can be encountered even in a simple basin and ridge setting, in which possible source rocks are reasonably constrained. Interpretations will obviously be more difficult in structurally complex regions and in areas with displaced terraines. We advise caution in the sole use of some classic provenance techniques which clearly have their limitations and are unlikely to detect all possible provenances.

We wish to thank the UK Department of Energy for funding and permitting publication of this work. The views expressed here are not necessarily those of the Department. Amerada Hess and their partners kindly agreed to permit mention of specific wells. We are grateful to Caleb Brett Laboratories and the Imperial College Reactor Centre for undertaking the geochemical anlayses. This paper was improved by the helpful comments of R. Knox, M. Leeder and an anonymous reviewer. This paper is published with the approval of the Director of the British Geological Survey, NERC.

References

BAIN, D. C. & RUSSELL, J. D. 1981. Swelling minerals in a basalt and its weathering products from Morvern, Scotland: II. swelling chlorite. *Clay Minerals*, **16**, 203–212.

BATTEN, D. J. 1991. Reworking of plant microfossils and sedimentary provenance. *This volume.*

BJØRLYKKE, K., DYPVIK, H. & FINSTAD, K. G. 1975. The Kimmeridge shale, its composition and radioactivity. *In: Jurassic Northern North Sea Symposium, Stavanger,* Norsk Petroleumforening *1975,* 12.1–12.20.

BLATT, H., MIDDLETON, G. & MURRAY, R. 1980. *Origin of sedimentary rocks.* Prentice-Hall, New Jersey.

CHANG, H. K., MACKENZIE, F. T & SCHOONMAKER, J. 1986. Comparisons between the diagenesis of dioctahedral and trioctahedral smectite, Brazilian offshore basins. *Clays and Clay Minerals*, **34**, 407–423.

CHURCHMAN, G. J. 1980. Clay minerals formed from micas and chlorites in some New Zealand soils. *Clay Minerals*, **15**, 59–76.

COWAN, G. 1989. Diagenesis of Upper Carboniferous sandstones: southern North Sea basin. *In:* WHATELEY, M. K. G. & PICKERING, K. T. (eds) *Deltas: sites and traps for fossil fuels.* Geological Society, London, Special Publication, **41**, 57–73.

DICKINSON, W. R. 1985. Interpreting provenance relations from detrital modes of sandstones. *In:* ZUFFA, G. G. (ed.) *Provenance of arenites.* D. Reidel, Dordrecht, 333–361.

DIXON, J. E. & FITTON, J. G. 1981. The tectonic significance of post-Carboniferous igneous activity in the North Sea basin. *In:* ILLING, L. V. & HOBSON, G. D. (eds) *Petroleum geology of the continental shelf of North-West Europe.* Heyden, London, 121–137.

ELDERS, C. F. 1987. The provenance of granite boulders in conglomerates of the northern and central belts of the Southern Uplands of Scotland.

Journal of the Geological Society, London, **144**, 853–863.

ESHET, Y., DRUCKMAN, Y., COUSMINER, H. L., HABIB, D. & DRUGG, W. S. 1988. Reworked palynomorphs and their use in determination of sedimentary cycles. *Geology,* **16**, 662–665.

FISHER, R. S. 1988. Clay minerals in evaporite host rocks, Palo Duro basin, Texas Panhandle. *Journal of Sedimentary Petrology,* **58**, 836–844.

FLOYD, P. A. & LEVERIDGE, B. E. 1987. Tectonic environment of the Devonian Gramscatho basin, south Cornwall: framework mode and geochemical evidence from turbiditic sandstones. *Journal of the Geological Society, London,* **144**, 531–542.

GUY-OHLSON, D., LINDQVIST, B. & NORLING, E. 1987. Reworked Carboniferous spores in Swedish Mesozoic sediments. *Geologiska Foreningens i Stockholm Forhandlingar,* **109**, 295–306.

HARTMAN, W. D., WENDT, J. W. & WIEDENMAYER, F. 1980. Living and fossil sponges (notes for a short course). GINSBURG, R. N. & REID, P. (Compilers) *Sedimenta VIII.* Comparative Sedimentology Laboratory, University of Miami, Florida.

HAUGHTON, P. D. W. 1988. A cryptic Caledonian flysch terrane in Scotland. *Journal of the Geological Society, London,* **145**, 685–703.

—— & FARROW, C. M. 1989. Compositional variation in Lower Old Red Sandstone detrital garnets from the Midland Valley of Scotland and the Anglo-Welsh Basin. *Geological Magazine,* **126**, 373–396.

HELMOLD, K. P. 1985. Provenance of feldspathic sandstones—the effect of diagenesis on provenance interpretations: a review. *In:* ZUFFA, G. G. (ed.) *Provenance of arenites.* D. Reidel, Dordrecht, 139–163.

—— & VAN DE KAMP, P. C. 1984. Diagenetic mineralogy and controls on albitization and laumontite formation in Paleogene arkoses, Santa Ynez Mountains, California. *In:* MCDONALD, D. A. & SURDAM, D. C. (eds) *Clastic diagenesis.* American Association of Petroleum Geologists Memoir, **37**, 239–276.

HUMPHREYS, B. & LOTT, G. K. 1990. An investigation into nuclear log responses of North Sea Jurassic sandstones using mineralogical analysis. *In:* HURST, A., LOVELL, M. A. & MORTON, A. C. (eds) *Geological Applications of Wireline Logs.* Geological Society, London, Special Publication, **48**, 223–240.

HURST, A. 1982. The clay mineralogy of Jurassic shales from Brora, NE Scotland. *In:* VAN OLPHEN, H. & VENIALE, F. (eds) *Developments in sedimentology,* International Clay Conference 1981, **35**, 677–684.

—— 1985. The implications of clay mineralogy to palaeoclimate and provenance during the Jurassic in NE Scotland. *Scottish Journal of Geology,* **21**, 143–160.

JACKSON, M. L. 1963. Interlaying of expansible layer silicates in soils by chemical weathering. *Clays and Clay Minerals,* **11**, 29–46.

JOHNSON, H. D., MACKAY, T. A. & STEWART, D. J. 1986. The Fulmar Oil-field (Central North Sea): geological aspects of its discovery, appraisal and development. *Marine and Petroleum Geology,* **3**, 99–125.

KARLIN, R. 1980. Sediment sources and clay mineral distributions off the Oregon coast. *Journal of Sedimentary Petrology,* **50**, 543–560.

MACKIE, W. 1927. The apatites in sedimentary rocks as indicators of the amount of atmospheric carbonic acid in the periods of deposit. *Transactions of the Geological Society of Glasgow,* **17**, 407–421.

MATTER, A. & RAMSEYER, K. 1985. Cathodoluminescence microscopy as a tool for provenance studies of sandstones. *In:* ZUFFA, G. G. (ed.). *Provenance of arenites.* D. Reidel, Dordrecht, 191–211.

MORTON, A. C. 1985. Heavy minerals in provenance studies. *In:* ZUFFA, G. G. (ed.). *Provenance of arenites.* D. Reidel, Dordrecht, 249–277.

—— 1986. Dissolution of apatite in North Sea Jurassic sandstones: implications for the generation of secondary porosity. *Clay Minerals,* **21**, 711–733.

—— 1991. Geochemical studies of detrital heavy minerals and their application to provenance research. *This volume.*

MUIR, M. D. 1967. Reworking in Jurassic and Cretaceous spore assemblages. *Review of Palaeobotany and Palynology,* **5**, 145–154.

REYNOLDS, R. C. JR. 1988. Mixed layer chlorite minerals. *In:* BAILEY, S. W. (ed.) *Hydrous phyllosilicates.* Mineralogical Society of America Reviews in Mineralogy, **19**, 599–629.

RITCHIE, J. D., SWALLOW, J. L., MITCHELL, J. G. & MORTON, A. C. 1988. Jurassic ages from intrusives and extrusives within the Forties igneous province. *Scottish Journal of Geology,* **24**, 81–88.

RÜTZLER, K. & MACINTYRE, I. G. 1978. Siliceous sponge spicules in coral reef sediments. *Marine Biology,* **49**, 147–159.

SAWYER, R. K., WIELAND, C. C. & GRIFFIN, G. M. 1988. The clay mineralogy of calcitic seat earth in the northern Everglades of Florida. *Journal of Sedimentary Petrology,* **58**, 81–88.

SCOTCHMAN, I. C. 1987. Clay diagenesis in the Kimmeridge Clay Formation, onshore UK, and its relation to organic maturation. *Mineralogical Magazine,* **51**, 535–551.

STREEL, M. & BLESS, M. J. M. 1980. Occurrence and significance of reworked palynomorphs. *Mededeelingen Rijks Geologische Dienst Nieuwe Serie,* **32**, 69–80.

TAYLOR, S. R. & MCLENNAN, S. M. 1981. The composition and evolution of the continental crust: rare earth element evidence from sedimentary rocks. *Philosophical Transactions of the Royal Society, London,* **A301**, 381–399.

TREVENA, A. S. & NASH, W. P. 1981. An electron microprobe study of detrital feldspar. *Journal of Sedimentary Petrology,* **51**, 137–150.

WILMOT, R. D. 1985. Mineralogical evidence for sediment derivation and ice movement within the Wash drainage basin, eastern England. *Clay Minerals,* **20**, 209–220.

WINDLE, T. M. F. 1979. Reworked Carboniferous spores: an example from the Lower Jurassic of northeast Scotland. *Review of Palaeobotany and Palynology,* **27**, 173–184.

Petrographic evidence of different provenance in two alluvial fan systems (Palaeogene of the northern Tajo Basin, Spain)

JOSÉ ARRIBAS & M. EUGENIA ARRIBAS

*Departamento de Petrología y Geoquímica, Universidad Complutense de Madrid,
28040 Madrid, Spain*

Abstract: Palaeogene detrital deposits of the northern Tajo Basin are coalescent alluvial fan systems interfingering distally with lacustrine carbonates. Non-carbonate extrabasinal clasts increase to the east while carbonate extrabasinal clasts decrease. Rock fragments increase to the west, while the feldspar/quartz ratio remains constant. Rock fragments define two sedimentary domains: the Iberian, in the east, was derived from Mesozoic rocks of the Iberian Range, and the Central System, to the west, was derived from Cretaceous cover and Palaeozoic metamorphic basement. Evolution of sandstone composition is related to erosion of the source areas and is different in the two domains. The tectonic setting is apparently 'recycled orogen', providing calcareous rock fragments are included in the total lithic clasts.

The Tajo Basin is located in central Spain, and was filled during the Tertiary by continental sediments (carbonate, evaporites and terrigenous). This basin is limited at the NW edge by the Central System, and at the NE edge by the Iberian Range. The Central System is a large exposure of Hercynian granites hosted by low to high-rank metamorphic rocks. The Iberian Range is a mountain belt with double vergence and developed from a depositional trough of the aulacogen type filled with Mesozoic deposits within the Iberian plate (Alvaro *et al.* 1979). Palaeogene deposits appear scattered along the border of the Tajo Basin. The northern Palaeogene outcrops are nearest to the area of interaction between the Iberian Range and the Central System (Fig. 1).

The base of the studied Palaeogene succession is apparently conformable over a Palaeogene evaporite unit, and the top is partially covered and eroded. The lower part of the Palaeogene succession contains a faunal association of macro- and micro-mammals indicating a Headonian age (Arribas *et al.* 1983). Within the succession two lithological units (Carbonate Unit and Detrital Unit) are differentiated (Arribas 1986) (Fig. 2). The Carbonate Unit, with a thickness between 200 m and 500 m, was formed in a variety of carbonate facies within a lacustrine–paludal environment. The Detrital Unit, which grades into the Carbonate Unit contains several detrital facies (lobes, channels, sheets and massive lutites) related to prograding alluvial fans (Arribas *et al.* 1983). The thickness of the Detrital Unit varies between 200 m and 340 m.

Thus, the Palaeogene succession reflects evolution from a lacustrine carbonate environment (Carbonate Unit) to a prograding alluvial fan environment (Detrital Unit).

The Palaeogene succession is a synorogenic unit, formed during the build-up of the Alpine chains in a compressive phase that formed the mountain belts of the Iberian Range and Central System. The synorogenic nature of the Palaeogene deposits is documented by the prograding alluvial fan system and the occurrence of important progressive unconformities.

The aim of this paper is to document the sandstone composition of the Palaeogene succession, and to analyse the role of Central System and Iberian Range as source areas during Palaeogene sedimentation.

Methods

Thirty-nine petrographic thin sections from five stratigraphic sections (Beleña de Sorbe, Membrillera, Torremocha de Jadraque, Negredo and Baides) have been analysed (Fig. 1). The selection of these samples has been made from a representative sampling in each stratigraphic section, sampling sandstones corresponding to the grain size interval 3–0ϕ. In each thin section a modal analysis of 300 points has been made using the petrographic groups defined by Zuffa (1980). Thus, it is possible to treat the data according to both the 'traditional' (or 'Indiana School' in Ingersoll *et al.* 1984) and Gazzi-Dickinson methods. Thin sections have been stained with sodium cobaltinitrite and alizarin-red-s solutions for feldspar and carbonate identifications, respectively.

From Morton, A. C., Todd, S. P. & Haughton, P. D. W. (eds), 1991,
Developments in Sedimentary Provenance Studies.
Geological Society Special Publication No. 57, pp. 263–271.

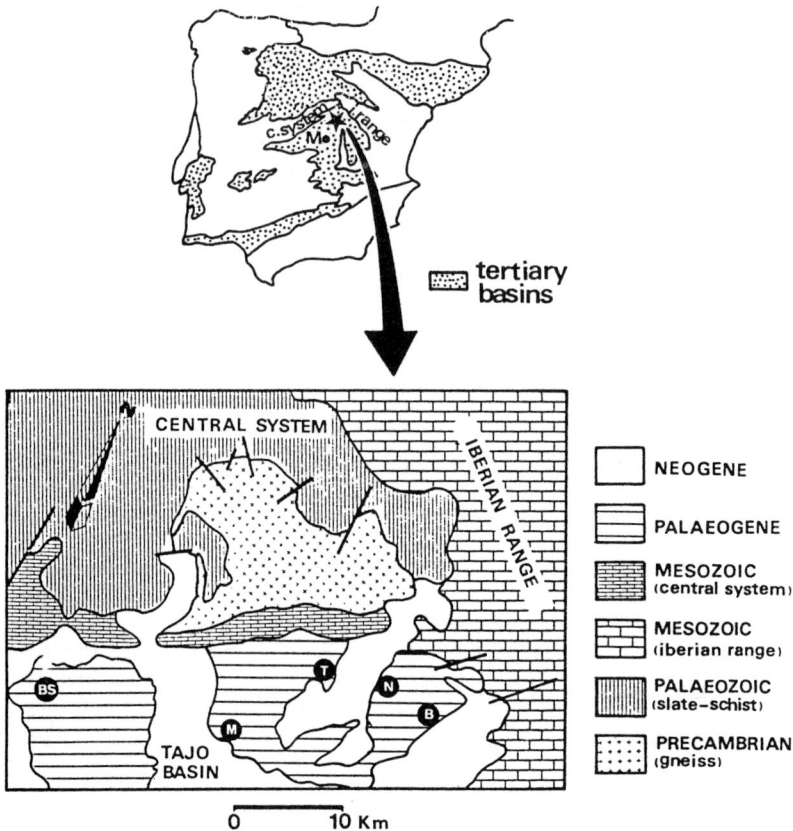

Fig. 1. Geological setting and location map of the stratigraphic sections. BS, Beleña de Sorbe section; M, Membrillera section; T, Torremocha de Jadraque section; N, Negredo section; B, Baides section.

Textures and components of arenites

Detrital arenites are fine- to very coarse-grained and are well to moderately-well sorted (So = 1.2–2.0). Two main components form the arenite framework: siliciclastic and carbonate grains. Carbonate grains are coarser than siliciclastic particles. Roundness is also controlled by clast compositions; carbonate grains are well rounded (Powers 1953) while quartz grains are subangular–subrounded.

Compaction has produced some pressure solution contacts between clasts of different composition (quartz–carbonate grains) and mechanical deformation of labile grains (intraclasts) generating a micritic pseudomatrix.

Generally, the arenite framework is grain supported. However, the tops of some channel-fill sequences are composed of matrix-supported sandstones with high contents of micritic matrix. The origin of this micritic matrix is related to palaeosols and appears to be associated with intrabasinal carbonate grains (pedogenetic intraclasts; Arribas 1986).

The cement is sparry pore-filling calcite with a mosaic texture or as overgrowths around monocrystalline calcite grains.

Sand grains have been divided into the four groups defined by Zuffa (1980): (1) non-carbonate extrabasinal (NCE), (2) carbonate extrabasinal (CE), (3) non-carbonate intrabasinal (NCI) and (4) carbonate intrabasinal (CI).

Non-carbonate extrabasinal grains (NCE)

Four categories have been distinguished: quartz, feldspar, metamorphic rock fragments, and micas and other minerals. Quartz grains are mainly monocrystalline (more than 80% of total quartz) with non-undulatory extinction. Monocrystalline quartz grains with rounded or irregular overgrowths are common, demonstrating a second-cycle origin from previous sandstones. Quartz grains with evaporite mineral inclusions are common. Plagioclase is absent, and K-feldspar occurs as orthoclase and microcline. Some single K-feldspar grains have inherited overgrowths (Fig. 3a). The content of feldspar is low, and never exceeds 15% of the total framework grains. Metamorphic rock fragments include mica-schist, slate and meta-arkose (Fig. 3b). Phyllosilicates (biotite, muscovite and chlorite) are as accessory components (less than 1%), and glauconite, phosphate and heavy minerals (tourmaline, zircon and titanite) have also been observed.

Carbonate extrabasinal grains (CE)

This group is represented by limestone and dolostone fragments. Limestone fragments are generally micritic showing a wide variety of microfacies (mudstones with equinoids, pelmicrites, biosparites, etc.) (Fig. 3c). Dolostone fragments are coarsely crystalline (dolosparites), sometimes partially dedolomitized. Dolomicritic grains are also present but in low percentages (Fig. 3d). Other CE grains include recrystallized bioclasts (mainly molluscs and echinoids).

Non-carbonate intrabasinal grains (NCI)

These clasts are very scarce and consist of silty-clayey grains, larger in size than associated extrabasinal siliciclastic particles. They are commonly squeezed between other clasts to form pseudomatrix.

Carbonate intrabasinal grains (CI)

Poorly lithified intraclasts and micritic grains have been distinguished (Fig. 3e). These grains coexist with carbonate extrabasinal grains. The main characteristics used to discriminate between both types of grains have been grain size, roundness and induration (Zuffa 1980, 1985). Reworked crystal-aggregates of *Microcodium* have been observed in the sand fraction. Because *Microcodium* develops in palaeosols, they are interpreted as intrabasinal. Also, some ostracode fragments appear as carbonate intrabasinal clasts (Fig. 3f).

Palaeogene arenites have a variable composition according to the CE–NCE–CI diagram (Zuffa 1980) (Fig. 4). Some have an important CE content, approaching 'calclithite' compo-

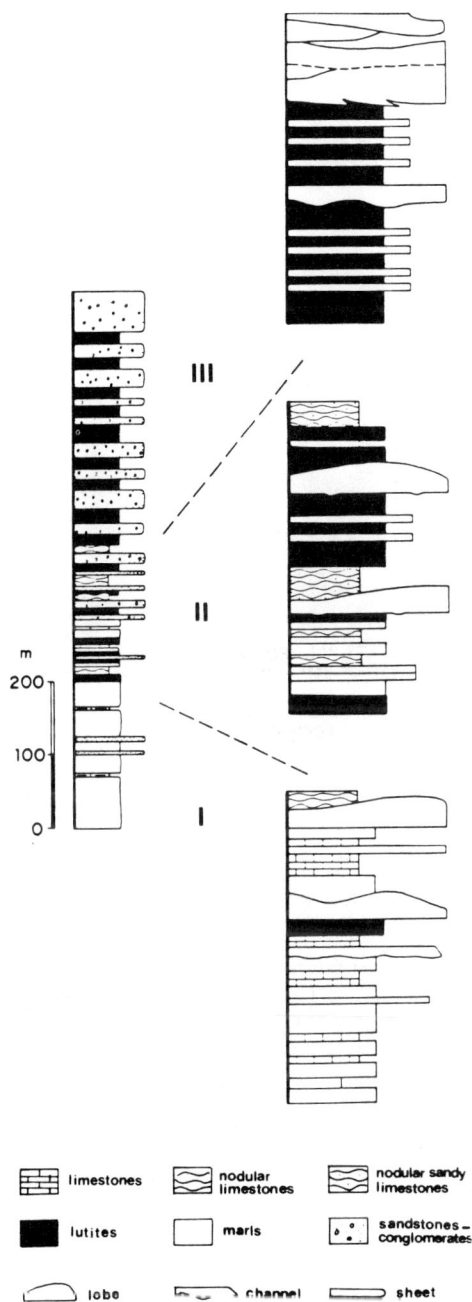

Fig. 2. The representative vertical profile of the Palaeogene succession. I, the Lower Carbonate Unit (lacustrine sediments); II, the Upper Carbonate Unit (lacustrine, paludal and alluvial fan sediments); III, the Detrital Unit (alluvial fan sediments).

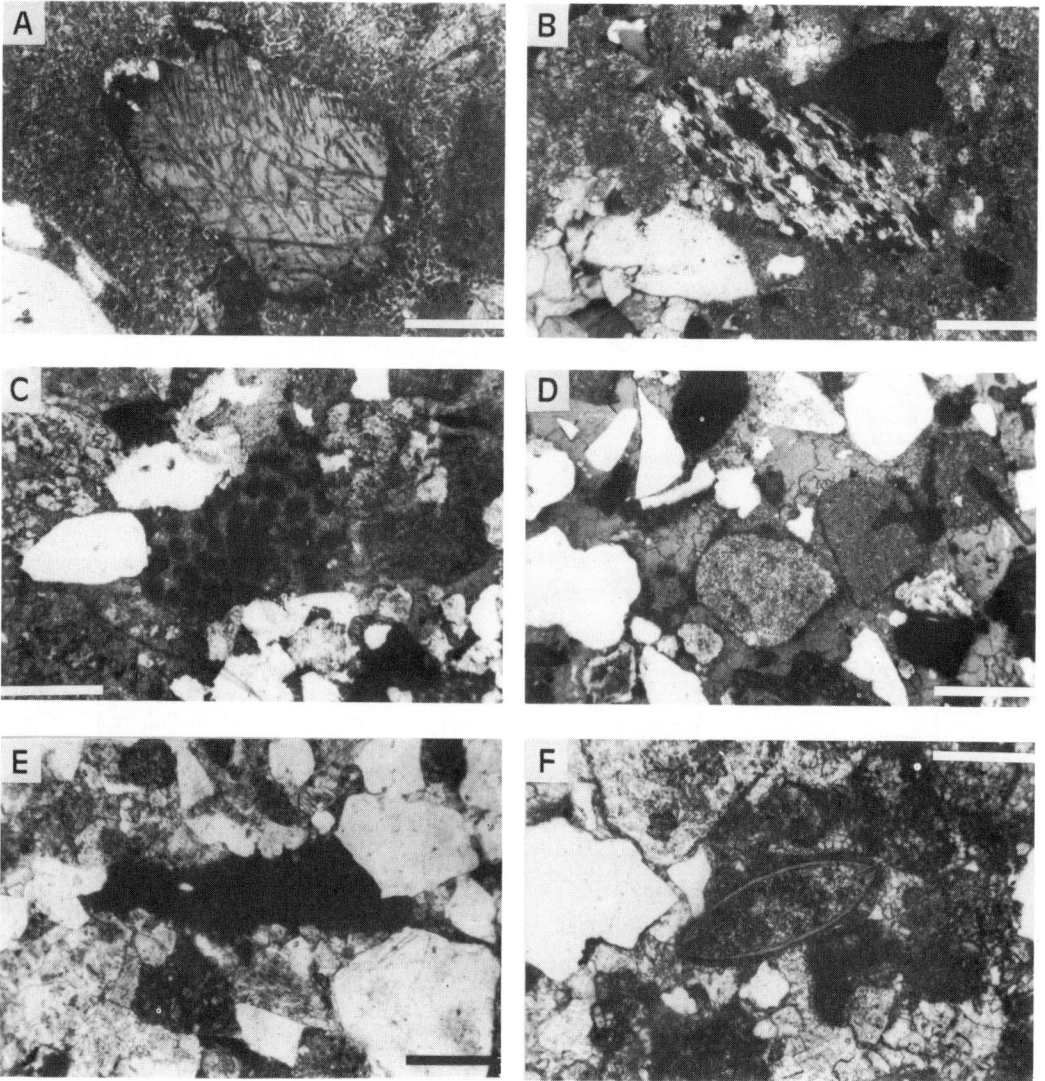

Fig. 3. Detrital components of Palaeogene sandstones. (**A**) K-feldspar grain with inherited overgrowth. Crossed polars. Scale bar, 0.2 mm. (**B**) Metamorphic rock fragment (mica-schist). Crossed polars. Scale bar, 0.5 mm. (**C**) Limestone fragment showing a pelloidal microfacies. Crossed polars. Scale bar, 0.5 mm. (**D**) Extrabasinal dolomitic grains. Plane light. Scale bar, 0.5 mm. (**E**) Intrabasinal micritic grain showing deformation by mechanical compaction. Plane light. Scale bar, 0.5 mm. (**F**) Intrabasinal ostracode grain. Plane light. Scale bar, 0.2 m.

sitions (Folk 1959), whereas others are predominantly of NCE type (sandstones *sensu stricto.*). Finally hybrid arenites are also present, with similar CE and NCE amounts and low percentages of CI.

According to Pettijohn *et al.* (1973), the classification of Palaeogene arenites into classic sandstone types is only feasible if only extrabasinal (CE and NCE) grains are considered (Fig. 5). Thus the terrigenous framework of the Palaeogene arenites is litharenite and sublitharenite. However, this is only valid if all CE grains are plotted at the R pole on the QFR diagram. Note that some CE grains are monocrystalline or fossil fragments, but are here considered rock fragments.

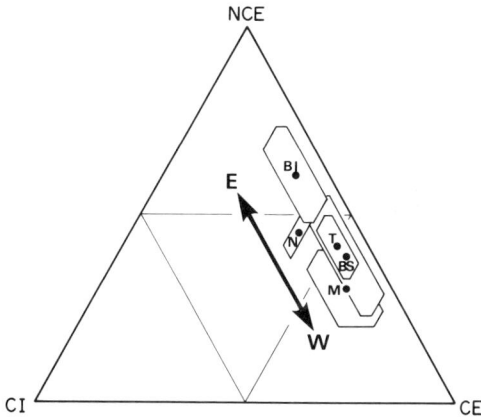

Fig. 4. Ternary plot of mean (point) and standard deviation (hexagons) values of Palaeogene sandstones, according to the criteria of Zuffa (1980). NCE: non-carbonate extrabasinal grains. CE: carbonate extrabasinal grains. CI: carbonate intrabasinal grains. BS, Beleña de Sorbe section; M, Membrillera section; T, Torremocha de Jadraque section; N, Negredo section; BI, Baides section.

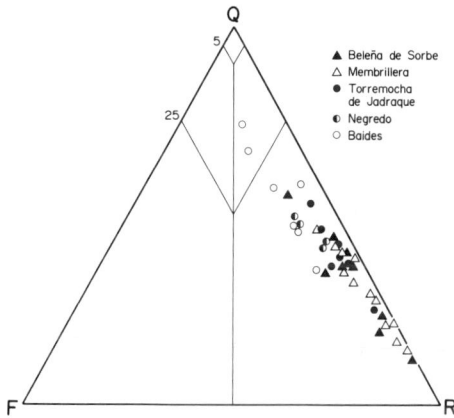

Fig. 5. Composition of terrigenous modes of Palaeogene sandstones in a QFR diagram (Pettijohn *et al.* 1973). Q, quartz grains; F, feldspar grains; R, rock fragments.

Provenance results

Petrographic parameters indicate that the Palaeogene sandstones are mainly sedimentoclastic (Ingersoll 1983). Most of the lithic fragments are sedimentary (carbonates) (Fig. 6); the presence of abraded quartz and feldspar overgrowths indicates they are second-cycle. However, a small proportion of metamorphic rock

fragments is also present, implying input of grains from epicrustal rocks. Sandstone composition varies laterally between different outcrop areas, as well as changing through time (up-section), in each area.

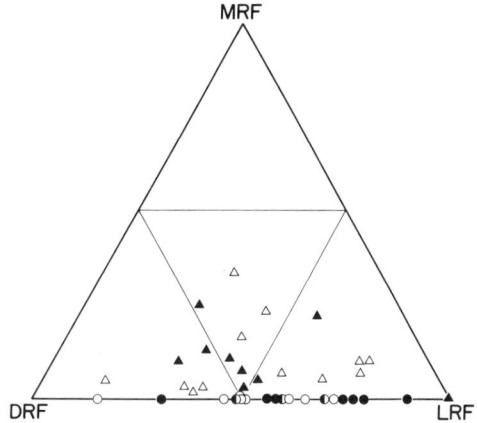

Fig. 6. Composition of total rock fragments in the Palaeogene sandstones. MRF, metamorphic rock fragments; DRF, dolostone rock fragments; LRF, limestone rock fragments. Legend as for Fig. 5.

Sandstone compositon and geographic distribution

The NCE–CE–CI diagram (Fig. 4) displays a diminution of NCE grains (siliciclastic) from east to west. This relates to the lithology of the source areas. The eastern Iberian Range has a greater potential to produce siliciclastic deposits from Permo-Triassic and Cretaceous sandstone formations compared with the western Cretaceous Central System cover. CI contents are similar in all studied areas, and are associated with cannibalistic erosion of interbedded paludal–lacustrine deposits.

Differences in sandstone compositions can also be observed on the QFR diagram (Fig. 5). The Palaeogene sandstones from the west (Beleña de Sorbe and Membrillera sections) are rich in rock fragments (Q_{25}-F_5-R_{70}), evolving to Q_{40}-F_{10}-R_{50} (Torremocha de Jadraque and Negredo) and to Q_{60}-F_{15}-R_{25} (Baides) toward the east. Thus, the Iberian Range extrabasinal contribution was mainly quartz and feldspar, while Central System provided more rock fragments. The Q/F ratio is very similar in all sandstones, and can be defined as a linear equation: $Q = 5F + 9$. This demonstrates that the quartz and the feldspar source must be similar. Siliciclastic Cretaceous formations (e.g. Arenas de

Utrillas Formation) found principally in the Iberian Range, are considered to be the most probable source of quartz and feldspar.

With regard to rock fragment composition, the relation between the three major categories (limestones, dolostones and metamorphic rock fragments) in Palaeogene sandstones has been represented in Fig. 6. In this diagram drastic differences between geographical areas can be observed. Rock fragments are exclusively carbonate in Palaeogene sandstones near to the Iberian Range in the eastern area (Baides, Negredo and Torremocha de Jadraque), which plot at the base of ternary diagram, since metamorphic rock fragments are virtually absent. However, metamorphic rock fragments are present in the framework of sandstones of the western area (Beleña de Sorbe and Membrillera), showing contributions from metamorphic complexes of the Central System. These rock fragments are low to medium rank metamorphic (slate and mica-schist), and form up to 35% of the total rock fragments.

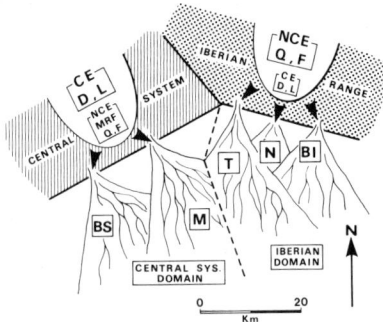

Fig. 7. Sandstone provenance and dispersal pattern sketch emphasizing the sedimentary domains.

These results have important inferences regarding the palaeogeography of Palaeogene sedimentation. It is possible to establish two geographically well-defined sedimentary domains, based on variations in source lithologies (Fig. 7): the Iberian domain in the east (Baides, Negredo and Torremocha de Jadraque) derived from Mesozoic sedimentary rocks of the Iberian Range, and the Central System domain, to the west (Beleña de Sorbe and Membrillera), derived from Cretaceous sedimentary cover and Palaeozoic metamorphic basement. The sharp change in MRF content (absent or present) of framework sandstones shows the clear boundary between the domains.

Evolution of sandstone composition

Vertical trends in sandstone composition have been determined for all five stratigraphic sections, using several compositional indices (Fig. 8).

NCE/NCE + CE values increase up-section in Central System domain (Beleña de Sorbe and Membrillera), interpreted as the result of stripping off Cretaceous carbonate cover and progressive unroofing of the underlying metamorphic complex. However, this index decreases on Iberian domain sections. The upper Mesozoic succession in the Iberian Range is a thick (c. 300 m) section of Jurassic carbonate deposits (dolostones and limestones at the top) and a Cretaceous siliciclastic formation with a dolomite unit above. Thus, the decrease of the NCE/NCE + CE index in Iberian Domain can be related to lithological properties of eroded source rocks, as an 'inverted' stratigraphy of upper Mesozoic succession. These differences in the evolution of sandstone composition verify the existence of previously defined sedimentary domains.

Vertical trends of D/D + L (dolostone fragments versus total carbonate fragments) are similar to those of NCE/NCE + CE. The upward decrease of dolostone fragments in sandstones of the Iberian domain is because the top of the Mesozoic section in the Iberian Range is dominantly dolomitic. Thus, an 'inverted' stratigraphic lithology of rock fragments takes place also in the Palaeogene succession.

The Q/Q + M relation (total quartz versus total quartz plus metamorphic rock fragments) remains more or less constant through Palaeogene sedimentation in Central System domain, whereas this index remains at 1 in the Iberian domain because of the absence of metamorphic rock fragments.

As discussed previously, a close relation between quartz and feldspar exists. In both sedimentary domains F/F + Qm ratio remains constant through time, never exceeding 0.3. However, there is a pronounced increase of this index in the Beleña de Sorbe section, indicating a local contribution from crystalline rocks. Gneissic rocks are exposed in the lower part of the metamorphic complex of the Central System (Soers 1972) (Fig. 1). Thus, the youngest Palaeogene sandstones in the Central System domain are probably related to the exposure of gneissic rocks in the source area.

Geotectonic setting and composition

Dickinson & Suczek (1979) and Dickinson et al.

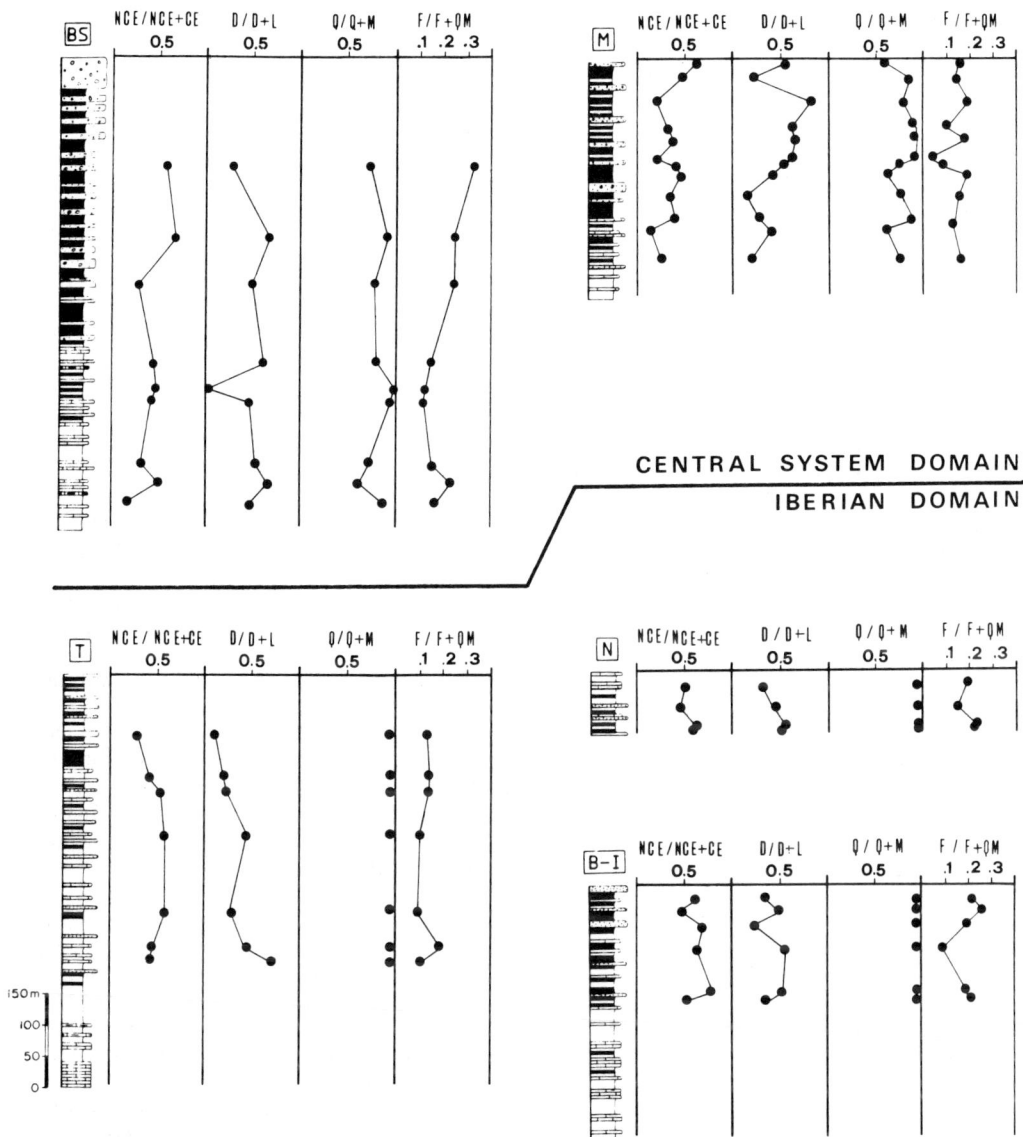

Fig. 8. Vertical trends in sandstone composition of the Palaeogene sections using several indices. BS, Beleña de Sorbe section; M, Membrillera section; T, Torremocha de Jadraque section; N, Negredo section; BI, Baides section.

(1983) pointed out that modal compositions of sandstone suites from specific geotectonic settings plot on QFL and QmFLt ternary diagrams. However, these authors exclude extrabasinal detrital carbonates in calculations of detrital modes. Later, Mack (1984), Zuffa (1980) and Ingersoll et al. (1987) have outlined the importance of including such detrital grains in the L pole.

When plotted on a QFLt diagram, according to Dickinson et al. (1983) criteria (carbonate extrabasinal excluded), all Palaeogene sandstones group together in a problematic area of imprecise provenance (Fig. 9a), within the recycled orogen, stable craton, and 'mixed' provenance fields. However, if carbonate extrabasinal grains (Lc) are included with total lithics (Fig. 9b), all studied Palaeogene sandstones plot

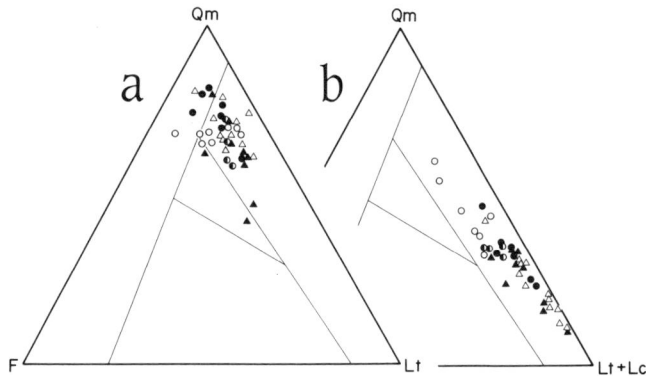

Fig. 9. Composition of Palaeogene sandstones. (**a**) Following criteria of Dickinson *et al.* (1983). (**b**) Including carbonate rock fragments (LC) in lithic pole.

within the recycled orogen provenance field. This inferred provenance is consistent with the regional geological setting. Palaeogene sandstones are syntectonic deposits produced by erosion of the Iberian Range sedimentary succession in the east, and sedimentary (Cretaceous cover) and low to medium rank metamorphic rocks of the Central System in the west, during Alpine tectogenesis.

Furthermore, this distribution of data on the QmF(Lt + Lc) diagram is similar to that of the QFR diagram, showing the same geographical distribution of sandstone composition, more lithic in the west.

Conclusions

Petrographic data on composition of Palaeogene sandstones framework in the northern Tajo Basin suggest that these sandstones are mainly sedimentoclastics (*sensu* Ingersoll 1983), with a litharenitic composition.

Sandstone composition changes with geographical distribution, with a progressive diminution of NCE grains from east to west. Rock fragments, mainly dolostone and limestone fragments (CE grains), increase to the west. Q/F values remain constant in all localities, indi-

cating a common source for these minerals, probably the Cretaceous sandy formations (e.g. Arenas de Utrillas Formation). Carbonate intrabasinal grains (CI) are also present, the result of erosion of contemporaneous lacustrine–paludal deposits.

A detailed analysis of the rock fragments nature permits the establishment of two sedimentary domains: (1) the Iberian domain in the east, derived from Mesozoic sedimentary rocks of the Iberian Range, and (2) the Central System domain in the west derived from Cretaceous cover (sedimentary rocks) and Palaeozoic metamorphic basement of the Central System.

Vertical trends in sandstone composition, on the basis of NCE/NCE + CE, D/D + L and Q/Q + M indices, reveal differences between both domains, related to the lithological properties of the eroded source terrains.

Finally, we conclude that the sandstone composition is consistent with a recycled orogen provenance (Dickinson *et al.* 1983). However, this is only clear if carbonate rock fragments (Lc) are included in the total lithic population.

The authors wish to acknowledge R. Valloni and M. A. Velbel for their valuable suggestions. We thank A. Morton for his collaboration in the English version of the text.

References

ALVARO, M., CAPOTE, R. & VEGAS, R. 1979. Un modelo de evolución geotectónica para la Cadena Celtibérica. *Acta Geológica Hispánica*, **14**, 172–177.

ARRIBAS, M. E. 1986. Petrologia y análisis secuencial de los carbonatos lacustres del Paleógeno del

sector N de la Cuenca Terciaria del Tajo. *Cuadernos de Geología Ibérica*, **10**, 295–334.
——, DIAZ-MOLINA, M., LOPEZ-MARTINEZ, N. & PORTERO, J. M. 1983. El abanico aluvial paleógeno de Beleña de Sorbe (Cuenca del Tajo): facies, relaciones espaciales y evolución. *X Congreso Nacio-*

nal de Sedimentologia, Menorca, Comunicaciones, **1**, 34–38.

DICKINSON, W. R. & SUCZEK, C. A. 1979. Plate tectonics and sandstone compositions. *American Association of Petroleum Geologists Bulletin*, **63**, 2164–2182.

—— & VALLONI, R. 1980. Plate settings and provenance of sands in modern ocean basins. *Geology*, **8**, 82–86.

——, BEARD, L. S., BRAKENBRIDGE, G. R., ERJAVEC, J. L., FERGUSON, R. C., INMAN, K. F., KNEPP, R. A., LINDBERG, F. A. & RYBERG, P. T. 1983. Provenance of North American Phanerozoic sandstones in relation to tectonic setting. *Geological Society of America Bulletin*, **94**, 222–235.

FOLK, R. L. 1959. Practical petrographic classification of limestones. *American Association of Petroleum Geologists Bulletin*, **43**, 1–38.

GARZANTI, E. 1986. Source rock versus sedimentary control on the mineralogy of deltaic volcanic arenites (Upper Triassic, northern Italy). *Journal of Sedimentary Petrology*, **56**, 267–275.

INGERSOLL, R. V. 1983. Petrofacies and provenance of late Mesozoic forearc basin, northern and central California. *American Association of Petroleum Geologists Bulletin*, **67**, 1125–1142.

INGERSOLL, R. V., BULLARD, T. F., FORD, R. L., GRIMM, J. P., PICKLE, J. D. & SARES, S. W. 1984. The effect of grain size on detrital modes: a test of the Gazzi-Dickinson point-counting method. *Journal of Sedimentary Petrology*, **54**, 103–116.

——, CAVAZZA, W. & GRAHAM, S. A. 1987. Provenance of impure calclithites in the Laramide foreland of southwestern Montana. *Journal of Sedimentary Petrology*, **57**, 995–1003.

MACK, G. H. 1984. Exceptions to the relationship between plate tectonics and sandstone composition. *Journal of Sedimentary Petrology*, **54**, 212–220.

PETTIJOHN, F. J., POTTER, P. E. and SIEVER, R. 1973. *Sand and Sandstones*. Springer-Verlag, Berlin.

POWERS, M. C. 1953. A new roundness scale for sedimentary particles. *Journal of Sedimentary Petrology*, **23**, 117–119.

SOERS, E. 1972. Stratigraphie et geologie structurale de la partie oriental de la Sierra de Guadarrama. *Studia Geologica*, **4**, 7–94.

SUTTNER, L. J. & DUTTA, P. K. 1986. Alluvial sandstone composition and paleoclimate, I. Framework mineralogy. *Journal of Sedimentary Petrology*, **56**, 329–345.

ZUFFA, G. G. 1980. Hybrid arenites: their composition and classification. *Journal of Sedimentary Petrology*, **50**, 21–29.

ZUFFA, G. G. 1985. Optical analysis of arenites: influence of methodology on compositional results. *In*: ZUFFA G. G. (ed.) *Provenance of Arenites*. Reidel, Dordrecht, 165–190.

Changes in the provenance of pebbly detritus in southern Britain and northern France associated with basin rifting

I. R. GARDEN

Badley, Ashton and Associates Limited, Winceby House, Winceby, Horncastle, Lincs. LN9 6PB, UK

Abstract: Stratigraphical and geographical variations in the composition of pebble suites in Upper Jurassic and Lower Cretaceous rocks in southern Britain and northern France provide a means for studying both local and regional changes in tectonic conditions associated with basin development. Local variations in pebble suites provide evidence of major fault-associated uplift and erosion of intrabasinal and basin-margin highs during earliest Cretaceous times, and the subsequent post-faulting subsidence and marine onlap in mid-Cretaceous times.

Regional provenance studies have demonstrated that the Upper Jurassic and Lower Cretaceous pebble suites of southern Britain and Normandy are separable into six assemblages.

Assemblage 1 is dominated by Carboniferous shelf chert, and is typical of Upper Jurassic pebble suites in Oxfordshire and Wiltshire, and Cretaceous pebble suites reworked from these older pebble beds. This detritus was probably derived from the western margin of the Anglo-Brabant Massif.

Assemblage 2 which is quartz dominated, includes Carboniferous basinal chert and tourmalinite, and is restricted to the Cretaceous pebble beds of Dorset. This assemblage was sourced principally from the Cornubian Massif.

Assemblage 3 contains subequal proportions of quartz and Carboniferous shelf chert and is present in the Lower Greensand pebble beds of the Isle of Wight, Hampshire, Wiltshire and Oxfordshire and Wealden pebble beds in the western and southern Weald. The distribution of this assemblage implies its derivation from the eastern margin of the Welsh Massif.

Assemblage 4 is characterized by a high proportion of sandstone clasts and is restricted to the Cretaceous pebble suites of Kent and East Sussex. Material in this assemblage was derived principally from the southern margin of the Anglo-Brabant Massif.

Assemblage 5 is compositionally akin to assemblage 3 but is restricted to the Albian pebble beds of the East Midlands Shelf. This material was derived from the northern margin of the Anglo Brabant Massif, with minor amounts of material from the Market Weighton High.

Assemblage 6 is quartz dominated and includes clasts of deformed sandstone and vein rock. This assemblage is restricted to the Aptian and Albian pebble beds of Normandy which are derived from the northern margin of the Armorican Massif.

In the Weald and Central Channel Basins, a change from chert-dominated assemblage 1 suites to quartz-rich pebble suites (assemblages 2 to 5) occurred at the beginning of the Cretaceous as a result of uplift and erosion of the marginal massifs adjacent to the rifting basin, accompanied by a regional fall in relative sea level. In contrast, on the East Midlands Shelf modifications to the detrital suite did not occur until the mid-Cretaceous, due to the Early Cretaceous reworking of Mesozoic detritus on the Shelf, probably from the uplift and erosion of the Sole Pit area. Subsequently, with the onset of subsidence in the Sole Pit area and a regional marine transgression during Albian times, the adjacent massifs were rejuvenated causing an input of fresh clastic material onto the Shelf.

The marine and continental rocks of Late Jurassic and Early Cretaceous age in southern Britain and northern France were deposited in a series of subsiding basins (Fig. 1) associated with the opening of the Bay of Biscay and the North Atlantic (Ziegler 1981). In conjunction with this subsidence, massif areas and intrabasinal highs were uplifted and eroded, shedding detritus into the adjacent basins. Petrographical analysis of such detritus provides a means for determining the extent and timing of the uplift and erosion and the resulting sediment distribution. Detailed

From Morton, A. C., Todd, S. P. & Haughton, P. D. W. (eds), 1991,
Developments in Sedimentary Provenance Studies.
Geological Society Special Publication No. 57, pp. 273–289.

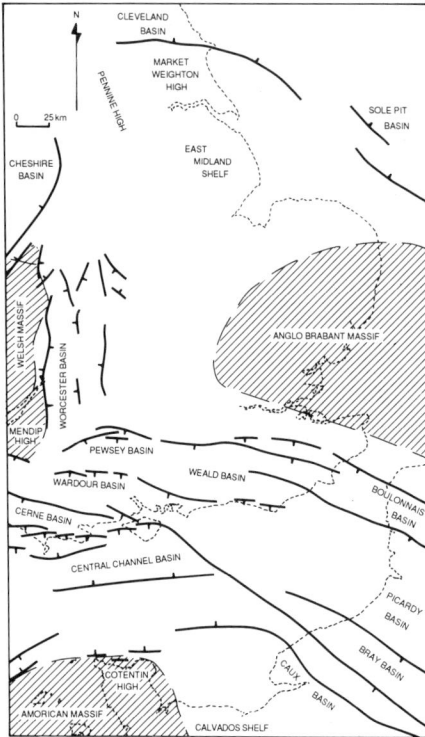

Fig. 1. Structural element map for southern Britain and northern France, illustrating the principal basin and massif areas (after Whittaker 1985 and Simpson *et al.* 1989).

provenance studies have been undertaken previously for the Wealden of the Weald and the Dorset areas (Allen 1954, 1960, 1961, 1967*b*, 1972) and for the Lower Greensand in southern England (Kirkaldy 1947; Wells & Gossling 1947; Middlemiss 1975) and northern France (Juignet *et al.* 1973). However, apart from Allen's work on the Wealden of northwest Europe (Allen 1959, 1965, 1967*a*) no systematic investigation of sediment provenance of the Upper Jurassic and Lower Cretaceous sequences in southern Britain and northern France has to date been attempted.

This paper presents the results of a study into regional and stratigraphical variations in the provenance of coarse-grained detritus from Oxfordian to Albian times in southern Britain and northern France. The principal aim of the study has been to determine the effects of basin rifting on sediment provenance. Samples were collected from 36 localities (Fig. 2) and 25 horizons (Fig. 3).

Analysis was undertaken on the greater-than-1 mm size fraction ($<0\varphi$) of pebble and granule-bearing intervals, with samples disaggregated and sieved into half φ size fractions. Identification of the clast types was made using a binocular microscope, with samples categorized according to their parental lithology and subdivided on textural features. Selected samples were examined using standard transmitted light microscopy, scanning electron microscopy and X-ray diffraction.

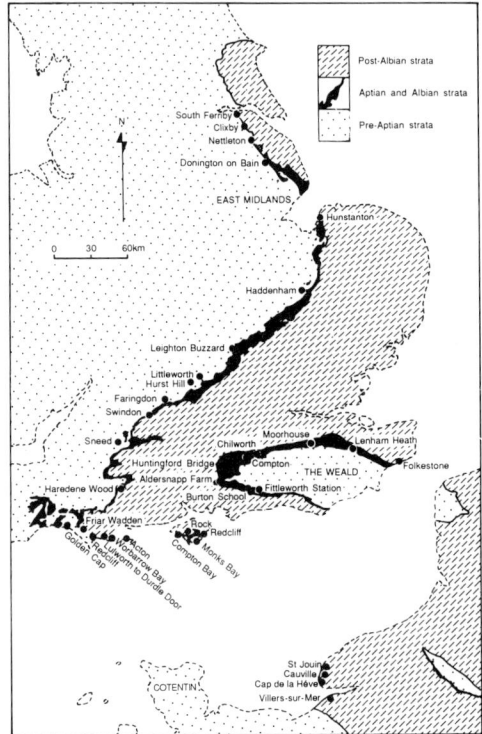

Fig. 2. Location map for the thirty-six principal sample sites used during this study.

Clast types

Ten principal clast types are recognized in the Upper Jurassic and Lower Cretaceous intervals studied. The majority of these are not specific to single source areas.

Quartz: ubiquitous and commonly the dominant clast type in pebble suites. These clasts

Fig. 3. Simplified Upper Jurassic and Lower Cretaceous lithostratigraphy for southern Britain and Normandy (after Rawson *et al.* 1978; Cope *et al.* 1980). The 25 principal sample horizons used in this study are numbered.

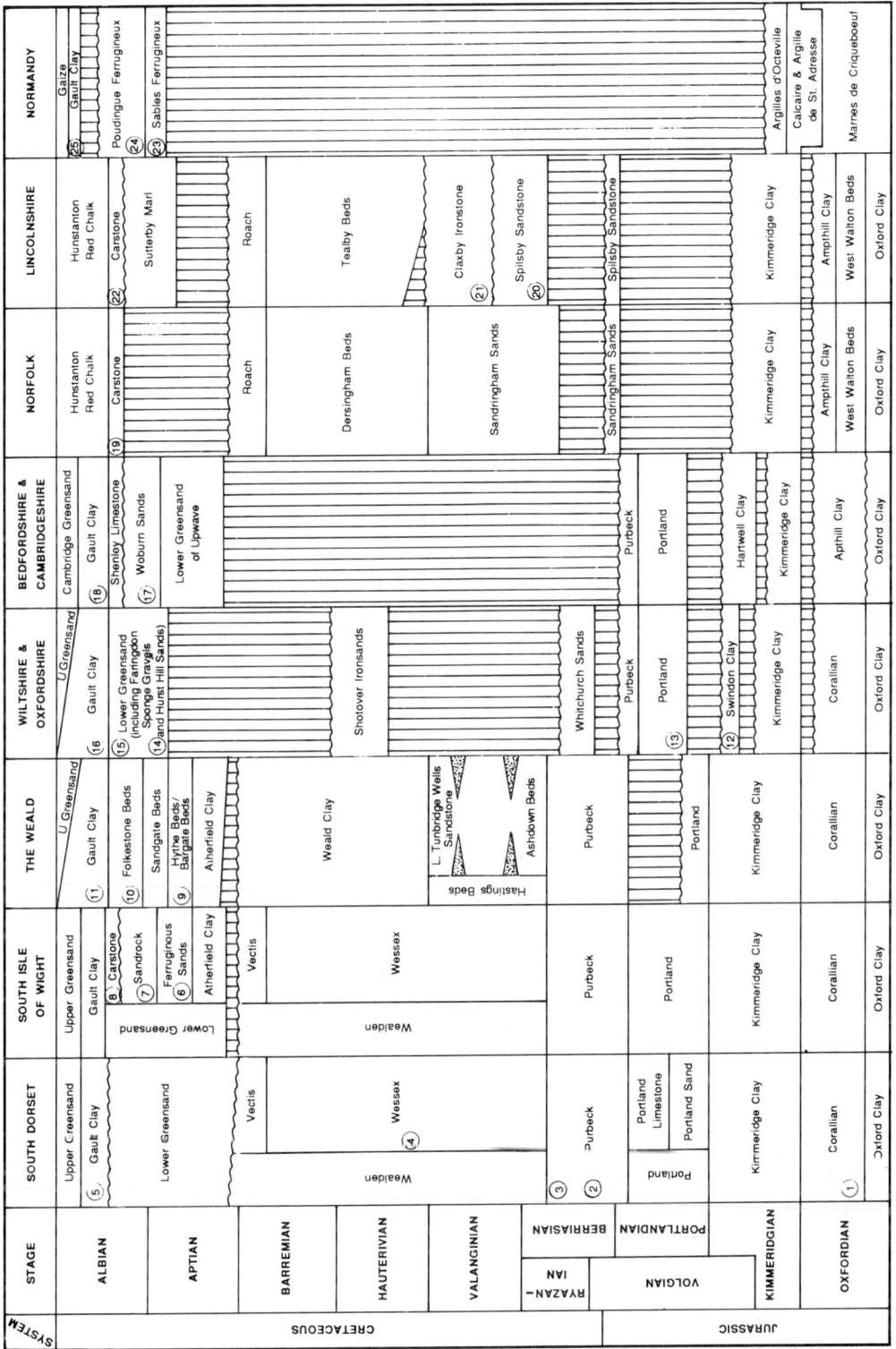

Stratigraphic correlation chart (read across: regional columns; read down: System / Stage). Circled numbers are locality/horizon references.

System	Stage	South Dorset	South Isle of Wight	The Weald	Wiltshire & Oxfordshire	Bedfordshire & Cambridgeshire	Norfolk	Lincolnshire	Normandy
CRETACEOUS	ALBIAN	Upper Greensand; Gault Clay (5)	Upper Greensand; Gault Clay	Gault Clay (11); U. Greensand	Gault Clay (16); U. Greensand	Cambridge Greensand; Gault Clay (18)	Hunstanton Red Chalk; Carstone (19)	Hunstanton Red Chalk; Carstone (22)	Gaize (25); Gault Clay
CRETACEOUS	APTIAN	Lower Greensand	Lower Greensand; Carstone (8); Sandrock (7); Ferruginous Sands (6); Atherfield Clay	Folkestone Beds (10); Sandgate Beds; Hythe Beds / Bargate Beds (9); Atherfield Clay	Lower Greensand (including Faringdon Sponge Gravels) (15) and Hurst Hill Sands (14)	Shenley Limestone; Woburn Sands (17); Lower Greensand of Upware	Roach; Sutterby Marl	Roach; Sutterby Marl	Poudingue Ferrugineux (24); Sables Ferrugineux (23)
CRETACEOUS	BARREMIAN	Vectis	Vectis	Weald Clay	Shotover Ironsands		Dersingham Beds	Tealby Beds	
CRETACEOUS	HAUTERIVIAN	Wessex (4)	Wessex	Weald Clay; L. Tunbridge Wells Sandstone	Shotover Ironsands		Dersingham Beds	Claxby Ironstone (21)	
CRETACEOUS	VALANGINIAN	Wealden	Wealden	Hastings Beds; Ashdown Beds	Whitchurch Sands		Sandringham Sands	Spilsby Sandstone (20)	
CRETACEOUS	RYAZAN- IAN / BERRIASIAN	Wealden	Wealden	Purbeck	Whitchurch Sands; Purbeck	Purbeck	Sandringham Sands	Spilsby Sandstone; Sandringham Sands	
JURASSIC	PORTLANDIAN / VOLGIAN	Purbeck (3); Portland Limestone; Portland Sand; Portland (2)	Purbeck; Portland	Purbeck; Portland	Purbeck; Portland (13)	Purbeck; Portland	Kimmeridge Clay	Kimmeridge Clay	
JURASSIC	KIMMERIDGIAN	Kimmeridge Clay	Kimmeridge Clay	Kimmeridge Clay	Swindon Clay (12); Kimmeridge Clay	Hartwell Clay; Kimmeridge Clay	Kimmeridge Clay	Kimmeridge Clay	Argiles d'Octeville; Calcaire & Argile de St. Adresse
JURASSIC	OXFORDIAN	Corallian (1); Oxford Clay	Corallian; Oxford Clay	Corallian; Oxford Clay	Corallian; Oxford Clay	Apthill Clay; Oxford Clay	Ampthill Clay; West Walton Beds; Oxford Clay	Ampthill Clay; West Walton Beds; Oxford Clay	Marnes de Criqueboeuf

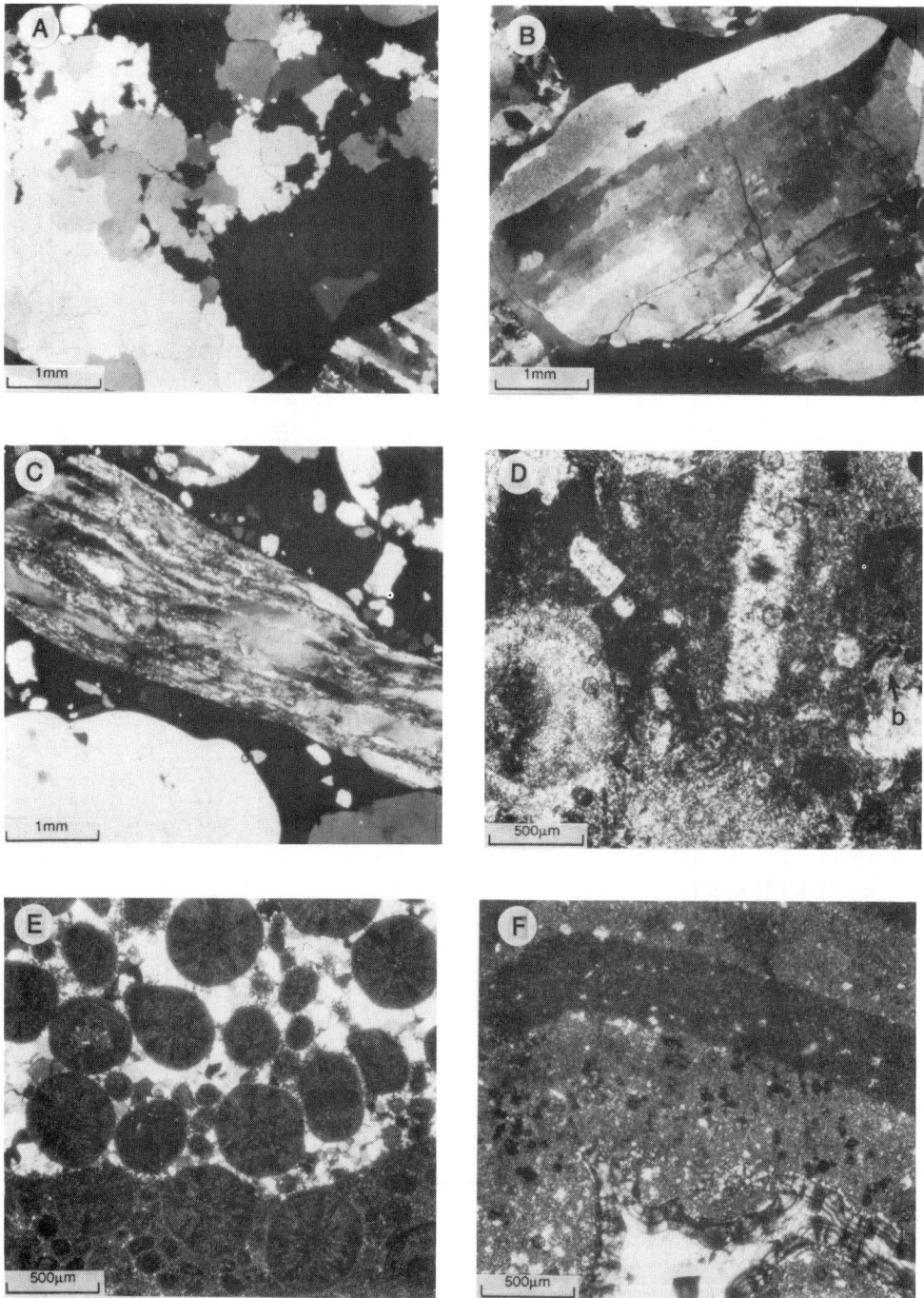

Fig. 4. Photomicrographs of selected quartz and Carboniferous shelf chert clast fabrics, with sample horizon indicated in parentheses. (**A**) Granoblastic quartz (Poudingue Ferrugineux, Cauville, Normandy). (**B**) Fibrous quartz (Gault Clay, Golden Gap, Dorset). (**C**) Mylonitized quartz (Poudingue Ferrugineux, Cauville, Normandy) (**D**) Bioclastic packstone with crinoid (a) and foraminifera (b) fragments (Gault Clay, Leighton Buzzard, Bedfordshire). (**E**) Oolitic grainstone with megaquartz cement (Carstone, Rock, Isle of Wight). (**F**) Silicified dolomitic mudstone (Carstone, Compton Bay, Isle of Wight). All the samples are in cross-polarized light.

typically have granoblastic fabrics and less frequently are fibrous or mylonitized (Figs 4A, B & C). The presence of quartz in all the pebble beds indicates that it is derived from each of the identified source massifs around southern Britain. It is noteworthy that fibrous quartz is in general only locally abundant and is present in greater abundance in immature pebble suites.

Carboniferous shelf chert: rounded, generally unweathered chert derived from silicified shallow marine limestones of Carboniferous age, as indicated by their included fauna and flora. This diverse group includes silicified bioclastic, oolitic, pelletal and dolomitic limestones (Figs 4D, E & F). Clasts from Lower Cretaceous rocks in southern England with similar fabrics were described in detail by Wells (1947) and Allen (1961). The cherts are principally replacements of packstones, wackestones and mudstones, but silicified oolitic, crinoidal and foraminiferal grainstones are also present. Significantly, calcareous material and chalcedonic cements are scarce in these clasts. This, together with the predominance of silicified packstones and wackestone clasts, shows that the suite is not representative of the chert in currently exposed Carboniferous limestone. Probably the suite was mechanically and chemically matured, removing the less stable types prior to deposition. Carboniferous chert is common to many of the studied pebble suites and is locally dominant. However, because of the maturation processes mentioned above it has not proved possible to trace this material to specific outcropping silicified limestones. The present distribution of silicified Carboniferous limestones in outcrop suggests that the principal sources of such clasts were on the Welsh and Anglo-Brabant Massifs, and possibly on the Pennine High.

Carboniferous basinal chert: rounded, unweathered clasts of silicified radiolarian and spicular mudrocks, with common cross-cutting quartz veins (Fig. 5A). The provenance of radiolarian clasts of this type in the Wessex Formation of Dorset was discussed by Oakley (1947). These clasts are readily matched with the Carboniferous chert formations of Devon and Cornwall from which they are likely to have been derived. They are common only in the Lower Cretaceous pebble beds of Dorset.

Sandstones: a diverse suite of clasts which includes quartz, lithic and micaceous arenites and siltites (Figs 5B & C). Details of sandstone clast compositions comparable with those identified in this study are given by Allen (1967b), Wells & Gossling (1947) and Juignet *et al.* (1973). All sandstone clasts are well cemented by quartz, or have illitic matrix and highly compacted fabrics. Lithic grains are predominantly silicified or illitized volcanic fragments.

Sandstone clasts are generally scarce in the majority of the pebble suites, which are supermature, being dominated by quartz and finely crystalline chert grains. In such suites, sandstone clasts are principally quartz arenites. Locally, lithic and micaceous arenites are moderately abundant, or form the dominant clast type (e.g. in the Lower Cretaceous of Kent), indicating a relatively immature clast suite.

The distribution of sandstone clasts indicates that they are derived from Devonian and Lower Palaeozoic rocks on the Anglo-Brabant Massif (e.g. Allen 1967b) and Palaeozoic and Precambrian rocks in Cornubia and Armorica. Additional material may also have been reworked from Triassic pebble beds and from Lower Palaeozoic strata outcropping on the Welsh Massif.

Tourmalinites: clasts containing abundant tourmaline. These include tourmalinized mudrocks and sandstones, pneumatolitic breccias and quartz–tourmaline vein rocks (Figs 5D & E). They are similar to rocks in the metamorphic aureoles of granites in Devon and Cornwall (Wells 1946; Goode & Taylor 1980) and to clasts in the Triassic pebble beds of southern and central England (Campbell-Smith 1963). However, Allen (1972) demonstrated by the use of K–Ar dating that some of the clasts were of significantly older age than tourmalinized rocks present in Devon and Cornwall and had ages more consistent with tourmaline-bearing rocks on the Armorican Massif. Thus, part of this group has been either derived directly from the Armorican Massif and/or reworked from Triassic pebble beds that were originally of Armorican derivation (e.g. Lamont 1946; Wills 1956). Tourmalinite clasts are scarce in the pebble beds of southern Britain and Normandy, except in Dorset where they form a minor but characteristic phase in Cretaceous pebble beds.

Silicified volcanics and jaspers: represent a minor but ubiquitous suite of clasts. This suite is dominated by laminated argillaceous chert with scattered siliceous phenocrysts (Fig. 5F) which are interpreted as felsic tuffs or aphanitic rhyolites. Other clasts include silicified porphyritic lavas, and jaspers with spherulitic and peloidal fabrics. Cross-cutting quartz veins are common. The wide distribution of the felsic tuff/aphanitic rhyolite clasts suggests derivation from a number of sources, the most likely of which are Lower Palaeozoic and Precambrian volcanic rocks on the Anglo-Brabant, Welsh and Cornubian Massifs. However, Devonian conglomerates could also provide a source. Jasper clasts are

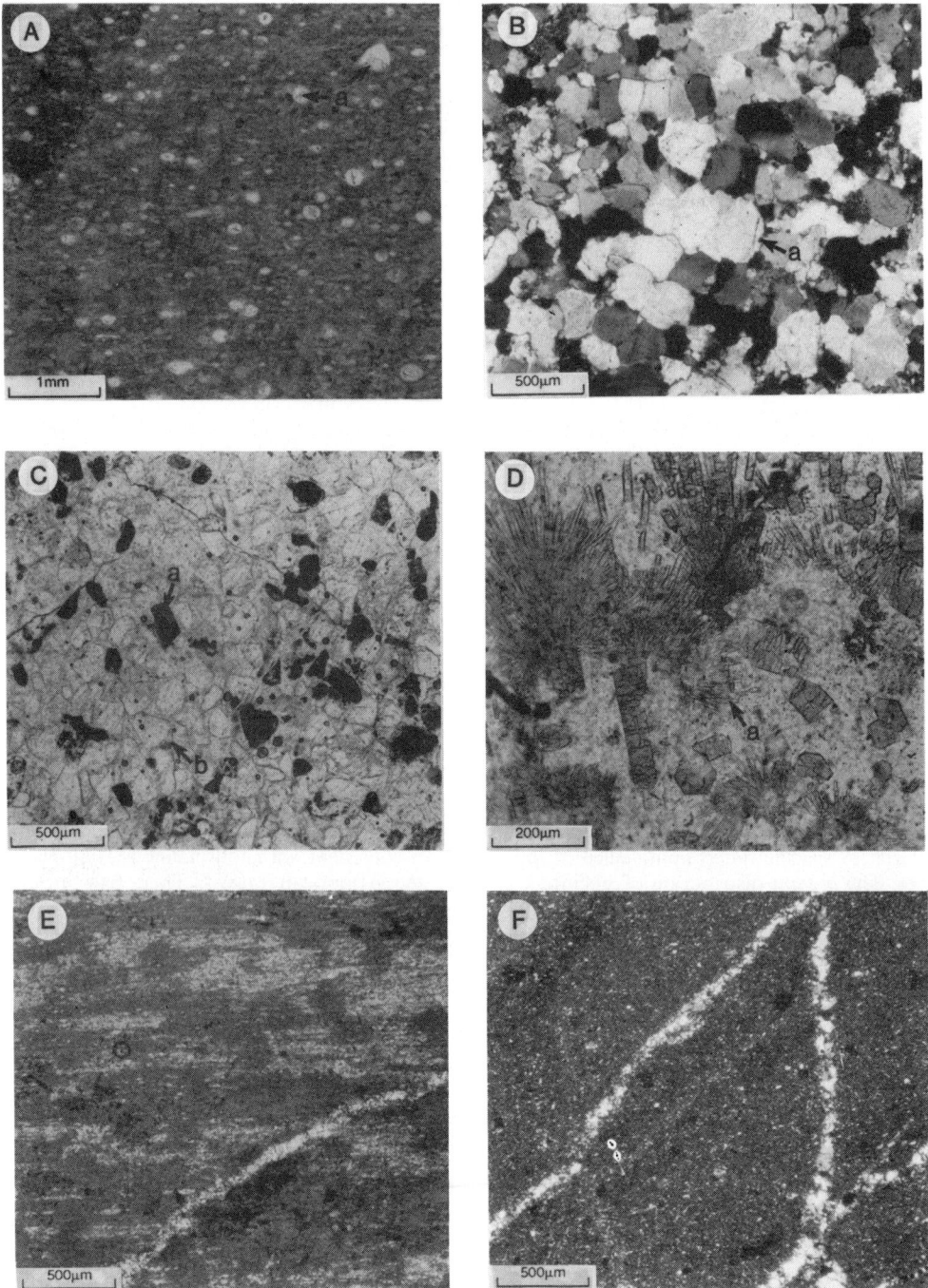

Fig. 5. Photomicrographs of Carboniferous basinal chert, sandstone, tourmalinite and silicified volcanic clast fabrics, with sample horizons indicated in parentheses. (**A**) Silicified mudstone with numerous radiolaria (a; Coarse Quartz Grit, Wessex Formation, Dorset). (**B**) Quartz arenite with pore-filling quartz overgrowths (a; Gault Clay, Haredene Woods, Wiltshire). (**C**) Lithic arenite with altered volcanic rock fragments (a) and pore-filling quartz cement (b; Folkestone Beds, Moorhouse, Kent). (**D**) Quartz–tourmaline vein rock with radiating tourmaline (a; Gault Clay, Golden Gap, Dorset), (**E**) foliated tourmalinized sediment (Gault Clay, Durdle promontory, Dorset). (**F**) Silicified aphenitic tuff (Coarse Quartz Grit, Wessex Formation, Durdle promontory, Dorset). Photomicrographs A, C D and E are in plane-polarized light, and B and F are in cross-polarized light.

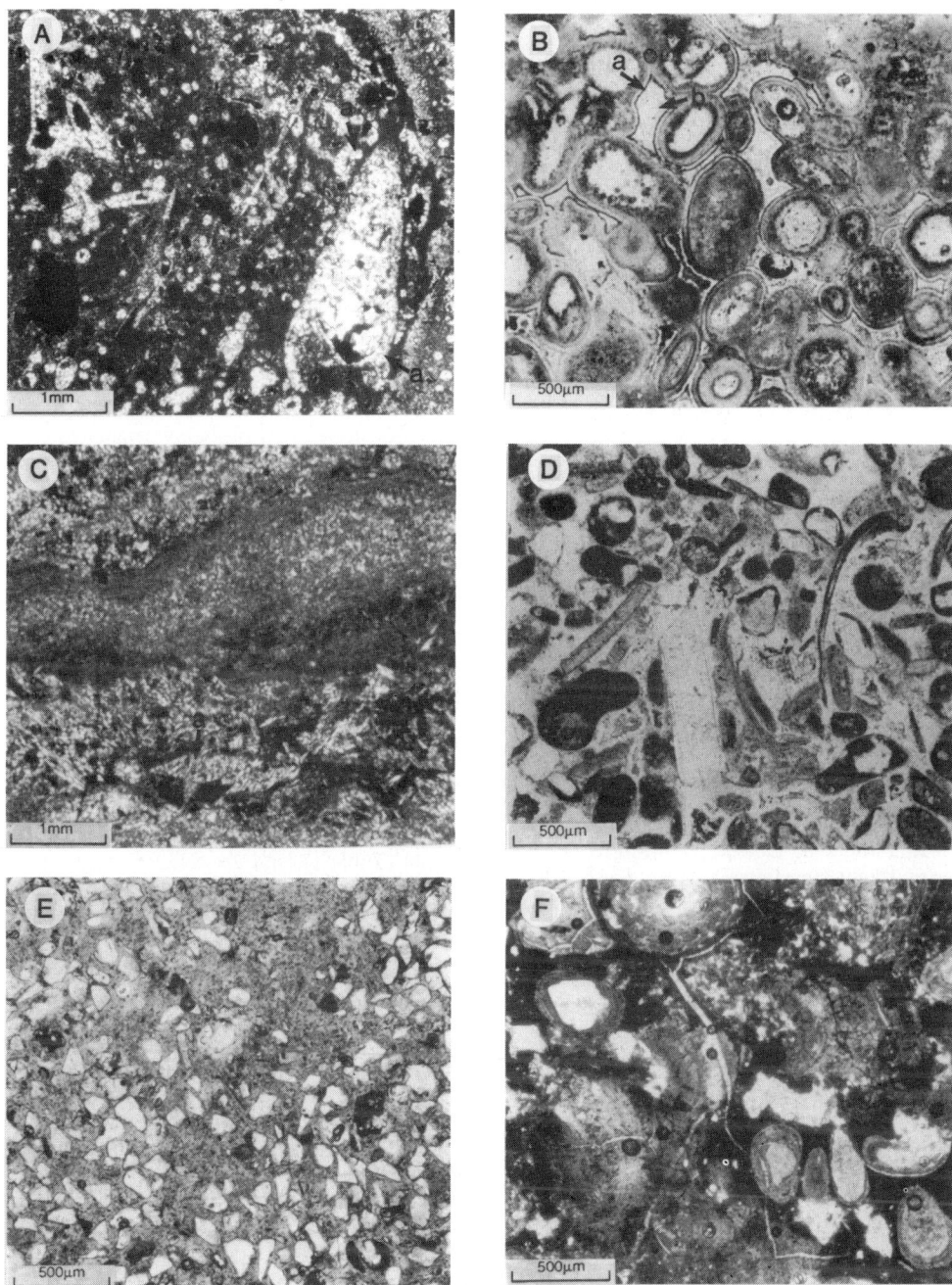

Fig. 6. Photomicrographs of Jurassic and Cretaceous chert, limestone, phosphorite and ironstone clasts, with source intervals defined and sample horizons indicated in parentheses. (**A**) Silicified packstone with *Rhaxella* and *Aptyxiella* (a) from the Portland Limestone (Coarse Quartz Grit, Wessex Formation, Stair Hole, Dorset). (**B**) Silicified oolitic grainstone with chalcedony (a) and megaquartz (b) cements, from the Portland Limestone (Coarse Quartz Grit, Wessex Formation, St Oswald's Bay, Dorset). (**C**) Silicified stromatolite with quartzine pseudomorphs after gypsum (a), from the basal Purbeck (Gault Clay, Durdle promontory, Dorset). (**D**) Calcareous oolitic and bioclastic grainstone with megaquartz pore-filling cement, from the Corallian (Faringdon Sponge Gravel, Lower Greensand, Faringdon, Oxfordshire). (**E**) Phosphate-cemented sandstone, probably from the Spilsby Sandstone (Carstone, Nettleton, Lincolnshire). (**F**) Oolitic ironstone, probably from the Lower Cretaceous (Carstone, Hunstanton, Norfolk). Photomicrographs A and C are in cross-polarized light, B, D, E and F are in plane-polarized light.

appreciably less common but also likely to be derived from Palaeozoic or Precambrian volcanic rocks or reworked through pebble beds. Clasts of jasper are present in moderate abundance in the Lower Cretaceous pebble suites of Dorset. Such material is likely to have been derived from Cornubia or Armorica. Similar clasts in the Hastings Beds were probably reworked from Devonian pebble beds on the Anglo-Brabant or Welsh Massifs (Allen 1967*b*).

Mesozoic cherts: derived from the Portland Limestone and Purbeck Formations, and the Corallian Group. Chert types are differentiated by their fauna and flora, and typically show white, porcellanous weathering fabrics. They include silicified bioclastic and oolitic grainstones and packstones, bioclastic wackestones and mudstones, silicified evaporites and silicified wood (Figs 6A, B & C). Portland and Purbeck cherts are locally common in the Wessex and Gault Clay Formations in Dorset and scarce in the Lower Greensand in the Isle of Wight. Allen (1960) reported Portland chert in the Hastings Group pebble beds of the Weald and West & Hooper (1969) noted Portland detritus in the Purbeck Formation in Dorset. Purbeck detritus is also present in the Gault Clay of the southwest Weald (Hampshire) and Corallian chert is present in the Lower Greensand of Oxfordshire. Many of the Portland and Purbeck clasts can be traced to local outcropping or subcropping silicified limestones in Dorset, Hampshire and the Isle of Wight. In contrast, there are no documented chert-bearing source rocks for Corallian chert in Oxfordshire where they were presumably entirely removed by erosion.

Jurassic limestones: derived from the Corallian, Portland Limestone and possibly the Kimmeridge Clay. These include bioclastic and oolite grainstones, packstones and wackestones, calcite-cemented sandstones and lime mudstones (Fig. 6D). Clasts of this type can be traced to nearby outcropping source rocks. Portland clasts are locally common in the *Unio* Beds of the Purbeck Formation in Dorset (West & Hooper 1969), and Corallian and Kimmerdige Clay clasts are present in the Lower Greensand in Oxfordshire (Arkell 1947).

Phosphorites: clasts of microcrystalline apatite which replaced mudrocks, and cemented glauconitic and goethite-ooid-bearing sandstones (Fig. 6E). Phosphatic, oolitic packstone clasts are also present but are scarce. Reworked fossil moulds are moderately common and locally abundant. These include phosphatized bivalves and ammonites derived from the Kimmeridge Clay, and belemnite phragmacones that have probably been reworked from the Upper Jurassic and

Lower Cretaceous marine sandstones of Lincolnshire. Phosphate clasts are locally common in the Portland Limestone (Neaverson 1925), Spilsby Sandstone (Kelly 1980), Wealden (Allen 1960), Lower Greensand (Arkell 1939) and Gault Clay, and are also present in the Purbeck Formation in Dorset (West & Hooper 1969; Ensom 1985).

Ironstone: clasts of microcrystalline goethite, including goethite-cemented sandstones and siltstones, laminated, peloidal and oolitic ironstones, and goethitic bioclast moulds (Fig. 6F). Bioclast moulds are derived from the Oxford Clay and are particularly abundant in the Bargate Beds of the northwest Weald (Arkell 1939). Other ironstone clasts probably originated from Upper Jurassic and Lower Cretaceous mudrocks. Ironstone clasts are particularly common in the Ferruginous Sands of the Isle of Wight, the Bargate and Folkestone Beds of the northwest Weald, and the Carstone of Lincolnshire.

Clast assemblages: stratigraphical and geographical variations

Six pebble assemblages can be distinguished in the Upper Jurassic and Lower Cretaceous of southern Britain and Normandy. They are geographically and, to some extent, stratigraphically distinct (Table 1), and are differentiated on the basis of their relative proportions of quartz, Carboniferous shelf chert and sandstone. Carboniferous basinal chert, tourmalinite, and locally-derived Jurassic and Cretaceous detritus allow further refinement.

The composition of the assemblages reflects the source-area geology at the time of deposition. Changes in assemblage type within an area therefore provide evidence for a change in source, which can result from either alteration in sediment transport and distribution or progressive denudation of the source area. Thus provenance variations reflect the evolution of both the basins and source massifs.

Assemblage 1

This is dominated by Carboniferous shelf chert, with subordinate quartz. Sandstone and silicified volcanics from a very minor component, and tourmalinite and Carboniferous basinal chert clasts are virtually absent. The chert principally comprises silicified mudstones with scarce bioclasts, peloids and silicified dolomite rhombs. Quartz pebbles are typically well rounded and

Table 1. *Definitions of the six principal pebble assemblages recognized in southern Britain and Normandy, together with their regional and stratigraphical distribution and source areas*

Assemblage	Assemblage characteristics	Distribution		Source areas
		Stratigraphical	Geographical	
1	Dominated by Carboniferous shelf chert with minor quartz and trace quantities of silicified volcanic rocks and quartzite. Reworked Jurassic ironstone and phosphate are moderately common locally.	Lower Greensand Group	Surrey & Bedfordshire	
		Hastings Beds & Weald Clay Groups	Surrey & Sussex	Principally reworked from older Cretaceous and Jurassic pebble beds
		Spilsby Sandstone & Claxby Ironstone Formations	Lincolnshire	
		Purbeck (Intermarine Beds) Formation	Dorset	
		Portland Group	Wiltshire & Oxfordshire	Derived from the western margin of the Anglo-Brabant Massif, prior to appreciable Mesozoic erosion.
		Corallian Group	Dorset	
2	Dominated by quartz with minor quantities of Carboniferous basinal chert, tourmalinite and sandstone, and trace amounts of jasper, Jurassic and Cretaceous chert, phosphate and limestone are locally abundant.	Purbeck (*Unio* Beds), Wessex & Gault Clay Formations	Dorset	Derived from the Cornubian Massif, possible minor Armorican influence.
3	Subequal proportions of quartz and Carboniferous shelf chert with trace amounts of silicified volcanic rocks, quartzite and tourmalinite. Jurassic chert, phosphate and limestone are locally abundant but generally scarce.	Lower Greensand Group & Gault Clay Formation	Isle of Wight, Wiltshire, Oxfordshire, Bedfordshire, Hampshire & Sussex	Principally derived from the Welsh Massif and its eastern margins. Some reworking of older Cretaceous and Jurassic pebble suites.
		?Whitchurch Sands & Hastings Beds	Wiltshire & Sussex	
4	Dominated by quartz with subequal to subordinate amounts of sandstone and Carboniferous shelf chert, and scarce silicified volcanic rocks. Jurassic phosphate has been recorded by previous workers.	Lower Greensand Group	Kent	Source rocks on the southern margin of the Anglo-Brabant Massif. Some reworking of Jurassic pebble beds.
		Hastings Beds Group	Kent	
5	Subequal quartz–Carboniferous shelf chert to quartz-dominated assemblage with trace amounts of silicified volcanic rocks and sandstone. Jurassic/Cretaceous phosphates and ironstone typically form a major component in this pebble suite.	Carstone Member	Lincolnshire, Norfolk & South Humberside	Principally from the northern margin of the Anglo-Brabant Massif. Minor amounts of material from the Market Weighton High.
6	Quartz-dominated assemblage with subordinate deformed sandstones and cherts, silicified volcanic rocks and tourmalinites. Scarce Palaeozoic and Jurassic chert and moderately abundant Jurassic phosphate.	Lower Greensand & Gault Clay (equivalents)	Normandy	Sourced from the Cotentin High (northeastern Armorican Massif).

fibrous types are very scarce. Jurassic phosphatic detritus is locally abundant in the pebble assemblage.

Occurrence. Pebbly units with this assemblage include: those of Oxfordian to Early Berriasian age, the Corallian and middle Purbeck limestones in Dorset, the Lower and Upper Lydite Beds in Wiltshire and Oxfordshire, the Middle Spilsby Sandstone nodule bed in Lincolnshire (Fig. 7A); Late Berriasian to Barremian pebble beds of the Hastings and Weald Clay Groups in the northwest Weald (Allen 1954, 1967*b*, 1981, this volume) and the basal Claxby Ironstone in Lincolnshire (Fig. 7B); Aptian and Albian pebble beds of the Lower Greensand in the northwest Weald and Cambridgeshire (Figs 7C & D).

Source. The composition of this assemblage indicates derivation of detritus from an area with extensive Carboniferous limestone cover and no significant quartzose rocks. The scarcity of quartz provides the best evidence of source. In most English and Welsh areas where Carboniferous limestone outcrops, it is overlain by Namurian sandstones and conglomerates, which typically include quartz-pebble conglomerates and quartz-pebble-bearing sandstones. Additionally, the Dinantian limestones of South Wales and the Anglo-Brabant Massif overlie Devonian rock which include quartz-pebble bearing fluvial sandstones and conglomerates, and those of the Pennines rest on metamorphosed and quartz-veined Lower Palaeozoic rocks. Therefore to derive a chert-rich assemblage without incorporating quartz pebbles requires a relatively distinct geology, namely a very limited Namurian–Westphalian sequence overlying Carboniferous limestone that largely blankets older rocks. Clearly, the Namurian sandstones and conglomerates of South Wales and the Pennines reduce the possibility that these areas were sources. In contrast, boreholes onto the Anglo-Brabant Massif have proved Carboniferous limestone at its margins without overlying Namurian sequences (Sellwood & Scott 1986; Foster *et al.* 1989). However, over large areas of the massif, the Carboniferous is totally absent and Devonian and Lower Palaeozoic rocks subcrop beneath the Cretaceous cover (Smith 1985).

The work of Allen (1954, 1959, 1967*b*) showed that during Early Cretaceous times the Anglo-Brabant Massif formed a major sediment source and erosion extended down into Devonian and Lower Palaeozoic rocks. In addition, subcrop mapping of the structure reveals negligible Carboniferous limestone beneath the Aptian and Albian cover (Whittaker 1985). Therefore the Anglo-Brabant Massif could have been the primary source for this assemblage but only prior to the Early Cretaceous phase of erosion.

Assemblage 1 pebble suites of Late Berriasian to Albian age are more restricted geographically than those of the Oxfordian to Early Berriasian age (Fig. 7B). Probably much of the Late Berriasian to Albian coarse-grained clastic material was second cycle and was reworked from older pebble beds, in particular those of the Portland and Corallian Groups.

Assemblage 2

This is dominated by quartz, including fibrous and vein types, together with subordinate amounts of sandstone, Carboniferous basinal chert, tourmalinite and jasper. Carboniferous shelf chert is notably absent. Jurassic detritus, including chert, phosphate and limestone, is locally abundant, and scarce clasts of Cretaceous chert are present in some horizons.

Occurrence. The assemblage is typical of the Wealden (Late Berriasian to Barremian; Fig. 7B) and Gault Clay (Albian, Figs 7C & D) pebble beds in Dorset. Scarce but distinctive quartz–tourmaline and quartz clasts are present in the *Unio* Beds of the Purbeck Formation (Late Berriasian) in Dorset, making the earliest known appearance of the assemblage (Fig. 7B).

Source. The predominance of relatively immature quartz clasts, associated with Carboniferous basinal cherts and tourmalinites reworked from matamorphic aureoles of granites, implies derivation at least in part from the Cornubian Massif. In addition, the distinct geographical restriction of this assemblage to the Central Channel Basin is consistent with a local source. The general absence of Carboniferous shelf cherts indicates no significant input from the north or east of Dorset, where such material is moderately common in pebble suites of comparable age (Figs 7B & D). The possibility of an Armorican or more remote western source was considered by Allen (1969, 1972) to account for some tourmalinite and jasper clasts, and cannot be discounted here. However, the textural immaturity of the assemblage and the ease with which the majority of the clasts can be traced to potential sources on the Cornubian Massif, indicates that this was the main source area. The more exotic clasts noted by Allen (1969, 1972) could poten-

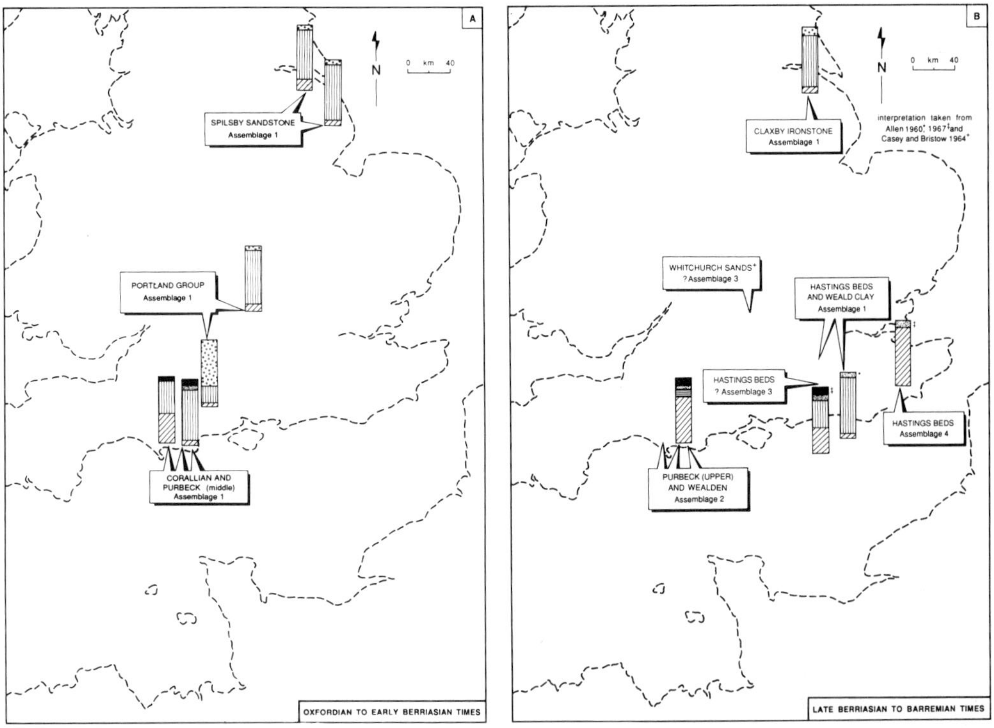

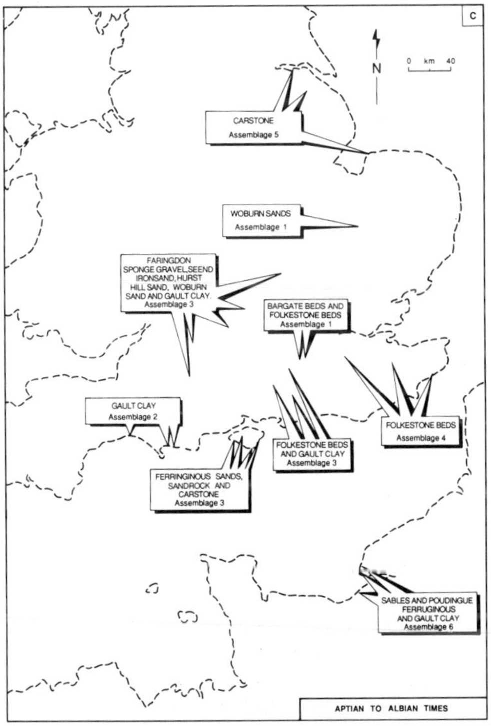

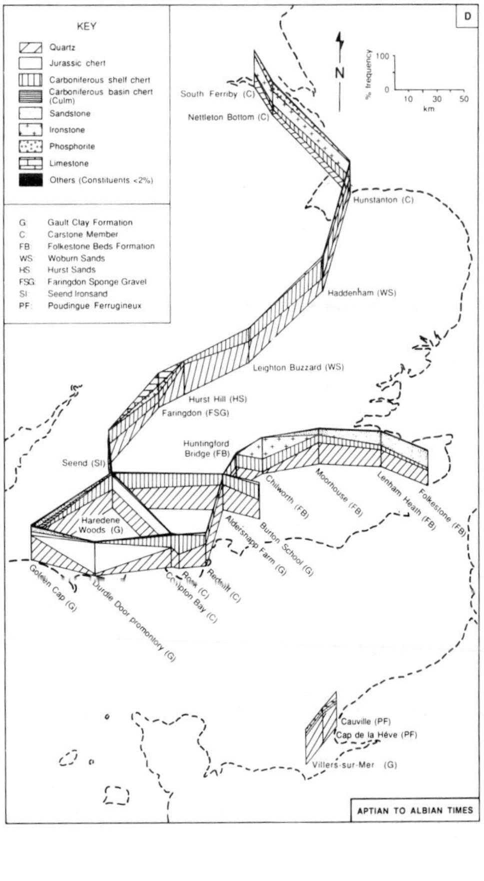

Fig. 7. Pebble assemblage distribution maps for (**A**) Oxfordian to Early Berriasian times; (**B**) Late Berriasian to Barremian times; (**C**) Aptian to Albian times; (**D**) a fence diagram illustrating the complex compositional variation in the Aptian and Albian pebble suites geographically.

tially have been derived from the margins of the Cornubian Massif, being reworked from Triassic pebble beds which have their origin, in part, on the Armorican Massif (Lamont 1946; Campbell-Smith 1963).

Assemblage 3

These pebble suites contain subequal amounts of quartz and Carboniferous shelf chert, with minor abundances of sandstone, tourmalinite and silicified volcanic rocks. The quartz is typically granoblastic, and the chert is dominated by wackestones and mudstones, with moderately abundant packstones and grainstones. Kimmeridge Clay phosphorites are relatively common, and Corallian and Kimmeridge Clay limestone and chert clasts are locally abundant.

Occurrence. Typically the Lower Greensand and basal Gault Clay pebble suites (Aptian to Albian) of the Isle of Wight, Wiltshire, Oxfordshire, Hampshire and Sussex have assemblages of this composition (Fig. 7C). Descriptions of the Berriasian Whitchurch Sands pebble suite (Casey & Bristow 1964) in Wiltshire and the Berriasian to Hauterivian Hastings Beds in the western and southern Weald (Allen 1967b) suggest that these fall into the same category (Fig. 7B).

Source. In Oxfordshire and Wiltshire, where Portlandian pebble suites are of assemblage 1 and the Aptian and Albian pebbles belong to assemblage 3 (Figs 7A, C), quartz-rich detritus must have been contributed during Early Cretaceous times. Suitable sources for the latter are present on the Welsh Massif where Namurian and Devonian pebble beds were exposed during the Early Cretaceous. The Anglo-Brabant Massif might also have been a source following partial removal of Carboniferous cover and exposure of Devonian and Lower Palaeozoic rocks. However, the repeated deposition of assemblage 1 pebble suites during Aptian and Albian times in the northwest Weald, between the Anglo-Brabant Massif and the area of deposition of the assemblage 3 suites, indicates that this massif is unlikely as a source. Hence the Welsh Massif is a more probable source.

Tourmalinite and Carboniferous basinal cherts are very minor components and unlikely to indicate a direct Cornubian influence. They were probably reworked from Triassic rocks in the West Midlands on the eastern margin of the Welsh Massif.

Assemblage 4

This contains subequal proportions of quartz and sandstone and subordinate Carboniferous shelf chert. Silicified volcanic rocks form a minor component. The sandstone clasts are diverse and immature, and include quartz, lithic and micaceous arenites and siltites.

Occurrence. The assemblage is typified by the Albian pebble suites of the Folkestone Beds in Kent (Figs 7C & D) and closely matches the Aptian and Albian suites described from the Hythe and Folkestone Beds of west Kent by Wells & Gossling (1947). In the Weald, the Ashdown and Lower Tunbridge Wells pebble beds (Valanginian to Hauterivian) described by Allen (1954, 1967b) have similar compositions (Fig. 7B).

Source. As Allen (1967b) demonstrated, the Wealden pebble beds with this assemblage are largely derived from Devonian and Lower Palaeozoic rocks subcropping beneath the Cretaceous on the Anglo-Brabant Massif, with Carboniferous limestone sourcing Carboniferous shelf chert, and Jurassic rocks supplying Carboniferous shelf chert pebbles and Jurassic clasts. The distribution of Lower Greensand pebble suites mimics that of the Wealden pebble beds, and similar sources are envisaged.

Assemblage 5

An assemblage composed of subequal proportions of quartz and Carboniferous shelf chert, minor amounts of sandstone and abundant Cretaceous and Jurassic ironstone and phosphorite.

Occurrence. Although superficially similar to assemblage 3, this is stratigraphically and geographically restricted to the Carstone (Albian) of the East Midlands (Figs 7C & D). It is therefore isolated from assemblage 3. The Carstone suites show a progressive increase in abundance of quartz pebbles and decrease in Carboniferous shelf cherts northwards (Fig. 7D).

Source. As thickness, grain size and clast abundance increase southwards along the outcrop of the Carstone the principal sediment supply is likely to have been from the northern margin of the Anglo-Brabant Massif. Quartz was probably derived from the Devonian and Silurian rocks subcropping beneath the Cretaceous cover, with Carboniferous shelf cherts derived from outliers of Carboniferous limestone and recycled through Upper Jurassic and Lower Cretaceous pebble beds having assemblage 1 suites. The increase in quartz abundance northwards may have resulted from dilution of the southerly-derived detritus by quartz clasts eroded from the Great Oolite on the Market Weighton High. This twofold sourcing for the Carstone concurs with the heavy mineral findings of Versey & Carter (1926) and Rastall (1930).

Assemblage 6

Quartz is dominant, and deformed and unde-formed vein-rocks, sandstones, tourmalinites and chert are minor components. Supplementing this assemblage are Jurassic phosphorites and cherts. Significantly, Carboniferous cherts (shelf or basinal) are lacking, and tourmalinites are relatively scarce.

Occurrence. This assemblage characterizes the Aptian and Albian, Sables and Poudingue Fer-rugineux and the basal Gault Clay pebble beds in Normandy (Figs 7C & D), and is distinct from assemblages 2 and 3 in the rocks of equivalent age in Dorset and the Isle of Wight respectively.

Source. Juignet *et al.* (1973) traced much of the detritus to the Palaeozoic and Precambrian rocks of the Cotentin High and this concurs with evidence found in this study. Jurassic rocks on the margin of the Cotentin High or an intra-basinal high are likely to have supplied the Jurassic detritus.

Discussion

The change in sediment provenance in southern Britain from the initially widespread pebble suites of assemblage 1 (Fig. 7A) to the diverse assemblages of 2, 3, 4 and 5 occurred at differ-ent times. Over much of the Weald and Central Channel basin the switch took place during the earliest Cretaceous (Fig. 7B), whereas on the East Midlands Shelf it did not occur until the Albian (Figs 7C & D). This difference reflects the differences in structural history, which was the principal control on the provenance of coarse-grained detritus during the Late Jurassic–Early Cretaceous interval. Consequently it is necessary to consider provenance in the two areas separ-ately.

Weald and Central Channel basins

The important change of provenance in the Weald and Central Channel basins coincides with a major tectonic phase in northwest Europe (Ziegler 1979), commonly referred to as the Late Cimmerian event, together with a phase of rela-tive sea level fall (Vail & Todd 1981; Haq *et al.* 1987). Subsequent changes reflected later tecto-nic events and further sea level changes. A com-bination of tectonic and eustatic factors influenced the rate and depth of erosion on the source massifs and dispersal of the sediment as intrabasinal highs partitioned the basins. The alteration in provenance patterns in the Weald and Central Channel basins can be considered in relation to three principal time intervals: Oxfor-dian to Early Berriasian, Late Berriasian to Barremian, and Aptian to Albian.

Oxfordian to Early Berriasian. The Late Jurassic and earliest Cretaceous pebble suites are dominated by coarse-grained detritus from the Anglo-Brabant Massif, although sands derived from Cornubia and Armorica may be present in Dorset (Neaverson 1925; Latter 1926). The assemblage 1 pebble suites are highly mature, with only the most resistant and compositionally homogeneous chert types derived overwhelm-ingly from Carboniferous limestones (Fig. 8A).

The latter part of this interval was marked in the Weald and Central Channel Basins by fault-ing and accelerated subsidence, with subsidence emphasized in hangingwalls to major faults (Chadwick 1986). Therefore isostatic uplift of footwall blocks within the basins and the massifs flanking them (*cf.* Jackson & McKenzie 1983; Barr 1987; Roberts & Yielding, 1991) resulted in enhanced flows of sediment into the adjacent half grabens. However, the high sea-level during the Kimmeridgian (Hallam 1978; Vail & Todd 1981; Haq *et al.* 1987) caused submergence of the source massifs and restricted the distribution of coarse-grained sediment.

Relative sea level began to fall in Portlandian times. Initially, the relief of the source massifs was too low for the transport of significant amounts of detritus into the basins (Allen 1981). Pebble suites were largely derived from older sequences undergoing erosion on intrabasinal structures (*e.g.* reworking of Corallian detritus into the middle Purbeck in Dorset), and from Carboniferous limestone which was probably the most significant massif cover at this time (Fig. 8A).

Late Berriasian to Barremian. Faulting continued during the Early Cretaceous resulting in the continued accumulation of thick sedimentary sequences, albeit in appreciably more confined basins than those of Kimmeridgian age

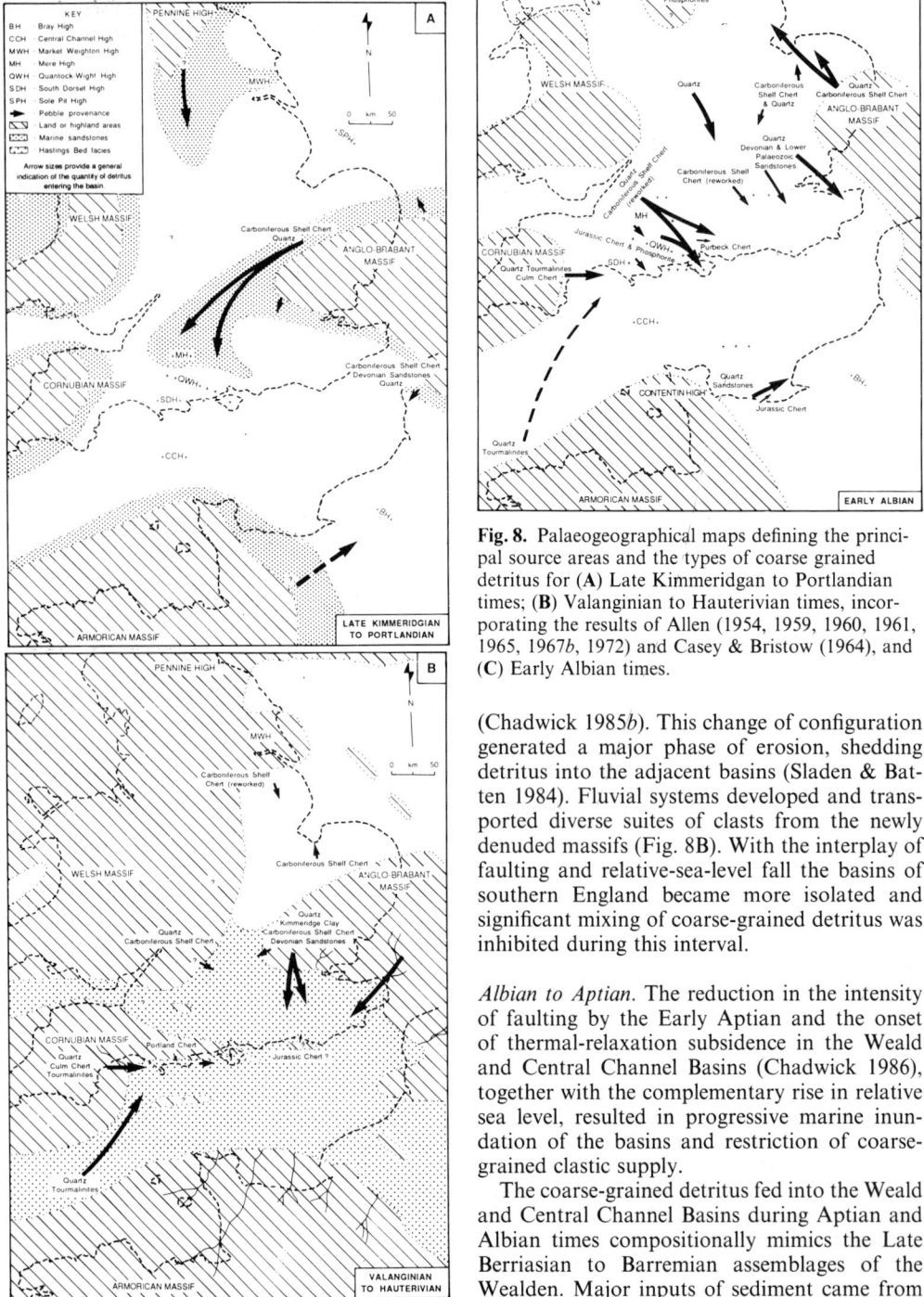

Fig. 8. Palaeogeographical maps defining the principal source areas and the types of coarse grained detritus for (**A**) Late Kimmeridgan to Portlandian times; (**B**) Valanginian to Hauterivian times, incorporating the results of Allen (1954, 1959, 1960, 1961, 1965, 1967*b*, 1972) and Casey & Bristow (1964), and (**C**) Early Albian times.

(Chadwick 1985*b*). This change of configuration generated a major phase of erosion, shedding detritus into the adjacent basins (Sladen & Batten 1984). Fluvial systems developed and transported diverse suites of clasts from the newly denuded massifs (Fig. 8B). With the interplay of faulting and relative-sea-level fall the basins of southern England became more isolated and significant mixing of coarse-grained detritus was inhibited during this interval.

Albian to Aptian. The reduction in the intensity of faulting by the Early Aptian and the onset of thermal-relaxation subsidence in the Weald and Central Channel Basins (Chadwick 1986), together with the complementary rise in relative sea level, resulted in progressive marine inundation of the basins and restriction of coarse-grained clastic supply.

The coarse-grained detritus fed into the Weald and Central Channel Basins during Aptian and Albian times compositionally mimics the Late Berriasian to Barremian assemblages of the Wealden. Major inputs of sediment came from all the principal source massifs (Fig. 8C). These influxes are associated locally with unconformities in the Weald (Casey 1961; Middlemiss 1967), Central Channel (Casey 1961) and Bristol Channel (Kamerling 1979) Basins, and include significant amounts of Jurassic detritus adjacent to eroded footwall blocks (*e.g.* Arkell 1939). Lake & Karner (1987) related the unconformities to faulting associated with rift/drift transitions in the Bay of Biscay and the North Atlantic. In conjunction with this, Lake & Karner (1987) recognized short-lived isostatic uplift of the basement massifs which they assigned to thermal effects associated with continental rifting. Such uplifts reactivated the source areas, increasing the flows of coarse-grained detritus into the basins.

Progressive marine incursion during the Aptian and Albian culminated in the Middle Albian transgressions across the massifs (Owen 1971) which finally cut off the sources of coarse-grained clastic sediment.

East Midlands Shelf

In contrast to the Weald and Central Channel basins, there is no evidence on the East Midlands Shelf of a change in sediment type at the beginning the Cretaceous. Instead, assemblage 1 suites persisted until Valanginian times (Claxby Ironstone), the detritus being reworked from older pebble beds. Only in the Albian (Carstone) is there evidence for an appreciable change in sediment provenance, recognized by the appearance of an assemblage 5 suite. The contrasting provenance history of the East Midlands Shelf with respect to the Weald and Central Channel basins can be related to the distinct structural development of the Shelf area during Late Jurassic and Early Cretaceous times.

Volgian to Barremian. The Upper Jurassic and Lower Cretaceous stratigraphical record for the East Midlands shelf indicates slow, punctuated sedimentation (Rawson & Riley 1982). The assemblage 1 pebble suite of the Spilsby Sandstone (late Volgian to Valanginian) implies derivation similar to that of the Lydite Beds suite (Kimmeridgian to Early Volgian) in the south Midlands. However, investigation into the distribution of Mesozoic rocks in the southern North Sea suggests that sandstones of Spilsby Sandstone age were derived from the Sole Pit and Winterton High areas (Glennie & Boegner 1981; van Hoorn 1987; Badley *et al.* 1989). Uplift and erosion of the Sole Pit area during latest Jurassic and earliest Cretaceous times therefore appears to have directed detritus onto the East Midlands

Shelf, the sediment being reworked from older Mesozoic rocks which included pebble beds having assemblage 1 compositions. This influx occurred at the expense of detritus from the northern margin of the Anglo-Brabant Massif, an area presumably of low relief.

The uplift of the Sole Pit area during Early Cretaceous times appears to have coincided with only subdued subsidence on the East Midlands Shelf (Chadwick 1985a). Negligible fresh sediment was introduced onto the shelf at this time with pebbles locally reworked from the Spilsby Sandstone at the northern margin of the shelf, probably in association with movements on the Market Weighton High (Kent 1978; Kirby & Swallow 1987).

Aptian to Albian. The interpretation of seismic lines crossing the South Hewett fault given by Badley *et al.* (1989) indicates that sedimentation in the Sole Pit area was renewed by Albian times. This was probably a significant event being coeval with the introduction of fresh detritus from the Anglo-Brabant Massif and Market Weighton High onto the East Midlands Shelf. The regional Aptian and Albian marine transgressions onto these areas are therefore assumed to have resulted in their rejuvenation and transport of detritus onto the Shelf.

Conclusions

Variations in composition of the coarse-grained detritus geographically and stratigraphically enable recognition of six principal clast assemblages in the Upper Jurassic and Lower Cretaceous rocks of southern Britain and northern France. The pebble assemblages are characteristic of their source areas, the principal basin-bounding massifs, and define the distribution of detritus derived from each. A systematic change in provenance of the detrital suites during Late Jurassic to Early Cretaceous times in the Weald and Central Channel Basins reflects uplift and erosion of the massifs in response to basin rifting and opening of the Bay of Biscay and the North Atlantic. On the East Midlands Shelf, however, subdued Early Cretaceous subsidence in conjunction with minor erosion at the shelf margins (*e.g.* Sole Pit and Market Weighton areas) caused only continued reworking of older pebble suites and severely restricted the introduction of fresh detritus until Albian times.

Work for this study was undertaken at Southampton University, but many of the ideas have been developed subsequently at Badley, Ashton & Associates. Discussion with A. Roberts, M. Ashton and

P. Allen proved beneficial and they are gratefully thanked. The financial support of BP Exploration Company and Badley, Ashton & Associates is also acknowledged.

References

ALLEN, P. 1954. Geology and geography of the London-North Sea Uplands in Wealden times. *Geological Magazine*, **91**, 498–508.

—— 1959. The Wealden environment: Anglo-Paris basin. *Philosophical Transactions of the Royal Society, London*, **B242**, 283–346.

—— 1960. Strand-line pebbles in the mid-Hastings Beds and the geology of the London Uplands. General features. Jurassic pebbles. *Proceedings of the Geologists' Association, London*, **71**, 156–168.

—— 1961. Strand-line pebbles in the mid-Hastings Beds and the geology of the London Uplands. Carboniferous pebbles. *Proceedings of the Geologists' Association, London*, **72**, 271–285.

—— 1965. L'age du Purbecko-Wealdien d'Angleterre. *Mémoires du Bureau Récherches Géologique et Minières*, **34**, 321–326.

—— 1967a. Origin of the Hastings facies in Northwestern Europe. *Proceedings of the Geologists' Association, London*, **78**, 27–105.

—— 1967b. Strand-line pebbles in the mid-Hastings Beds and the geology of the London Uplands. Old Red Sandstone, New Red Sandstone and other pebbles. Conclusion. *Proceedings of the Geologists' Association, London*, **78**, 241–276.

—— 1969. Lower Cretaceous sourcelands and the North Atlantic. *Nature*, **222**, 657–658.

—— 1972. Wealden detrital tourmaline: implications for northwestern Europe. *Journal of the Geological Society, London*, **128**, 273–294.

—— 1981. Pursuit of Wealden models. *Journal of the Geological Society, London*, **138**, 375–405.

—— 1990. Provenance research: Torridonian and Wealden examples. This volume.

ARKELL, W. J. 1939. Derived ammonites from the Lower Greensand of Surrey and their bearing on the tectonic history of the Hog's Back. *Proceedings of the Geologists' Association, London*, **50**, 22–25.

—— 1947. *The Geology of Oxford*, Clarendon, Oxford.

BADLEY, M. E., PRICE, J. D. & BACKSHALL, L. C. 1989. Inversion, reactivated faults and related structures: seismic examples from the Southern North Sea. *In*: COOPER, M. A. & WILLIAMS, G. D. (eds) *Inversion Tectonics*. Geological Society, London, Special Publication, **44**, 201–219.

BARR, D. 1987. Lithospheric stretching, detached normal faulting and footwall uplift. *In*: COWARD, M. P., DEWEY, J. F. & HANCOCK, P. L. (eds) *Continental Extensional Tectonics*. Geological Society, London, Special Publication, **28**, 75–94.

CAMPBELL-SMITH, W. 1963. Description of the igneous rocks represented amongst pebbles from the Bunter pebble beds of the Midlands of England. *Bulletin of the British Museum (Natural History) (Mineralogy)*, **2**, 3–17.

CASEY, R. 1961. The stratigraphical palaeontology of the Lower Greensand. *Palaeontology*, **3**, 487–621.

—— & BRISTOW, C. R. 1964. Notes on the ferruginous strata in Buckinghamshire and Wiltshire. *Geological Magazine*, **101**, 116–128.

CHADWICK, R. A. 1985a. Permian, Mesozoic and Cenozoic structural evolution of England and Wales in relation to the principles of extension and inversion tectonics. *In*: WHITTAKER, A. (ed.) *Atlas of Onshore Sedimentary Basins in England and Wales: Post-Carboniferous Tectonics and Stratigraphy*. British Geological Survey, Blackie, Glasgow, 9–25.

—— 1985b. End Jurassic-early Cretaceous sedimentation and subsidence (late Portlandian to Barremian), and the late-Cimmerian unconformity. *In*: WHITTAKER, A. (ed.) *Atlas of Onshore Sedimentary Basins in England and Wales: Post-Carboniferous Tectonics and Stratigraphy*. British Geological Survey, Blackie, Glasgow, 52–56.

—— 1986. Extensional tectonics in the Wessex Basin, southern England. *Journal of the Geological Society, London*, **143**, 465–488.

COPE, J. C. W., DUFF, K. L., PARSONS, C. F., TORRENS, H. S., WIMBLEDON, W. A. & WRIGHT, J. K. 1980. *A Correlation of the Jurassic rocks in the British Isles. Part 2, Middle and Upper Jurassic*. Geological Society, London, Special Report **15**.

ENSOM, P. C. 1985. Derived fossils in the Purbeck Limestone Formation, Worbarrow Tout, Dorset, *Proceedings of the Dorset Natural History and Archaeological Society*, **106**, 166.

FOSTER, D., HOLLIDAY, D. W., JONES, C. M., OWEN, B. & WELSH, A. 1989. The concealed Upper Palaeozoic rocks of Berkshire and South Oxfordshire. *Proceedings of the Geologists' Association, London*, **100**, 395–407.

GLENNIE, K. W. & BOEGNER, P. L. E. 1981. Sole Pit inversion tectonics. *In*: ILLING, L. V. & HOBSON, G. D. (eds) *Geology of the Continental Shelf of North-west Europe*. Institute of Petroleum, Heyden, London, 110–120.

GOODE, A. J. J. & TAYLOR, R. T. 1980. Intrusive and pneumatolitic breccias in the south-west of England. *Report of the Institute of Geological Science*, **80/2**, 1–16.

HALLAM, A. 1978. Eustatic cycles in the Jurassic. *Palaeogeography, Palaeoclimatology, Palaeoecology*, **23**, 1–32.

HAQ, B. V., HARDENBOL, J. & VAIL, P. R. 1987. Chronology of fluctuating sea levels since the Triassic. *Science*, **235**, 1156–1167.

JACKSON, J. & MCKENZIE, D. 1983. The geometrical evolution of normal fault systems. *Journal of Structural Geology*, **5**, 471–482.

JUIGNET, P., RIOULT, M. & DESTOMBES, P. 1973. Boreal influences in the Upper Aptian-Lower Albian beds of Normandy, northwest France. *In*: CASEY, R. & RAWSON, P. F. (eds) *The Boreal Lower Cretaceous*. Geological Journal Special Issue, **5**, 303–326.

KAMERLING, P. 1979. The geology and hydrocarbon habitat of the Bristol Channel basin. *Journal of Petroleum Geology*, **2**, 75–93.

KELLY, S. R. A. 1980. *Hiatella*—a Jurassic bivalve squatter? *Palaeontology*, **23**, 769–781.

KENT, P. E. 1978. Subsidence and uplift in east Yorkshire and Lincolnshire: a double inversion. *Proceedings of the Yorkshire Geological Society*, **42**, 505–524.

KIRBY, G. A. & SWALLOW, P. W. 1987. Tectonism and sedimentation in the Flamborough Head region of north-east England. *Proceedings of the Yorkshire Geological Society*, **46**, 301–309.

KIRKALDY, J. F. 1947. The provenance of the pebbles in the Lower Cretaceous rocks. *Proceedings of the Geologists' Association, London*, **58**, 223–241.

LAKE, S. D. & KARNER, G. D. 1987. The structure and evolution of the Wessex Basin, Southern England: an example of inversion tectonics. *Tectonophysics*, **137**, 347–378.

LAMONT, A. 1946. Fossils from Midland Bunter Pebbles collected in Birmingham. *Geological Magazine*, **83**, 39–44.

LATTER, M. P. 1926. The petrography of the Portland Sand of Dorset. *Proceedings of the Geologists' Association, London*, **37**, 73–91.

MIDDLEMISS, F. A. 1967. Analysis of structure in a region of gentle *en echelon* folding. *Jahrbuch für Mineralogie, Geologie und Palaeontologie Abhandlungen*, **129**, 137–156.

—— 1975. Studies in the sedimentation of the Lower Greensand of the Weald. 1875–1975: a review and commentary. *Proceedings of the Geologists' Association, London*, **86**, 457–473.

NEAVERSON, E. 1925. The petrography of the Upper Kimmeridge Clay and the Portland Sand in Dorset, Wiltshire, Oxfordshire and Buckinghamshire. *Proceedings of the Geologists' Association, London*, **36**, 240–256.

OAKLEY, K. P. 1947. A note on Palaeozoic radiolarian chert pebbles found in the Wealden Series of Dorset. *Proceedings of the Geologists' Association, London*, **58**, 255–258.

OWEN, H. G. 1971. Middle Albian stratigraphy in the Anglo-Paris basin *Bulletin of the British Museum (Natural History) (Geology) Supplement*, **8**, 1–164.

RASTALL, R. H. 1930. The petrography of the Hunstanton Red Rock. *Geological Magazine*, **67**, 436–458.

RAWSON, P. F. & RILEY, L. A. 1982. Latest Jurassic-early Cretaceous events and the "Late Cimmerian Unconformity" in the North Sea area. *Bulletin of the American Association of Petroleum Geologists*, **66**, 2628–2648.

——, CURRY, D., DILLEY, F. C., HANCOCK, J M., KENNEDY, W. J., NEAL, J. W., WOOD, C. J. & WORSSAM, B. C. 1978. A correlation of Cretaceous rocks in the British Isles. Geological Society, London, Special Report, **9**.

ROBERTS, A. M. & YIELDING, G. 1991. Deformation around basin-margin faults in the North Sea/Mid-Norway rift. *In*: ROBERTS, A. M., YIELDING, G. & FREEMAN, B. (eds) *The Geometry of Normal Faults*. Geological Society, London, Special Publication (in press).

SELLWOOD, B. W. & SCOTT, J. 1986. A geological map of the sub-Mesozoic floor beneath Southern England. *Proceedings of the Geologists' Association, London*, **97**, 81–85.

SIMPSON, I. R., GRAVESTOCK, M., HAM, D., LEACH, H. & THOMPSON, S. D. 1989. Notes and cross-sections illustrating inversion tectonics in the Wessex Basin. *In*: COOPER, M. A. & WILLIAMS, G. D. (eds) *Inversion Tectonics*. Geological Society, London, Special Publication, **44**, 123–129.

SLADEN, C. P. & BATTEN, D. J. 1984. Source-area environments of Late Jurassic and Early Cretaceous sediments in southeast England. *Proceedings of the Geologists' Association, London*, **95**, 149–163.

SMITH, N. J. P. 1985. The pre-Permian subcrop map. *In*: WHITTAKER, A. (ed.) *Atlas of Onshore Sedimentary Basins in England and Wales: Post-Carboniferous Tectonics and Stratigraphy*. British Geological Survey, Blackie, Glasgow, 6–8.

VAIL, P. R. & TODD, R. G. 1981. Northern North Sea Jurassic unconformities, chronostratigraphy and sea-level changes from seismic stratigraphy. *In*: ILLING, L. V. & HOBSON, G. D. (eds) *Petroleum Geology of the Continental Shelf of North-west Europe*. Institute of Petroleum, Heyden, London, 216–235.

VAN HOORN, B. 1987. Structural evolution, timing and tectonic style of the Sole Pit inversion. *Tectonophysics*, **137**, 239–294.

VERSEY, H. C. & CARTER, C. 1926. The petrography of the Carstone and associated beds in Yorkshire and Lincolnshire. *Proceedings of the Yorkshire Geological Society*, **20**, 349–365.

WELLS, A. K. 1947. On the origin of the oolitic, spherulitic and rhomb-bearing cherts. *Proceedings of the Geologists' Association, London*, **58**, 242–255.

—— & GOSSLING, F. 1947. A study of the pebble beds in the Lower Greensand in east Surrey and west Kent. *Proceedings of the Geologists' Association, London*, **58**, 194–222.

WELLS, M. K. 1946. A contribution to the study of luxullianite. *Mineralogical Magazine*, **27**, 186–194.

WEST, I. M. & HOPPER, M. J. 1969. Detrital Portland chert and limestone in the Upper Purbeck Beds, at Friar Waddon, Dorset. *Geological Magazine*, **106**, 277–280.

WHITTAKER, A. (ed.) 1985. *Atlas of Onshore Sedimentary Basins in England and Wales. Post-Carboniferous Tectonics and Stratigraphy*. British Geological Survey, Blackie, Glasgow.

WILLS, L. J. 1956. *Concealed Coalfields. A Palaeogeographical Study of the Stratigraphy and Tectonics of Mid-England in Relation to Coal Reserves*. Blackie, Glasgow.

ZIEGLER, P. A. 1978. North-west Europe: tectonics and basin development. *Geologie en Mijnbouw*, **57**, 589–626.

—— 1981. Evolution of sedimentary basins in north-west Europe. *In*: ILLING, L. V. & HOBSON, G. D. (eds) *Petroleum Geology of the Continental Shelf of North-west Europe*. Institute of Petroleum, Heyden, London, 3–39.

Sandstones of arc and ophiolite provenance in backarc basin, Halmahera, eastern Indonesia

GARY NICHOLS[1], KUSNAMA[2] & ROBERT HALL[2]

[1] *Department of Geology, Royal Holloway and Bedford New College, University of London, Egham, Surrey TW20 0EX, UK*

[2] *Department of Geological Sciences, University College London, Gower Street, London WC1E 6BT, UK*

Abstract: Analyses of sedimentary rocks from a late Neogene backarc basin on the island of Halmahera in eastern Indonesia show that the detrital mineral assemblages in the sandstones have distinctive characteristics. Quartz is extremely rare in the entire sequence of up to 4 km of basin sediments, clearly indicating that there has been no input from continental sources throughout the basin history. Sediments derived from two distinct provenance areas are recognized: sandstones dominated by volcanic rocks occur in the western half of the basin and in the eastern part there are black sands composed largely of ultrabasic debris interbedded with carbonate mudstones. These sandstone petrographies reflect the nature of the terrains which have bordered the Halmahera Basin throughout its history and other small basins formed in an 'oceanic' setting may show similar characteristics. Temporal changes in the volcaniclastic components are attributed to stages in the evolution of the adjacent volcanic arc and do not reflect the degree of dissection of an arc massif.

Numerous sedimentary basins of variable sizes have formed during the Cenozoic in the complex mosaic of small plates at the west Pacific margin (Fig. 1). These basins lie in forearc and back-arc collisional settings and in strike-slip zones in areas of oblique collision. The nature of the fill of these sedimentary basins is controlled by the equatorial latitude (and hence high biogenic productivity) and the nature of the surrounding terranes. These include small pieces of continental crust detached by rifting and strike-slip faulting, old arc volcanic rocks and related sedimentary rocks, pieces of oceanic crust structurally imbricated with these arc massifs and active volcanic arcs. Each of these types of source area provide distinctive suites of mineral grains and lithic clasts to the adjacent basins and their contribution can be assessed from petrographic studies of the basin-fill sediments.

Sandstones with low quartz contents and a high proportion of volcanic lithic fragments are characteristic of undissected arc provenances (Dickinson & Suczek 1979). Dickinson (1982) demonstrates that forearc sandstones of the Circum-Pacific belt have similar compositions; variations in the detrital modes are interpreted to reflect varying degrees of dissection of the volcano-plutonic complexes of arc massifs. Dickinson (1982) notes that these sands have a consistent feldspar to quartz ratio of between 1:1 and 1:2 and that the petrology of these sands

may serve as a reliable guide for the recognition of analogous sequences in other arcs.

In this paper, a small Neogene basin in a backarc setting adjacent to the Halmahera island arc in Eastern Indonesia is considered (Fig. 1). Throughout its history this basin received detritus only from arc and ophiolitic terranes. These terranes originated in the west Pacific, far from any continental crust and throughout the history of the basin there has been no influx of material from the Australian continental margin which lies to the south. The arenites of the Halmahera Basin are remarkable and distinctive because they are quartz-free; the clastic components are dominated by material derived from island arc volcanic and ultrabasic or gabbroic ophiolitic rocks. The sedimentary sequence in this basin is an interesting example of a basin formed in an 'oceanic' setting which may later become incorporated into an orogenic belt as collision proceeds.

Tectonic setting

The island of Halmahera lies in the middle of a number of small plates at the convergent triple junction where the Pacific, Eurasian and Indian/Australian plates meet (Hamilton 1979). It presently lies at the southwest corner of the Philippine Sea Plate (Nichols *et al.* 1990). To the

From Morton, A. C., Todd, S. P. & Haughton, P. D. W. (eds), 1991,
Developments in Sedimentary Provenance Studies.
Geological Society Special Publication No. 57, pp. 291–303.

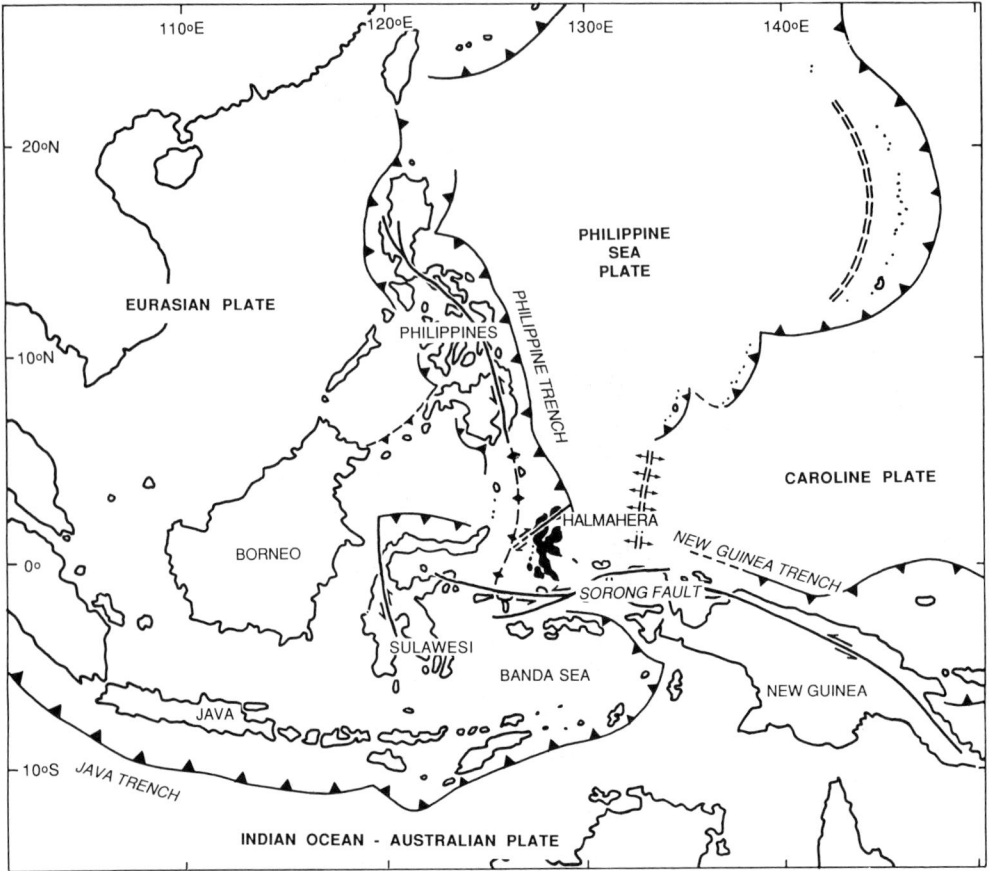

Fig. 1. The tectonic setting of Halmahera in the centre of a mosaic of plates at the triple junction between the Pacific, Eurasian and Indian/Australian plates (from Nichols *et al.* 1990).

south, the Sorong Fault Zone is a strike-slip plate boundary with the Australian Plate; to the west convergence between the Philippine Sea Plate and the Eurasian margin is taken up in the dual subduction systems of the Molucca Sea Collision Zone (Silver & Moore 1978; Moore & Silver 1983).

During the late Neogene, the Halmahera Trench was the site of eastward subduction of the Molucca Sea Plate beneath Halmahera and there is still an associated active volcanic arc along the west side of the island (Hatherton & Dickinson 1969). Subduction at the Halmahera Trench commenced in the Late Miocene (Hall 1987; Nichols & Hall 1990). Up until this time, the Miocene had been a period of shallow marine carbonate sedimentation over most of Halmahera and adjacent areas. In the late Miocene uplift along the western edge of the island and downwarp to the east led to the formation of a basin, termed the 'Halmahera Basin' (Nichols &

Hall 1990). The Halmahera Basin was a broad (at least 200 km perpendicular to the arc) back-arc basin from the Late Miocene until it was modified by east–west compression across the island at the end of the Pliocene (Nichols & Hall 1990).

The island of Halmahera has a distinctive K-shape of four arms linked by a central zone (Fig. 2). This study has concentrated on the south-eastern and southwestern arms of the island where late Miocene to Pliocene sedimentary rocks of the Halmahera basin form extensive areas of outcrop. Exposure on the island is limited to sections in deeply incised rivers in a rugged area which is totally covered by tropical rain forest and largely without tracks.

Stratigraphy

The stratigraphy of Halmahera is summarized in

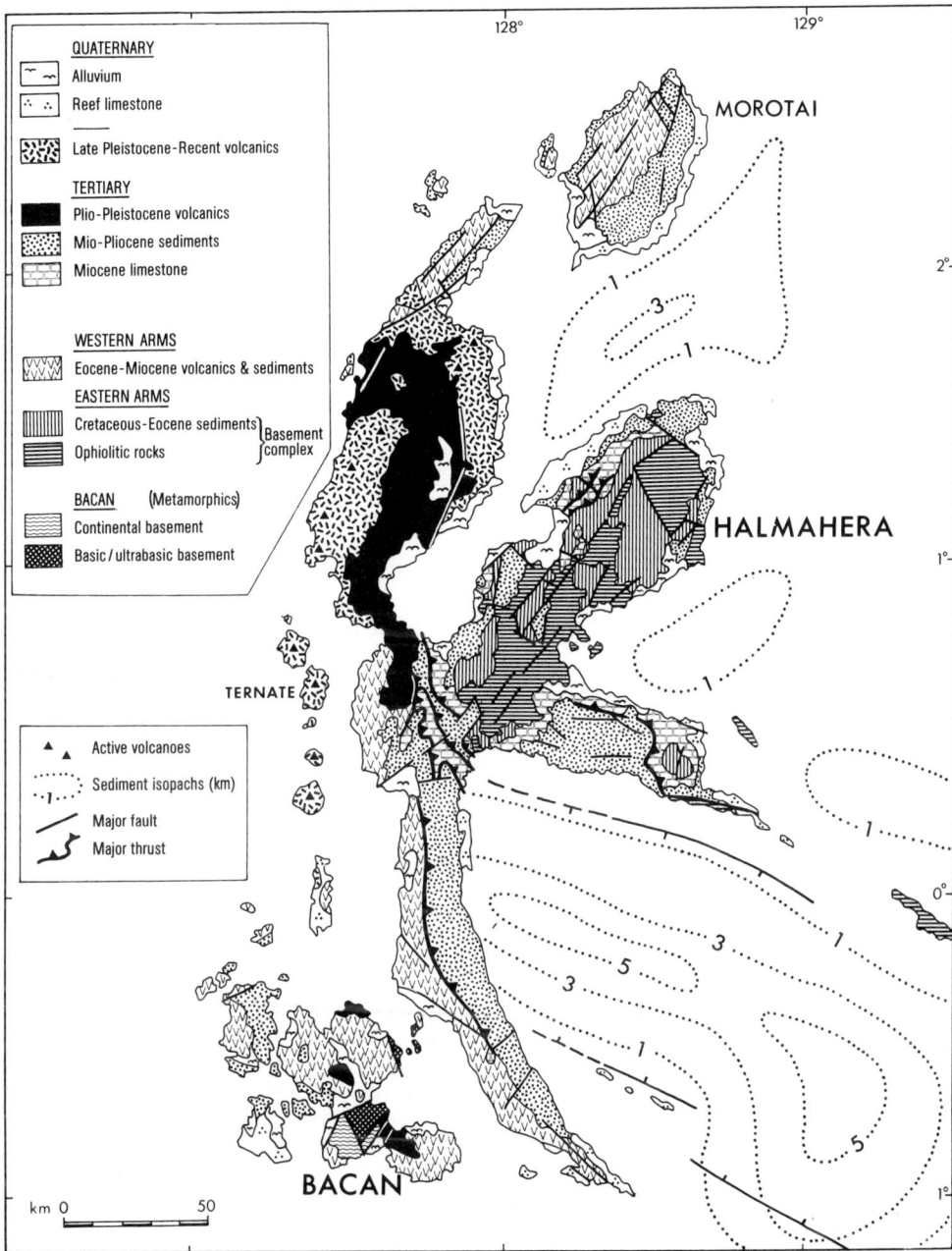

Fig. 2. Outline of the geology of Halmahera The Mio-Pliocene sedimentary rocks in the southern part of Halmahera form the Weda Group. The basement sediments in eastern Halmahera are the Buli Group and the volcanic basement of western Halmahera is the Oha Volcanic Formation. The central fold zone lies north–south across the centre of the island.

Fig. 3. The basement of the eastern part of the island is made up of a Mesozoic ophiolite imbricated with a complex of Upper Cretaceous and Eocene volcaniclastic rocks and limestones deposited in a forearc setting, the Buli Group (Hall *et al.* 1988a, b). This basement was deformed, uplifted and partly eroded before the beginning of the Miocene; basal conglomerates

CENTRAL AND SW ARM (WEST SIDE)	SW ARM (EAST SIDE)	SE AND NE ARMS

QUATERNARY VOLCANICS AND VOLCANICLASTICS   QUATERNARY REEF LIMESTONES

TAFONGO VOLCANIC FORMATION
Late Pliocene - mid-Pleistocene
volcanics and volcaniclastics

KULEFU FORMATION
(?Pliocene - Pleistocene)
Tuffaceous sandstone
200m +

GOLA FORMATION
(Late Miocene -) Pliocene
Limestone and calcareous mudstone
300m +

DUFUK FORMATION
(Late Miocene -) Early Pliocene
Shallow marine mudstone, sandstone, conglomerate
1000m - 1500m

SAOLAT FORMATION
Late Miocene - Pliocene
Marlstone, redeposited reef limestone and
littoral sandstone
1000m +

WEDA GROUP

AKELAMO FORMATION
Late Miocene (- Early Pliocene)
Carbonaceous mudstone
0 - 500m

LOKU FORMATION
Late Miocene (- Early Pliocene)
Turbidite and debris flow deposits
1000m +

SUPERAK FORMATION
Late Miocene (- Early Pliocene)
Shallow marine conglomerate and sandstone
1000m - 1500 m

SUBAIM LIMESTONE FORMATION
Early - Late Miocene (? Early Pliocene)
Reef and reef-associated limestone
500m +

OHA VOLCANIC FORMATION
(?Late Cretaceous - Eocene)
Arc volcanics and volcaniclastics

**JAWALI & GEMAF CONGLOMERATE
FORMATIONS**
(?Early Miocene) Fluvial & littoral
conglomerate and sandstone. 0-400m

OPHIOLITIC BASEMENT COMPLEX
(a) BULI GROUP - forearc volcanics,
volcaniclastics and sediments, Late Cretaceous and
Eocene
(b) Ultrabasic and basic rocks of ophiolitic
character

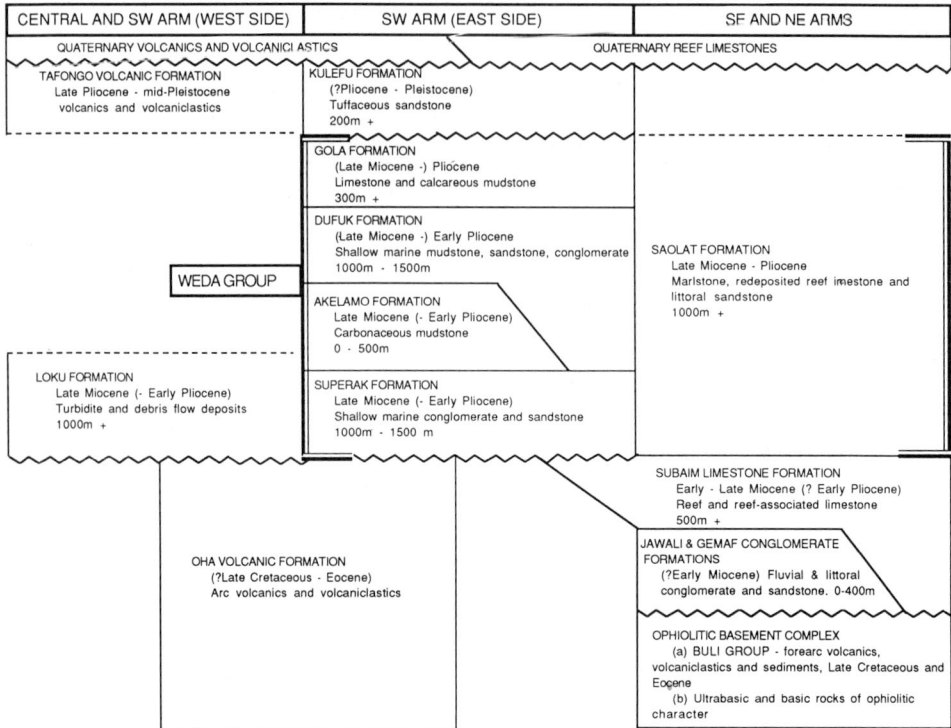

Fig. 3. A summary of the stratigraphy of central, southern and eastern Halmahera. Possible extensions to the age ranges are shown in parentheses. Thicknesses of the formations are shown where they can be estimated.

of fluvial and littoral origin are found underlying a Miocene carbonate sequence up to 500 m thick. These carbonates from the Subaim Limestone Formation (Hall *et al.* 1988*c*) and they are dominantly shallow marine reef and reef-related deposits.

In western Halmahera the basement is a deformed association of volcanic and coarse volcaniclastic rocks of basaltic to andesitic composition, the Oha Volcanic Formation (Hall *et al.* 1988*b*; Hakim 1989). These rocks have not been dated, but they contain pebbles of Eocene limestone and they show similarities to the Upper Cretaceous to Eocene rocks of the Buli Group in Eastern Halmahera; the volcanic rocks of the Oha Volcanic Formation are very similar to the volcanic component of the Buli Group in their petrology, textures, mineralogy and chemistry (Hakim 1989). This suggests that western and eastern Halmahera formed an arc–forearc complex in the Late Cretaceous and Eocene.

Miocene limestones are found as abundant clasts within the Neogene rocks of the western arms of the island; a small outlier of limestone on the southwest arm may also be Miocene in age. These occurrences suggest that the Miocene limestone cover was largely stripped off when the western side of the island was uplifted in the Late Miocene. The Loku Formation lies unconformably on the basement (the Oha Volcanic Formation) and is exposed on the west side of the western arm of Halmahera. It is a sequence of turbiditic sandstones and siltstones which is strongly deformed and overlain by largely undeformed sedimentary rocks of the Weda Group. The nature of the contact is uncertain; it may be a thrust or an unconformity. The Loku Formation has not been included in this study.

Sedimentary rocks of late Neogene age on Halmahera were previously assigned to the Weda Series (Bessho 1944) and to the Weda Formation by Apandi & Sudana (1980), Supriatna (1980) and Yasin (1980). Most of these rocks are now assigned to the Weda Group (Fig. 3) which is subdivided into a number of formations by Hall *et al.* (1988*b, c*) and Nichols & Hall (1990). These formations have been dated as late Miocene to Pliocene in age on the basis of foraminifers and nannofossils. In western Halmahera the Superak and Akelamo Formations are Late Miocene in age; the Dufuk and

Gola Formations are early Pliocene in age. The Saolat Formation of eastern Halmahera is the stratigraphic equivalent of the western arm formations and is late Miocene to early Pliocene in age. In the Central Zone of Halmahera Upper Miocene–Lower Pliocene and Pliocene sedimentary rocks are mapped as undifferentiated Weda Group because the tight folding and the nature of the exposure in this belt precludes division of the sequence into separate formations.

At the northern end of the SW arm rocks assigned to the Kulefu Formation rest unconformably upon the folded rocks of the Weda Group. No fauna has been found in the Kulefu Formation and the formation is considered likely to be Pliocene in age (Nichols & Hall 1990).

Weda Group and younger sedimentary rocks

The Weda Group constitutes the fill of the Halmahera Basin which developed behind the Halmahera Arc by crustal downwarp during the late Neogene (Nichols & Hall 1990). The basin was at least 200 km across from east to west and up to 400 km long, parallel to the Halmahera Trench (Fig. 4). Reconstructions of the Halmahera Basin take into account at least 60 km of late Pliocene east–west shortening across the fold and thrust belt of the Central Zone (Hall *et al.* 1988*b*; Nichols & Hall 1990). The samples discussed in this study are arenites from the Superak, Dufuk and Saolat Formations of the Weda Group, undifferentiated Weda Group of the Central Zone, and the Kulefu Formation.

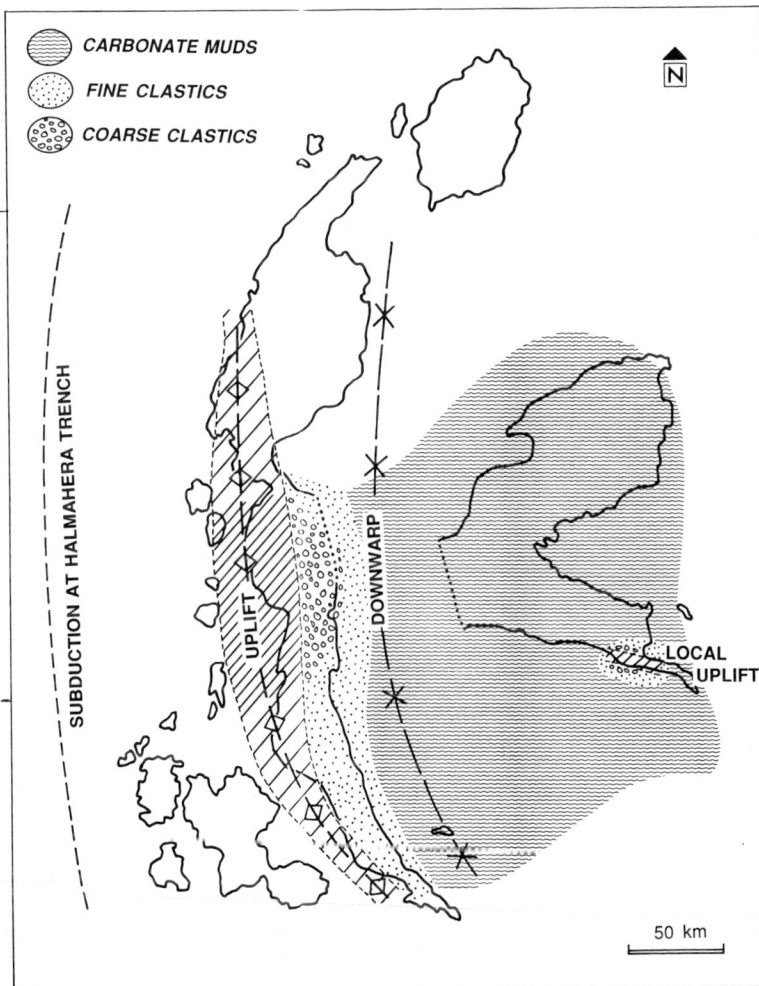

Fig. 4. Palaeogeographic reconstruction of the Halmahera Basin in the late Miocene to Pliocene. 60 km of later Pliocene shortening, calculated by restoring balanced cross-sections through the central fold zone, has been taken into account.

Table 1. *Percentages of the detrital components of 84 sandstones from the Halmahera Basin calculated from point-counting of thin sections*

Sample	Volc:P	Volc:A	Feld	Pyrox	Hblnd	L.UB	L.Sed	Biocl	Organ	Matrix	Miscel
Superak Formation											
HG 2	41.9	11.3	0.0	5.0	0.0	0.0	24.2	3.4	0.0	9.8	4.4
HG 44	1.7	0.8	26.0	1.9	0.0	0.0	15.4	35.0	3.7	10.2	5.2
HG 89	15.3	5.2	42.0	0.0	0.0	0.0	15.6	4.4	2.9	6.7	7.9
HG 97	12.9	5.4	28.4	0.0	0.0	0.0	20.0	7.9	0.0	21.9	3.5
HG 99	39.1	9.3	10.2	0.0	0.0	0.0	25.4	4.2	0.9	9.3	1.6
HG 166	37.3	14.3	5.3	0.0	3.9	0.0	29.7	0.0	0.0	7.6	2.0
HR 13	26.8	1.7	44.7	0.8	9.0	0.0	3.0	0.3	0.0	8.3	5.3
HR 14	35.0	9.8	5.4	6.0	3.2	0.0	12.0	7.1	0.0	17.9	3.7
HR 16	9.5	3.1	20.5	0.3	2.4	0.0	33.8	13.8	1.9	9.8	4.8
HR 18	19.8	7.4	9.1	3.1	7.0	0.0	11.0	19.0	0.0	20.4	3.3
HR 20	27.3	4.8	35.7	1.6	10.6	0.0	4.5	0.0	0.4	8.2	6.9
HR 35	34.6	3.3	38.8	0.4	5.6	0.0	4.3	0.0	0.0	9.5	3.4
HR 37	28.9	5.0	0.0	0.0	8.7	0.0	14.7	23.8	0.0	16.5	2.4
HR 49	41.3	14.3	15.1	13.3	0.0	0.0	2.8	0.0	0.0	9.2	3.9
HR 67	27.0	4.3	29.1	4.4	1.6	0.0	17.8	2.4	1.8	7.0	4.6
HR 79	17.6	8.0	2.1	0.0	0.0	0.0	21.7	36.7	0.0	11.4	2.5
HR 88	67.4	0.0	7.8	5.0	0.0	0.0	0.0	0.0	0.0	19.1	5.6
HR 89	30.0	10.2	10.6	5.0	0.0	0.0	30.1	4.7	0.7	7.5	1.2
HR 90	18.4	4.0	39.3	1.3	0.6	0.0	19.5	5.7	0.0	8.4	2.9
HR 91	46.9	9.8	3.1	3.7	0.0	0.0	24.0	0.0	0.0	8.6	4.0
HR 93	39.9	4.4	12.4	3.9	0.3	0.0	27.4	0.0	0.0	8.3	3.3
HR 95	18.0	4.5	28.3	0.0	0.0	0.0	10.8	25.4	1.5	9.6	1.9
HR 107	12.4	3.5	10.6	0.8	0.0	0.0	28.4	31.8	3.0	5.5	4.1
HR 144	6.9	0.0	6.1	0.0	0.0	0.0	4.9	64.9	3.0	11.8	2.4
Dufuk Formation											
HG 11	15.7	4.0	20.2	0.0	3.0	0.0	17.3	19.8	0.0	17.7	2.3
HG 14	3.4	0.0	15.2	1.1	1.4	0.0	31.1	21.6	8.2	12.3	5.6
HG 19	0.3	0.0	25.1	0.0	0.8	0.0	15.0	29.1	5.5	16.8	7.5
HG 20	8.0	3.1	35.6	1.5	3.3	0.0	18.8	11.3	0.0	15.4	3.1
HG 22	23.3	2.3	38.1	1.9	1.1	0.0	19.7	3.2	1.3	6.8	2.3
HG 103	17.3	0.0	7.1	1.1	0.0	0.0	37.6	3.1	3.2	23.2	7.3
HR 23	21.7	3.1	42.3	8.8	14.1	0.0	0.0	0.0	0.0	6.2	3.9
HR 28	14.9	2.4	19.9	0.2	5.4	0.0	33.7	13.2	0.0	6.7	3.6
HR 31	17.7	6.6	13.2	8.3	1.7	0.0	27.1	6.5	0.0	11.0	7.7
HR 38	0.0	0.0	6.5	0.0	0.0	0.0	11.7	57.4	0.0	19.0	5.4
HR 39	2.3	0.0	14.8	0.2	0.0	0.0	24.2	29.9	2.7	21.4	4.6
HR 40	0.0	0.0	4.5	0.0	0.0	0.0	19.5	55.6	0.0	17.9	2.5
HR 44	20.2	2.6	36.1	3.1	0.0	0.0	18.8	0.0	2.4	8.6	8.2
HR 110	23.3	8.5	39.9	3.5	8.0	0.0	4.2	0.0	0.0	9.2	3.4
HR 111	4.1	0.0	17.1	3.0	5.0	0.0	18.0	32.1	3.8	12.2	4.9
HR 115	8.4	1.9	19.0	0.0	3.8	0.0	12.8	39.0	3.5	6.8	4.7
HR 116	6.6	0.0	30.6	0.0	6.8	0.0	18.7	0.0	8.8	23.3	5.2
HR 117	11.9	0.0	29.7	2.0	6.6	0.0	12.7	22.0	3.6	7.1	4.4
HR 125	30.4	9.3	31.4	3.1	1.9	0.0	13.1	0.6	0.0	9.6	0.7
HR 147	31.2	9.1	23.9	0.0	0.0	0.0	17.2	7.3	0.0	8.6	2.7

Fold zone

Sample											
HG 149	56.4	3.4	10.0	0.8	5.8	0.0	10.4	2.1	2.7	6.8	1.5
HG 156	19.8	6.3	11.0	1.4	1.7	0.0	14.6	18.7	1.6	19.2	5.5
HG 157	5.4	12.7	37.2	4.6	2.8	0.0	18.5	0.0	1.1	8.4	9.3
HG 158	20.7	5.5	20.8	12.6	2.6	0.0	20.0	1.1	0.0	10.6	6.0
HG 165	25.4	7.2	14.5	6.1	2.1	0.0	24.9	5.0	1.2	8.3	5.3
HG 183	12.9	4.1	32.4	2.6	0.0	0.0	23.0	6.1	1.5	15.8	1.7
HG 184	27.3	6.4	30.8	4.1	1.9	0.0	15.8	0.0	0.0	11.5	2.2
HG 186	10.6	1.1	15.9	1.4	1.4	0.0	20.2	28.8	0.0	15.1	5.6
HG 188	13.2	4.1	39.2	3.7	1.2	0.0	16.6	2.7	2.1	10.0	7.1
HG 190	22.6	7.6	46.1	4.4	3.2	0.0	3.8	0.0	0.0	9.7	2.5

Saolat Formation

Sample											
HG 202	1.4	1.7	7.8	3.7	0.0	34.6	27.8	8.1	1.0	10.4	3.4
HG 208	0.0	0.0	0.0	0.0	0.0	6.1	35.8	29.4	0.0	22.3	6.4
HG 216	0.0	0.0	2.9	2.2	0.0	59.7	25.0	0.0	0.0	7.6	2.5
HG 217	1.0	0.0	0.6	0.0	0.0	61.7	20.5	0.0	0.0	8.7	8.6
HG 218	0.3	0.0	0.4	0.0	0.0	61.3	14.6	1.3	0.0	19.2	2.2
HG 222	0.0	0.0	3.1	1.4	0.7	44.5	11.9	13.5	0.0	19.9	4.5
HG 246	0.0	0.0	0.0	0.0	0.0	0.0	8.7	70.5	0.0	19.4	1.3
HG 252	2.9	0.0	1.4	1.6	0.0	47.0	20.3	3.6	0.0	22.0	4.2
HG 259	19.8	5.0	0.0	1.0	0.0	52.8	19.0	0.0	0.0	16.0	8.4
HG 261	0.0	0.0	0.5	1.3	0.0	35.9	18.3	1.7	0.7	10.6	6.2
HG 264	0.0	0.0	3.0	3.3	1.7	46.1	24.4	0.0	0.0	15.4	6.1
HG 267	0.0	0.0	0.0	0.0	0.0	57.2	31.0	0.0	0.0	8.4	3.4
HL 34	0.0	0.0	0.4	8.7	0.0	70.7	4.5	0.3	0.0	9.4	6.1
HL 39	0.0	0.0	0.0	10.9	0.0	48.4	10.4	12.4	0.0	12.5	5.3
HL 41	0.0	0.0	2.6	9.0	0.1	49.9	14.0	3.3	0.0	11.4	9.7
HP 76	0.0	0.0	1.3	0.0	0.0	77.0	11.6	0.0	0.0	7.1	2.9
HP 78	0.0	0.0	0.0	0.0	0.0	66.2	15.5	0.0	0.0	9.3	9.0
HR 216	0.0	0.0	0.0	3.4	0.3	80.3	3.9	0.0	0.0	6.9	5.5
HR 246	0.0	0.0	0.0	2.4	0.0	63.4	19.7	0.0	0.0	9.8	4.4
HR 247	0.8	0.0	0.0	0.0	0.0	78.3	10.3	0.0	0.0	7.8	2.8
HR 334	0.0	0.0	1.0	1.2	0.0	70.3	10.5	1.9	0.7	10.5	5.2
HR 341	0.0	0.0	2.1	0.3	0.0	71.7	13.4	0.0	0.0	7.2	4.5
HR 342	0.8	0.0	0.0	0.3	0.0	74.9	14.5	0.0	0.0	6.0	3.4
HR 343	0.0	0.0	0.6	0.0	0.0	73.0	22.0	0.0	0.0	2.6	1.6
HR 367	0.0	0.0	8.9	0.0	0.0	19.1	3.8	41.2	0.0	23.6	3.5

Kulefu Formation

Sample											
HG 27	4.8	0.0	32.2	0.5	1.2	0.0	11.9	14.6	1.4	28.5	5.0
HG 28	37.0	4.5	31.7	2.6	11.3	0.0	3.8	0.0	0.0	7.9	1.2
HG 137	2.5	1.7	31.7	0.8	20.4	0.0	10.6	0.0	0.0	25.9	4.0
HG 138	1.3	2.4	40.2	3.3	13.9	0.0	8.1	0.0	2.2	23.5	4.3
HG 139	18.5	1.0	47.4	3.8	9.5	0.0	6.5	0.0	0.0	9.3	1.4

The Superak Formation is a sequence of conglomerates and sandstones interpreted as fan-delta and shallow marine deposits (Hall *et al.* 1988*b*). The Akelamo Formation is a poorly exposed sequence of calcareous mudstones found in discontinuous outcrops above the Superak Formation and is devoid of arenaceous material. Overlying the Superak and Akelamo Formations is the Dufuk Formation; this is a shallow marine sequence of bioturbated and cross-bedded sandstones, siltstones, calcareous mudstones and intraformational conglomerates. The Gola Formation is poorly exposed and is composed of limestone and calcareous mudstones without any sandy beds. The Saolat Formation is exposed over a large area on the southeastern arm of Halmahera. It is a mainly thick bedded calcareous mudstone and marl with some limestone beds. There are also beds of distinctive black sandstone and fine conglomerate. The arenites and rudites are well sorted, the clasts are well rounded and the beds display low angle cross stratification; this suggests deposition in a littoral environment (Hall *et al.* 1988*b*; Nichols & Hall 1990). The Kulefu Formation includes cross-bedded and bioturbated tuffaceous sandstones.

Data set

The data presented in Table 1 and used in the generation of the triangular plots and graphs in this paper were obtained by point counting thin sections of sedimentary rocks. The material under the cross wires of the microscope was identified at 0.2 mm intervals in ribbons across the slide and for each thin section between 500 and 600 counts were made. Fixed interval counting was used in preference to grain counting to reflect the relative volume of each constituent; as the selected arenites are all moderately to well sorted, the differences between the results obtained by the two methods are likely to be minor. The data presented are from 80 fine- to medium-grained sandstones selected from the Weda Group. The sandstone samples selected are generally fresh with only very minor alteration of feldspars and ferromagnesian minerals. In places the edges of the grains have been slightly corroded and replaced by calcite. With the exception of some of the aphyric volcanic lithic clasts the lithic fragments are usually easily identifiable.

Points in the count were classified into the 11 categories described below. These were chosen after petrographic examination of samples from each of the formations. As quartz is very rare it is not identified as a separate category; lithic fragments are divided into four groups to reflect the types of rock fragments present. Cement and voids are excluded from the totals.

(i) Porphyritic volcanic lithic clasts are an important component of many of these sandstones. They are volcanic rocks which occur in pieces ranging from fine sand in the sandstones to boulders in conglomerates. They are typically pyroxene and hornblende andesites with large plagioclase phenocrysts.

(ii) Aphyric volcanic lithic clasts are most common in Superak Formation arenites. They are typically basaltic in composition. Volcanic glass, which occurs in the Kulefu Formation, is included in this category.

(iii) Plagioclase feldspars with compositions in the range of andesine to labradorite (determined optically) are very common. Alkali feldspar is absent. The degree of alteration of the feldspars varies; many feldspar grains show complex zoning and inclusion patterns. The freshest feldspars occur in the Kulefu Formation.

(iv) Dark green to brown pleochroic hornblende grains are common, particularly in the Kulefu Formation, where the crystals are very fresh.

(v) Pyroxenes in these sedimentary rocks are principally clinopyroxenes with smaller amounts of orthopyroxenes: they occur in small amounts in many of the samples.

(vi) Sedimentary lithic clasts are mainly mudstones, fine siltstones and limestones and they occur in almost all the samples examined.

(vii) Clasts of serpentinite and other ultrabasic lithic fragments form an important component of the sandstones of eastern Halmahera, making up over 70% of some of the arenites from the Saolat Formation. Ultrabasic lithic clasts are absent from arenites of the SW arm and Central Zone.

(viii) Planktonic and benthic foraminifers are a very common bioclastic component of the sandstones examined. In the Saolat Formation there are thick beds of foraminiferal mudstone, wackestone and packstone. Other bioclastic components in the arenites are fragments of coralline algae, corals, echinoids, polyzoans and molluscs.

(ix) Organic material (wood, spores and resin) constitutes up to 10% of the sand size clasts, but occurs more commonly as finer material disseminated in the matrix.

(x) Clay minerals and micrite are the main components of the matrix to these sandstones. Fine silt-sized feldspars, ferromagnesian minerals, lithic fragments and disseminated organic matter also form part of the matrix.

(xi) The final category of miscellaneous material includes a wide variety of heavy minerals, including chrome spinel in the Saolat Formation, olivine and opaque minerals.

Data presentation

It is usual to present modal analyses of arenites in the form of triangular diagrams. The most commonly used diagram is the quartz–feldspar–lithic (QFL) plot as these three components usually make up the bulk of the clasts. However, because of the extreme scarcity of quartz, either as mineral grains or in lithic fragments, this standard plot is inappropriate for the arenites of the Halmahera Basin. A plot of lithic components is used by Dickinson & Suczek (1979) and Dickinson (1982) for comparison of circum-Pacific sandstones; a similar plot, using sedimentary, volcanic and ultrabasic rock fragments as the three components, is shown in Fig. 5 for the arenites of the Halmahera Basin. A drawback of this method of presentation is that other important and abundant components, particularly feldspar and other mineral grains, are excluded; this plot is hence not truly representative of the composition of the rock. As an alternative to triangular diagrams, a series of stacked bar charts (Fig. 6) has been used to present the

modal compositions of arenites from the Weda Group of the SW arm and the Central Zone. These diagrams allow the variations in more than three components (or groups of components) to be displayed together. In Fig. 6 hornblende and pyroxene crystals have been grouped together and organic debris has been grouped with bioclastic material. The samples in each formation have been ranked in order of abundance of porphyritic volcanic lithic clasts, which is the commonest grain type in these arenites.

Spatial and temporal variations in the framework modes

Spatial variations

The lithic components plot (Fig. 5) shows clearly that the arenites from the eastern part of the island, the Saolat Formation, have a different provenance from those of western Halmahera. Lithic clasts of serpentinite, the main ultrabasic lithic component, are abundant in the Saolat Formation, constituting up to 80% of the rock for some samples. Notably lacking in the arenites of the Saolat Formation in the SE arm is the volcanic mineral and lithic debris typical of the Weda Group in the SW arm, such as zoned

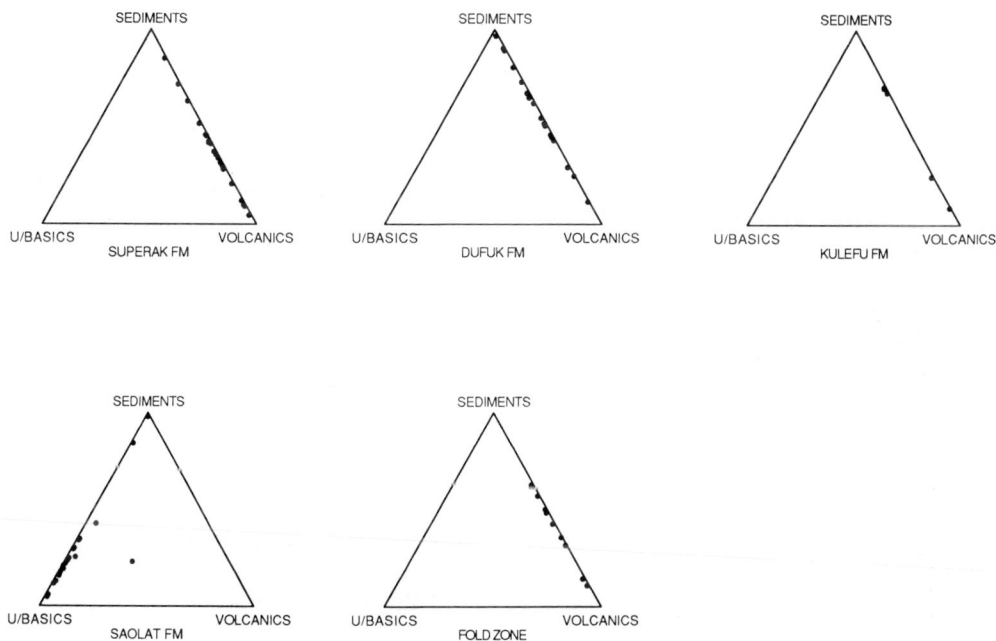

Fig. 5. Ternary plots of the lithic components (lithic fragments of sedimentary, ultrabasic and volcanic rocks) for arenites from the Weda Group and the Kulefu Formation.

SUPERAK FORMATION

DUFUK FORMATION

FOLD ZONE   **KULEFU FM**

- ■ PORPHYRITIC VOLCANICS
- ▨ APHYRIC VOLCANICS
- ⊟ FELDSPARS
- ▤ HORNBLENDE & PYROXENES
- ☐ LITHIC SEDIMENTS
- ☐ BIOGENIC CLASTS
- ▥ MISCELLANEOUS
- ▨ MATRIX

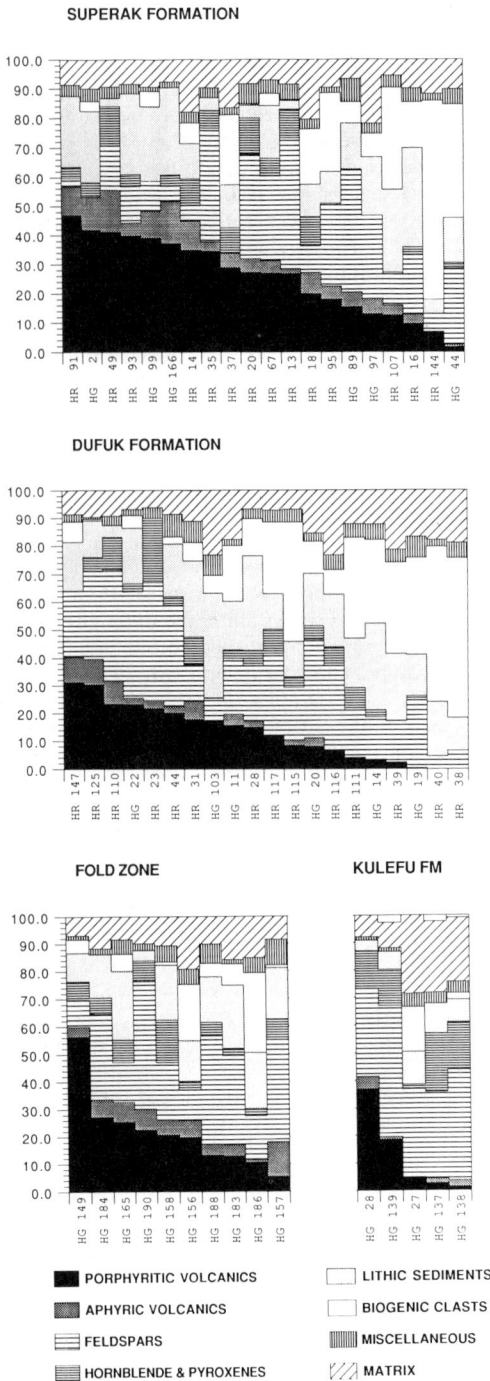

Fig. 6. The principal components of the main arena-ceous formations in the SW arm and the Central Zone presented as stacked bar charts. See text for discussion.

plagioclase feldspars, fresh hornblende, and por-phyritic volcanic lithic clasts.

In contrast, ultrabasic lithic clasts are entirely absent from the Weda Group arenites of the Central Zone and southwest arm. The Superak arenites are dominated by volcanic debris (crys-tal and volcanic lithic grains) with considerable variation in the proportions of each (Fig. 6). The Dufuk arenites are composed mainly of vol-canic debris (crystal and volcanic lithic grains), sedimentary lithic grains and bioclastic debris. Tightly folded and thrust faulted Upper Mio-cene to Pliocene sedimentary rocks in the Cen-tral Zone of the island have the same provenance as the Weda Group arenites of western Halma-hera. The Central Zone arenites are very rich in volcanic mineral and lithic grains with some variation in the proportions of crystals and vol-canic lithic grains. On the basis of their clastic components the Central Zone arenites may be equivalent to either the Superak Formation or the Dufuk Formation or both.

Temporal variations

The main differences between the Superak, Dufuk and Kulefu Formations of the SW arm can be seen on Fig. 6. The older sedimentary rocks (the Superak Formation) are generally richer in volcanic lithics than the overlying Dufuk Formation which has a higher crystal component. The volcanic lithic fragments in the Superak Formation include both porphyritic and aphyric rocks of andesitic and basaltic com-positions. In the Dufuk Formation aphyric vol-canic lithic clasts are less abundant. Compared to the Superak Formation, the Dufuk Forma-tion includes samples with greater proportions of matrix and greater proportions of non-vol-canic debris, reflecting the overall fining-upward trend of the Weda Group. Organic and bioclas-tic material occur in variable proportions in both the Superak and Dufuk Formations. Arenites from the Kulefu Formation are dominated by volcanic material and they are strikingly differ-ent from the older SW arm arenites of the Weda Group. Distinctive features of the Kulefu For-mation arenites include the abundance of fresh ferromagnesian minerals, the presence of small quantities of biotite and quartz, abundant pumice and rare dacite clasts. The implications of these temporal variations in the framework components in the western Halmahera sedimen-tary rocks are discussed in the following section.

Source areas

Eastern Halmahera

The principal source of detritus for the eastern part of the Halmahera Basin were ultrabasic, volcanic and sedimentary rocks from the Ophiolitic Basement Complex. Many of the sandstones in the Saolat Formation are almost exclusively composed of grains of serpentinite. These sandstones are an interesting example of a sharp contrast between the textural and compositional maturities of the sediment. Texturally many of these sandstones are very mature: the grains are well-rounded, they are well to very well-sorted into layers and contain little or no matrix. However, the absence of quartz, the low proportion of feldspars and the high proportion of lithic fragments would normally indicate a very immature sediment on compositional criteria. Serpentine, as a mineral species, is not even mentioned as a possible component of sandstone by other authors (e.g. Pettijohn 1975; Pettijohn *et al.* 1973). In the Saolat Formation the textural maturity is determined by the environment of deposition (a littoral setting), whereas the composition is determined by the available material, in this case a terrain composed of the ultrabasic part of an ophiolite.

Western Halmahera

The sandstones in the western part of the Halmahera Basin contain a very high proportion of material derived from a volcanic source area, including volcanic lithic fragments, plagioclase feldspars, pyroxenes and amphiboles. There are two likely sources for this detritus, the Late Cretaceous to Eocene volcanic basement of the Oha Volcanic Formation or contemporary volcanic rocks from the evolving Halmahera Arc. These two volcanic sources can be distinguished on both textural and compositional grounds (Hakim 1989). The Oha Volcanic Formation rocks are typically aphyric basalts; plagioclase feldspar and clinopyroxene are the principal phenocrysts. The Neogene arc volcanics are generally porphyritic andesites; hornblende and plagioclase are the most common phenocrysts together with minor amounts of clinopyroxene and orthopyroxene.

Conglomerates and sandstones in the Superak Formation include basaltic lithic clasts derived from the Oha Volcanic Formation, but the younger sedimentary rocks of the Weda Group contain many volcanic lithic clasts with an andesitic composition. This indicates that the older volcanic basement was supplying detritus in the early stages of development of the Halmahera Basin, but contemporaneous volcanism in the Halmahera Arc was the principal source of volcaniclastic detritus for most of the Weda Group sedimentary rocks.

A change in the volcanic character of the Halmahera Arc is indicated by the composition of arenites of the Kulefu Formation. The presence of biotite and quartz, pumice and dacite clasts indicates that they are the products of a distinct phase of volcanism in the Halmahera Arc. They are exceptionally rich in volcanic lithic and crystal debris, and the character of the debris suggests an explosive interval of acid volcanism.

Late Miocene to Pliocene palaeogeography

Two provenance areas are recognized for the clastic detritus in the Halmahera Basin in the late Neogene (Fig. 4). Sediments in the western part of the basin were mainly derived from a volcanic terrain of andesitic rocks. The coarsest sediments lay in the central part of western Halmahera with finer sediments to the south and east. This indicates that the source area lay to the west or northwest of the central part of the Halmahera. This area is currently an area of intense volcanic activity related to subduction at the Halmahera Trench.

The western Halmahera basement consists of island arc basalts belonging to the Oha Volcanic Formation which has been uplifted since the Miocene, resulting in the erosion of Miocene limestone cover and the elevation of the basement rocks to over 1000 m above sea level. However, the volcanic lithic fragments in the Weda Group sedimentary rocks are typically andesitic in character and are more similar to the volcanic rocks which form the present day Halmahera Arc (Hakim 1989). This suggests that the volcanism associated with the subduction of the Molucca Sea Plate generated the volcanic rocks which were eroded into the adjacent parts of the south Halmahera sedimentary basin. Once the Halmahera Arc became established, volcanic and volcaniclastic rocks of the arc mantled the Oha Volcanic Formation and these basement rocks were not an important source of detritus after the deposition of the Superak Formation.

Episodes in the activity in the Halmahera Arc and changes in the chemistry of the material erupted can be identified in the Weda Group sedimentary rocks. A period of volcanism in the Pliocene which produced an assemblage of minerals characteristic of more silicic rocks is represented by the Kulefu Formation. Periods of low clastic input are represented by the muddy and calcareous Akelamo and Gola Formations. These are interpreted as periods when the volcanic activity in the southern part of the Halmahera Arc was reduced and uplift to the west of the basin was not sufficient to generate large quantities of detritus.

Although large quantities of volcanic material were shed into the Halmahera Basin at its western edge, very little (if any) of this material was transported as far east as the present outcrops of the Saolat Formation. Taking the late Pliocene east–west shortening into account, these outcrops would have been around 100 km away from the western margin of the basin. In the eastern part of the Weda Group Basin most of the late Miocene to Pliocene sedimentary rocks are intrabasinal carbonates of bioclastic material, principally foraminiferal mudstones and wackestones. Beds of sandstone and conglomerate interbedded with these carbonates were derived from an uplifted block of ophiolite and deposited in a littoral environment. These uplifted blocks within the basin must have been similar to the islands of ophiolitic rocks seen at the present day on the eastern coast of Halmahera.

This pattern of sedimentation persisted until the Halmahera Basin was modified by late Pliocene east–west compression.

Discussion

A sedimentary basin filled by clastic sediments which are almost devoid of quartz, and with serpentinite as an important clastic component in some areas, is clearly unusual but may not be unique in terms of the detrital composition. The west Pacific has been a region of subduction and arc activity throughout the Cenozoic. The oceanward shift of the subduction system currently at the Mariana Trench has resulted in linear remnant arcs within the Philippine Sea Plate. These old arc massifs are now being transported westward as the Philippine Sea Plate converges on the Eurasian margin. Small ocean basins bounded by older arc terrains have also formed to the north and east of New Guinea. As convergence between the Pacific and Eurasian plates has proceeded, some of these arc massifs have become amalgamated with each other and with slices of ophiolitic rocks to form complexes like the basement of Halmahera. Halmahera has clearly not been close enough to continental crust to receive any quartz detritus at any time in its history, and other regions of the west Pacific have had similar histories. Continental crust has only arrived adjacent to Halmahera (to the south) relatively recently as a result of the oblique convergence between the Pacific and Australian plates.

Basins have formed within the convergent zone of the west Pacific by spreading or downwarp in the backarc region, as forearc basins, intra-arc basins or as pull-apart basins between strike-slip faults. Volcaniclastic debris from contemporaneous or older arc terrains will be the principal clastic component of these basins and quartz may be absent where they are not close to any continental margin or continental fragment. Ultrabasic rocks will also contribute detritus where ophiolitic rocks have been imbricated with the arc massifs. When these basins become incorporated into an orogenic belt at the Eurasian or Australian margin their sedimentary infill will clearly show whether the basin formed (or resided) close to continental crust or if it was always surrounded by terrains of exclusively 'oceanic' origin.

The data presented here have implications for the interpretation of arc-related basin sediments. Dickinson (1982) suggested that an upward trend in arc-related sediments from volcanic lithic sandstones to feldspathic sandstones indicates a progressive dissection of the arc. However, the changing volcanic component of arenites in the Halmahera Basin reflects stages in the evolution of the volcanic arc. A pre-existing volcanic basement was uplifted and eroded in the early stages of subduction at the Halmahera Trench, but as the volcanic arc became established, the basement was mantled by younger volcanic rocks; this relationship can be seen today in the north west arm of Halmahera where Plio-Quaternary volcanic rocks overly an older volcanic basement (Fig. 2). Variations in the proportions of volcanic lithic fragments and feldspar crystals in the Halmahera Basin sediments reflect changes in the contemporaneous arc volcanism and in this case does not indicate the degree of dissection of the arc massif.

The authors are grateful for the financial assistance of Amoco, Enterprise Oil, Total Indonesia, Union Texas, and the British Council. Our thanks to D. Pirrie and P. Stone for their helpful reviews.

References

APANDI, T. & SUDANA, D. 1980. *Geological map of the Ternate quadrangle, North Maluku.* Geological Research and Development Centre, Bandung, Indonesia.

BESSHO, B. 1944. [Geology of the Halmahera islands.] *Geographical Journal (LVI),* **664**, 145–203 [in Japanese].

DICKINSON, W. R. 1982. Compositions of Sandstones in Circum-Pacific Subduction Complexes and Fore-Arc Basins. *American Association of Petroleum Geologists Bulletin,* **66**, 121–137.

—— & SUCZEK, C. A. 1979. Plate Tectonics and Sandstone Compositions. *American Association of Petroleum Geologists Bulletin,* **63**, 2164–2182.

HAKIM, A. S. 1989. *Tertiary volcanic rocks from the Halmahera Arc, Indonesia: petrology, geochemistry and low temperature alteration.* MPhil. Thesis, University of London.

HALL, R. 1987. Plate boundary evolution in the Halmahera region, eastern Indonesia. *Tectonophysics,* **144**, 337–352.

——, AUDLEY-CHARLES, M. G., BANNER, F. T., HIDAYAT, S. & TOBING, S. L. 1988*a*. The basement rocks of the Halmahera region, east Indonesia: a Late Cretaceous-Early Tertiary forearc. *Journal of the Geological Society, London,* **145**, 65–84.

——, NICHOLS, G. J., BALLANTYNE, P. D., MILSOM, J. S., CARTER, D. J., HAKIM, A. S. & KUSNAMA, 1988*b*. *UCL-GRDC Halmahera expedition 1987, Third report.* University College London internal report.

——, AUDLEY-CHARLES, M. G., BANNER, F. T., HIDAYAT, S. & TOBING, S. L. 1988*c*. Late Palaeogene-Quaternary Geology of Halmahera, Eastern Indonesia: initiation of a volcanic island arc. *Journal of the Geological Society, London,* **145**,

577–590.

HAMILTON, W. 1979. *Tectonics of the Indonesian region.* US Geological Survey Professional Paper **10 78**.

HATHERTON, T. & DICKINSON, W. R. 1969. The relationship between andesitic volcanism and seismicity in Indonesia, the Lesser Antilles, and other island arcs. *Journal of Geophysical Research,* **74**, 5301–5310.

MOORE, G. F. & SILVER, E. A. 1983. Collision processes in the northern Molucca Sea. *In:* HAYES, D. E. (ed.) The Tectonic and Geologic Evolution of Southeast Asian Seas and Islands, Part 2. American Geophysical Union Monograph **27**, 360–372.

NICHOLS, G. J. & HALL, R. 1990. Basin formation and Neogene sedimentation in a backarc setting, Halmahera, eastern Indonesia. *Marine and Petroleum Geology,* (in press).

——, HALL, R., MILSOM, J. S., MASSON, D. G., PARSON, L., SIKUMBANG, N., DWIYANTO, B. & KALLAGHER, H. J. 1990. The southern termination of the Philippine Trench. *Tectonophysics,* (in press).

PETTIJOHN, F. J. 1975. *Sedimentary Rocks* (Third Edition). Harper and Row, London.

——, POTTER, P. E. & SIEVER, R. 1973. *Sand and Sandstone.* Springer-Verlag, New York.

SILVER, E. A. & MOORE, J. C. 1978. The Molucca Sea collision zone, Indonesia. *Journal of Geophysical Research,* **83**, 1681–1691.

SUPRIATNA, S. 1980. *Geologic map of the Morotai quadrangle, North Maluku.* Geological Research and Development Centre, Bandung, Indonesia.

YASIN, A. 1980. *Geologic map of the Bacan quadrangle, North Maluku.* Geological Research and Development Centre, Bandung, Indonesia.

Nature and record of igneous activity in the Tonga arc, SW Pacific, deduced from the phase chemistry of derived detrital grains

PETER A. CAWOOD

Department of Earth Sciences, Memorial University of Newfoundland, St John's, Newfoundland, Canada A1B 3X5

Abstract: Sedimentary rocks in dredge samples from the Tonga arc, SW Pacific, are simple two-component admixtures of volcanogenic detritus and bioclastic material. The volcanic detritus consists largely of lithic and plagioclase grains, but also includes, in decreasing order of abundance, significant minor amounts of clinopyroxene, orthopyroxene and pigeonite. In addition, amphibole, olivine and opaque oxide minerals are accessory phases and trace amounts of monocrystalline quartz occur in a few samples. Plagioclase grains have high An and low Or contents and fall largely in the bytownite field. Pyroxene grains show progressive Fe enrichment and low Ti contents. Volcanic glasses also show Fe enrichment and have low K_2O contents. These features indicate derivation of the volcanic detritus from a low-K tholeiite source. The uniform compositional range and Oligocene to Holocene age range of the detritus is consistent with their derivation from the adjacent Tongan arc, and indicate that Pliocene rifting of the arc and associated back-arc basin formation did not affect the composition and affinities of the magma source.

Provenance studies are concerned with establishing the source area for the detrital components of sedimentary rocks. Standard techniques for determining provenance are modal analysis of detrital framework components (e.g. Crook 1974; Dickinson & Suczek 1979; Dickinson & Valloni 1980; Dickinson *et al.* 1983; Valloni & Mezzadri 1984) or whole rock sediment geochemistry (Bhatia & Taylor 1981; Bhatia 1983; Bhatia & Crook 1986; Roser & Korsch 1986). Both techniques have established a first-order relationship between sediment composition and tectonic setting, although exceptions to this relationship can occur through the tectonic juxtaposition of contrasting source terrains, such as that taking place at oceanic trenches (Mack 1984; Velbel 1985; Underwood 1986; Cawood 1990a).

Modal and geochemical analysis are whole rock techniques that produce an average of the selected sample. For these techniques to succeed in establishing provenance they need (and often assume) a single homogeneous source terrain. They can delineate inter-sample variations in composition and provenance but they cannot determine if intra-sample provenance mixing has taken place. Whole rock analyses also assume that, during any alteration to which the rock samples may have been subjected (compaction, lithification, diagenesis, metamorphism, deformation), at least some element abundances and ratios have remained unaffected and thus provide a guide to provenance.

A method for determining sediment provenance which avoids any assumptions about source homogeneity and degree of sample alteration is the study of relict detrital grain compositions (Cawood 1983). This technique establishes the provenance of individual grains and is, therefore, a unique method for determining intra-sample provenance mixing, particularly from lithologically similar source terrain (Cawood 1990a). This paper determines the phase chemistry, by electron microprobe analysis, of detrital grains in volcaniclastic sediments from the intra-oceanic Tonga arc, SW Pacific.

Volcanic activity at intra-oceanic and continental margin magmatic arcs is characterized by a large ratio of fragmental to effusive material, with the volume of fragmental volcanic debris being generally greater than 85% (Garcia 1978). In contrast, volcanic activity at mid-ocean ridge and ocean island settings contains generally less than 20% by volume fragmental material, and often less than 5% fragmental material. Fragmentation of convergent plate margin magmas is a function of their high volatile content and interaction with external water. Fragmental volcanic material derived from the magmatic arc forms extensive pyroclastic and epiclastic deposits in the adjacent forearc, back-arc and intra-arc sedimentary basins. Thus, the deposits within these basins, rather than the volcanic rocks within the magmatic arc, provide the most complete record of igneous activity along the plate margin. The aim of this paper is to outline

From Morton, A. C., Todd, S. P. & Haughton, P. D. W. (eds), 1991, *Developments in Sedimentary Provenance Studies.* Geological Society Special Publication No. 57, pp. 305–321.

the nature and record of magmatism in the Tongan arc in the southwest Pacific from the study of derived volcaniclastic debris.

Tectonic setting

The Tonga arc lies along the boundary between the Indo-Australia and Pacific plates in the southwest Pacific (Fig. 1). The arc system is divisible into three morpho-tectonic elements. Extending west from the Tonga trench these are: Tonga Ridge, constituting the active arc and forearc; Lau Basin, representing the actively spreading back-arc basin; Lau Ridge, forming the remnant arc. Subduction of the Pacific plate beneath the Tonga arc commenced in the Eocene (Packham 1978), and prior to the initiation of arc rifting and back-arc spreading in the Pliocene (Malahoff *et al.* 1982), the Lau and Tonga ridges formed a single magmatic arc forearc succession (Packham 1978; Hawkins *et al.* 1984). The active volcanic arc along the western margin of the Tonga Ridge is termed the Tofua arc.

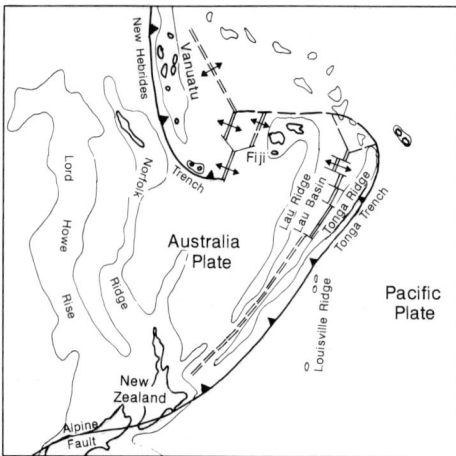

Fig. 1. Tectonic setting of Tonga arc system at boundary of Indo-Australia and Pacific plates. Major tectonic elements outlined by the 2000 m bathymetric contour.

In 1984, the USGS research vessel S. P. Lee collected an assemblage of volcanic and sedimentary rocks from 23 dredge sites around the Tongan arc system. The majority of sites were located along the Tonga Ridge (site nos 3–12 and 15–21). Remaining dredge sites were located either on the Lau Ridge (site nos 22, 23, 25), along the active spreading ridge in the Lau Basin (site nos 1 and 2; Jenner *et al.* 1987; Vallier *et al.* 1991), or on a seamount from the eastern flank

of the South Fiji Basin (site no. 26). Three dredge sites failed to recover any material (site nos 13, 14, 24). Dredge sites from the Tonga and Lau ridges collected a variety of arc-derived sedimentary rocks of Tertiary age. In addition, Dredge 20 which is located on the lower slope of the Tonga Ridge at the along-strike extension of the Louisville Ridge hotspot chain, also contained rocks which yielded late Cretaceous and mixed late Cretaceous and late Tertiary ages (Ballance *et al.* 1989*a*; Cawood 1990*a*). The Cretaceous material is believed to represent fragments of a seamount from the Louisville hot spot chain which is being subducted below, and in part accreted to, the Tongan arc (Ballance *et al.* 1989*b*). This study provides a detailed analysis of Tertiary age, arc-derived volcanic debris from the Tonga and Lau ridge dredge hauls, including samples from the Dredge 20 site. The provenance of the older Cretaceous and mixed Cretaceous–Tertiary samples is discussed by Cawood (1990*a*). Table 1 gives a list of samples selected for detailed study and Fig. 2 shows dredge haul locations. Samples range in age from Oligocene to Pleistocene and cover approximately three quarters of the entire range over which the arc system has been active.

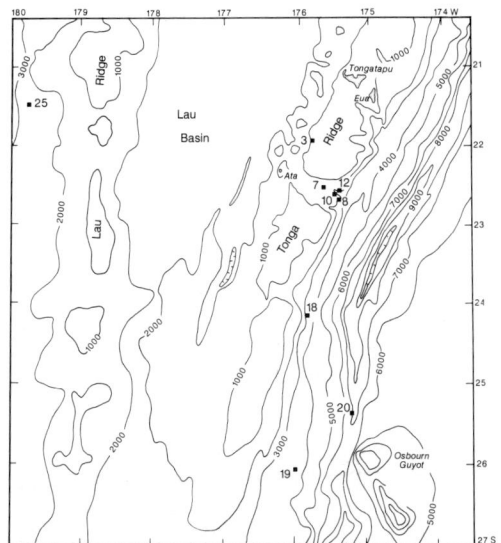

Fig. 2. Location of Tonga arc dredge samples selected for study.

Petrography

Sedimentary rocks dredged from the Tonga forearc are simple two-component admixtures of volcaniclastic debris and pelagic sediment

Table 1. *Age range of samples selected for detailed microprobe and modal analysis*

Sample	Age
3-3*	Mid-Pleistocene to Holocene
7-3*	Early Pleistocene
8-2*	Late Miocene
10-3	N/F
12-4	Neogene
12-5*	Late Oligocene
12-9	Early Pliocene
18-3b†	Late Miocene to Early Pliocene
18-5†	Early Pleistocene
18-10	Late Oligocene–Early Miocene
18-11	Early Miocene
19-2†	Mid Miocene–Early Pliocene
19-4	Early Pleistocene
20-28	N/D
20-39	N/D
20-60	Pliocene to Pleistocene
20-63*	Late Tertiary
20-65	Late Tertiary
20-66	Eocene or younger
20-67	Late Miocene to Pliocene
20-68	barren
20-69	barren
21-3†	?Early Pleistocene
25-7	N/D (?late Oligocene)

* probe and modal analysis of sample
† only modal analysis of sample
N/F, no foraminifera or coccoliths; N/D, age not determined; barren, no fossils recovered.
Sample numbers give both dredge haul location and the rock number for each haul e.g. 8–2 is rock sample 2 from dredge haul 8; location of dredge hauls given in Fig. 1.

(Cawood 1985; Exon *et al.* 1985; Ballance *et al.* 1990). Proportions of these two end-member components vary and cover the spectrum of rock types from ash-bearing lime mudstones to carbonate-free volcaniclastic sandstones and granule conglomerates. Samples are grey to brown and often have a speckled (salt and pepper) appearance, reflecting the mixing of light and dark coloured detrital phases. Sorting is moderate to poor with some sandstone and granule conglomerate samples containing outsized pumice clasts up to 40 cm long. Degree of induration varies from poor to moderate.

The dominant volcanic components within the Tonga dredge samples are volcanic lithic grains and plagioclase crystals. Calcic clinopyroxene is a ubiquitous minor phase and calcium-poor pyroxene, olivine, hornblende, opaque oxide minerals and quartz are rare accessory phases in some samples. The biogenic component of the samples consists mainly of foraminifera and calcareous nannofossils but also includes minor and vari-

able proportions of radiolaria, sponge spicules and pteropods.

Volcanic lithic grains are dominated by microlitic and vitric textural varieties with only rare lathwork and felsitic grains. The character of the volcanic glass ranges from brown, high relief grains to colourless, low relief material suggesting the igneous source rocks range from basic to silicic compositions (cf. Cawood 1985).

All feldspar grains are of plagioclase composition; alkali feldspar is absent. Quartz is only present in monocrystalline form, and shows an embayed and resorbed outline, typical of grains of volcanic origin. Colourless to pale green augite is the principal calcium clinopyroxene phase and colourless to pink–green hypersthene is the main calcium-poor pyroxene, although pigeonite is present in a number of samples. Amphibole grains are generally brown green or reddish brown hornblende.

Framework grains are embedded in a matrix composed of a variable mixture of lime mud and fine pyroclastic ash. The relative proportion of the lime mud to ash is approximately directly proportional to the amount of framework biogenic grains. Samples in which biogenic grains are common have a matrix rich in lime mud whereas samples in which the biogenic component is either absent or only present in trace amounts have a carbonate free matrix. These latter rock types probably include some reworked tuffs.

Modal analyses

Modal analyses for nine representative specimens are listed in Table 2. Analyses are based on total grain count of 400 points.

Modally analysed samples are of Miocene or younger age (Table 1) and show a uniform composition characterized by trace amounts of quartz, plagioclase as the only feldspar phase ($P/F = 1$) and all lithic fragments of volcanic provenance ($L_v/L_t = 1$). These features, together with the high proportion of fragmental pyroclastic debris, are indicative of an intra-oceanic magmatic arc source. This is emphasized on the QFL plot (Fig. 3) where all samples plot along the F–L tie, towards the L apex. The modal range of the Tonga dredge samples is similar to other intra-oceanic arc derived sandstones (Maynard *et al.* 1982; Packer & Ingersoll 1986), and analyses plot in the undissected arc field of Dickinson *et al.* (1983; Fig. 3). Although metamorphic and clastic sedimentary lithic fragments are absent, samples from dredge 6, which were unsuitable for modal analysis, contain rock fragments of highly altered vesicular lava, granitoid clasts and

Table 2. *Modal analyses of selected dredge haul samples from the Tonga Arc*

Sample	3-3	7-3	8-2	12-5	18-3	18-5	19-2	20-63	21-3
Quartz	—	—	—	—	0.2	—	—	—	—
Feldspar	9.5	6.5	7.2	8.2	9.7	4.2	12.2	3.5	14.7
Lithic clasts									
lath.	1.2	0.7	1.0	1.5	0.7	0.2	0.7	0.2	2.5
micro.	19.5	8.5	16.7	58.7	11.5	7.7	29.5	5.0	18.2
fels.	0.2	0.2	1.5	0.7	1.5	0.2	3.5	—	1.7
vitr.	10.7	9.4	7.5	15.0	34.5	11.1	38.3	40.8	4.0
%Lv.	31.6	18.8	26.7	75.9	48.2	19.2	72.0	46.0	26.4
Biog.	3.5	20.7	5.2	2.5	0.2	33.7	0.5	0.2	7.2
Cpx.	6.1	1.7	1.5	1.2	2.0	0.5	2.2	1.7	6.2
Cppx.	0.4	0.4	1.2	0.4	—	—	—	0.2	0.5
Olv.	0.2	—	—	0.4	0.2	—	—	—	—
Hbl.	—	0.2	—	0.2	—	—	—	—	—
Opq.	0.2	0.2	—	0.2	—	0.2	0.4	0.4	0.5
Matrix	48.5	51.5	58.2	11.0	39.5	42.2	12.7	48.0	44.5
Total	100.0	100.0	100.0	100.0	100.0	100.0	100.0	100.0	100.0

Abbreviations: lath., lathwork; micro., microlitic; fels., felsitic; vitr., vitric; %Lv, total volcanic detritus; Bio., biogenic grains; Cpx., clinopyroxene; Cppx., Ca-poor pyroxene (includes orthopyroxene and pigeonite); Olv., Olivine; Hbl., hornblende; Opq., opaque oxide minerals.

?quartzose rocks, presumably derived from the pre-middle Eocene basement of the Tongan arc system (Ballance *et al.* 1990).

The absence of K-feldspar and biotite from all Tertiary age dredge samples and the presence of pigeonite, in at least some samples, indicates that the magmatic arc source for the volcanic detritus was of low-K tholeiitic affinities (Cawood 1985, 1990*b*).

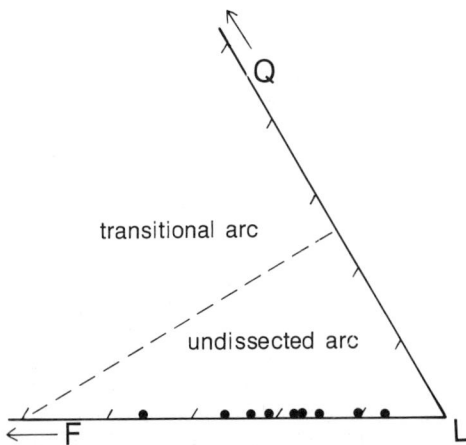

Fig. 3. QFL plot of detrital modes of Tonga arc dredge samples. Dashed line separates the undissected and transitional arc fields of Dickinson *et al.* (1983).

Phase chemistry

Detrital phases within the dredge samples were analysed by electron microprobe to delineate the character and magmatic affinities of the volcanic source terrain(s) and to determine the intra-sample homogeneity of the source. Mineral and glass analyses were carried out at Memorial University of Newfoundland on the JEOL JXA-50A electron probe microanalyser with wavelength dispersive spectrometers. Operating conditions were an accelerating voltage of 15 kV, beam current of around 22 nA, beam diameter of 1–2 μm and maximum count times of 30 s. For volcanic glass analyses the microprobe beam was defocused to a diameter of around 5 to 10 μm. Analyses were corrected to standard minerals using the factors of Bence & Albee (1968). Over 700 microprobe analyses on plagioclase, calcium-rich clinopyroxene, orthopyroxene, pigeonite, olivine, opaque oxide minerals, and volcanic glass were carried out on 21 dredge samples (Table 1). Representative analyses are presented in Tables 3 to Table 9 and summarized graphically in Figs 4 to 16.

Plagioclase

Plagioclase feldspar analyses show a broad range of compositions extending from An_{98} to An_0 (Figs 4–6; Table 3). However, approximately 95% of plagioclase grains have an An content of greater than 50 mol %. This relationship is emphasized on the histogram plot in Fig. 5 in which plagioclase composition is expressed in terms of variation in anorthite and albite

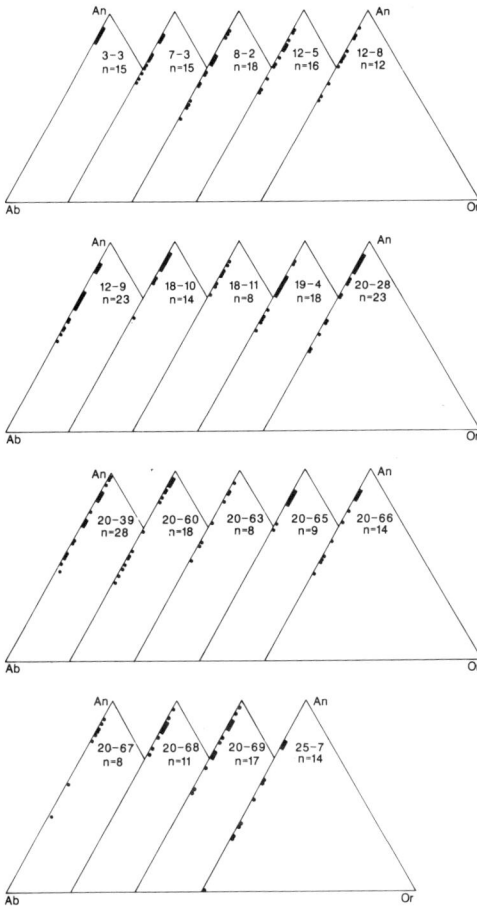

Fig. 4. Plagioclase compositions from Tonga arc dredge samples. Abbreviation: n, number of analyses.

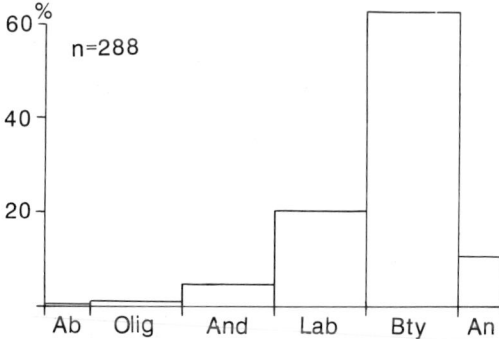

Fig. 5. Histogram showing compositional range of plagioclase feldspar analyses. Abbreviation: n, number of analyses.

albite (combined). Plagioclase grains of albite and oligoclase composition are restricted to sample 25-7 from the Lau Ridge. The presence of albite in this sample, along with petrographic evidence for secondary zeolites, reflects the effects of low grade metamorphism.

All plagioclase grains have low Or contents, generally less than 0.5 mole percent and this value is constant with respect to variations in the An to Ab ratio (Figs 4 and 6).

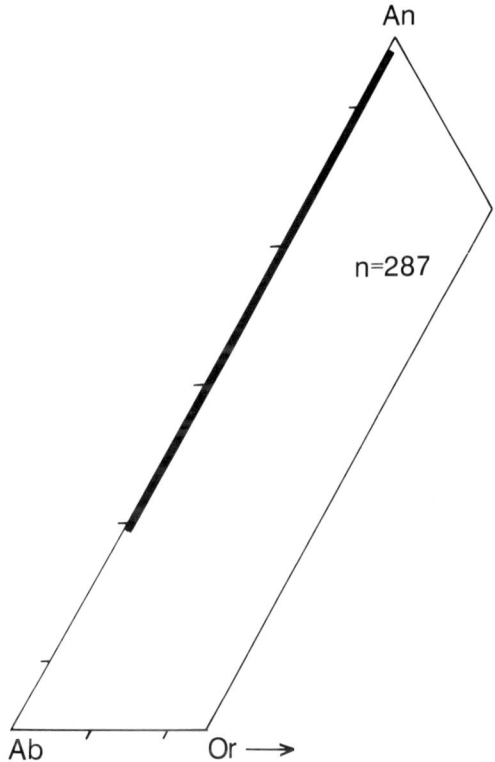

Fig. 6. Compositional field of plagioclase grains for the Tonga arc dredge samples. Analysis of secondary albite from sample 25-7 is excluded from plot. Abbreviation: n, number of analyses.

components. Of the 288 analysed grains 11% are anorthite, 63% bytownite, 21% labradorite, 5% andesine and approximately 1% oligoclase and albite (combined).

The compositional range of the detrital Tonga arc plagioclase grains is similar to that recorded by Cawood (1985) from Miocene and younger volcaniclastic detritus from the Tonga Ridge immediately south of Eua, and is also similar to the composition of plagioclase grains in volcanic rocks from the active Tofua magmatic arc (cf. Ewart 1982; Ewart et al. 1973, 1977). The overall low Or and high An content of the detrital plagioclase grains indicates derivation from volcanic rocks of low-K tholeiitic compositions.

Table 3. *Selected microprobe analyses of plagioclase grains from the Tonga arc dredge samples*

| Sample | 8-02 | 8-02 | 12-08 | 12-08 | 12-09 | 19-04 | 19-04 | 20-28 | 20-28 | 20-28 | 25-07 | 25-07 |
Grain	05	06	07c	08	05c	11	12	10r	16	18	06	11
SiO_2	51.18	48.28	49.30	45.21	55.00	46.70	51.04	57.60	46.07	48.77	61.13	48.24
TiO_2	0.03	0.01	0.02	0.00	0.01	0.05	0.06	0.05	0.07	0.11	0.00	0.00
Al_2O_3	31.36	32.43	31.58	33.79	27.83	33.43	31.68	25.25	32.37	31.68	23.77	32.50
FeO	0.77	0.89	0.63	0.55	0.45	0.83	0.59	0.58	0.71	0.70	0.27	0.58
MnO	0.04	0.00	0.00	0.00	0.01	0.02	0.03	0.03	0.04	0.06	0.02	0.03
MgO	0.11	0.14	0.01	0.09	0.04	0.09	0.02	0.01	0.13	0.16	0.01	0.02
CaO	14.11	17.02	15.32	18.17	11.17	17.15	14.39	8.80	18.08	16.52	5.88	15.66
Na_2O	2.93	1.45	2.62	0.98	5.49	1.21	2.89	6.16	0.62	2.09	8.06	2.36
K_2O	0.03	0.02	0.02	0.02	0.07	—	0.02	0.07	0.02	0.08	0.07	—
Cr_2O_3	—	0.03	0.04	0.02	0.04	—	—	—	0.02	—	0.01	—
NiO	na	na	na	na	na	0.03	0.01	0.05	0.01	—	0.02	0.03
Total	100.56	100.27	99.54	98.83	100.11	99.51	100.73	98.60	98.14	100.17	99.24	99.42

Grain number refers to the analysed grain within the sample; a letter suffix to the grain number indicates that there is more than one analysis of the grain; c, core; r, rim. Abbreviation: na, not analysed.

Table 4. Selected microprobe analyses of clinopyroxene grains from the Tonga arc dredge samples

Sample Grain	02-03 05	03-03 06c	07-03 04c	07-03 04r	07-03 07	12-05 05c	12-05 06c	12-05 06r	18-10 01	18-10 02	18-10 03	18-11 06
SiO_2	52.32	53.66	54.30	53.81	51.59	50.48	51.37	51.77	49.71	49.71	50.90	51.24
TiO_2	0.23	0.09	0.08	0.08	0.41	0.61	0.32	0.38	0.66	0.49	0.51	0.34
Al_2O_3	2.38	0.69	0.72	1.26	2.27	1.21	2.57	2.61	2.57	0.93	2.76	2.18
FeO	6.70	2.60	2.81	3.82	11.86	21.96	8.14	8.37	16.24	23.66	12.82	10.18
MnO	0.16	0.07	0.09	0.06	0.34	0.63	0.23	0.17	0.30	0.48	0.34	0.21
MgO	17.05	18.80	18.61	17.95	14.62	11.93	15.80	15.97	13.78	11.39	15.54	15.78
CaO	20.45	23.17	22.54	21.41	18.45	13.73	20.32	20.09	15.39	11.55	16.28	18.97
Na_2O	0.25	0.12	0.11	0.15	0.25	0.27	0.29	0.12	0.30	0.60	0.29	0.25
K_2O	0.02	0.01	—	0.02	—	0.03	—	0.01	0.01	0.01	0.01	0.01
Cr_2O_3	0.39	0.57	0.37	0.62	0.04	0.03	0.05	0.07	0.05	—	0.04	0.13
NiO	0.05	0.05	—	—	—	—	—	—	0.03	0.02	0.05	0.05
Total	100.00	99.83	99.63	99.18	99.83	100.88	99.09	99.56	99.04	98.84	99.54	99.34

Grain numbering procedure and abbreviations as for Table 3.

Pyroxene

Pyroxene compositions fall in the clinopyroxene, orthopyroxene and pigeonite fields (Fig 7). Clinopyroxene analyses show a wide compositional variation and on the pyroxene quadrilateral plot in the diopside, endiopside, salite, augite, subcalcic augite and ferroaugite fields between Ca_{18}–Ca_{49}, Mg_{30}–Mg_{57} and Fe_4–Fe_{41} (Figs 7 and 8; Table 4).

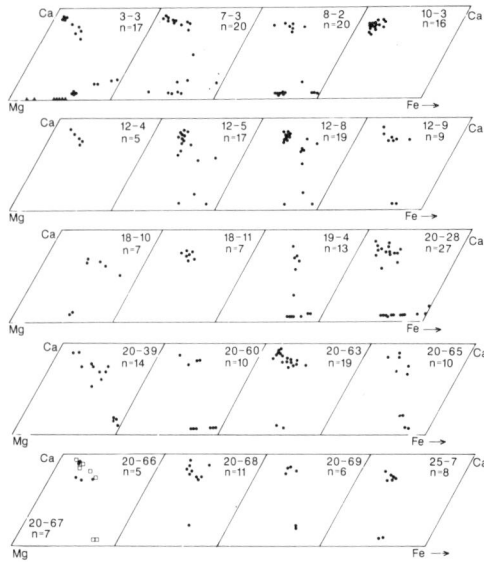

Fig. 7. Composition of pyroxene and olivine from the Tonga arc dredge samples plotted in terms of Ca–Mg–Fe. Coordinates of parallelogram are Ca–Mg–Fe 0–100–0, Ca–Mg–Fe 50–50–0, Ca–Mg–Fe 50–0–50, Ca–Mg–Fe 0–0–100. Symbols: solid circles, pyroxene analyses (except for sample 20-67 which is shown by open squares; triangles, olivine analyses. Abbreviation: n, number of analyses.

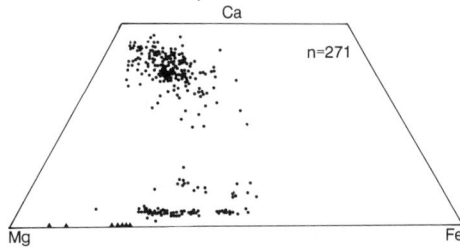

Fig. 8. Compositional fields of pyroxene and olivine grains for the Tonga arc dredge samples. Symbols: solid circles, pyroxene analyses; triangles, olivine analyses. Abbreviation; n, number of analyses.

Studies of igneous rocks have shown that pyroxene composition is directly related to magma type and tectonic setting (e.g. Kushiro 1960; Le Bas 1962; Verhoogen 1962; Nisbet &

Pearce 1977; Leterrier *et al.* 1982). The high Si and low Al and Ti contents of the Tonga detrital clinopyroxene grains are indicative of pyroxenes crystallizing from subalkaline magmas (Fig. 9). On the eigenvector-based discrimination diagram of Nisbet & Pearce (1977) clinopyroxenes fall largely within the field for volcanic arc basalts (Fig. 10). Similarly on the Ti–Ca + Na plot of Leterrier *et al.* (1982) analyses fall in the field for tholeiitic and calc-alkali basalts (Fig. 11). On the Ti–Al$_{tot}$ diagram of Leterrier *et al.* (1982) for discriminating clinopyroxenes from tholeiitic and calc-alkali basalts, the Tonga analyses fall mainly in the tholeiitic field but with scatter into the calc-alkali field.

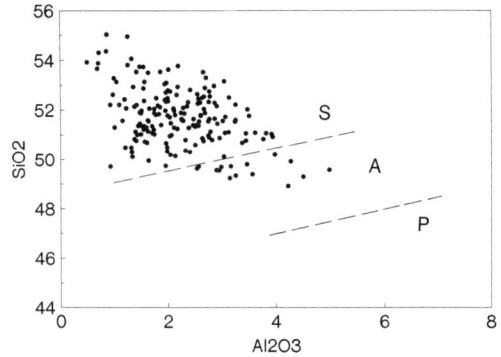

Fig. 9. SiO_2–Al_2O_3 covariation diagram for detrital Tonga arc clinopyroxene grains. Boundaries taken from Le Bas (1962): S, subalkaline; A, alkaline; P, peralkaline.

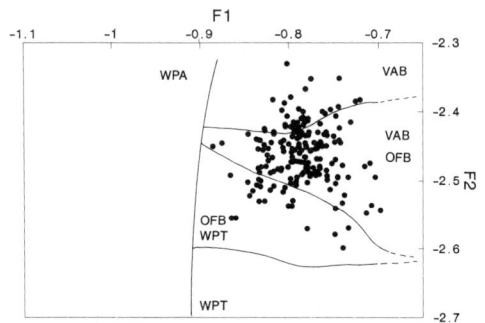

Fig. 10. Plot of discriminant functions, F1 versus F2, for detrital Tonga arc clinopyroxene analyses. Fields and discriminant functions are from Nisbet & Pearce (1977): VAB, volcanic arc basalt; OFB, ocean floor basalt; WPT, within plate tholeiitic basalt; WPA within plate alkali basalt.

A problem in applying the above types of discrimination diagrams to detrital mineral grains, and possibly some of the reason for the scatter of the Tongan analyses on these diagrams, is that they were primarily designed for

Table 5. *Selected microprobe analyses of orthopyroxene grains from the Tonga arc dredge samples*

Sample Grain	07-03 02r	07-03 03r	08-02 02c	08-02 03	08-02 04c	08-02 07c	20-28 08	20-28 09	20-28 11	20-28 12	20-28 13
SiO_2	52.08	52.74	53.72	53.45	51.53	54.48	49.67	52.38	52.63	52.02	53.16
TiO_2	0.19	0.19	0.19	0.16	0.28	0.06	0.15	0.20	0.20	0.05	0.16
Al_2O_3	1.27	1.37	1.31	1.00	0.74	0.67	0.95	0.46	0.71	0.43	1.29
FeO	20.84	18.29	18.88	19.82	27.96	18.60	31.53	24.11	21.39	28.14	18.94
MnO	0.52	0.41	0.56	0.53	1.43	0.52	1.01	0.05	0.99	1.34	0.80
MgO	23.39	24.71	24.36	23.91	16.79	24.49	15.62	20.55	22.65	17.23	23.49
CaO	1.91	1.80	1.65	1.78	1.59	1.90	1.54	1.55	1.76	1.50	1.53
Na_2O	0.02	0.00	0.00	0.05	0.04	0.09	0.01	0.09	0.00	0.14	0.04
K_2O	0.02	0.01	0.01	0.01	0.00	0.01	0.01	—	0.01	—	0.03
Cr_2O_3	0.05	0.08	0.04	0.04	0.04	0.01	—	—	—	—	—
NiO	na	na	na	na	na	na	—	0.06	0.06	0.02	0.03
Total	100.29	99.60	100.72	100.75	100.40	100.83	100.49	99.45	100.40	100.87	99.47

Grain numbering procedure and abbreviations as for Table 3.

differentiating pyroxenes from mafic lavas, although the volcanic arc basalt field in the Nisbet & Pearce (1977) diagram does include pyroxene compositions from arc andesites (Nisbet, pers. comm.). For detrital sediments the silica content of the source lavas for the framework grains is generally unknown. Thus, unless an independent means is available for estimating silica content of the source rocks, this limitation must be borne in mind when applying these diagrams to provenance studies. For the Tongan samples, detrital glass compositions suggest a range of SiO_2 values encompassing mafic to acidic compositions (Cawood 1985, 1990b, see below). However, glass compositions were determined on detrital fragments which lacked pyroxene inclusions, and therefore it cannot be assumed that pyroxene was on the liquidus and crystallizing throughout the entire range of SiO_2 values recorded from the glasses. Recently, Morris (1988) has developed a clinopyroxene discriminant function plot for distinguishing between basalt, andesite, and dacite plus rhyolite from volcanic arc sources. On this diagram, detrital clinopyroxenes from the Tonga dredge cover the basalt to rhyolite compositional range (Fig. 12). This suggests that pyroxene was crystallizing throughout the range of glass compositions and that the two are consanguineous.

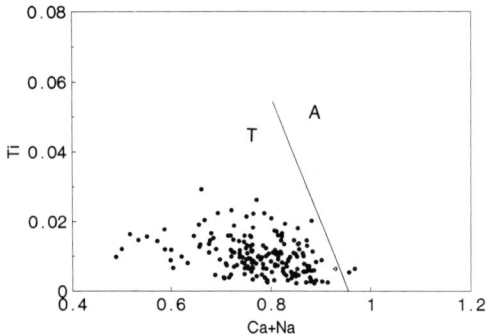

Fig. 11. Plot of Ti versus Ca + Na for detrital Tonga arc clinopyroxene analyses. Solid line discriminates field of alkalic basalts (A) from tholeiitic and calc-alkali basalts (T; Leterrier *et al.* 1982).

Calcium-poor detrital pyroxene phases in the Tonga dredge hauls include both orthopyroxene and pigeonite. Orthopyroxene compositions range from Fe_{26}–Fe_{51} (Table 5; Figs 7 and 8). They are mainly hypersthene but extend into the bronzite and ferrohypersthene fields. In addition, sample 7-3 contains a relatively Mg-rich bronzite (Fe_{15} mol %; Fig. 7). Pigeonite analyses range from Fe values of 27–51 mol % and on the pyroxene quadrilateral fall in the Mg-rich pigeo-

nite and pigeonite fields (Table 6; Figs 7 and 8). Pigeonite and orthopyroxene compositions show a similar range in their Fe/Mg ratios. One pigeonite analysis contains 14 mol % Ca and merges with clinopyroxenes in the subcalcic augite field. The proportion of non-quadrilateral components Al, Ti, and Na show a general decrease from calcium-rich to calcium-poor pyroxenes. In the calcium-rich pyroxenes Ti shows a tendency to increase with increasing Al, which suggests that the $CaTiAl_2O_6$ component of the pyroxene structure influences composition.

The strong Fe-enrichment trend of both the calcium-rich and calcium-poor pyroxene grains (Fig. 8) is characteristic of pyroxenes crystallizing from a low-K tholeiitic magma (Gill 1981; Ewart 1982).

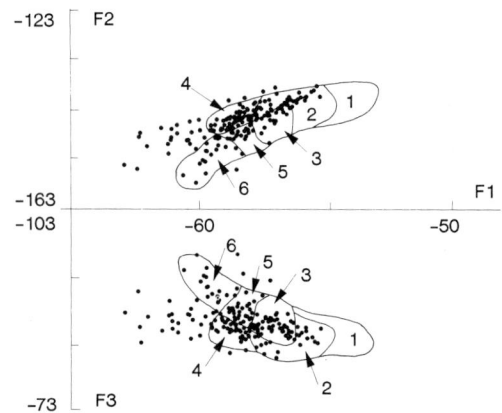

Fig. 12. Clinopyroxenes discriminant function plot of Morris (1988) for determining parent magma type. Fields are for clinopyroxenes from modern arc volcanic rocks: 1, basalt; 2, basalt + andesite; 3, basalt + andesite + (dacite + rhyolite); 4, andesite; 5, andesite + (dacite + rhyolite); 6, dacite + rhyolite.

Olivine, amphibole and opaque oxide minerals

No systematic study of detrital olivine, amphibole or opaque oxide minerals from the Tonga dredge hauls was undertaken. Olivine analyses are from sample 3-3 and are presented in Table 7. Olivine compositions are plotted on Figs 7 and 8, and range from Fe_8–Fe_{27} (Fo_{73}–Fo_{92}; Table 7). Apart from the one Mg-rich bronzite, orthopyroxene compositions are antithetic to those of olivine (Fig. 8), suggesting that olivine was the initial Ca-poor ferromagnesium phase to crystallize from the source magmas and was replaced in more fractionated magmas by orthopyroxene and pigeonite.

Table 6. *Selected microprobe analyses of pigeonite grains from the Tonga arc dredge samples*

Sample	03-03	03-03	07-03	12-08	20-28	20-39	20-65	20-68
Grain	03c	04c	02	02	07	04	01b	11
SiO_2	49.48	51.57	53.31	51.53	49.92	51.52	51.48	51.78
TiO_2	0.35	0.28	0.18	0.30	0.29	0.17	0.29	0.21
Al_2O_3	0.84	1.28	1.00	0.89	0.72	0.43	4.49	0.82
FeO	28.22	22.91	17.53	23.52	30.97	27.84	19.69	20.15
MnO	0.63	0.52	0.44	0.62	0.69	1.12	0.40	0.47
MgO	13.75	18.28	23.16	18.00	14.37	16.21	18.27	19.74
CaO	5.07	4.14	4.75	4.80	3.55	3.52	5.10	5.66
Na_2O	0.02	0.11	0.08	0.01	0.13	0.02	0.25	0.12
K_2O	—	0.03	0.01	—	—	—	0.02	0.01
Cr_2O_3	0.02	—	0.06	—	0.02	—	0.02	—
NiO	0.01	0.03	na	na	0.02	0.01	na	na
Total	98.39	99.15	100.52	99.67	100.68	100.84	100.01	98.96

Grain numbering procedure and abbreviations as for Table 3.

Table 7. *Selected microprobe analyses of olivine grains from the Tonga arc dredge samples*

Sample	03-03	03-03	03-03	03-03	03-03	03-03
Grain	01	02	03c	04	05	06
SiO_2	37.87	37.71	40.40	39.05	39.05	37.64
TiO_2	0.01	0.04	0.01	0.03	0.02	0.01
Al_2O_3	0.06	0.04	0.04	0.11	0.07	0.07
FeO	23.24	23.57	11.92	22.38	20.85	22.38
MnO	0.32	0.35	0.17	0.36	0.33	0.29
MgO	38.54	37.91	47.62	38.66	40.10	40.25
CaO	0.22	0.18	0.20	0.23	0.27	0.18
Na_2O	—	0.02	0.07	0.01	0.08	0.05
K_2O	—	0.02	0.03	0.04	0.01	0.01
Cr_2O_3	—	0.02	0.05	0.05	0.02	0.04
NiO	0.07	0.05	0.15	0.08	0.05	0.04
Total	100.33	99.91	100.66	101.00	100.85	100.96

Grain numbering procedure and abbreviations as for Table 3.

Amphibole is calcic in composition with $(Ca + Na)_B \geqslant 1.34$ and $Na_B < 0.67$ and belong to the hornblende series (Table 8). TiO_2 contents are around 2% or less and in terms of the classification scheme of Leake (1978) the analyses fall in the magnesio-hornblende $(Na + K)_A < 0.50;\ Ti < 0.50)$ and edenite $(Na + K)_A \geqslant 0.50; Ti < 0.50; Fe^3 \leqslant Al^{VI})$ fields. Compositions are typical of those found in convergent plate magmatic arc settings (Gill 1981; Ewart 1982).

Opaque oxide minerals are spinel group titanomagnetites. Compositions are similar to those reported by Cawood (1985) for similar age samples from the Tonga Platform to the north and are also similar in composition to those reported by Ewart (1976) from the Tofua arc.

Volcanic glass

Analyses of detrital volcanic glass provide a direct means of assessing magmatic affinities of the source terrane. Volcanic glass is a common product of convergent plate magmatism and reflects the explosive nature of the igneous activity. Glass occurs in all of the Tonga dredge haul samples, but only a few samples contain material fresh enough for microprobe analysis. Even in samples containing optically fresh glass, analysis totals are generally low, ranging from 90% to 100%, with most falling between 93% to 97% (Table 8). This reflects some degree of alteration, to which glass is particularly susceptible because of its amorphous character and its high surface area to volume ratio (e.g. pumice, glass shards). Alteration, however, is probably limited to simple hydration, which although diluting absolute elemental abundances, does not appear to have affected primary igneous trends or element ratios (Fujioka *et al.* 1980; Cawood 1985).

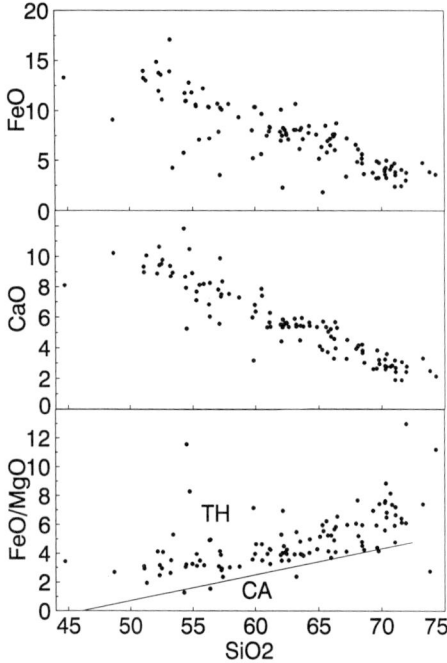

Fig. 13. Plot of CaO and FeO oxide abundance and FeO/MgO ratio versus silica for detrital Tonga arc volcanic glass analyses (n = 156). Solid line separates calc-alkali (CA) and tholeiitic (TH) rock series of Miyashiro (1974).

The SiO$_2$ content of the volcanic glass ranges from 45% to 75%, covering the compositional range basalt to rhyolite. Analyses show normal igneous differentiation trends of decreasing FeO, MgO, TiO$_2$ and CaO, and increasing K$_2$O and FeO/MgO with increasing SiO$_2$ content (e.g. Fig. 13), although FeO and MgO show some scatter, particularly at lower SiO$_2$ values. Inter- and intra-sample scatter of major elements at given values of SiO$_2$ probably reflects temporal and spatial changes in magma composition of the igneous source rocks, although alteration associated with glass hydration may be an additional component of at least some sample scatter. Each sedimentary rock sample is prob- ably derived from a source region containing a record of numerous (rather than one) igneous events, so some scatter of detrital glass compo- sitions is to be expected. Given the possibilities

for source diversity, the overall coherent igneous trends shown by the glass compositions are remarkable. This is emphasized on the plot of K$_2$O versus SiO$_2$ (Figs 14 and 15), with the majority of analyses falling in the low-K island arc tholeiite field. Some analyses have slightly higher K$_2$O contents and plot in the mildly, medium-K tholeiite (calc-alkaline) field (Fig. 15). These slightly more potassic compositions merge with the low-K tholeiite compositions and probably represent diversity within a single source area rather than input from multiple source areas. Analyses show the characteristic tholeiitic differentiation trend of overall increas- ing FeO/MgO ratio with increasing SiO$_2$ (Fig. 13) and almost all samples fall in the tholeiitic field of Miyashiro (1974) on the FeO/MgO versus SiO$_2$ plot. Similarly on the AFM plot (Fig. 16) analyses fall in the tholeiitic field of Irvine & Baragar (1971) and follow the general differentiation trend on the pigeonitic series of Kuno (1968).

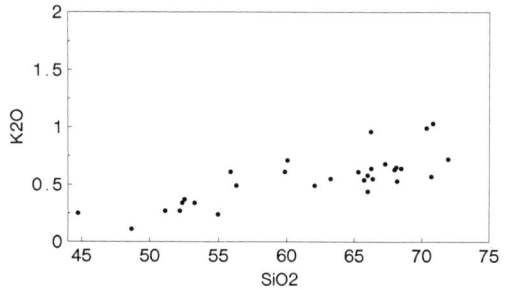

Fig. 14. Plot of K$_2$O versus SiO$_2$ for volcanic glass from Tonga dredge sample 20-39 showing intra- sample variability in oxide abundances.

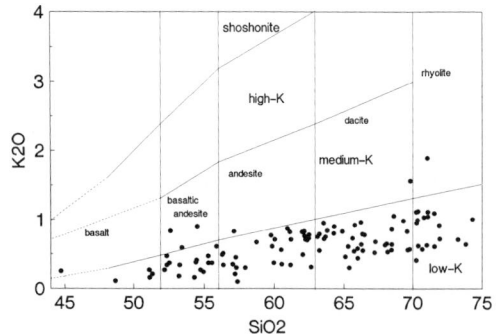

Fig. 15. Composite plot of K$_2$O versus SiO$_2$ for detrital volcanic glass from Tonga Ridge dredge samples (19-4, 20-28, 20-39, 20-65 and 20-69). The fields for low-K (island arc) tholeiite, medium-K tholeiite (calc-alkaline series), high-K tholeiite and shoshinite series are from Peccerillo & Taylor (1976) as modified by Ewart (1982) and Gill (1981).

Table 8. *Selected microprobe analyses of volcanic glass from the Tonga arc dredge samples*

| Sample | 19-04 | 19-04 | 19-04 | 19-04 | 20-39 | 20-39 | 20-39 | 20-39 | 20-39 | 20-28 | 20-28 | 20-28 | 20-28 |
Grain	04	18b	24	30	11	15	16	22	25	03	09	11	18
SiO_2	60.96	73.81	64.27	57.26	65.99	51.12	71.95	66.00	59.86	70.39	51.15	60.55	57.93
TiO_2	0.94	0.45	0.94	1.11	0.73	1.10	0.33	0.59	0.70	0.56	1.18	0.96	0.92
Al_2O_3	12.57	10.86	12.46	13.27	12.70	14.65	11.11	12.36	17.52	12.02	13.09	13.29	13.37
FeO	7.48	3.84	7.78	10.29	7.17	13.96	3.77	6.52	5.23	5.00	13.26	9.68	10.68
MnO	0.20	0.18	0.20	0.24	0.20	0.23	0.18	0.23	0.10	0.07	0.20	0.12	0.18
MgO	1.89	1.39	1.93	3.63	1.36	4.50	0.29	1.76	0.73	0.73	4.50	2.75	3.52
CaO	5.35	2.51	5.66	7.36	5.24	9.34	2.80	4.76	3.18	3.62	8.98	7.42	7.54
Na_2O	2.71	2.05	2.87	2.61	2.66	2.37	1.56	2.97	2.53	2.63	3.71	2.49	2.90
K_2O	0.87	0.64	0.80	0.45	0.58	0.27	0.72	0.44	0.61	0.62	0.16	0.35	0.30
Cr_2O_3	0.05	—	—	—	—	0.02	0.02	0.00	0.04	0.01	0.04	0.03	0.04
NiO	0.03	na	na	na	0.19	—	0.02	0.16	0.08	—	0.15	0.09	0.12
Total	93.05	95.73	96.91	96.22	96.82	97.56	92.75	95.79	90.58	95.65	96.42	97.73	97.50

Grain numbering procedure and abbreviations as for Table 3.

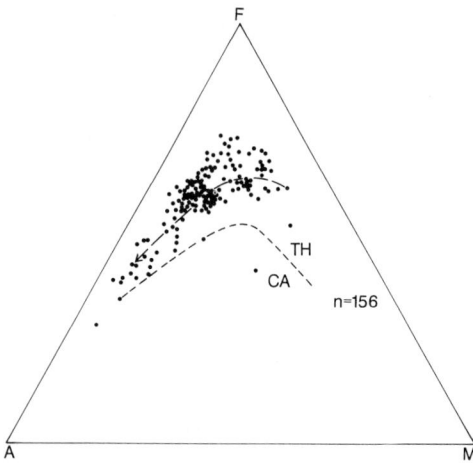

Fig. 16. AFM ternary plot for detrital volcanic glass from Tonga Ridge dredge samples. Line of short dashes separates tholeiitic (TH) and calc-alkali (CA) fields of Irvine & Baragar (1971); line of long dashes marks pigeonite series differentiation trend of Kuno (1968).

Discussion

Petrography and detailed phase chemistry analysis of Eocene to Pleistocene volcanic detritus in sediments from the Tonga arc indicates derivation from a magmatic arc of relatively uniform low-K tholeiitic affinities. The widespread presence of pyroclastic debris, particularly glass shards and pumice, is characteristic of convergent plate margin magmatic activity (Garcia 1978). Plagioclase grains have high An and low Or contents indicating crystallization from a magma low in potassium. Both the calcium-rich and calcium-poor ferromagnesium phases (clinopyroxene, pigeonite, orthopyroxene and olivine) have low Al and Ti contents, and show an overall Fe-enrichment trend indicating deviation from a subalkaline magma of low-K tholeiitic affinities. Pigeonite which is present within at least some of the samples is a characteristic component of igneous rocks of tholeiitic affinities (Kuno 1968). Hornblende and opaque oxide mineral compositions are within the range of intra-oceanic island arcs (Ewart 1982). Volcanic glass compositions are characterized by low K_2O contents and an FeO/MgO enrichment trend, typical of low-K tholeiitic arcs. In addition, the absence of alkali feldspar and biotite, and the presence of only rare amphibole is consistent with derivation from a low-K tholeiitic source.

Conclusions

The Oligocene to Pleistocene age of volcaniclastic material dredged from the southern part of the Tonga arc spans almost three quarters of the entire history of convergent plate interaction along this segment of Indo-Australia plate and Pacific plate boundary. Phase chemistry analysis of this volcanic debris indicates that the magmatic arc source maintained consistent low-K tholeiitic affinities throughout this time period. The age range of this material overlaps the Pliocene to Recent age for both initiation and opening of the Lau backarc basin and establishment of the Tofua magmatic arc. The low-K tholeiitic affinities of the Tofua arc (Ewart et al. 1973, 1977; Gill 1976; Ewart & Hawkesworth 1987) indicate that it, along with the pre-Pliocene arc preserved on the Lau Ridge (remnant arc), was the source for the volcaniclastic debris in the Tonga dredge hauls.

The uniform low-K tholeiitic affinities of the volcaniclastic debris during the Pliocene rifting of the magmatic arc indicates that rifting of the arc did not affect the geochemical character of the magma source. This conclusion contrasts with data from the northern part of the arc system which suggests the period of arc rifting corresponds with a change from Miocene calc-alkaline (medium-K tholeiite) magmatism, now preserved on the northern part of the remnant arc (Lau Ridge), to the Pliocene to Recent low-K tholeiitic magmatism of the active Tofua arc (Gill 1976; Cole et al. 1985).

The medium-K tholeiite affinities for the Miocene volcanic arc along the northern segment of the Lau Ridge and the low-K tholeiite affinities of the Miocene volcaniclastic detritus in the southern Tonga arc probably reflects along strike variations in the geochemical character of the arc (Cawood 1985). Kay et al. (1982) showed that the geochemical affinities of the Aleutian magmatic arc change abruptly from tholeiitic to calc-alkaline (medium-K) along strike due to changes in the crustal structure of the arc. A similar scenario may account for the inferred along strike variations in the nature of the Miocene magmatic arc on the Lau Ridge. The current bathymetry of the Lau Ridge changes dramatically along strike (Cawood 1985). North of 20°S, the ridge is relatively shallow (< 1000 m) and characterized by exposure of the numerous small volcanic islands of the Lau Group. South of the 20°S, the ridge is considerably deeper (> 1000 m) and largely unexposed. Variations in bathymetry must reflect along strike changes in crustal structure.

The cruise of the United States Geological Survey research vessel S. P. Lee to the Tonga arc was funded by the Tripartite agreement between the Australian, New Zealand and United States Governments in association with the United Nations Committee for the coordination of Joint Prospecting for Mineral Resources in the South Pacific Offshore Areas (CCOP/SOPAC). My participation in the cruise was facilitated by a grant from the Australian Development Assistance Bureau, Canberra. I am grateful to the captain and crew of the R/V S. P. Lee, co-chief scientists D. Scholl and R. Herzer, and co-cruise scientists P. Ballance and T. Vallier for their help during the cruise. The manuscript benefited from comments by J. Pearce and an anonymous reviewer.

References

BALLANCE, P. F., BARRON, J. A., BUKRY, D., CAWOOD, P. A., CHAPONIERE, G. C. H., FRISCH, R., HERZER, R. H., NELSON, C. S., QUINTERNO, P., RYAN, H., SCHOLL, D. W., STEVENSON, A. J., TAPPIN, D. R. & VALLIER, T. L. 1989a. Late Cretaceous pelagic sediments, volcanic ash and biotas from near the Louisville hotspot, Pacific Plate, paleolatitude ~42°S. *Palaeogeography, Palaeoecology and Palaeoclimatology*, **71**, 281–299.

——, SCHOLL, D. W., VALLIER, T. L., STEVENSON, A. J., RYAN, H. & HERZER, R. H. 1989b. Subduction of a Late Cretaceous seamount of the Louisville Ridge at the Tonga Trench: a model of normal and accelerated tectonic erosion. *Tectonics*, **8**, 953–962.

——, VALLIER, T. L., NELSON, C. S. & FRISCH, R. S. 1990. Petrology and sedimentology of mixed volcaniclastic and pelagic sedimentary rocks from the southern Tonga platform, trench slope and trench, and the Lau Ridge, *In*: BALLANCE, P. F., HERZER, R. H. & VALLIER, T. L. (eds) *Geology of the Tonga Ridge, Lau Basin and Lau Ridge region, southwest Pacific*. Circum-Pacific Council for Energy and Mineral Resources Earth Science Series, Houston, Texas, (in press).

BENCE, A. E. & ALBEE, A. L. 1968. Empirical correction factors for the electron micro analyses of silicates and oxides. *Journal of Geology*, **76**, 382–401.

BHATIA, M. R. 1983. Plate tectonics and geochemical composition of sandstones. *Journal of Geology*, **91**, 611–627.

BHATIA, M. R. & CROOK, K. A. W. 1986. Trace element characteristics of graywackes and tectonic setting discrimination of sedimentary basins. *Contributions to Mineralogy and Petrology*, **92**, 181–193.

—— & TAYLOR, S. R. 1981. Trace-element geochemistry and sedimentary provinces: a study from the Tasman geosyncline, Australia. *Chemical Geology*, **33**, 115–125.

CAWOOD, P. A. 1983 Modal composition and detrital clinopyroxene geochemistry of lithic sandstones from the New England Fold Belt (east Australia): A Paleozoic forearc terrane. *Geological Society of America Bulletin*, **94**, 1199–1214.

—— 1985. Petrography, phase chemistry and provenance of volcanogenic debris from the Southern Tonga Ridge: implications for arc history and magmatism, *In*: SCHOLL, D. W. & VALLIER, T. L. (eds) *Geology and offshore resources of Pacific island arcs—Tonga region*. Circum-Pacific Council for Energy and Mineral Resources Earth Science Series, Houston, Texas, **2**, 149–169.

—— 1990a. Provenance mixing in an intraoceanic subduction zone: Tonga Trench—Louisville Ridge collision zone, S. W. Pacific. *Sedimentary Geology*, **67**, 35–53.

—— 1990b. Volcaniclastic debris from the southern Tonga forearc: petrology and phase chemistry, *In*: BALLANCE, P. F., HERZER, R. H. & VALLIER, T. L. (eds) *Contributions to the Geology of the Tonga–Lau regions of the southwest Pacific*, Circum-Pacific Council for Energy and Mineral Resources Earth Science Series, Houston, Texas (in press).

COLE, J. W., GILL, J. B. & WOODHALL, D. 1985. Petrologic history of the Lau Ridge, Fiji, *In*: SCHOLL, D. W. & VALLIER, T. L. (eds) *Geology and offshore resources of Pacific island arcs—Tonga region*. Circum-Pacific Council for Energy and Mineral Resources Earth Science Series, Houston, Texas, **2**, 379–414.

CROOK, K. A. W. 1974. Lithogenesis and geotectonics: the significance of compositional variations in flysch arenites (graywackes), *In* DOTT, R. H. & SHAVER, R. H. (eds) *Modern and Ancient Geosynclinal Sedimentation*. Society of Economic Paleontologists and Mineralogists, Special Publication, **19**, 304–310.

DICKINSON, W. R. & SUCZEK, C. A. 1979. Plate tectonics and sandstone compositions. *American Association of Petroleum Geologists Bulletin*, **63**, 2164–2182.

—— & VALLONI, R. 1980. Plate settings and provenance of sands in modern ocean basins. *Geology*, **8**, 82–86.

——, BEARD, L. S., BRAKENRIDGE, G. R., ERJAVEC, J. L., FERGUSON, R. C., INMAN, K. F., KNEPP, R. A., LINDBERG, F. A. & RYBERG, P. T. 1983. Provenance of North American Phanerozoic sandstones in relation to tectonic setting. *Geological Society of America Bulletin*, **94**, 222–235.

EWART, A. 1976. A petrologic study of the younger Tongan andesites and dacites, and the olivine tholeiites of Niua Fo'ou Island, south-western Pacific. *Contributions to Mineralogy and Petrology*, **14**, 36–64.

—— 1982. The mineralogy and petrology of Tertiary–Recent orogenic volcanic rocks: with special reference to the andesite-basaltic compositional range, *In*: THORPE, R. S. (ed.) *Andesites*. John Wiley and Sons, New York, 25–95.

—— & HAWKESWORTH, C. J. 1987. The Pleistocene–Recent Tonga–Kermadec Arc Lavas: Interpretation of new isotopic and rare earth data in terms of a depleted mantle source model. *Journal of Petrology*, **28**, 495–530.

——, BROTHERS, R. N. & MATEEN, A. 1977. An outline of the geology and geochemistry, and the possible petrogenetic evolution of the volcanic rocks of the Tonga–Kermadec–New Zealand island arc. *Journal of Volcanology and Geothermal Research*, **2**, 205–250.

——, BRYAN, W. B. & GILL, J. B. 1973. Mineralogy and geochemistry of the younger volcanic islands of Tonga, S. W. Pacific. *Journal of Petrology*, **14**, 429–465.

EXON, N. F., HERZER, R. H. & COLE, J. W. 1985. Mixed volcaniclastic and pelagic sedimentary rocks from the Cenozoic southern Tonga Platform and their implications for petroleum potential, *In*: SCHOLL, D. W. & VALLIER, T. L. (eds) *Geology and offshore resources of Pacific island arcs—Tonga region*. Circum-Pacific Council for Energy and Mineral Resources Earth Science Series, Houston, Texas, **2**, 75–107.

FUJIOKA, K., FURUTA, T. & ARAI, F. 1980. Petrography and geochemistry of volcanic glass: Leg 57, Deep Sea Drilling Project, *In*: SCIENTIFIC PARTY, *Initial Reports of the Deep Sea Drilling Project*, **56** & **57**, Pt 2, Government Printing Office, Washington, DC, US, 1046–1066.

Garcia, M. O. 1978. Criteria for the identification of ancient volcanic arcs. *Earth Science Reviews*, **14**, 147–165.

GILL, J. B. 1976. Composition and age of Lau Basin and Ridge volcanic rocks: implications for evolution of an interarc basin and remnant arc. *Geological Society of America Bulletin*, **87**, 1384–1395.

—— 1981. *Orogenic andesites and plate tectonics*. Springer-Verlag, Berlin.

HAWKINS, J. W., BLOOMER, S. H., EVANS, C. A. & MELCHIOR, J. T. 1984. Evolution of intra-oceanic arc-trench systems. *Tectonophysics*, **102**, 175–205.

IRVINE, T. N. & BARAGAR, W. R. A. 1971. A guide to the chemical classification of common igneous rocks. *Canadian Journal of Earth Science*, **8**, 523–548.

JENNER, G. A., CAWOOD, P. A. RAUTENSCHLEIN, M. & WHITE, W. M. 1987. Composition of back-arc basin volcanics, Valu Fa Ridge, Lau Basin: Evidence for a slab-derived component in their mantle source. *Journal of Volcanology and Geothermal Research*, **32**, 209–222.

KUNO, H. 1968, Differentiation of basalt magmas, *In*: HESS, H. H. & POLDEVAART, A. (eds) *Basalts*, **2**, Wiley Interscience, New York, 624–688.

KUSHIRO, I. 1960. Si–Al relation to clinopyroxenes from igneous rocks. *American Journal of Science*, **258**, 548–554.

KAY, S. M., KAY, R. W. & CITRON, G. P. 1982. Tectonic controls on tholeiitic and calc-alkaline magmatism in the Aleutian Arc. *Journal of Geophysical Research*, **87**, 4051–4072.

LE BAS, M. J. 1962. The role of aluminium in igneous clinopyroxenes with relation to their parentage.

American Journal of Science, **260**, 267–288.

LEAKE, B. E. 1978. Nomenclature of amphiboles. *Canadian Mineralogist*, **16**, 501–520.

LETERRIER, J., MAURY, R. C., THORON, P., GIRARD, D. & MARCHAL, M. 1982. Clinopyroxene composition as a method of identification of the magmatic affinities of paleo-volcanic series. *Earth and Planetary Science Letters*, **59**, 139–154.

MACK, G. H. 1984. Exceptions to the relationship between plate tectonics and sandstone composition. *Journal of Sedimentary Petrology*, **54**, 212–220.

MALAHOFF, A., FEDEN, R. H. & FLEMING, H. S. 1982. Magnetic anomalies and tectonic fabric of marginal basins north of New Zealand. *Journal of Geophysical Research*, **87**, 4109–4125.

MAYNARD, J. B., VALLONI, R. and YU, H. 1982. Composition of modern deep-sea sands from arc-related basins, *In*: LEGGETT, J. K. (ed.) *Trench-forearc geology*. Geological Society London, Special Publication, **10**, 551–561.

MIYASHIRO, A. 1974. Volcanic rock series in island arcs and active continental margins. *American Journal of Science*, **274**, 321–355.

MORRIS, P. A. 1988. Volcanic arc reconstruction using discriminate function analysis of detrital clinopyroxene and amphibole from the New England Fold Belt, eastern Australia. *Journal of Geology*, **96**, 299–311.

NISBET, E. G. & PEARCE, J. A. 1977. Clinopyroxene composition in mafic lavas from different tectonic settings. *Contributions to Mineralogy and Petrology*, **63**, 548–554.

PACKER, B. M. & INGERSOLL, R. V. 1986. Provenance and petrology of Deep Sea Drilling Project sands and sandstones from the Japan and Marian forearc and backarc regions. *Sedimentary Geology*, **51**, 5–28.

PACKHAM, G. H. 1978. Evolution of a simple island arc—The Lau-Tonga Ridge. *Australian Society of Exploration Geophysicists Bulletin*, **9**, 133–140.

PECCERILLO, A. & TAYLOR, S. R. 1976. Geochemistry of Eocene calc-alkaline volcanic rocks from Kastamonu area, northern Turkey. *Contributions to Mineralogy and Petrology*, **58**, 63–81.

POLDERVAART, A. & HESS, H. H. 1951. Pyroxenes in the crystallization of basaltic magma. *Journal of Geology*, **59**, 472.

ROSER, B. P. & KORSCH, R. J. 1986. Determination of tectonic setting of sandstone–mudstone suites using SiO_2 content and K_2O/Na_2O ratio. *Journal of Geology*, **94**, 635–650.

UNDERWOOD, M. B. 1986. Sediment provenance within subduction complexes—an example from the Aleutian forearc, *In*: BILODEAU, W. L. (ed.) *Plate Tectonics and Petrologic Suites. Sedimentary Geology*, **51**, 57–73.

VALLIER, T. L., JENNER, G. A., FREY, F., GILL, J. DAVIES, A. S., HAWKINS, J. W., MORRIS, J. D., CAWOOD, P. A., MORTON, J., SCHOLL, D., RAUTENSCHLEIN, M., WHITE, W. M., WILLIAMS, R. W., VOLPE, A. M., STEVENSON, A. J. & WHITE, L. D. 1991. Subalkaline andesite from Valu Fa Ridge, a back arc spreading center in southern

Lau Basin: Petrogenesis, comparative chemistry and tectonic implication. *Chemical Geology*, (in press).

VALLONI, R. & MEZZADRI, G. 1984. Compositional suites of terrigenous deep-sea sands of the present continental margins. *Sedimentology*, **31**, 353–364.

VELBEL, M. A. 1985. Mineralogically mature sandstones in accretionary prisms. *Journal of Sedimentary Petrology*, **55**, 685–690.

VERHOOGEN, J. 1962. Distribution of titanium between silicates and oxides in igneous rocks. *American Journal of Science*, **260**, 211–220.

The provenance of sediments in the Barrême thrust-top basin, Haute-Provence, France

MARTIN J. EVANS[1] & MARIA A. MANGE-RAJETZKY[2]

[1]*Department of Earth Sciences, University of Liverpool, Liverpool L69 3BX, UK*
Present Address: BP Exploration, Britannic House, Moor Lane, London EC2Y 9BU, UK
[2]*Department of Earth Sciences, University of Oxford, Parks Road, Oxford OX1 3PR, UK*

Abstract: This paper records the results of an integrated analysis of heavy minerals, sedimentary facies, palaeocurrents and structural data. The objective of the study was to evaluate the provenance of the sediments of the Barrême thrust-top basin. This basin contains a Late Eocene to Late Oligocene succession of shallow marine, alluvial and lacustrine sediments. Late Eocene marine transgression resulted in starvation of clastic supply to the basin hence the Calcaires Nummulitiques received only a limited influx of heavy minerals derived from local sedimentary sources, augmented by the first appearance of assemblages with andesitic affinities at *c.* 42–41 Ma. Clastic members associated with the overlying Marnes Bleues record a significant diversification of provenance and sediment transport paths. Clastic detritus supplied to the shallow marine Grès de Ville were derived from Permo-Triassic sediments to the south. Conglomeratic submarine channels of the La Poste member mark the introduction of Alpine ophiolite–blueschist-derived minerals at *c.* 37–35 Ma, eroded from the Embrun–Ubaye thrust complex to the NE. The deltaic St Lions member is dominated by volcaniclastic assemblages. Completing the marine fill, the shoreface Grès de Senez received detritus from the southerly Maures-Esterel crystalline massif.

The succeeding Oligocene continental sediments record an increasing supply of minerals from Alpine lithologies. The alluvial Molasse Rouge contains reworked sedimentary detritus, including material cannibalized from the underlying marine Tertiary, and first-cycle Alpine minerals. The succeeding lacustrine Série Saumon–Série Grise show the increasing diversity of Alpine suites. The youngest sediments, the fluvial Grès Verts, signal a major influx of Alpine-derived assemblages related to reactivation of the Embrun–Ubaye thrusts between *c.* 28–27 and 25–24 Ma.

The sedimentary successions of foreland basins record the uplift and unroofing history of their adjacent mountain fold-thrust belts, but this stratigraphic record is often incomplete. The use of heavy mineral analyses in the study of foreland basin successions permits evaluation of the lithology and tectonic evolution of the source regions from which they were derived. This technique, combined with data on sequence arrangements, petrology, sedimentary facies and structural geology can be used to constrain the timing of uplift and erosion of the hinterland and facilitates reconstruction of the dynamic palaeo-drainage patterns of foreland basins.

The geological setting of the Barrême thrust-top basin is ideal for a study of time-varying provenance, controlled by tectonic episodes in the hinterland. This basin is located in the external zones of the SW Alps, in Alpes-de-Haute-Provence, between the Grès d'Annot turbidite basins to the east and the Digne–Valensole foreland basin to the west (Fig. 1).

Eleven stratigraphic units have been recognized within the basin, each characterized by distinctive heavy mineral assemblages. These as-semblages become increasingly diverse upwards in the stratigraphic column, reflecting the evolving complexity of lithologies exposed in the hinterland.

The evaluation of sediment transport pathways and reconstruction of the petrography of source areas are among the more testing problems encountered in basin analysis. This may result, for example, from a paucity of palaeo-current indicators, or the lack of present day exposure of potential parent lithologies. These problems are particularly acute in thrust belts where the geometrical relationships of source areas and receiving sedimentary basins undergo constant and often dramatic changes; moreover, active provenance areas may be subsequently buried beneath thrust sheets or eroded completely.

Additional complications are introduced in orogens where the foreland basin is dissected by thrusts and isolated segments of the basin are tectonically transported as thrust-top basins. Such an example is the Barrême syncline which has been displaced towards the SW as a thrust-top basin (Vann *et al.* 1986).

From Morton, A. C., Todd, S. P. & Haughton, P. D. W. (eds), 1991,
Developments in Sedimentary Provenance Studies.
Geological Society Special Publication No. 57, pp. 323–342.

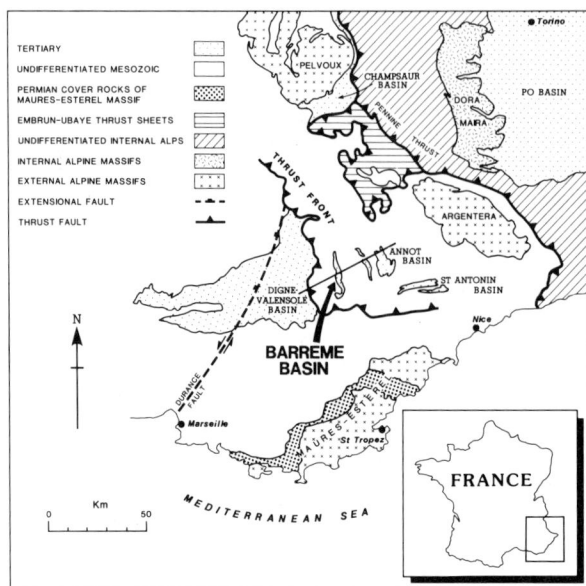

Fig. 1. Location map of the Barrême basin.

Sampling and analytical procedures

Each lithostratigraphic unit was sampled at regular intervals with the objective of obtaining heavy mineral assemblages representative of the particular unit. Depending on thickness and lithological inhomogeneity, 5–15 samples were collected from each unit. Samples were taken from well cemented beds in order to minimize the effects of post-depositional dissolution and present-day weathering.

For sample preparation standard laboratory disaggregation techniques were employed, followed by heavy mineral concentration in bromoform, using the centrifuge and partial freezing method (Mange & Maurer 1990). The grain-size distribution in the sample dataset varied from fine to coarse and it was necessary to select a correspondingly wide grain-size range for the heavy mineral analysis, in order to reduce the risk of leaving diagnostic mineral species unnoticed. For this reason the 0.063–0.300 mm size interval was extracted and analysed.

Grain mounts were made in liquid Canada balsam for microscopic identification. The results presented here are based on 200 grain counts per sample, using the ribbon counting method (Galehouse 1971). Micas, opaque grains and authigenic minerals were recorded but not included in the grain counts. Garnet was counted separately, thereby avoiding the masking effect of fluctuating garnet quantities on the associated minerals.

Electron probe microanalysis was performed on garnets, vesuvianite and volcaniclastic hornblende to reveal their chemistry and the nature of their host rocks.

Structural framework

The external Alps of Haute-Provence have for a long time been interpreted as a foreland fold-and-thrust belt. Present understanding of the complex structural geometries of this area owes much to comprehensive regional studies, completed over fifty years ago (Goguel 1936; de Lapparent 1938). More recently, the area has been interpreted as a predominantly thin-skinned thrust belt developed in Mesozoic rocks (Fig. 2) (Gigot et al. 1974; Siddans 1979; Vann et al. 1986; Apps 1987). The thrust system is a SW-directed, foreland propagating piggy-back sequence (Fig. 2) which deformed and inverted former Mesozoic extensional continental margin sequences (Lemoine et al. 1986; Hayward & Graham 1989).

Remnants of a formerly more extensive Tertiary foreland basin have been preserved in the thrust belt to the north and east of the mountain thrust front (Fig. 1). During its early marine phase the foreland basin was extensive and only partly compartmentalized into sub-basins (Apps 1987). However, the continued growth of thrust-related structures resulted in uplift and isolation of small remnants of the basin.

The Barrême basin is the most westerly of these basins (Fig. 1) and Vann et al. (1986) calculated that it has been transported a mini-

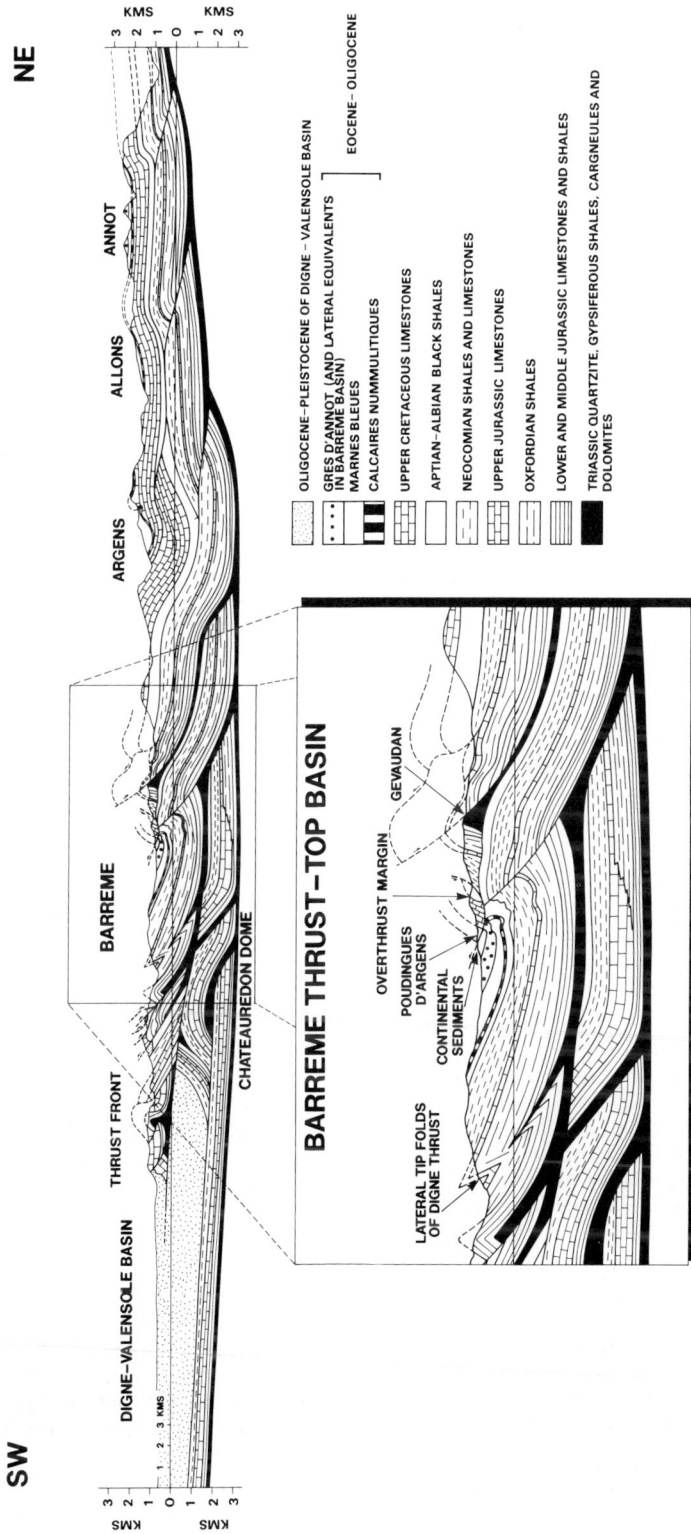

Fig. 2. Structural cross-section (by R. H. Graham) through the Provençal foreland thrust belt (location shown in Fig. 1), with total shortening of 55–58 km. The structure to the east of the Barrême thrust-top basin is a major duplex with the floor thrust in Triassic gypsiferous shales and cargneueles and the roof thrust in Aptian–Albian black shales. Late Cretaceous limestones occur above the roof thrust. To the east of Barrême the structure is an emergent imbricate stack; Late Cretaceous limestones are absent.

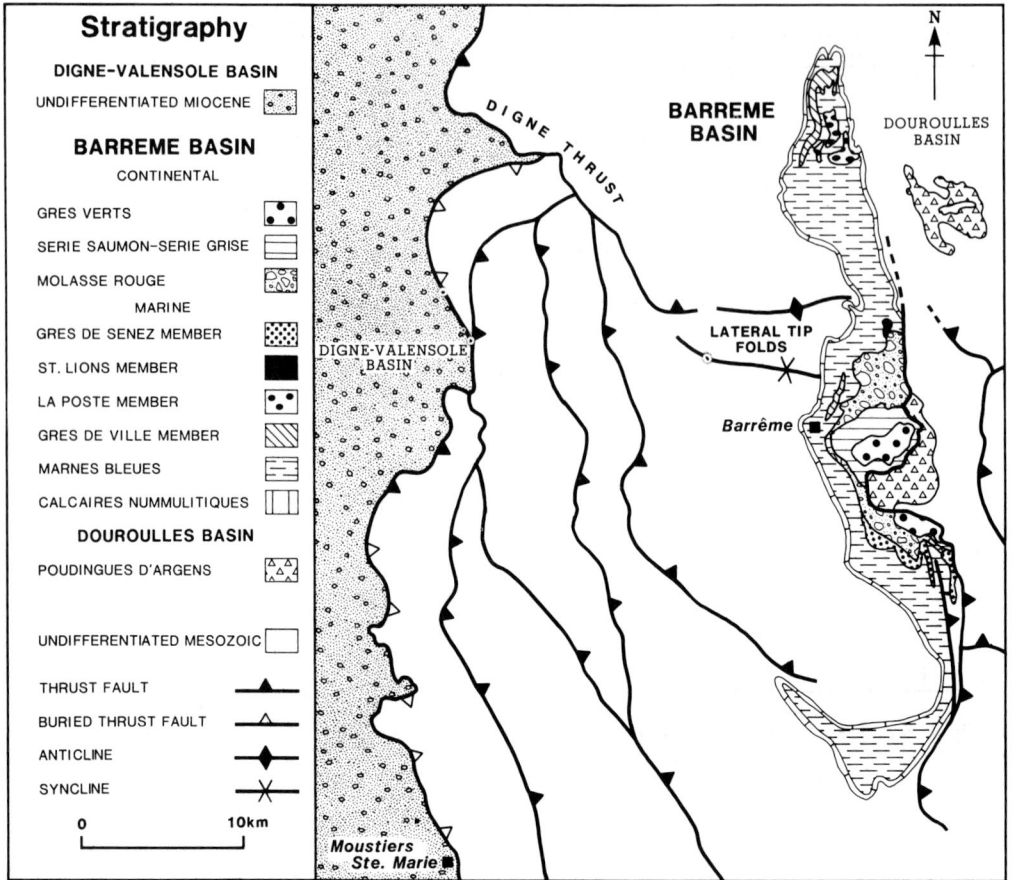

Fig. 3. Geological map of the Barrême thrust-top basin.

mum of 43 km towards the SW as a thrust-top basin. It is a narrow elongate syncline, aligned parallel to the NNW–SSE strike of thrusts (Fig. 3). It is 22.5 km in length, with a maximum width of 4 km in the vicinity of Barrême village, covering a total area of approximately 50 km².

Viewed in an E–W cross-section (Fig. 2) the Barrême syncline is strongly asymmetric, with a steeply-dipping to overturned eastern limb, and a relatively gentle dipping (up to 20°) western limb. Deformation is concentrated along the inner, thrust-bounded eastern margin (Chauveau & Lemoine 1961; Bodelle 1971; de Graciansky 1972), and east of Barrême village the basin is overthrust. The thrust fault places Upper Cretaceous limestones and ?Eocene Poudingues d'Argens conglomerates over the youngest Oligocene continental sediments (Fig. 3).

Lithostratigraphy

The sedimentary fill of the Barrême basin comprises approximately 750 m of marine and continental sediments (Fig. 4). Their age ranges from Eocene to Oligocene, thus linking the Eocene deposits of the Grès d'Annot basins to the East with the Oligocene to Plio-Pleistocene sediments of the Digne–Valensole basin to the west. This stratigraphic younging towards the west demonstrates that the depocentre migrated westwards as subsidence, driven by crustal loading, progressed towards the foreland margin.

Eleven stratigraphic units are recognized in the basin, resting unconformably on Lower Cretaceous shales and limestones (Fig. 4). The coarse-grained clastic units, the focus of this study, occur predominantly on the eastern margin of the basin (Fig. 5).

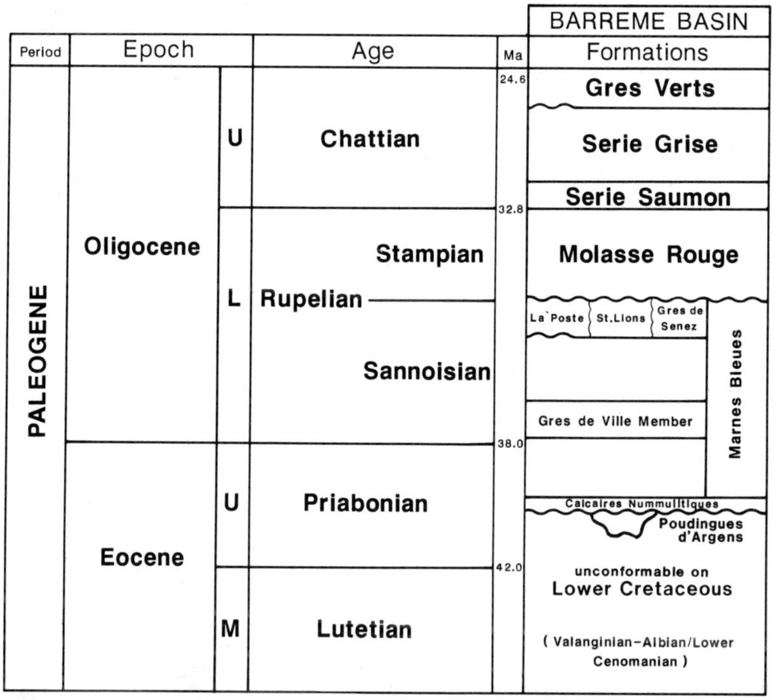

Period	Epoch		Age	Ma	BARREME BASIN Formations
PALEOGENE	Oligocene	U	Chattian	24.6	Gres Verts
					Serie Grise
				32.8	Serie Saumon
		L	Stampian / Rupelian		Molasse Rouge
			Sannoisian	38.0	La´Poste / St.Lions / Gres de Senez / Gres de Ville Member / Marnes Bleues
	Eocene	U	Priabonian	42.0	Calcaires Nummulitiques / Poudingues d'Argens
		M	Lutetian		unconformable on Lower Cretaceous (Valanginian–Albian/Lower Cenomanian)

Fig. 4. Stratigraphic table, based on biostratigraphic data from Bodelle (1971). Heavy mineral analyses indicate that the La Poste, St Lions and Grès de Senez members are not time-equivalent. Our revised stratigraphic interpretation is shown in Fig. 5. Age picks (Ma) were adopted from Harland *et al.* 1982.

The earliest Tertiary sediments

Poudingues d'Argens

The Poudingues d'Argens are the earliest Tertiary sedimentary rocks preserved in the Provençal thrust belt, although the precise age of the formation has yet to be established accurately. Bodelle (1971) concluded that the lower age limit is probably post-Senonian (on the basis of K–Ar geochronology) and the upper age limit is pre-Calcaires Nummulitiques (Fig. 4). In the Barrême basin the formation rests unconformably on Lower Cretaceous shales and limestones.

The Poudingues d'Argens consist predominantly of conglomerates, with minor intervals of colour-mottled mudstones, siltstones and thin sheet sandstones. This facies association is interpreted in terms of localized alluvial palaeovalley-fill deposits (Gubler 1958; Chauveau & Lemoine 1961; Bodelle 1971; Evans 1987). The conglomerates occur both as lenticular channel fills and as more laterally extensive bodies, the products of coalesced channels infilling the topographic relief which developed on the underlying uncon-

formity surface (Fig. 5). Such infill is particularly well exposed in the small Douroulles syncline, located 1.5 km to the east of the Barrême basin (Fig. 3).

Heavy minerals. Because the Poudingues d'Argens consist of nearly 100% locally derived Turonian–Senonian limestone clasts the heavy mineral yield is very poor. The assemblages are dominated by zircon with subordinate tourmaline, apatite and epidote (Fig. 6). The zircon population is diverse, with metamorphic, volcanic, metasedimentary and intrusive granitoid types present. Their morphology suggests both first cycle and polycyclic provenance.

Late Eocene marine transgression

Calcaires Nummulitiques

The Barrême basin contains the most westerly occurrence of the Calcaires Nummulitiques formation in the western Alpine foreland (Chauveau & Lemoine 1961), where it rests unconfor-

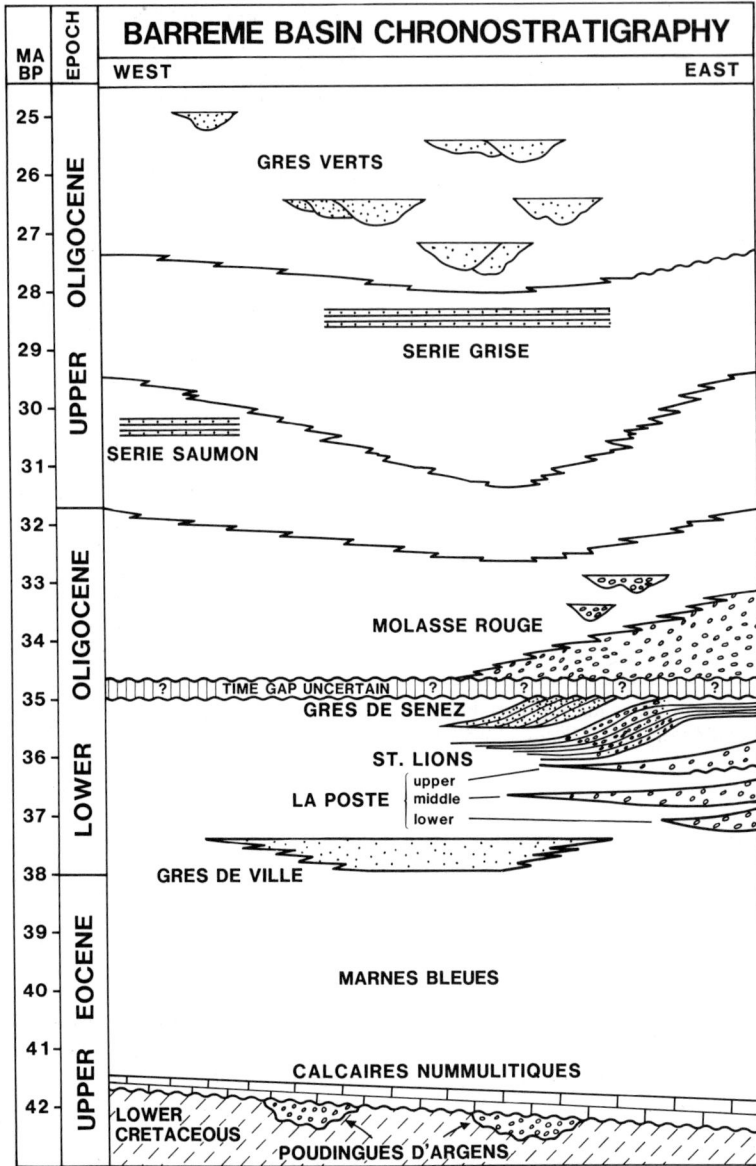

Fig. 5. Schematic chronostratigraphic diagram. Note the preferential development of coarse clastic units on the eastern margin of the basin, the exceptions being the axial inputs of the Grès de Ville (palaeoflow towards the NNW) and the Grès Verts (southerly flowing palaeocurrents).

mably on the Lower Cretaceous (Fig. 5). This formation is younger and thinner here than in the basins to the east. The extent of this diachronism is considered to range from Middle Eocene (Lutetian, *c.* 45 Ma) in the east, to Priabonian (late Eocene, *c.* 42 Ma) in the west (Espitalie & Sigal 1961).

The Calcaires Nummulitiques are transgressive shoreface deposits, comprising bioclastic packstones with abundant shallow marine fauna (Bodelle 1971). At Barrême the contact with the underlying Poudingues d'Argens fluvial conglomerates is marked by a bored clast horizon which is interpreted as a shoreface erosion surface (Evans 1987).

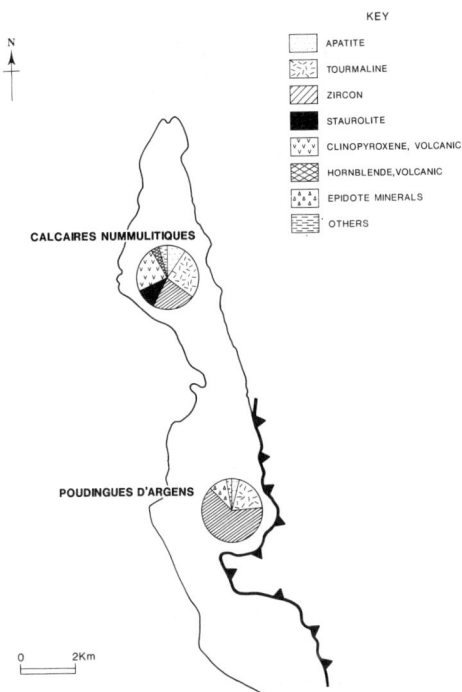

Fig. 6. Representative heavy mineral assemblages of the Poudingues d'Argens and Calcaires Nummulitiques.

Heavy minerals. The shallow Calcaires Nummulitiques carbonate shelf was starved of clastic input resulting in a low detrital content. The heavy mineral fractions are dominated by organic phosphatic debris (collophane and francolite), but small quantities of staurolite, zircon and tourmaline (Fig. 6) occurring together with clinopyroxene (green augite), hornblende and biotite (euhedral) were detected in each sample. The morphologies of the former (stable, ultrastable) suite indicate that these grains were derived from older sediments. The latter assemblage (Fig. 7A) points to a volcanic parentage.

Marine basin fill

Marnes Bleues

The Marnes Bleues formation consists of 350–400 m calcareous silty mudstones which show a rapid upwards transition from a benthonic to a pelagic fauna (Bodelle 1971), representing deepening associated with continued marine transgression. Four coarser clastic units are intercalated within and at the top of the Marnes Bleues

(Figs 4 & 5), and are termed members in this study. These comprise the Grès de Ville, La Poste, St Lions and Grès de Senez members. Heavy mineral analyses have focused on these coarser grade sediments.

Grès de Ville Member

The Grès de Ville represents the first significant input of sand grade detritus into the Marnes Bleues basin (Fig. 5). The sequence is 50–60 m thick, Sannoisian in age (Espitalie & Sigal 1961; Fig. 4) and interpreted to be laterally equivalent to the youngest Grès d'Annot sandstones found in basins to the east (Stanley 1961).

The dominant facies comprises thin (generally <20 cm), erosive based turbiditic sandstones, with axially-directed palaeocurrents towards the NNW and NW (Fig. 8). The sandstones are interbedded with grey silty mudstones and commonly exhibit a vertical sequence of sedimentary structures analogous to the Bouma turbidite sequence, although a storm-generated origin for at least some of these beds is indicated by hummocky cross-stratification and wave-modified current ripples.

Heavy minerals. The older Grès de Ville sediments are characterized by an apatite–staurolite–tourmaline assemblage, with abundant garnet and associated volcaniclastic titanite, euhedral biotite, green augitic clinopyroxene and green–brown hornblende. Staurolite, kyanite and garnet display dissolution features (sawtooth terminations and various etch facets). Their chemical stability is considerably higher than that of the associated pyroxenes and amphiboles. The coexistence of these minerals suggests that dissolution of the chemically stable species occurred during diagenesis in precursor sandstones. The survival of the delicate etch features suggests rapid transportation from the source area to the depositional site. The presence of unstable pyroxenes and hornblende indicates limited dissolution during diagenesis.

The dominantly rounded and subrounded apatites of the lower Grès de Ville are replaced in the younger beds by abundant euhedral and subhedral morphologies. These grains are accompanied by abundant euhedral biotite and such a marked change may be the result of increased volcanic activity (Fig. 7B).

La Poste Member

This member occurs as three discrete units interbedded within the Marnes Bleues (Fig. 5), herein termed the lower, middle and upper units. Maximum unit thickness is 50 m. It is Sannoisian in age and on the basis of biostratigraphic correlation at the top of the Marnes Bleues it is considered to be the lateral equivalent of the St Lions and Grès de Senez members (Bodelle 1971; Fig. 4). A revised lithostratigraphic framework based on index heavy mineral assemblages is shown in Fig. 5.

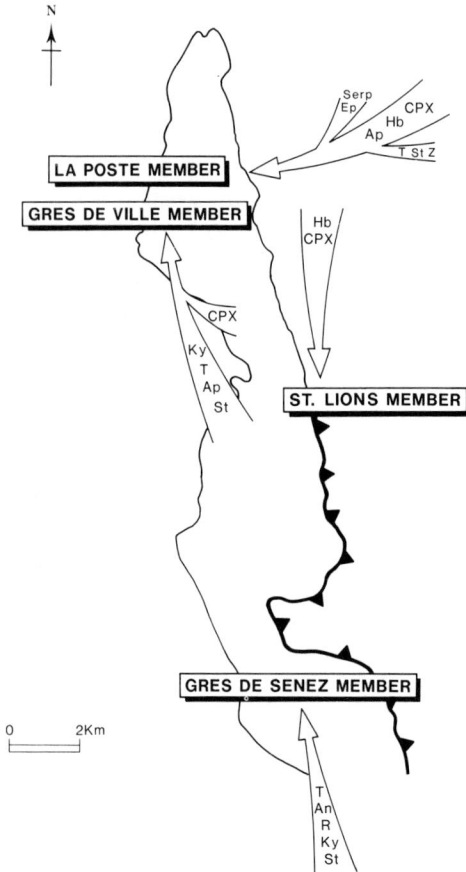

Fig. 8. Summary of palaeocurrents and heavy mineral species for clastic members of the marine fill. For explanation of abbreviations see Fig. 15.

Each unit exhibits a broadly coarsening-upwards sequence, passing from thin, erosive-based sandstones, resembling turbidites, and small isolated sandy channels, interbedded with Marnes Bleues sandstones, into coarse channelised and sheet conglomerates. The latter are associated with massive, trough cross-bedded and pebbly sandstones. Palaeocurrents in the La Poste are predominantly towards the WSW, perpendicular to the basin axis (Fig. 8).

The La Poste conglomerates and sands were deposited in a marine basin with a background suspended sediment fall-out of silty muds (Marnes Bleues). The dominant depositional processes are recognized as high density gravelly and sandy turbidity currents with sandy debris flows making a minor contribution. Density-flow deposition, at least in the lower two units, is interpreted to have occurred in large scale submarine channels, up to 1200 m wide, incised into the eastern margin of the Barrême basin (Evans 1987). The influx of coarse clastic sediments into the Marnes Bleues is linked to thrust activity east of the basin, resulting in oversteepening of the eastern margin which promoted channel incision, and culminated in the development of a syn-tectonic unconformity at the eastern margin of the basin (Fig. 5). These events at Barrême can be correlated with the emplacement of the 'Schistes á Blocks' melange further east in the Annot basins (Apps 1987).

Conglomerate clast types are dominantly Upper Cretaceous limestones, with rare clasts of Jurassic limestones and, according to Bodelle (1971), Upper Cretaceous Helminthoid flysch. The La Poste also records the first appearance of exotic clasts with Alpine affinities, which increase in abundance upwards. The wide spectrum of lithologies include serpentinites, serpentinised mafic rocks (e.g. basalt, gabbro, dolerite) as well as dolerite, gabbro, migmatite, prasinite, various granitoids, granophyre, andesite, rhyolite and metamorphosed quartzites (Gubler 1958; Chauveau & Lemoine 1961; Bodelle 1971). The metagabbros contain blue amphiboles, lawsonite, pumpellyite, jadeite and epidote (de Graciansky *et al.* 1971).

Heavy minerals. The heavy mineral spectrum of the lower La Poste unit is similar to that of the Upper Grès de Ville, being rich in both euhedral apatite and biotite (Fig. 9). A small amount of green augite, green–brown and basaltic hornblende complement the volcaniclastic suite. The proportion of augite increases upwards and it becomes predominant in the upper unit. The augite grains display etch features but the amphiboles are only slightly affected by dissolution. Non-volcanic detritus is represented by zircon, rutile, staurolite, kyanite and garnet, occurring in negligible quantities. The light fraction is dominated by well preserved zoned, intermediate to basic plagioclase (Fig. 7C).

The La Poste member is significant since it records the first appearance in the Barrême basin fill of high pressure index blue sodic amphibole together with traces of orthopyroxene (enstatite), diallage, serpentine and blue-green calcic amphibole, all with Alpine affinities (Fig. 9).

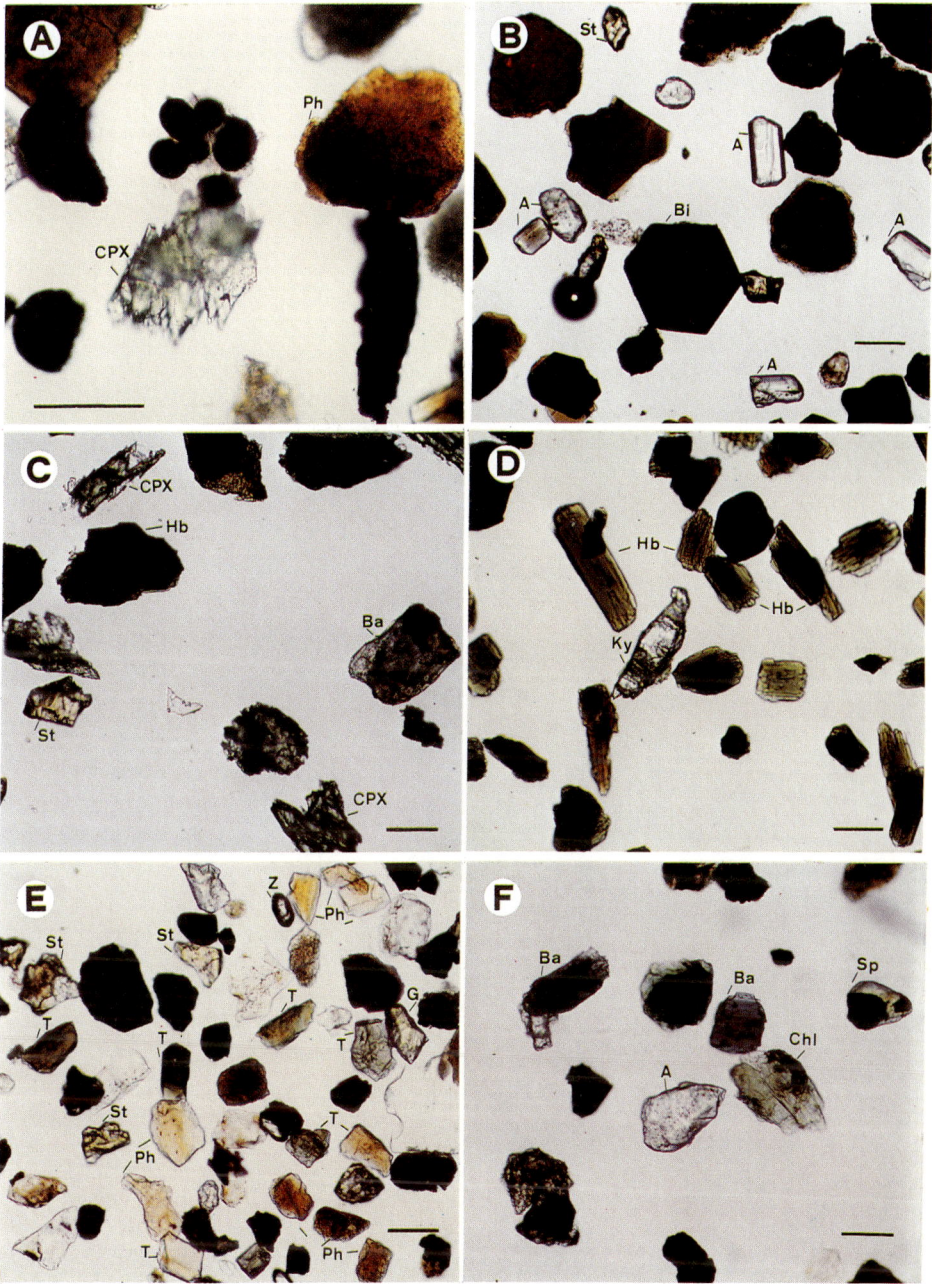

Fig. 7. Photomicrographs illustrating characteristic assemblages (**A**) Calcaires Nummulitiques: Clinopyroxene, phosphatic fragments and foraminifera infill; (**B**) Marnes Bleues–Grès de Ville: volcaniclastic hexagonal biotites and euhedral apatite; (**C**) Marnes Bleues La Poste member: etched staurolite, blue amphibole, volcaniclastic clinopyroxene and hornblende; (**D**) Marnes Bleues–St Lions member: hornblende and etched kyanite; (**E**) Marnes Bleues–Grès de Senez member: prismatic and angular tourmalines, staurolite, zircon with overgrowth, garnet, phosphatic debris and mica, note pronounced angularity of the grains; (**F**) Grès Verts: blue amphibole, serpentine, apatite, chlorite. A, apatite; Ba, blue amphibole; Bi, biotite; Chl, chlorite; CPX, clinopyroxene; Hb, hornblende; G, garnet; Ky, kyanite; Ph, organic phosphatic fragments; Sp, serpentine; St, staurolite; T, tourmaline; Z, zircon. Scale bar represents 100 μm in each photomicrograph.

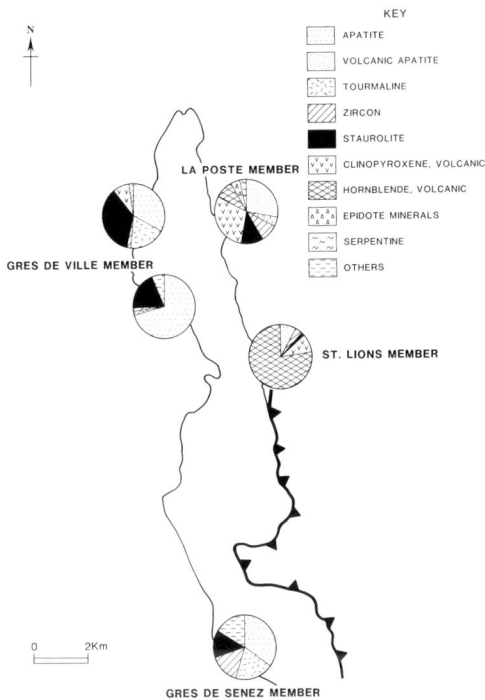

Fig. 9. Representative heavy mineral assemblages for clastic members of the marine fill.

St Lions Member

The St Lions member is late Sannoisian (Figs 4 & 5) and is interpreted to be laterally equivalent to the La Poste and Grès de Senez (Bodelle 1971). It occurs as an isolated outcrop in the central part of the basin (Fig. 3) and comprises a 55–60 m succession. This is made up of four smaller-scale coarsening-upwards sequences, stacked vertically with a pronounced offlap to the south. Their most striking feature is the development of large-scale, unidirectional conglomeratic foreset beds, 3–4 m high and dipping at 15–20°. Each coarsening-upwards package represents progradation of a small Gilbert-type delta into a marine setting (Evans 1987).

The conglomerate clasts are Upper Cretaceous and Upper Jurassic limestones, calcareous sandstones and radiolarian chert. Igneous and metamorphic constituents form approximately 15%; of these, 3% are andesite while the remainder are granite, serpentinized dolerite and gabbro, as well as microdiorite and microgranodiorite (Bodelle 1971) and abundant pumpellyite-bearing dolerite (de Graciansky *et al.* 1971).

Heavy minerals. The heavy mineral fractions of the St Lions member are composed almost entirely of volcaniclastic detritus. The dominant species is green–brown ferro-hastingsitic hornblende (Figs 7D & 9), and this is associated with smaller amounts of hastingsite, basaltic hornblende, euhedral apatite, titanite and biotite. Augite is present in minor quantities. The non-volcanic minerals include a few grains of blue sodic amphibole, kyanite and staurolite.

Grès de Senez member

The occurrence of this member is limited to the southern part of the basin (Fig. 3). It comprises a 40 m thick coarsening-upwards sequence with calcareous silty mudstones (Marnes Bleues) at the base, passing gradationally upwards into an intensely bioturbated sandstone-dominated interval. Strongly-ribbed bivalves and *Ophiomorpha* burrows become increasingly common upwards and the complete sequence is interpreted to be the product of a prograding sandy shoreline (Evans 1987).

Heavy minerals. The Grès de Senez heavy mineral assemblages contrast with those of its lateral equivalents, the La Poste and St Lions members. Staurolite, kyanite and andalusite, minerals of medium- to high-grade regional metamorphic parageneses, characterize the Grès de Senez. Volcaniclastic debris is absent as are Alpine-derived minerals, apart from traces of blue amphibole in one sample (Fig. 9). Apatite is abundant and exhibits angular, subangular or subrounded morphologies. First cycle euhedral and sharp prismatic tourmaline is relatively common. Many zircons have overgrowths which generally form in high grade metamorphics, migmatites or anatexites (Speer 1980). Kyanite and staurolite display etched or sharp angular morphologies (Fig. 7E) indicating primary and reworked sedimentary parentages.

Oligocene continental sedimentation

Molasse Rouge

The Molasse Rouge formation represents an important change in the depositional history of the Barrême basin. It marks the onset of continental sedimentation which persists throughout the remainder of the basin-fill. The sedimentary record of the transition is not complete as a syn-tectonic unconformity separates the underlying marine deposits from the Molasse Rouge (Fig. 5). This formation is 150 m thick and crops out in the central and southern areas of the basin (Fig. 3). It is barren of fauna and flora, and is assigned a Stampian age by bracketing between formations of known age above and below (Fig. 4).

The Molasse Rouge consists of an alluvial red-bed sequence which shows an overall fining-upwards trend. It is dominated by red-brown siltstones and silty mudstones with abundant palaeosols. Conglomeratic facies are common near the base of the formation (Fig. 5), representing ephemeral channels. Palaeocurrents are directed perpendicular to the basin axis, which flow towards the SW and W. The overall facies association is interpreted as a retreating alluvial apron, or a series of laterally coalesced alluvial fans deposited in an intermontane, semi-arid setting (Evans 1987).

Conglomerate clasts are composed predominantly of Upper Cretaceous limestone with occasional clasts showing inherited *Lithophaga* borings indicating derivation from older marine conglomerates. Granite, andesite, basalt and dolerite clasts constitute approximately 5%.

Heavy minerals. Both the heavy mineral yield and grain size distribution are highly variable. A multiple provenance of the Molasse Rouge sediments is indicated by three contrasting suites (Fig. 10):

(1) a zircon–tourmaline–staurolite assemblage with rounded and etched morphologies of polycyclic origin;

(2) rounded hornblende and green augite, together with slightly abraded euhedral biotite and apatite, as well as zoned plagioclase (reworked from the underlying marine Tertiary);

(3) traces of blue sodic amphibole, lawsonite, non-volcanic calcic amphibole, serpentine, epidote and euhedral garnet of first cycle origin.

Série Saumon–Série Grise

These formations occur in the central and southern parts of the Barrême basin (Fig. 3), where the Série Saumon is 40 m thick and the Serie Grise attains 110–120 m. The sequence is dated as Chattian on the basis of mammal teeth (Carbonnel *et al.* 1972).

The Série Saumon is characterized by varicoloured, colour-mottled mudstones with incipient calcrete development. Thin molluscan

wackestones with rootlet traces, charophyte oogonia and freshwater molluscs occur at the top of the sequence. The facies association suggests deposition in a lake–marginal marsh.

The Série Grise is dominated by grey calcareous mudstones with thin micritic limestones and wave-rippled siltstones. Macrofauna include *Helix*, *Unio* and *Lymnea*, and an interval from near the base of the formation yielded a rich mammal fauna (Carbonnel *et al.* 1972). The facies and palaeontological evidence indicate a shallow lacustrine depositional environment

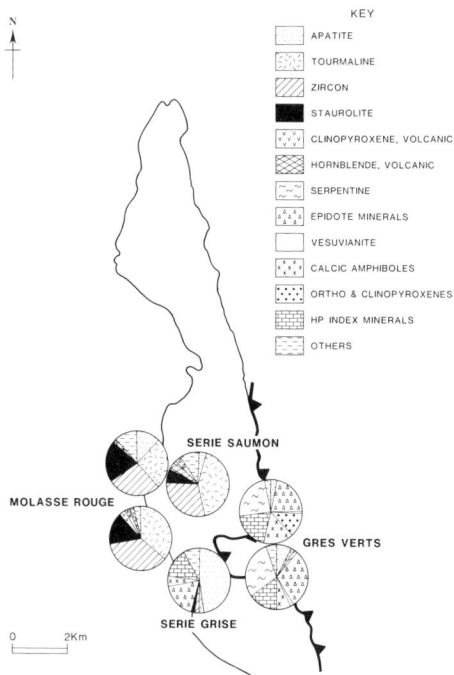

Fig. 10. Representative heavy mineral assemblages for the continental formations.

Heavy minerals. These formations show transitional mineral suites between the Molasse Rouge and the overlying Grès Verts (Fig. 10). The Série Saumon contains reworked volcaniclastic material, some polycyclic species, and a range of first cycle minerals, representing the increasing diversity of Alpine assemblages e.g. epidote minerals, serpentine, blue sodic amphibole, lawsonite and pumpellyite. These are associated with garnet and various calcic amphiboles (pale blue–green hornblende, tremolite and actinolitic amphibole). Schistose rock fragments, composite grains comprising metamorphic minerals and various micas complement the first cycle assemblages.

Grès Verts

The Grès Verts is the youngest formation, dated as Chattian by mammal remains (Fig. 4; Carbonnel *et al.* 1972). It is 60–70 m thick and has an areal distribution similar to that of the underlying continental sediments (Fig. 3). Locally the contact with the Série Grise is a syn-tectonic unconformity at the eastern margin of the basin (Fig. 5), (Pairis 1971; de Graciansky 1972).

The facies of the Grès Verts represent a dramatic contrast to the underlying, low-energy lake deposits, comprising coarse-grained fluvial channel sandstones and associated fine-grained colour-mottled overbank deposits (Fig. 5). The channels are the product of bedload deposition in vigorous, laterally stable streams with an ephemeral discharge pattern (Evans 1987). Sets of trough cross-bedding (up to 1.5 m set height) indicate flow from N to S (Fig. 11).

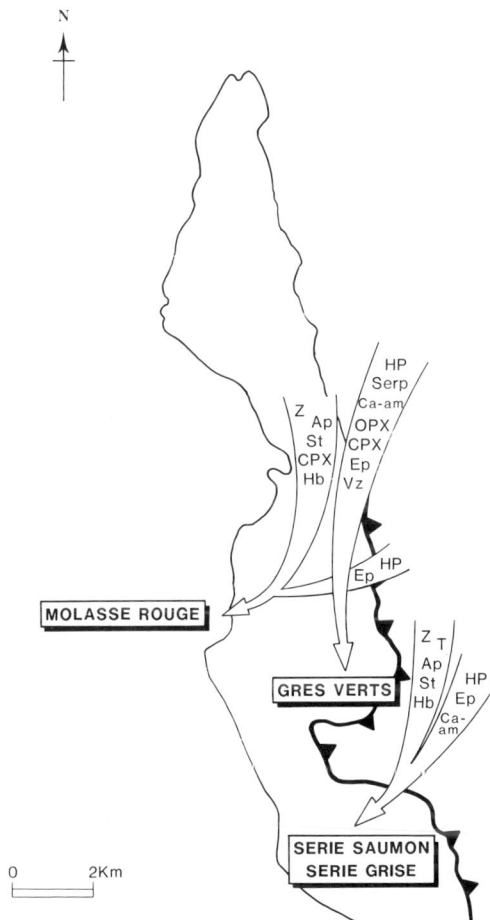

Fig. 11. Summary of palaeocurrents and heavy mineral species for the continental formations. For explanation of abbreviations see Fig. 15.

Heavy minerals. The Grès Verts contain the most complex assemblage of heavy minerals in the Barrême basin-fill (Fig. 7F). In addition to the Alpine species present in the Série Grise (see above), jadeite, ophiolite-derived hypersthene, enstatite, diopside, diallage, chromian spinel and abundant serpentine flakes occur together with various calcic amphiboles and some garnet (Fig. 10). Serpentines (which give the sandstones their distinctive green colouration) are either serpentine rock fragments or serpentinized olivine, pyroxenes and amphiboles. A diagnostic mineral, vesuvianite, is present in small amounts in each sample. It has a prismatic, more rarely irrregular or rounded habit and exhibits distinctive strong purple–blue and dull honey–yellow anomalous interference colours, negative elongation and parallel extinction (Mange & Maurer 1990). Sector twinned and strongly birefringent grossular garnet is a common associate.

The provenance of the sediments

Information on the lithostratigraphic units, depositional environments and distinctive heavy mineral suites is summarized in Table 1.

The Poudingues d'Argens

The continental Poudingues d'Argens are conglomerates composed of limestone clasts with a low heavy mineral yield. This reflects the physiographic setting of the Provençal region during the earliest phases of thrusting and foreland basin development. Regional topography was low, resulting from Palaeocene–Lower Eocene erosion (Baudrimont & Dubois 1977). The Poudingues d'Argens was locally derived from Mesozoic carbonates and shales, with a very low supply of siliciclastic detritus.

Marine sediments

Priabonian marine transgression resulted in low rates of siliciclastic supply to the shelf during deposition of the Calcaires Nummulitiques and Marnes Bleues. Of the limited number of detrital heavy minerals staurolite, garnet and some zircons were probably derived from pre-existing sediments. The Permo-Triassic cover rocks of the external crystalline massifs contain abundant staurolite, kyanite, apatite and garnet (Stanley 1965), and the inherited etch features observed in the Tertiary sediments are signatures of their recycled origin.

Table 1. *Summary table providing information on lithostratigraphic units, depositional environments, heavy mineral suites and source areas*

Lithostratigraphic units	Depositional environment	Heavy mineral assemblages	Provenance-index minerals	Source areas
Grès Verts	Fluvial channels and interchannel floodplain	Epidote minerals, Ca-amphibole, OPX, CPX, blue amphibole, lawsonite, chrome spinel, serpentine, vesuvianite	OPX, CPX (non-volcanic), high pressure index minerals, vesuvianite	Embrun–Ubaye thrust sheets incorporating rocks with Penninic affinities
Série Saumon Série Grise	Lake-marginal marsh and shallow lake	Epidote minerals, Ca-amphibole, blue amphibole, lawsonite, tourmaline, zircon, staurolite, recycled volcanic hornblende	Epidote minerals, Ca-amphibole, high pressure index minerals, reworked volcanic hornblende	Embrun–Ubaye thrust sheets, pre-existing Marnes Bleues sediments
Molasse rouge	Semi-arid alluvial apron/fans	Zircon, tourmaline, staurolite, apatite; reworked volcanic CPX; Ca-amphibole, high pressure-index minerals	Reworked volcanic CPX, Ca-amphibole, high pressure-index minerals	Embrun–Ubaye thrust sheets, cannibalized Marnes Bleues sediments
Marnes Bleues Gres de Senez Member	Progradational shoreface	Apatite, zircon, tourmaline, staurolite, kyanite, titanite, rutile	Staurolite, kyanite	High-grade metamorphics of the Maures-Esterel Massif
St Lions Member	Stream-dominated Gilbert-type deltas	Zircon, tourmaline, staurolite; volcanic hornblende, CPX and apatite	Volcanic suite and recycled sedimentary minerals	Local pre-existing sediments and andesitic volcanism
La Poste Member	Large scale submarine channels	Zircon, tourmaline, staurolite; volcanic hornblende and CPX; blue amphibole, serpentine	Volcanic suite and recycled minerals, Alpine serpentine and blue amphibole	Local pre-existing sediments, andesitic volcanism and Alpine thrust sheets
Grès de Ville Member	Storm-influenced shallow marine	Zircon, tourmaline, rutile; volcanic apatite and biotite	Recycled minerals, volcaniclastics	Permo-Triassic cover of the Maures-Esterel Massif, andesitic volcanism
Calcaires Nummulitiques	Transgressive shoreface	Apatite, zircon, tourmaline, staurolite; volcanic hornblende, CPX and apatite	Recycled minerals, volcaniclastics	Local Permo-Triassic sediments, andesitic volcanism
Poudingues d'Argens	Alluvial palaeovalley fills	Apatite, zircon, tourmaline, rutile, epidote	Recycled minerals, first cycle zircon and epidote	Underlying Cretaceous sediments

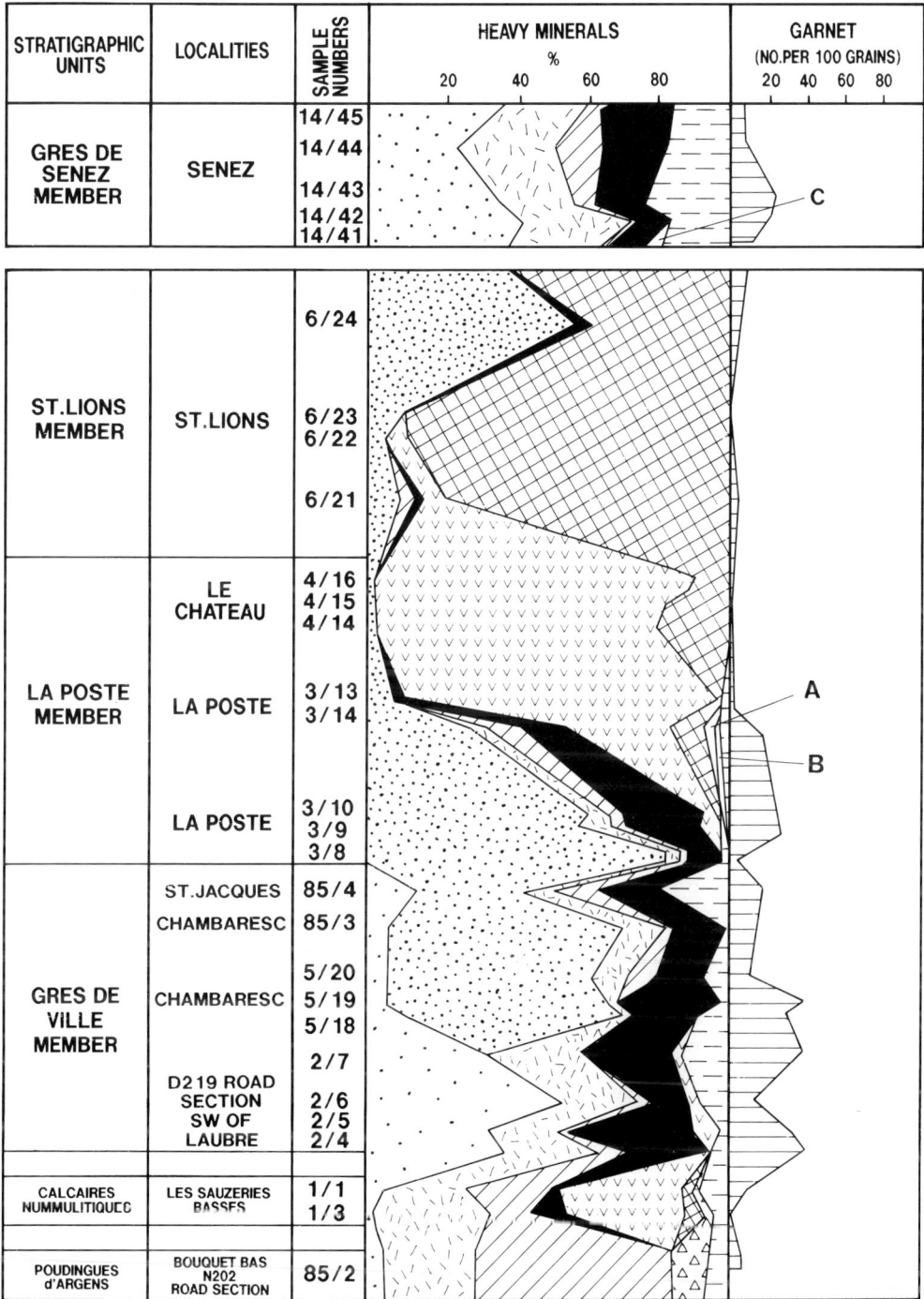

Fig. 12. Vertical distribution plot of heavy minerals for the marine fill, illustrating the stratigraphic evolution of assemblages. Key is the same as for Figs 9 and 10. A, high pressure index minerals; B and C, epidote minerals.

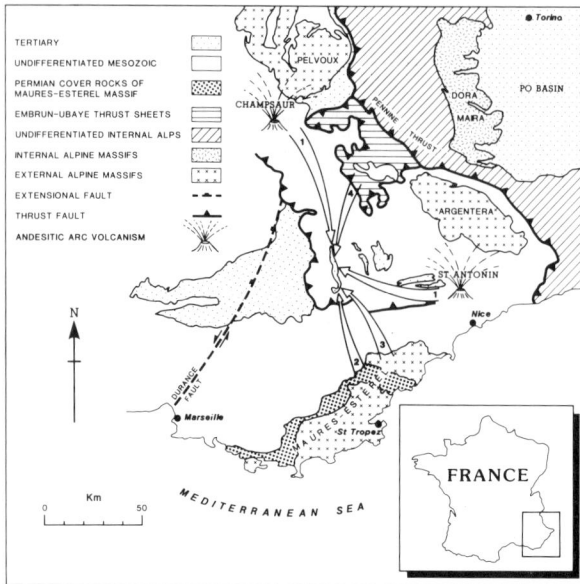

Fig. 13. Interpretation of provenance and major sediment delivery routes to the Barrême basin. 1, Andesitic volcaniclastics; 2, Permian sedimentary cover of the Maures–Esterel massif; 3, Maures–Esterel foreland basement massif; 4, Ophiolitic and blueschist detritus from the Embrun–Ubaye thrust sheets.

The Calcaires Nummulitiques records the first appearance of volcaniclastic material in the Barrême basin-fill (Fig. 12) at *c.* 42–41 Ma. The well preserved euhedral morphologies of delicate biotite flakes indicate an air-borne origin for at least some of the volcanic detritus, resulting from explosive activity. This implies the existence of contemporaneous andesitic volcanic centres during the late Eocene Nummulitic marine transgression, coinciding with the initial subsidence phase of the SW Alpine foreland basin.

The existence of late Eocene to early Oligocene andesitic arc volcanism in the Swiss and French Alps is recorded in several turbiditic sequences characterized by abundant volcaniclastic material. These volcaniclastic greywackes are generally called the Taveyannaz and Champsaur sandstones. Several hypotheses have been proposed to explain the presence of volcaniclastic debris in these sediments (Martini & Vuagnat 1968; Beuf *et al.* 1961; Sawatzki 1975; Homewood 1983). Recently, this volcanism has been attributed to deep northward-trending fractures in the autochthonous foreland basement, associated with extension (Homewood & Caron 1982; Homewood 1983). However, in the case of the Barrême basin deposition of the Calcaires Nummultiques occurred in a foreland basin setting associated with compressional tectonics.

The exact location of the volcanic centres remains unclear. There are no exposed Tertiary volcanic complexes in the western Alps,

although it is probable that they were buried beneath later thrusts. In addition to the Taveyannaz–Champsaur volcaniclastics, proximal andesitic breccias occur in the late Eocene–Oliocene of the St Antonin basin (Alsac *et al.* 1969; Bodelle 1971). Bodelle (1971) interpreted the breccias as gravity flow deposits generated on the flanks of a volcano and suggested two possible centres of emission in the St Antonin basin. The fine-grained volcanic detritus at Barrême could therefore have been derived from either the Champsaur area to the north or from St Antonin to the south, or both (Fig. 13). It is also possible that a currently unexposed centre existed in close proximity to the basin.

The Sannoisian Grès de Ville heavy minerals and palaeocurrent data indicate dual provenance (Fig. 12). Polycyclic detritus was supplied by the Permo-Triassic cover of the Maures–Esterel massif to the south of the basin (Fig. 13), and this was mixed with arc-derived volcaniclastics. Sediment transport from the south raises the question whether detritus from the Alpine and crystalline basement lithologies of Corsica was also delivered to the basin. There is ample palaeomagnetic, plate tectonic and petrological evidence suggesting that the Corso-Sardinian land mass and Provence were once contiguous (amongst others Stanley & Mutti 1968; de Jong *et al.* 1973; Alvarez 1976; Le Pichon 1984; Rehault *et al.* 1985). The lower time-limit of the rotation of these blocks (followed by the opening of the Gulf of Lion), is well established as being Late Oligocene/Early Miocene. Although northerly palaeoflows suggest the possibility of contribution from Corso-Sardinian sources, the absence of minerals typical of Alpine lithologies indicates that Corso-Sardinian supply to the Grès de Ville depositional system was unlikely.

Air-borne euhedral apatites and biotites, common in the younger Grès de Ville sediments indicate explosive activity. Clinopyroxene and hornblende are virtually absent in the explosive material (Fig. 12).

Products of the explosive phase can be found in the Lower La Poste units but diminish in the upper unit, where the decline of air-borne debris is paralleled by a marked increase in clinopyroxene (Fig. 12). By contrast, its lateral equivalent, the St Lions member, is rich in hornblende (Fig. 12), a feature which cannot be ascribed to dissolution of pyroxenes. The explanation considered most likely is a change in the andesitic source rock composition. Two types of andesite clasts were described by Siedon (1981), one containing olivine, abundant augite and subordinate hornblende and a second type, rich in hornblende but with little pyroxene. This possible compositional evolution in the andesitic volcanic products may have controlled the contrasting heavy mineral assemblages of the La Poste and St Lions members. Implications for the validity of biostratigraphic correlation of the La Poste and St Lions members are discussed below.

The switch from axial, northerly palaeocurrents of the Grès de Ville to the lateral WSW-directed input of the La Poste, and the southerly prograding St Lions deltas (Fig. 8), is associated with the first appearance of both exotic clasts and heavy mineral species of Alpine parageneses. This event is linked to the unroofing of the Embrun–Ubaye thrust sheets during the Sannoisian between c. 37–35 Ma (Fig. 13). According to Fry (1989), the lower Embrun–Ubaye thrust sheets were emplaced over a time span ranging from Lutetian to Stampian. The Sannoisian erosive event can be related to a major phase of uplift, subaerial exposure and erosion as the Embrun–Ubaye thrusts overrode the Grès d'Annot, following emplacement of their precursor 'Schistes à Blocs' olistostromes (Kerckhove 1969; Apps 1987; Fry 1989).

Pebbles and heavy minerals of uplifted Alpine subduction complexes, occurring in the La Poste and St Lions members, are significant in the timing of structural episodes in the hinterland. Their presence indicates that Alpine blueschists and meta-ophiolite subduction complexes occurring as slices in the Embrun–Ubaye thrust sheets had already been thrust to surface by the Sannoisian.

Only a negligible contribution of sand-grade detritus is inferred from the Late Cretaceous Helminthoid Flysch, owing to their fine grain size and limited content of heavy minerals (small quantities of apatite, our observations). Chauveau & Lemoine (1961) and Bodelle (1971) document clasts of Helminthoid Flysch in the Oligocene conglomerates of Barrême, linking their provenance to the Embrun–Ubaye thrust sheets. It is most likely that these were derived from the Nappe de l'Autapie, emplaced during the Late Eocene–Early Oligocene. The emplacement of the structurally higher Parpaillon nappe commenced in the Miocene (Fry 1989) which postdates deposition of the youngest sediments (Upper Oligocene) in the Barrême basin.

Detrital minerals of the Grès de Senez are predominantly first cycle, derived from medium to high grade metamorphic rocks, and they show affinities to the Maures-Esterel massif to the south (Fig. 13). The lithology of this massif is extremely complex and rock types include granitoids, migmatites, medium- to high-grade crystalline schists and Permian volcanic rocks (Gueirard 1962; Boucarut 1971; Campredon & Boucarut 1975; Pupin 1976). The area has been a prominent sediment source since the Early Cretaceous (Bordet 1951) and experienced a regional uplift during the Oligocene (Aubouin & Mennessier 1963). The entire Grès de Senez heavy mineral spectrum shows a strong resemblance to that of the Lower Miocene sandstones of the Nerthe Chain in the south of France, known to have been derived from the Maures–Esterel massif (Colomb 1972; Lay & Parfenoff 1972, and our own investigations) thus providing supporting evidence for the provenance of the Grès de Senez. Similarly to the Grès de Ville, the absence of typical Alpine minerals precludes contribution from the Corso-Sardinian region. The lack of andesitic volcanic detritus, significant in the La Poste and St Lions sediments (Fig. 12), indicates an apparently abrupt cessation of the volcaniclastic supply to the basin.

Implication of mineralogical data for correlation of the marine sediments

The marked contrast of heavy mineral compositions between the La Poste and St Lions members, and the lack of volcaniclastic grains in the Grès de Senez have important implications which modify the biostratigraphic correlation scheme for the Barrême basin established by Bodelle (1971) and de Graciansky et al. (1982). It appears that compositional evolution of the

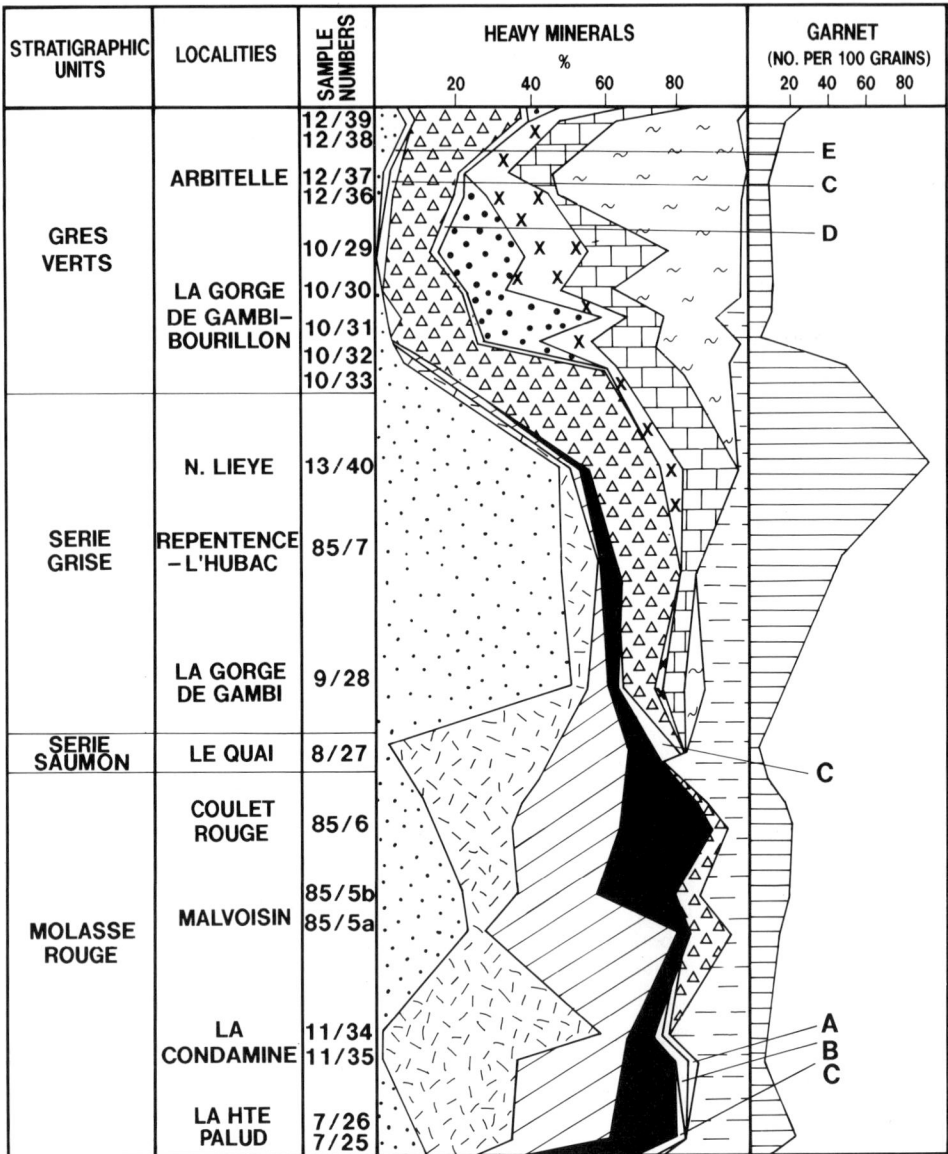

Fig. 14. Vertical distribution plot of heavy minerals for the continental fill. Key is the same as Figs 9 and 10. A, high pressure index minerals; B, reworked volcanic clinopyroxene; C, reworked volcanic hornblende; D, vesuvianite, E, tourmaline.

volcaniclastic supply towards hornblende-rich andesites may have controlled the contrasting heavy mineral suites of the La Poste and St Lions members, indicating that the St Lions sediments may be slightly younger. The Grès de Senez is devoid of volcaniclastic minerals. Although selective sorting during transportation may reduce or remove some species resulting from differing size, shape and density of the various minerals (controlling their hydraulic be-

haviour), hornblende and pyroxenes have different hydraulic properties and it is unlikely that the entire hornblende and pyroxene populations were eliminated during transport. Biotite together with euhedral apatite are dominantly of air-borne origin not influenced by surface transport. The effects of diagenetic dissolution may be considered in the elimination of the unstable pyroxenes and hornblende, but these grains are present in older formations in the basin fill, and it appears that these sediments have experienced a similar diagenetic history. The absence of stable apatite (volcanigenic) and biotite cannot be ascribed to dissolution. The most plausible explanation of their absence is that explosive andesitic volcanism had ceased to the south of the basin (the provenance area) before the deposition of this shallow marine sequence. This suggests that the Grès de Senez is younger than the La Poste and St Lions with which it was correlated on the basis of biostratigraphic data. Figure 4 shows the previously accepted stratigraphic framework for the basin, whereas our interpretation, based on mineralogical data, is indicated in Fig. 5.

Continental sediments

The continental Molasse Rouge received a limited supply of siliciclastic detritus. The coarse grained material was sourced locally from Upper Cretaceous limestones and the Poudingues d'Argens (Gubler 1958). Nevertheless the three contrasting heavy mineral suites (Fig. 14) indicate a multiple provenance for the sand size grains:

(1) Permo-Triassic sediments, which were ultimately sourced from medium to high-grade metamorphic rocks;
(2) cannibalization of underlying Marnes Bleues clastic members, signalled by rounded volcanic hornblendes, pyroxene and titanite;
(3) small amounts of material from Alpine terranes, and the progressive abundance of epidote indicates the increasing importance of this source.

During deposition of the Série Grise the renewed thrusting of Alpine structural units between c. 31–28 Ma is reflected by the complexity of heavy mineral suites (Fig. 14). High pressure index species, together with those derived from ophiolites, indicate the availability of meta-ophiolite-blueschist rocks of Penninic affinities.

Detritus from Alpine sources, first observed in the Sannoisian La Poste member, reaches its maximum diversity in the Chattian Grès Verts (Fig 14). The overwhelming abundance of Alpine material in the sediment dispersal system between 28–27 and 25–24 Ma is related to a major phase of thrusting and uplift, resulting in rejuvenation of hinterland relief coupled with intense erosion.

Diagnostic minerals (e.g. chrome spinel, serpentine, lawsonite, blue amphibole and vesuvianite) signal derivation from ophiolites, blue-schists and rhodingites. These occur as dismembered and metamorphosed slices in the Haute-Ubaye valley (Kerckhove 1969; Steen 1972, 1975). Despite a retrograde greenschist facies metamorphic overprint, earlier blueschist assemblages are well preserved. Serpentinites are abundant in the sequence; the contact between the serpentinites and the wall rock is characterized by reaction rims of rhodingitic affinity in which calc-silicates, including vesuvianite, were formed (Steen 1975). These data, combined with N to S palaeoflow directions recorded from channel sandstones, dictate that the provenance of the Grès Verts can be confidently linked with a major phase of out-of-sequence displacement and uplift of the Embrun-Ubaye thrust sheets (Fig. 13).

Conclusions

(1) The Palaeogene fill of the Barrême thrust-top basin preserves a complex evolutionary pattern of major sediment transport routes (Fig. 13) and sand provenance which has been reconstructed by the study of heavy minerals (Fig. 15). The evolution of provenance can be summarized as an overall transition from pre-Tertiary foreland sedimentary cover rocks, andesitic volcaniclastics, foreland crystalline basement, to a final overwhelming influx of detritus derived from Alpine parageneses.

(2) During the Late Eocene and Early Oligocene, marine sedimentation was accompanied by andesitic volcanism with volcaniclastic debris first appearing in the Calcaires Nummulitiques at c. 42–41 Ma. Volcanism probably ceased before the deposition of the Grès de Senez around c. 36–35 Ma.

(3) Alpine high pressure index minerals first appear in the coarse-grained La Poste member at c. 37–36 Ma, implying that the initial unroofing of ophiolitic slices associated with the Embrun–Ubaye thrust sheets occurred at this time.

(4) The abundance of Alpine ophiolite- and blueschist-derived assemblages in the Grès Verts between c. 28–27 Ma records major reactivation and uplift of the Embrun–Ubaye thrust sheets.

(5) Despite sediment supply from the south for the Grès de Ville and Grès de Senez members, the absence of diagnostic Alpine species suggests that there was no appreciable contribution from the Corso-Sardinian landmass. By contrast, minerals of Alpine affinities do occur in the La Poste and Grès Verts, but palaeocurrent data, indicating sediment supply from the N and NW, precludes the derivation of material from Corso-Sardinian sources.

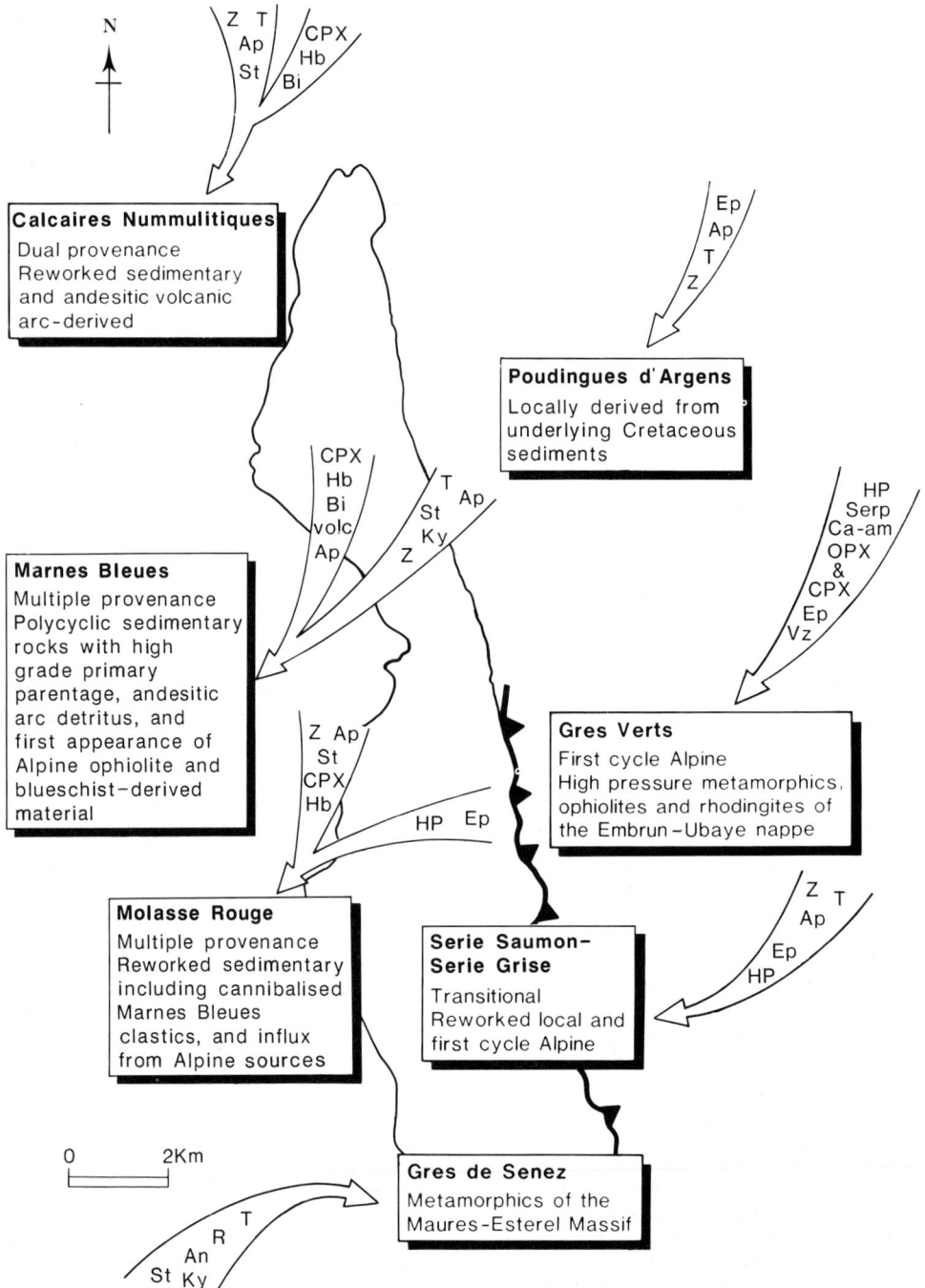

Fig. 15. Summary of heavy mineral assemblages and provenance. Refer to Fig. 5 for basin stratigraphy. Abbreviations: An, andalusite; Ap, apatite; Bi, (volc) volcanic biotite; Ca-am, calcic amphiboles; CPX, clinopyroxene; Ep, epidote; Hb, hornblende; HP, high pressure index minerals; Ky, kyanite; OPX, orthopyroxene; R, rutile; Serp, serpentine; St, staurolite; T, tourmaline; Vz, vesuvianite; Z, zircon.

(6) The contrasting heavy mineral assemblages of the La Poste, St Lions and Grès de Senez members suggests that Bodelle's (1971) correlation of these clastic sequences, based on biostratigraphy, may be invalid. Mineralogical data presented herein favour the chronological interpretation shown in Fig. 5.

This paper forms an extension of a PhD research project undertaken by M. J. E. at University College Swansea and Liverpool University. M. J. E. is indebted to T. Elliott for his guidance and encouragement, and for his constructive criticism of the manuscript. The paper has benefitted from valuable comments by P. A. Allen, G. M. Apps, J. Bellamy, N. Fry, H. G. Reading, J. A. Scott, C. P. Sladen and G. G. Zuffa and further improved by useful suggestions from the reviewers A. C. Morton and K. Stattegger. S. Petrie is gratefully acknowledged for drafting the figures, with financial assistance provided by BP. Analytical work was carried out at the University of Berne, Switzerland and M. A. M-R is grateful for the support of A. Matter. H. Haas provided kind assistance with the heavy mineral separation.

References

ALSAC, C., BOCQUET, J. & BODELLE, J. 1969. Les roches volcaniques tertiaires du synclinal de Saint-Antonin (Alpes Maritimes). *Bulletin Bureau Recherches Géologiques et Minières*, **3**, 45–56.

ALVAREZ, W. 1976. A former continuation of the Alps. *Geological Society of America Bulletin*, **87**, 891–896.

APPS, G. M. 1987. *Evolution of the Grès d'Annot Basin, SW Alps.* PhD thesis, University of Liverpool.

AUBOUIN, J. & MENNESSIER, G. 1963. Essai sur la structure de la Provence. *In: Livre à la mémoire du Professeur P. Fallot.* Société géologique de France, **11**, 45–98.

BAUDRIMONT, A. F. & DUBOIS, P. 1977. Un bassin Mésogéen du domaine péri-Alpin: Le sud-est de la France. *Bulletin Centres Recherches Exploration-Production Elf Aquitaine*, **1**, 261–308.

BEUF, S., BIJU-DUVAL, B. & GUBLER, Y. 1962. Les formations volcano-détritiques du tertiaire de Thônes (Savoie), du Champsaur (Hautes-Alpes) et Clumanc (Basses-Alpes). *Travaux du Laboratoire Géologie, Grenoble*, **37**, 142–156.

BODELLE, J. 1971. *Les Formations nummulitiques de l'arc de Castellane.* PhD thesis, University of Nice.

BORDET, P. 1951. *Étude géologique et pétrographique de l'Esterel.* Mémoire Explicative de la Carte Géologique de la France.

BOUCARUT, M. 1971. *Étude volcanologique et géologique de l'Esterel (Var France).* PhD thesis, University of Nice.

CAMPREDON, R. & BOUCARUT, M. 1975. *Alpes-Maritimes, Maures, Esterel.* Guides géologiques régionaux. Masson, Paris.

CARBONNEL, G., CHATEAUNEUF, J-J., FEIST-CASTEL, M., DE GRACIANSKY, P-J. & VIANEY-LIAUD, M. 1972. Les apports de le paléontologie (spores et pollens, charophytes, ostracodes, mammifères) et à la paléogeographie des molasses de l'oligocène supérieur de Barrême (Alps de Haute-Provence). *Compte Rendus de l'Academie des Sciences, Paris*, **275D**, 2599–2602.

CHAUVEAU, J-C. & LEMOINE, M. 1961. Contribution à l'étude géologique du synclinal du tertiaire de Barrême (moitié nord). *Bulletin Service Carte Géologique de la France*, **58**, 147–178.

COLOMB, E. 1972. Étude des minéraux lourds et de leur provenance. Contribution à l'étude de l'aquitanien La coupe de Carry-le-Ruet (Bouches-du-Rhône, France). *Bulletin Bureau Recherches Géologiques et Minières (deuxième série) Section 1*, 91–95.

DE JONG, K. A., MANZONI, M., STAVENGA, T. & VAN DIJK, F. 1973. Palaeomagnetic evidence for rotation of Sardinia during the Early Miocene. *Nature*, **243**, 281–283.

ESPITALIE, J. & SIGAL, J. 1961. Microstratigraphie des marnes Bleues des bassins tertiaires des Alpes méridionales; le genre Caucasina (foraminifères). *Revue Micropaleontologie*, **3**, 201–206.

EVANS, M. J. 1987. *Tertiary sedimentology and thrust tectonics in the southwest Alpine foreland basin, Alpes de Haute-Provence, France.* PhD thesis, University of Wales.

FRY, N. 1989. Southwestward thrusting and tectonics of the western Alps. *In:* COWARD, M. P., DIETRICH, D. & PARK, R. G. (eds) *Alpine tectonics.* Geological Society London, Special Publication, **45**, 83–109.

GALEHOUSE, J. S. 1971. Point counting. *In:* CARVER, R. E. (ed.) *Procedures in sedimentary petrology.* Wiley, New York, 385–407.

GIGOT, P., GRANDJACQUET, C. & HACCAREL, D. 1974. Evolution tectonosédimentaire de la septentrionale du bassin tertiaire de Digne depuis l'éocene. *Bulletin de la Société géologique de France*, **16**, 128–139.

GOGUEL, J. 1936. *Description tectonique de la bordure des Alpes de la Bléone au Var.* Mémoire Explicative de la Carte Géologique de la France.

GRACIANSKY, P-C. DE 1972. Le bassin tertiaire de Barrême (Alpes de Haute-Provence): relations entre déformation et sédimentation; chronologie des plissements. *Compte Rendus de l'Academie des Sciences, Paris*, **275D**, 2825–2828.

——, LEMOINE, M. & SALIOT, P. 1971. Remarques sur la présence de minéraux et de paragenèses du métamorphisme alpin dans les galets des conglomérats oligocènes de synclinal de Barrême (Alpes de Haute-Provence). *Compte Rendus de l'Academie des Sciences, Paris*, **272D**, 3243–3245.

——, DUROZOY, G. & GIGOT, P. 1982. *Notice explicative de la feuille Digne à 1 : 50,000.* Carte Géologique de France, Bureau de Recherches Géologique et Minières.

GUBLER, Y. 1958. Étude critique des sources du matériel constituant certaines séries détritiques dans le tertiaire des Alpes françaises du sud: formations détritiques de Barrême, flysch Grès d'Annot. *Eclogae geologicae Helvetiae*, **51**, 942–977.

GUEIRARD, S. 1962. *Le Massif des Maures. Guide géologique*. Hermann, Paris.

HARLAND, W. B., COX, A. V., LLEWELLYN, P. G., PIKTON, C. A. G., SMITH, A. G. & WALTERS, R. 1982. *A geologic time scale*. Cambridge University Press.

HAYWARD, A. B. & GRAHAM, R. H. 1989. Some geometrical characteristics of inversion. *In*: COOPER, M. A. & WILLIAMS, G. D. (eds) *Inversion Tectonics*. Geological Society, London, Special Publication, **44**, 17–39.

HOMEWOOD, P. 1983. Palaeogeography of Alpine Flysch. *Palaeogeography, Palaeoclimatology, Palaeoecology*, **44**, 169–184.

—— & CARON, C. 1982. Flysch of the western Alps. *In*: Hsü, K. J. (ed.) *Mountain building processes*. Academic Press, London.

KERCKHOVE, C. 1969. La "zone du flysch" dans les nappes de l'Embrunais-Ubaye (Alpes occidentales). *Géologie Alpine*, **45**, 5–204.

LAPPARENT, A-F, DE 1938. Études géologiques dans les regions provençales et alpines entre le Var et la Durance. *Bulletin Service Carte Géologique de la France*, **XL 198**, 1–301.

LAY, J. & PARFENOFF, A. 1972. Études des minéraux lourds. Contribution à l'étude de l'aquitanien La coupe de Carry-le-Ruet (Bouches-du-Rhône, France). *Bulletin Bureau Recherches Géologiques et Minières, (deuxième série) Section 1*, 85–90.

Lemoine, M., BAS, T., ARNAUD, H., DUMONT, T., GIDON, M., BOURBON, M., GRACIANSCKY, P-C, DE, RUDKIEWITZ, J-L., MEGARD-GALLI, J. & TRICART, P. 1986. The continental margin of the Mesozoic Tethys in the western Alps. *Marine and Petroleum Geology*, **3**, 179–199.

LE PICHON, X. 1984. The Mediterranean Seas. *In*: *Origin and History of Marginal and Inland Seas*. Proceedings of the 27th International Geological Congress, **23**. VNU Science Press, 189–222.

MANGE, M. A. & Maurer, H. F. W. 1990. *Heavy minerals in colour*. Unwin Hyman, London. (in press).

MARTINI, J. & VUAGNAT, M. 1968. Considérations sur le volcanisme post-ophiolitique dans les Alpes occidentales. *Geologische Rundschau*, **57**, 264–276.

PAIRIS, J. L. 1971. Tectonique et sédimentation tertiaire sur la marge orientale du bassin de Barrême (Alpes de Haute-Provence) *Géologie Alpine*, **47**, 203–214.

PUPIN, J. P. 1976. *Signification des caractères morphologiques du zircon commun des roches en pétrologie. Base de la méthode typologique—Applications*. PhD thesis, University of Nice.

REHAULT, J-P, BOILLOT, G. & MAUFFRET, A. 1985. The Western Mediterranean Basin. *In*: STANLEY, D. J. & WESEL, C. F. (eds) *Geological evolution of the Mediterranean Basin*. Springer-Verlag, New York, 101–129.

SAWATZKI, G. G. 1975. Étude géologique et minéralogique des flyschs à grauwackes volcaniques du synclinal de Thônes (Haute- Savoie, France). Grès de Taveyanne et grès du val d'Illiez. *Archives des Sciences, Genève*, **28**, *fasc. 3*, 265–368.

SIDDANS, A. W. B. 1979. Arcuate fold and thrust patterns in the subalpine chains of southeast France. *Journal of Structural Geology*, **7**, 117–126.

SIEDON, T. 1981. *Données nouvelles sur les formations détritiques de l'arc de Castellane, sud-est de la France (conglomerats tertiaires des synclinaux de St. Antonin, Barrême, Majastre); pétrographie des galets des roches endogènes; application de la typologie du zircon accessoire; approche paléogéographique*. Doctoral thesis, 3e cycle. University of Nice.

SPEER, J. A. 1980. Zircon. *In*: RIBBE, P. H. (ed.) *Reviews in mineralogy. Vol. 5. Orthosilicates*. Mineralogical Society of America, Washington DC, 67–112.

STANLEY, D. J. 1961. Études sedimentologiques des Grès d'Annot et de leurs équivalents latéraux. *Revue Institut Français de Pétrole*, **16**, 1231–1254.

—— 1965. Heavy minerals and provenance of sands in flysch of central and southern French Alps. *Bulletin of American Association of Petroleum Geologists*, **49**, 22–40.

—— & MUTTI, E. 1968. Sedimentological evidence for an emerged land mass in the Ligurian Sea during the Palaeogene. *Nature*, **218**, 32–36.

STEEN, D. M. 1972. *Étude géologique et pétrographique du complexe ophiolitique de la Haute-Ubaye (Basses-Alpes, France)*. Mémoire Départment Minéraux Université Genève.

—— 1975. Géologie et métamorphism du complexe ophiolitique de la Haute-Ubaye. *Schweizerische Mineralogische und Petrographische Mitteilungen*, **55**, 523–566.

VANN, I. R., GRAHAM, R. H. & HAYWARD, A. B. 1986. The structure of mountain fronts. *Journal of Structural Geology*, **8**, 215–227.

Evolution of the Devonian Hornelen Basin, west Norway: new constraints from petrological studies of metamorphic clasts

SIMON J. CUTHBERT

Department of Geology and Applied Geology, The University, Glasgow G12 8QQ UK
Present address: Department of Civil Engineering, Paisley College of Technology, High Street, Paisley PA1 2BE, UK

Abstract: The Middle Devonian Hornelen Basin of western Norway is a continental fault-basin which formed during the late stages of the Caledonian orogeny, closely following the Late Silurian crustal thickening and high pressure metamophism responsible for the formation of eclogites in the adjacent Western Gneiss Region. Clasts from alluvial fan conglomerates which fringe the basin provide important constraints on the type of basin tectonism and the uplift history of this part of the orogen.

The clasts lack a petrographic linkage with the Western Gneiss Region (from which they are separated by a fault), but show a strong petrographic linkage with units of the allochthon (which lie unconformably below the basin fill). The clast sourceland had a similar metamorphic evolution to the allochthon, and both lack evidence for the high pressures recorded in the Western Gneiss Region. Both clasts and allochthon cooled through muscovite argon-retention temperatures about 40 Ma earlier than the Western Gneiss Region. Depth-time curves derived from isotopic and petrological results show that prior to basin initiation the Western Gneiss Region was uplifting faster than the clast source, but still lay at about 10 km depth during basin formation.

A tectonic scenario is suggested in which much of the uplift of the Western Gneiss Region took place before the deposition of the Hornelen Basin sediments, aided by crustal thinning along a major extensional mylonitic shear-zone. However, final exposure of this high-pressure metamorphic terrain only took place after the Hornelen Basin sediments were deposited, aided by uplift in the footwall of the basin sole fault.

Petrographic studies of conglomerate clasts can provide useful evidence for the nature of crustal blocks which are no longer accessible to observation due to displacement, removal or burial. They may act as indicators of tectonic activity (Ballance 1980), palaeogeology (e.g. Haughton 1988) or igneous activity (e.g. Haughton *et al.* 1990) in their sourcelands. This contribution explores the usefulness of provenance studies in yet another context; the evolution of metamorphic belts. Clasts shed from uplifting orogenic belts might be expected to record the exhumation history of metamorphic terrains whose overburden has otherwise been removed by erosion and tectonic processes. The importance of exhumation history in the evolution of metamorphic belts is now widely recognized (England & Richardson 1977); this type of provenance study may shed more light on the nature and thermal evolution of the overburden, and the exhumation mechanism.

Conglomerate clasts from the Hornelen Basin of western Norway have been examined in an attempt to reconstruct the metamorphic evolution of their sourcelands for comparison with presently exposed Caledonian metamorphic terrains in the area. The Hornelen Basin is the largest of five outcrops of Middle Devonian sandstones and conglomerates lying along the coast of Norway between Bergen and Trondheim (Fig. 1). They lie in the most internal exposed part of the Scandinavian Caledonides (Gee *et al.* 1985) in fault contact with the Western Gneiss Region, the lowest exposed structural level of the orogen. The Western Gneiss Region has suffered a late Silurian high-pressure (eclogite-facies) metamorphism (Griffin & Brueckner 1980; Griffin *et al.* 1985) indicating tectonic burial to depths in excess of 80 km below a thick assemblage of thrust sheets (Cuthbert *et al.* 1983; Cuthbert & Carswell 1990). The depositional age of the sediments (about 380 Ma; Harland *et al.* 1982) is close to isotopic cooling ages from the Western Gneiss Region (about 425–370 Ma; e.g. Brueckner 1972; Lux 1985; compilation of Kullerud *et al.* 1986). The specific aim of this study was to investigate the uplift history of the Western Gneiss Region using the clastic sample in the Devonian basins. The Hornelen Basin was chosen for particular attention because it is the best described in terms of its stratigraphy and sedimentology, and contains the largest strati-

From Morton, A. C., Todd, S. P. & Haughton, P. D. W. (eds), 1991,
Developments in Sedimentary Provenance Studies.
Geological Society Special Publication No. 57, pp. 343–360.

graphic section. The results described here show that these clasts were not sourced from the Western Gneiss Region, but come from higher crustal levels which have subsequently been removed during erosional and tectonic exhumation.

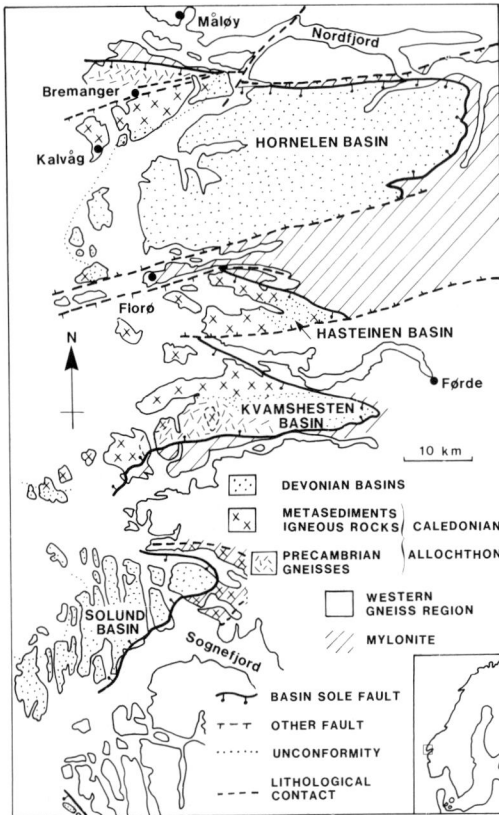

Fig. 1. Geological map of western Norway between Nordfjord and Sognefjord, showing the locations of the Devonian Old Red Sandstone basins, the Western Gneiss Region and the overlying allochthon, and the distribution of mylonites (after Kildal 1970, with modifications by Norton 1986 and this author). Fig. 2 occupies the part of the map to the north of Førde.

Main features of the Hornelen Basin

The Hornelen Basin lies to the south of Nordfjord in the county of Sogn og Fjordane (Figs 1 & 2). It has a trapezoidal outcrop area about 50 km E–W and 25 km N–S. The sediments are dominated by grey–green and red sandstones and minor siltstones with a narrow fringe of conglomerates and breccias up to 2 km wide (Steel

1976; Steel *et al.* 1977; Steel & Gloppen 1980; Bryhni 1978). These sediments comprise the Hornelen Group (Bryhni 1978). Rare plant and arthropod fossils in some of the youngest sandstones and siltstones are of mid-Middle Devonian age (Hoeg 1936).

The sedimentology of the Hornelen Group has been comprehensively described by Steel and his co-workers (Steel *et al.* 1985 and references therein), Bryhni (1978) and Larsen *et al.* (1987). The basin fill has a stratigraphic thickness of 25 km, organized into about 100 coarsening-upward megacycles. The marginal conglomerates and breccias are debris-flow and streamflow deposits formed on alluvial fans, with palaeoflows directed away from the present basin margins (Larsen *et al.* 1987). They interfinger with axial fluvial sandstones and floodplain siltstones with dominant palaeoflows toward the west. The megacycles for the axial and marginal systems are exactly in phase even though they apparently originated from independent drainage systems. Also, the depocentre of each alluvial fan cycle shows a consistent eastward shift relative to the one below it, creating a total eastward onlap across the basin of about 50 km. This cyclicity and onlap has been attributed to repeated lowering and lateral shift of the basin floor relative to the clast sourceland (Steel 1976). The sense of onlap may be either due to a stationary basin floor with an eastward shift of depocentre (Steel 1976), or a westward motion of the basin floor with a stationary depocentre (Hossack 1984). The present vertical thickness of the Hornelen Group is probably not more than 8 km. The sediments are essentially unmetamorphosed except for local cleavage development in siltstones.

The Hornelen Group is fault-bounded on its N, S and E sides (Fig. 2). In the extreme west it lies unconformably on units of the Caledonide allochthon. The bounding fault is scoop-shaped, dipping to the west at about 30° in the E and curving around to strike E–W on the north and south margins, where it dips basinwards at about 40° (Norton *et al.* 1990). This fault has been re-activated by later E–W faults along the N and S margins, but the preservation of the fringe of alluvial-fan conglomerates suggests that these later (Mesozoic?) faults lie close to the original base margin (Norton *et al.* 1990). Below the bounding fault is a zone of mylonite 1.5–2.5 km thick, capped by a thin zone of cataclasite and a chloritised tectonic breccia (Norton *et al.* 1990). Below the mylonites are gneisses of the Western Gneiss Region with common eclogites (Bryhni 1966; Griffin & Mørk 1981). The strike of the mylonitic foliation follows the form of the basin margin, and kinematic indicators show a consistent top-down-to-the-west shear sense, which is opposite in polarity to that of the Caledonide ('Scandian') thrust motions (Seranne 1988; Norton et al. 1990). Brittle structures at the fault contact between the Hornelen Group and the mylonites give a similar sense of slip (Seranne 1988).

The structural and sedimentological features of the Hornelen Basin outlined above have recently been incorporated into tectonic models for basin formation involving low-angle faulting on a scoop-shaped low-angle detachment, with deposition in the resultant half-graben basin on the hangingwall (Hossack 1984; Norton 1987; Seranne & Seguret 1987). According to this model the Western Gneiss Region has been uplifted in the footwall of the detachment fault in a manner similar to metamorphic core complexes (Davis & Coney 1979; McClay et al. 1986; Norton 1986). Uplift was therefore controlled not by erosion alone, but by excision of a very large crustal section by the detachment fault and its associated mylonitic shear-zone. The magnitude of the loss of crustal section is indicated by the juxtaposition across the detachment of high-pressure metamorphic rocks below unmetamorphosed sediments. The basin should have sampled detritus shed from the rising footwall as it emerged at the hangingwall cut-off.

Geology of the basement

Two main groups of rocks can be differentiated in the basement surrounding the Hornelen Basin (Fig. 2); those of the Western Gneiss Region, which are in fault contact with the Hornelen Group and occur mostly on the mainland, and those which lie unconformably below the Hornelen Group on the islands of the west coast.

The Western Gneiss Region in the Nordfjord area has previously been described by Bryhni (1966); Kildal (1970); Bryhni & Grimstad (1971), and Bryhni et al. (1981). To the south and east of the Hornelen Basin three sub-groups can be recognized.

(1) Feldspathic quartzites and muscovite schists ($\pm$ garnet) with minor dolomitic marbles and some large bodies of garnetiferous amphibolite (Lykkjebø Group of Bryhni et al. 1981).

(2) Dark, biotite–epidote gneisses, commonly containing bodies of meta-anorthosite, meta-gabbro, serpentinite and pink augen gneiss and common relics of granulite-facies mineral assemblages. A few bodies of eclogite are also found within these rocks (Eikefjord Group of Bryhni et al. 1981).

(3) Granitic and granodioritic gneisses with common pegmatites and numerous eclogite bodies (Jostedal Complex of Bryhni 1966).

To the north of the Hornelen Basin similar lithologies to those in the south are present, but the major groups are not as easily differentiated from each other due to intense deformation. In addition, the northwestern part of this region is characterized by banded granodioritic gneisses and augen gneisses, with minor pelitic garnet–phengite–kyanite gneisses. Eclogites are very common here; indeed this is one of the classic areas for these striking rocks (Bryhni et al. 1969). Ultramafic rocks are common in this area also, including dark serpentinites and distinctive green dunites. With the exception of the Jostedal Complex, most of the Western Gneiss Region rocks immediately adjacent to the Hornelen Basin bounding faults are mylonitic.

Basement rocks unconformably overlain by the Hornelen Group are most extensively exposed on the island of Bremangerlandet (Fig. 2). The following description is based on recent mapping by S. Cuthbert, I. Bray, R. Johnson and M. G. Norton. A large body of quartz diorite intrudes a pelitic melange, now largely hornfelsed, containing blocks of banded chert and greywacke several hundred metres across (this unit correlates with the Kalvåg melange, described by Bryhni & Lyse 1985). Near Kalvåg the melange is intruded by a large body of leucogabbro, which also underlies the archipelago of small islands to the west (Kildal 1970). To the north the hornfels passes gradationally into pelitic greenschists (chlorite–mica–garnet schists) whose schistosity overpirnts the hornfels fabric. These schists appear to correlate with those on the islands of Batalden and Skorpa (Fig. 2), where a pale chloritic gneiss with a well-developed spaced cleavage is also found locally (Spinnanger 1975). Below the schists is a complex zone of interleaved feldspathic quartzites and gneisses, including a distinctive very coarse pink augen gneiss, underlain by extensive banded granodioritic two-mica gneisses. Bodies of garnet-free amphibolite are commonly found in this unit, associated with white muscovite-bearing pegmatites. The gneisses are entirely non-mylonitic, but they are separated by a narrow zone of chloritic breccia and fault-gouge from the mylonites which occupy the northernmost part of the island.

The gneisses of Bremangerlandet have previously been correlated with those of the Western Gneiss Region (Kildal 1970), but the southern zone of non-mylonitic gneisses lacks eclogite. All the metabasic bodies are amphibolites with no sign of a previous eclogite mineralogy. The

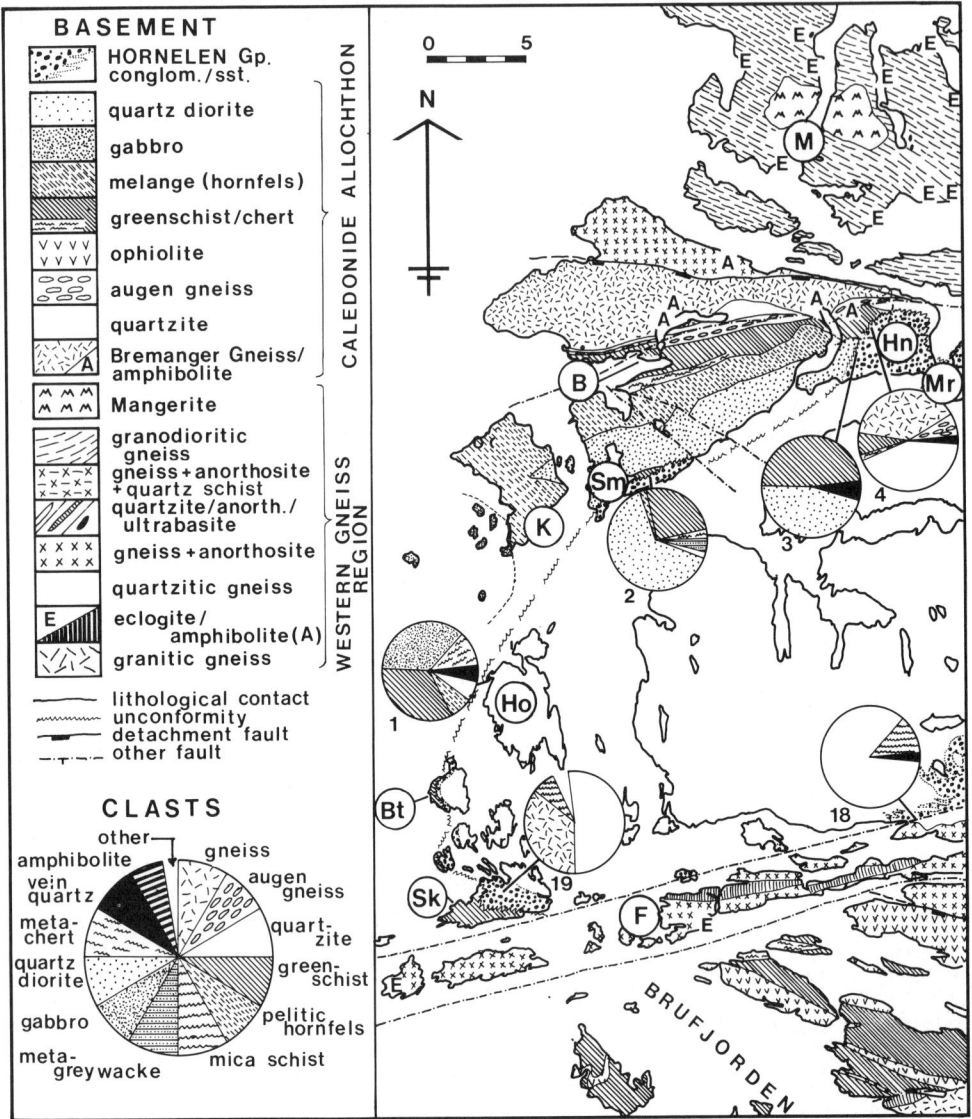

(Fig. 1)

gneissic banding is more coarsely differentiated than that found to the north of Nordfjord. The overlying quartzites and schists, which by analogy with similar rocks on Atløy to the south of this area (Brekke & Solberg 1987) have only suffered low-grade Caledonian metamorphism, locally display pre-metamorphic unconformable relationships to the gneiss. On this basis the gneisses are probably at the same tectonic level as the Dalsfjord 'nappe' and the Jotun nappe of the Caledonian allochthon (Fig. 1). The thick mylonites in the north of the island contain lithologies more typical of the Western Gneiss Region on the mainland, including meta-anorthosite and metagabbro. The zone of brecciation and fault-gouge separating the two major gneiss units on Bremangerlandet seems to correspond to the fault bounding the Hornelen Group on the mainland to the east. It follows from this interpretation that both the Hornelen Group and its depositional basement are everywhere separated from rocks of the Western Gneiss Region by faults.

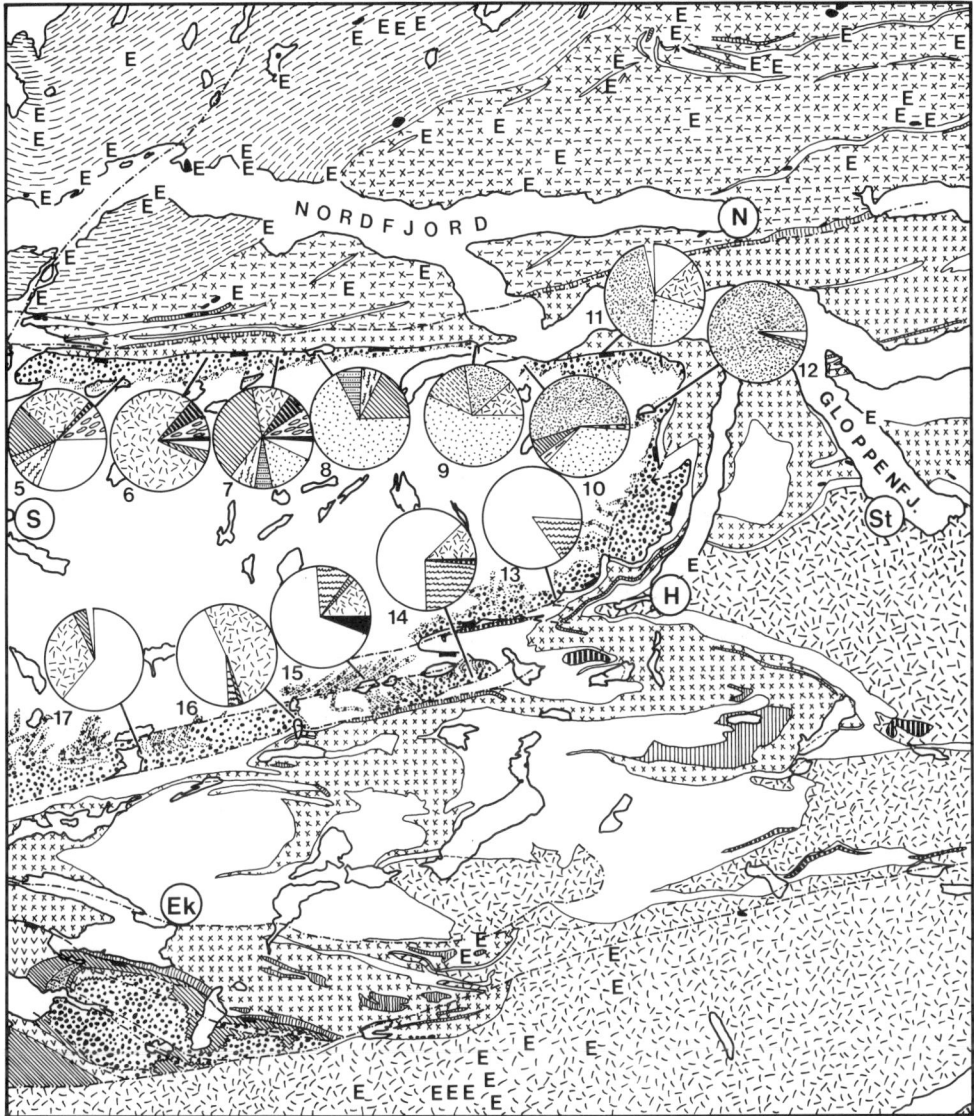

Fig. 2. Lithological map of the Hornelen Basin area, showing the distribution of clast types in marginal alluvial fan conglomerates of the Hornelen Group (pie charts, marked with site numbers mentioned in the text). Data for sites 11, 16, 17 & 19 are from Evensen (1978), the rest are the author's own. Place names (encircled): B, Bremangerlandet; Bt, Batalden; Ek, Eikefjord; F, Florø; H, Hyen; Hn, Hornelen; Ho, Hovden; K, Kalvåg; M, Maløy; Mr, Marøy; N, Nordfjordeid; Sm, Smørhamn; St, Stranda. Compiled from Kildal (1970); Spinnangr (1975); Maehle (1975); Evensen (1978); Larsen & Steel (1978); Griffin & Mørk (1981); Bryhni & Lyse (1985); Seranne (1988); Norton *et al.* (1990); unpublished preliminary 1:50 000 scale map sheets 'Nordfjordeid', 'Fimlandsgrend' and 'Naustdal' of the Norwegian geological survey (Norges Geologiske Undersøkelse), and unpublished mapping of the author, I. Bray, R. Johnson, M. Norton & W. Wilks.

Analytical and petrological methods

Mineral chemistry was determined on the Microscan 5 electron microprobe at Glasgow University using a Link Systems energy dispersive analyser, and the Microscan 5 electron microprobe at the Grant Institute of Geology, Edinburgh, with wavelength dispersive spectrometers. Both natural minerals and pure metals

Table 1. *Details of samples analysed for this study*

Sample	Lithology	Locality	Grid Ref. (series M711)	Clast site	Geological unit
Hcg20	muscovite–quartz schist	Haukå	1118 II 001395	18	Hornelen Group
Hcg85	mica–chlor–garnet schist	Vingevatnet	1118 I 108611	5	Hornelen Group
Hcg86	mica–chlor–garnet schist	Vingevatnet	1118 I 108611	5	Hornelen Group
Hbm53	pheng–garnet–kyan gneiss	Krokenakken	1118 I 085706	—	W. Gneiss Region
Hbm58	pheng–bio–garnet gneiss	Verpeneset	1118 I 006694	—	W. Gneiss Region
Hbm112	sill–garnet–hornfels	Rylandsvatan	1118 IV 865607	—	Kalvåg melange
Hbm113	mica–chlor–garnet schist	Marøy	1118 IV 040625	—	Bremanger schists

Note: clast site numbers refer to sampling stations and pie diagrams shown on Fig. 2.

were used as standards in each case. Analyses of minerals are displayed in Table 2.

Metamorphic temperatures (Table 3) were determined using Fe–Mg exchange geothermometry for garnet and mica pairs. The garnet–biotite thermometer of Indares & Martignole (1985) was used for all samples except clast Hcg85 (Tables 1 & 3), which lacked biotite, so the Green & Hellman (1982) garnet–phengite thermometer was used instead. Analyses were not corrected for ferric iron content, which is standard practice for garnet–biotite thermometry, but may lead to large errors in garnet–phengite thermometry due to the small total iron content of phengites. Results from garnet–phengite were up to 100°C different to those for garnet–biotite in the same sample, possibly due to the lack of correction for ferric iron, differences in the calibration of the geothermometers or even variation in post-crystallization re-equilibration of Fe and Mg. An error of 100°C will produce an extra uncertainty in pressure for sample Hcg85 of 1 kbar. The Indares & Martignole thermometer incorporates a correction for Ca and Mn in garnet. Sample Hbm58 (Table 1) contains garnets with high Mn concentrations (Table 2) which may not be adequately corrected for, leading to underestimation of temperature. However, these garnets are compositionally zoned, and at temperatures significantly higher than those calculated for this sample (Table 3) one might expect such zoning to have been erased, so any error is assumed to be small. For Hbm58 an error of 100°C would result in a pressure error of 1.3 kbar.

Metamorphic pressures (Table 3) were determined using the thermodynamic dataset and computer program THERMOCALC of Powell & Holland (1988), which calculates average pressures and uncertainties for a set of independent reactions which can be written between the end members in the minerals present in a rock. Activities for phase components were calculated assuming ideal mixing on sites for micas and chlorite, the Ganguly & Saxena (1984) formulation for garnet and the Hodges & Royden (1984) formulation for plagioclase.

K–Ar isotopic analyses (Table 4) were made on muscovite separates obtained by standard heavy liquid and magnetic separation techniques at Glasgow University, with final purification to better than 99% by hand-picking. Potassium concentrations were determined in duplicate on a Corning-EEL 450 flame photometer at the Scottish Universities Research and Reactor Centre, East Kilbride. ^{40}Ar concentrations were determined on AEI MS10 spectrometers at SURRC, East Kilbride and at the University of Leeds, using radio frequency heaters, standard purification methods and isotope dilution. Results for samples run at both laboratories were identical within analytical error. Decay constants used in age determinations are those recommended by Steiger & Jäger (1977) and analytical uncertainties are quoted at two sigma.

Metamorphic evolution of the basement

The Western Gneiss Region is well known for its abundant bodies of eclogite. Detailed discussions of these rocks can be found in Cuthbert *et al.* (1983); Griffin *et al.* (1985), and Cuthbert & Carswell (1990). Most of the eclogites appear to have been formed by high-pressure metamorphism of a variety of basic protoliths during the

Table 2. *Microprobe analyses of minerals used in geothermobarometry (see Table 1 for sample details)*

Sample	Garnet–mica schist clast Hcg85				Garnet–mica schist clast Hcg86			
Phase	grt	musc	chlor	plag	grt	musc	bio	plag
SiO$_2$	37.85	46.97	25.80	64.29	37.72	45.52	32.68	63.17
TiO$_2$	0.02	0.48	0.10	0.00	0.03	0.89	1.39	0.00
Al$_2$O$_3$	21.01	31.86	21.54	21.81	20.85	34.11	16.04	22.92
FeO	30.53	1.66	21.99	0.07	31.97	1.36	24.45	0.46
MnO	0.72	0.01	0.10	0.01	3.38	0.00	0.62	0.00
MgO	2.52	1.80	17.20	0.01	3.09	0.98	10.02	0.14
CaO	7.20	0.00	na	3.35	3.02	0.00	0.00	3.39
Na$_2$O	na	1.21	na	9.54	na	1.20	0.04	8.48
K$_2$O	na	9.23	na	0.10	na	9.51	7.75	1.08
Total	99.88	93.23	86.73	99.17	100.08	93.56	92.99	99.65

Sample	Phengite–garnet gneiss Hbm58				Sillimanite hornfels Hbm112		
	grt	musc	bio	plag	grt	bio	plag
SiO$_2$	37.52	46.49	35.83	62.84	37.90	36.63	61.61
TiO$_2$	0.03	0.70	2.15	0.00	0.00	1.48	0.00
Al$_2$O$_3$	20.44	28.82	15.92	22.91	21.62	18.33	24.54
FeO	18.04	5.08	20.22	0.07	34.10	15.35	0.30
MnO	12.52	0.16	0.84	0.00	1.67	0.03	0.00
MgO	1.18	1.80	10.02	0.01	4.19	13.54	0.00
CaO	10.22	0.05	0.05	4.30	1.44	0.00	5.48
Na$_2$O	na	0.28	0.28	9.05	na	0.09	7.93
K$_2$O	na	10.80	9.75	0.11	na	8.61	0.17
Total	99.96	94.23	94.90	99.32	100.96	94.07	100.10

Sample	Garnet–mica schist Hbm113				
	grt	musc	bio	chlor	plag
SiO$_2$	37.54	46.03	36.02	25.64	63.07
TiO$_2$	0.07	0.48	1.65	0.08	0.00
Al$_2$O$_3$	21.03	32.70	17.48	21.60	22.80
FeO	21.03	1.84	18.89	21.98	0.19
MnO	2.48	0.01	0.15	0.08	0.00
MgO	2.38	1.35	10.50	16.22	0.04
CaO	5.62	0.00	0.00	na	4.33
Na$_2$O	na	1.04	0.08	na	8.98
K$_2$O	na	9.31	9.32	na	0.11
Total	100.39	92.78	94.09	85.61	99.53

na, not analysed.

Scandian phase (late Silurian) of the Caledonian oregeny. Examples from the Nordfjord area record pressures of about 19 kbars at 650–750°C (Carswell *et al.* 1985), although the recent discovery of coesite at one of the localities (Smith 1984) may indicate that pressures as high as 28 kbar were attained. A Sm-Nd mineral isochron from one of these eclogites gives an age of 423 ± 12 Ma (Griffin & Brueckner 1985; see Table 5), corresponding to a closure temperature of about 750°C (Mørk & Mearns 1986).

In contrast to the eclogites, high-pressure mineral assemblages in the gneisses are rare due to pervasive, preferential retrogression to the

Table 3. *Results of geothermobarometry*

Sample	Clast/ Basement	Temp (°C)	Method	Pressure (kbar)	Mineral assemblage
A Hcg85	Clast	570 ± 50	GH	6.9 ± 2.5	phe + chl + pl + qz + grt
B Hcg86	Clast	730 ± 50	IM	9.4 ± 2.0	phe + bt + pl + qz + grt
C Hbm112	Bsmt (Brem)	510 ± 40	IM	4.1 ± 1.8	bt + pl + qz + sil
D Hbm113	Bsmt (Brem)	560 ± 50	IM	9.2 ± 1.4	phe + chl + bt + pl + qz + grt
E Hbm58	Bsmt (WGR)	745 ± 50	IM	11.4 ± 2.0	phe + bt + pl + qz + grt

Letters in left-hand column refer to boxes in Fig. 3c. IM, garnet/biotite geothermometer of Indares & Martignole (1985); GH, garnet/phengite geothermometer of Green & Hellman (1982). Pressures calculated using computer program THERMOCALC (Powell & Holland 1988). See Table 1 for sample details. Brem, Bremangerlandet; WGR, Western Gneiss Region.

Table 4. *Results of K–Ar dating of muscovites*

Sample	Clast/ Basement	K(Wt%)	$^{40}Ar_{Rad}$ (STPcm^3 × 10^{-5} gm^{-1})	% Rad	Age (Ma)	Lab
6 Hcg20	clast	7.52	13.971	99.5	423 ± 8	East Kilbride
7 Hbm113	bsmt (Brem)	8.65	16.052	99.0	423 ± 8	Leeds
3 Hbm53	bsmt (WGR)	7.95	13.247	97.1	385 ± 8	East Kilbride

Numbers in left-hand column refer to error bars in Fig. 3a. Uncertainties quoted as 2σ STPcm^3 × 10^{-5} gm^{-1} $^{40}Ar_{Rad}$. See Table 1 for sample details.
Brem, Bremangerlandet; WGR, Western Gneiss Region.

Table 5. *Summary of published radiometric ages for the Western Gneiss Region in the Måløy area, outer Nordfjord*

Method	Age (Ma)	Locality	T_c(°C)	Authors
1 Sm–Nd grt–cpx	423 ± 12	Raudeberg, Måløy	~750	Griffin & Brueckner (1985)
2 $^{40/39}$Ar Hornbl	410 ± 1	Sorpollen, Måløy	500 ± 25	Lux (1985)
4 $^{40/39}$Ar Biotite	375 ± 4	Stadlandet, Måløy	300 ± 50	Lux (1985)
5 Rb–Sr WR–Bio	374 ± 2	Eldevik, Nordfjord	300 ± 50	Mearns, in Kullerud et al. (1986)

Numbers in left-hand column refer to brackets in Fig. 3a. Closure temperatures (T_c) are from Harrison (1981) for K–Ar hornblende; Jäger (1979) for K–Ar & Rb–Sr biotite, and Mørk & Mearns (1986) for Sm–Nd garnet–omphacite.

amphibolite facies. Close to the Hornelen basin this retrogression is related to mylonitisation. A gneiss sample (Hbm58) from the north shore of Nordfjord gives a pressure of 10–12 kbar at 700–800°C (Table 3), indicating decompression from peak eclogite facies conditions through at least 10 kbar with little temperature change. Below 10 kbar the temperature began to decline (Cotkin et al. 1988), reaching 500°C at 410 Ma (Table 5) and 300°C at about 375 Ma (Tables 4 & 5). A reconstruction of the P–t curve from these data (Fig. 3) shows uplift rates of 20 mm a^{-1}, slowing gradually to 2 mm a^{-1} above 30 km depth. An important corollary of this is that the Western Gneiss Region was still buried at about 10 km during the Middle Devonian, when the Hornelen Group was being deposited.

The age of the metamorphism in the Bremangerlandet gneisses is not known. The overlying metasediments were probably deposited in the late Ordovician or early Silurian, by analogy with similar rocks on Atløy to the south,

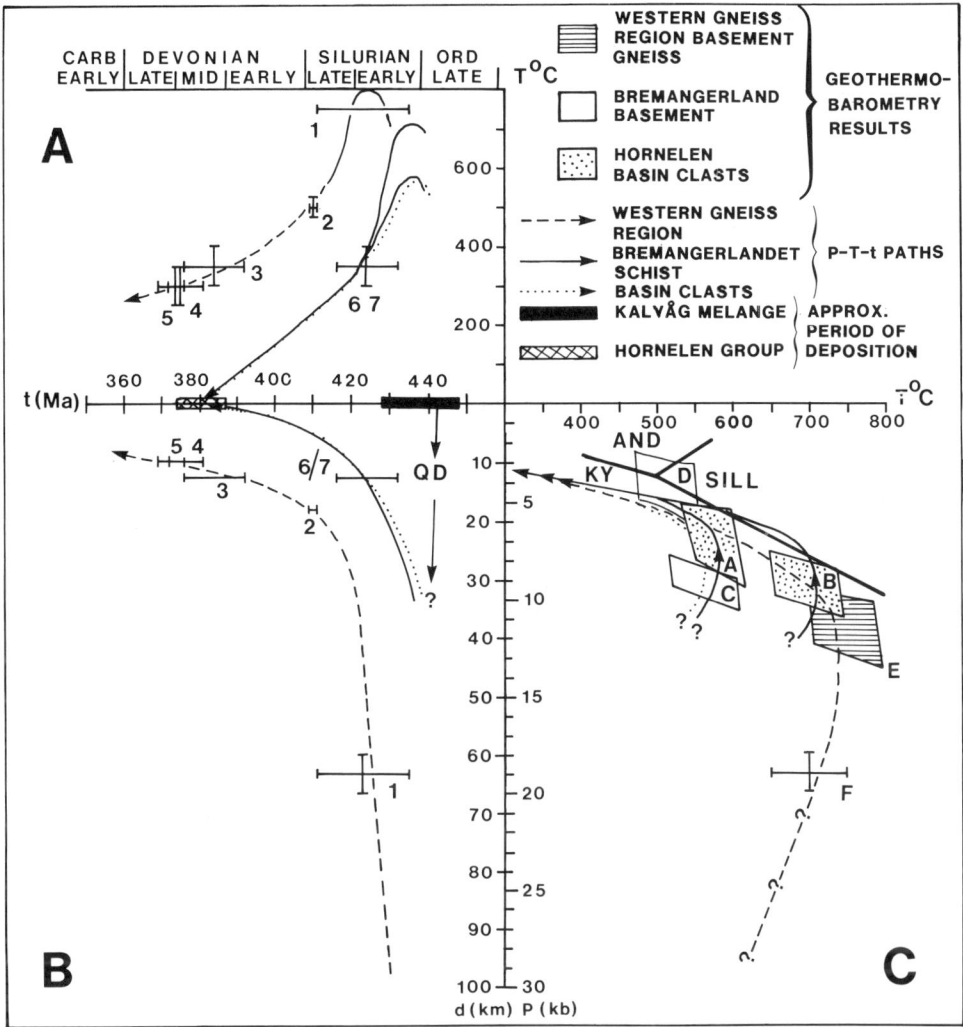

Fig. 3. Pressure–temperature–time (*P–T–t*) curves for clasts from the Hornelen Group, the Western Gneiss Region, and basement rocks from below the Hornelen Group unconformity on Bremangerlandet. Letters and numbers in error bars and boxes refer to samples in Table 1, except F, orthopyroxene eclogites from the outer Nordfjord area (Carswell *et al.* 1985). (**A**), *T–t* curves based on radiometric cooling ages (Tables 4 & 5) and published closure temperatures (*T_c*). (**B**) *P–t* (= depth–time) curves constructed from *P–T* and *T–t* curves; error bars correspond to cooling ages. QD, quartz diorite intrusion. (**C**), *P–T* curves based on geothermobarometry. The Western Gneiss Region *P–T* curve is based on Cotkin *et al.* (1988), extrapolated up-pressure through the eclogite field to the stability field of coesite, tentatively following Smith (1984). Thick lines are boundaries to Al–silicate polymorph stability fields.

(Brekke & Solberg 1987). A pressure estimate from a hornfels sample (Hbm112, Tables 1 & 3) indicates burial to 10–16 km during intrusion of the quartz diorite. The hornfelses were subsequently remetamorphosed to form the greenschists during a regional metamorphism (Cuthbert, unpublished results 1988); geothermometry on a garnet–two mica–chlorite schist from Marøy (sample Hbm 113, Tables 1 & 3) gives

560 ± 50°C at 9 ± 1 kbar. This pressure may not have been the maximum reached by the rock, as demonstrated by thermal models of the type developed by England & Richardson (1977), but a pressure of more than about 13 kbar is unlikely in view of the lack of relics of eclogite facies mineralogy in this block. A K–Ar muscovite age of 423 ± 8 Ma from the same Marøy mica schist (Table 4) gives the time of

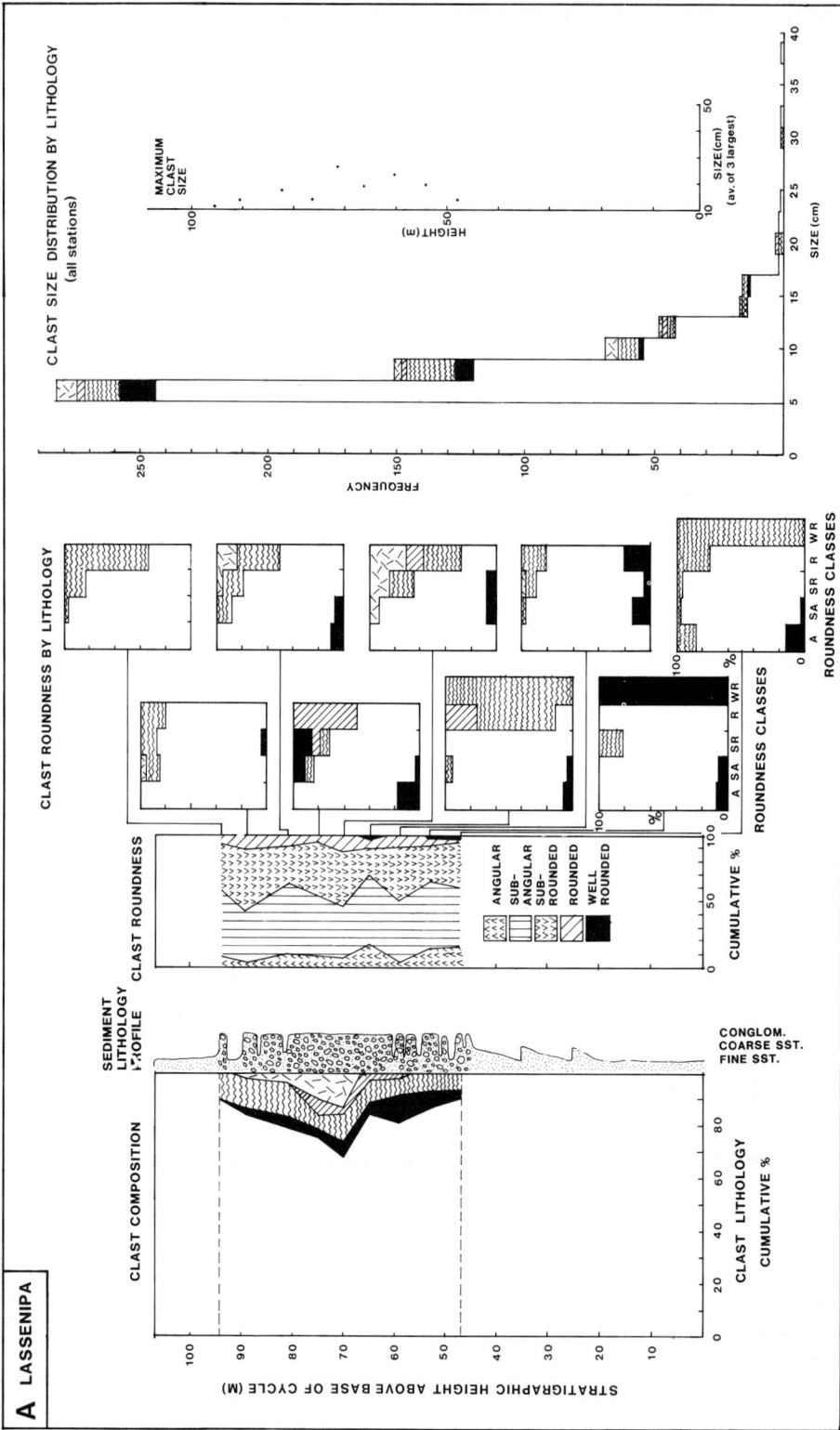

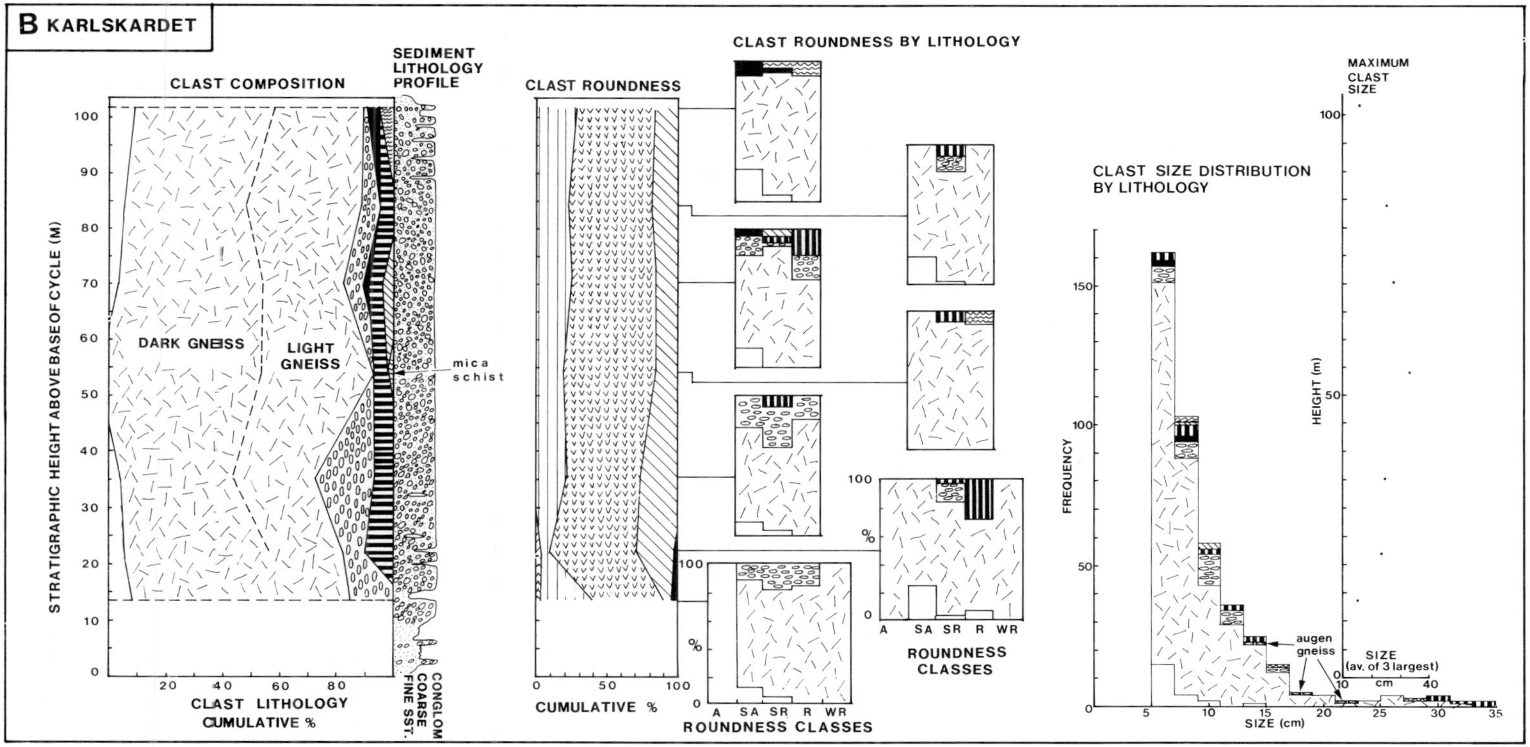

Fig. 4. Stratigraphic variation in clast lithology, roundness and size in two alluvial fan cycles of the Hornelen Group. (A) Lassenipa cycle, south margin (site 16, Fig. 2). (B) Karlskardet cycle, north margin (site 6, Fig. 2). Ornament for clast compositions are the same as those used in the pie charts in Fig. 2.

cooling through about 350°C (Jäger 1979), and surface exposure must have occurred by the Middle Devonian. The resultant P–t curve in Fig. 3 shows that this block lay at a depth of less than 15 km when the Western Gneiss Region was still buried at more than 60 km.

Clast lithologies

Approach to clast sampling

Studies of clast populations were restricted to the marginal fringe of alluvial fan breccias and conglomerates as they are most likely to represent proximal, first-cycle samples of the clastic sourcelands. Clast assemblages were logged at stations in fifteen different upward coarsening megacycles, spaced fairly evenly around the basin (Fig. 2). Stations were sited as close to the basin boundary as was practical in order to sample the most proximal sediments, hence minimising the effects of sample bias due to depositional processes within the basin. At each station data for about one hundred adjacent clasts within a single bed were logged for lithology, size (maximum exposed dimension) and roundness (visual estimation on scale of Pettijohn 1975, p. 57). In order to maximize confidence in lithology recognition only clasts larger than an arbitrary limit of 5 cm were logged.

Two cycles (site 15, Lassenipa, south margin and site 6, Karlskardet, north margin; Fig. 2) were selected for more detailed investigation in order to assess any bias in the clast populations resulting from drainage-basin processes or recycling of sediment. The sedimentology of both these units has previously been described in detail (Gloppen & Steel 1981; Larsen & Steel 1978; Larsen et al. 1987). Vertical sections were logged through the units and profiles constructed for clast lithology, size and roundness. The results are summarized in Figs 2, 4a & b.

In both profiles the diversity of the clast assemblage increases from the base of the conglomeratic part of the section into the most massive conglomerates, which are assumed to have been deposited at the time of maximum progradation of these alluvial fans. This may be due to progressive expansion of the drainage basin over different lithologies. Size sorting in relation to changing drainage basin capacity may also have affected diversity, but this is less likely as there seems to be no significant relationship between clast size and lithology.

The majority of the clasts are subangular and subrounded. At Lassenipa (Fig. 4a), where the assemblage is dominated by hard quartzites, about 40% of the clasts are subangular, and angular clasts are consistently present. Well-rounded clasts are rare, and tend to be restricted to the lowest parts of the section. Rounding shows a normal relationship with rock hardness, with quartzites and vein quartz dominating the more angular classes and schists usually being more rounded. Anomalously well-rounded, hard clasts are rare and no rebroken rounded clasts were observed. This is in sharp contrast with conglomerates in the axial, braided-alluvial system such as those forming the Hovden sandstone formation on the island of Hovden, where most clasts, including quartzites, are well-rounded (Spinnangr 1975; Cuthbert, unpublished data).

The absence of well-rounded clasts indicates transport distances of only a few kilometres (Pettijohn 1975, pp. 59–60). This is consistent with the rather small radius of the marginal conglomerate bodies (usually less than 2 km), which would be expected to have had drainage areas of similar magnitude (Steel & Gloppen 1980). On Hornelen mountain, Bremangerlandet (Fig. 2), a semi-coherent body of amphibolite and pegmatite with a maximum extent of 100 m lies within the conglomerates, which is interpreted as a rockslide. Such a body of rock is unlikely to have travelled more than a few hundred metres and must have had a local source. The lack of anomalously rounded and rebroken clasts implies that these sediments are predominantly first-cycle, and are therefore likely to represent the bedrock in the drainage basins from which they were sourced. The most diverse clast assemblages are found in the most massive conglomerates in each cycle, and these would give the best representations of palaeogeology in the adjacent basement during deposition. Sampling sites in all other cycles were selected to lie in the upper parts of such massive conglomerate sequences.

Stratigraphic distribution of clast assemblages

Clast assemblages have been logged at fifteen sites around the Hornelen Basin in marginal alluvial fan conglomerates. The results are displayed on the geological map of the area (Fig. 2), along with those from four sites logged by Evensen (1978).

Twelve main clast types have been differentiated; these are:

  (1) gneiss (plagioclase + quartz + 2 mica

or biotite + plagioclase + quartz + epidote)

(2) K-feldspar augen gneiss (K-feldspar + biotite + plagioclase + quartz + epidote)

(3) amphibolite (hornblende + plagioclase + epidote + biotite)

(4) feldspathic quartzite (quartz + K-feldspar + muscovite + plagioclase)

(5) mica-schist (muscovite + quartz + 2 feldspars ± biotite ± garnet)

(6) greenschist (muscovite + chlorite + biotite + plagioclase + quartz ± garnet)

(7) banded metachert (quartz + biotite + garnet + oxide)

(8) pelitic hornfels (biotite + feldspar + quartz ± sillimanite ± garnet)

(9) metagreywacke (quartz + feldspar ± biotite)

(10) gabbro (plagioclase + 2 pyroxenes ± amphibole)

(11) quartz diorite (plagioclase + quartz + biotite)

(12) vein quartz.

There is some gradation between certain pebble-types (e.g. quartzite and mica-schist, greenschist and metachert). Rare clasts of dolomitic marble and orthoclase pegmatite were also found.

Feldspathic quartzites dominate the southern margin of the basin (sites 13–19, Fig. 2), usually comprising more than 50% of the clasts. Muscovite schists are common here also, and a complete gradation between these and the quartzites can be found. Included with these types are distinctive, but rare muscovite–chlorite-plagioclase–quartz gneisses. All these lithologies commonly have a well-developed crenulation cleavage. Gneisses are usually dark, biotite-bearing varieties. Small amounts of greenschist are persistenly found along the southern margin.

The quartzite/schist dominated clast assemblage on the southern margin is superficially similar to the adjacent basement lithologies; indeed, they have been directly linked in previous provenance studies (Evenson 1978; Steel & Gloppen 1980). However, there are significant discrepancies between the clasts and the basement. Despite the close proximity of abundant anorthosite within the gneisses, no anorthosite clasts have been found. Also, none of the clasts, including the dark gneisses, is mylonitic, in strong contrast to the basement. Clasts of chloritic breccia and cataclasite, commonly found by the sole fault in the basement, have never been observed. Conversely, the only greenschist and chloritic gneiss are found in the basement in the extreme west, just below the unconformity. The balance of evidence points to a mismatch between clasts and basement at all sites investigated along this margin.

Clast assemblages along the northern margin (sites 1–12, Fig. 2) are more diverse than those to the south. Quartzites are less common and igneous rocks make an important contribution. Systematic stratigraphic variations occur along the northern margin: The oldest exposed sediments, in the extreme west on the island of Hovden (site 1) contain mainly clasts of greenschists and gabbros, with smaller amounts of pelitic hornfels, quartzite, metachert and vein quartz. At Smørhamn and the southwest slopes of Hornelen on Bremangerlandet (sites 2 and 3), greenschists are still important, but gabbros disappear and large amounts of white quartz diorite are found, with minor amounts of a hard, hornfelsic meta-greywacke at Smørhamn. Moving northeastwards along the unconformity on Hornelen there is an abrupt change to an assemblage dominated by quartzite and a muscovite–biotite gneiss. Gabbro and diorite are both absent here, although a few metachert and greenschist clasts perisist. Minor amounts of a distinctive coarse-grained pink augen-gneiss are found here for the first time. Passing up-section to the east similar assemblages, with varying proportions, are found at Vingevatnet and Karlskardet (sites 5 and 6), with the notable addition of clasts of amphibolite. At Ørneriervatent (site 7) greenschists increase in importance again and white quartz diorite clasts reappear. From there eastwards gneiss, augen gneiss and amphibolite become less prominent as first diorite, then gabbro dominates the assemblages. Hornfelsed greywacke and pelite are important components at Saeterdalsvatnet and Ålfoten (sites 8 and 9). In the youngest examined sediments, at Svartevatnet (site 12) the sediments are almost monomict gabbro conglomerates.

The basement immediately adjacent to the northern margin on the mainland (Fig. 2) consists entirely of pink augen gneisses, varying texturally from protomylonites to ultramylonites. These outcrop along a fault-bounded strip about 1 km wide, north of which are various types of non-mylonitic gneiss with large bodies of quartzite, anorthosite and serpentinite, and common bodies of eclogite. Diorite, gabbro, hornfels, greenschist, metachert and metasandstone are entirely absent in this part of the basement, hence these clast lithologies are clearly exotic. Anorthosite, serpentinite and eclogite are not found in the clasts, despite the proximity of such rocks in the basement. Steel &

Gloppen (1980) recorded serpentinite clasts along this margin, but none was found either in this study or that of Evenson (1978); it seems likely that the dark hornfelsic metapelite clasts were mis-identified as serpentinites.

The gneiss clasts fall into two groups (not differentiated on Fig. 2); a dark mica-rich type and a light, quartzo-feldspathic type, both of which are occasionally found in a single clast, indicating that they come from a common lithology which was quite coarsely differentiated into the two types. Similar gneisses occur in the basement, but the latter are highly variable, making comparisons difficult. Coarse-grained pink augen gneisses are common to both clasts and basement, but the clasts lack the high strains found in the strip of land just north of the boundary fault. The only metabasites found in the clasts are ordinary, granoblastic plagioclase amphibolites, in contrast to the abundant eclogites in the basement.

At site 4, just above the unconformity (Fig. 2) the clast assemblage is similar to that on the mainland nearby. The greenschist in the basement is very similar to the same lithology in the clasts, but most of the clasts are more similar to lithologies about 1 km further north. The gneisses show a good petrographic match with the main body of gneiss on the island, as do the augen gneisses. Moreover, at sites 2 and 3 there is an excellent match between clasts and basement. No basement is exposed adjacent to the most westerly sampling site on Hovden, but by extrapolation of the geology of nearby islands there seems to be a good clast-basement match here also.

Returning to the mainland, many of the clasts found along the northern boundary are very similar to those found in the basement below the unconformity on Bremangerlandet. Hornfelses, metacherts, metagreywackes, gabbros, diorites and greenschists are all quite distinctive, and despite the poorer diagnostic value of the gneisses and quartzites, the whole *clast assemblage* from this margin can be confidently correlated with a source having the same geology as the modern-day Bremangerlandet area. Conversely, the characteristic lithological *assemblage* found in the Western Gneiss Region is not represented in the clasts.

Tectonic and palaeogeological inferences

In order for the Bremangerlandet-type source to shed detritus into the basin, this terrain must have been broken by a fault and partly transported past the depocentre. The relative dis-

placement between the gabbro body near Kalvåg and the first appearance of gabbro clasts at Ålfoten (site 9, Fig. 2) indicates about 50 km of dextral displacement by the time that unit was deposited. This shear-sense is consistent with the sense of stacking of the fan units (Steel & Gloppen 1980). Following deposition of all the exposed sediments, the source terrane was entirely removed from the basin margin and replaced by the Western Gneiss Region rocks.

The presence of exotic greenschists as far west as sampling site 15 indicates a similar scale of displacement (sinistral in this case) on the southern margin. This is consistent with the extensional basin model of Hossack (1984), Seranne & Seguret (1987) and Norton (1987), in which the basin forms in a scoop-shaped half graben with the Hornelen Group and its depositional basement lying in the hangingwall, and the Western Gneiss Region forming the footwall (Fig. 5b).

The clast sourceland clearly had a complex metamorphic history, but in the absence of the field relationships from which to determine the relative timing of events only inferences can be made. By inspecting the distribution of clast assemblages in Fig. 2 it can be seen that certain lithologies are more closely associated with each other than others; for instance, hornfels and metachert are more likely to occur where gabbro and quartz diorite are abundant, while amphibolite and augen gneiss are more closely associated with gneiss. Quartzite and greenschist are almost ubiquitous, but quartzite (and mica schist) is most abundant where gneiss is common, and greenschist is more abundant where quartz diorite is present. Two contiguous terrains can therefore be defined; one dominated by gneisses, amphibolite, quartzite and mica schist, and another dominated by igneous rocks, hornfels and greenschists. It follows that the hornfels was probably a result of the intrusion of the gabbro and the quartz diorite. 'Greenschist' clasts often show evidence of development by recrystallization of pelitic hornfels, suggesting that the greenschist metamorphism postdated the intrusion. This scenario shows similarities with field relationships on Bremangerlandet below the basal unconformity to the Hornelen Group.

The lack of eclogite clasts in the Hornelen Basin has already been noted. Eclogite clasts are commonly found in local Quaternary river gravels and glacial drift, and one would confidently expect to find them in alluvial fan sediments if they occurred in the sourceland. However, the only metabasic clasts found in association with gneisses were hornblende–plagioclase amphibolites, so an eclogite facies metamorphism in the sourceland is thought to have been unlikely. The lack of mylonite and cataclasite clasts indicates that the shear-zone capping the Western Gneiss region was not breached during the Middle Devonian (see Miller & John 1988).

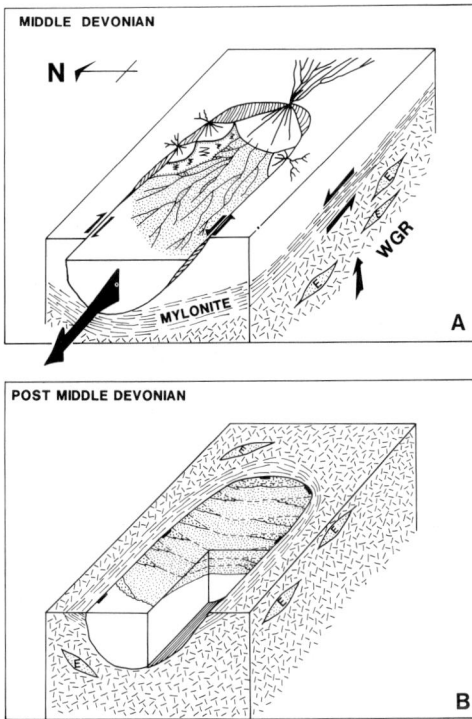

Fig. 5. Block diagrams showing relationships between the Hornelen Basin and its basement. (**A**) Syn-depositional geometry with a scoop-shaped half-graben. Detritus from a Bremanger-type basement is being shed into the basin to form alluvial fans which interfinger with axial floodplain sediments. The Western Gneiss Region has been uplifted by intra-crustal excision along a major extensional shear zone following high-pressure metamorphism, but still lies at mid-crustal levels below the basin. (**B**) Present-day geometry: the Western Gneiss Region and the mylonitic shear-zone have been captured by the sole fault to the Hornelen Basin, and brought to the surface to lie in juxtaposition with the Hornelen Group sediments and the Bremanger-type basement.

Metamorphic evolution of clasts

Two garnet–mica schist clasts from Vingevatnet (site 5) were selected for geothermometry (see Table 1). The two schists from site 5 gave overlapping pressure estimates of 6.9 ± 2.5 kbar and 9.4 ± 2.0 kbar at $570 \pm 50°C$ and $730 \pm 50°C$ respectively. The pressures are similar to that

determined for the basement schist from Marøy (Table 1 and Fig. 2), but it is clear that the clasts span a range of temperatures, which may correspond to a true metamorphic field gradient.

A mica schist clast from site 18 (Hcg20, Table 1) yielded enough fresh muscovite for K–Ar dating, giving an age of 423 ± 7 Ma (Table 4), identical to that from the Marøy schist and corresponding to a closure temperature of $350°C$ (Jäger 1979). Assuming that this sample shared the same uplift history as those from site 5, the P–T–t paths (Fig. 3) are very similar to that for the Bremangerlandet block, which is consistent with their lithological similarities.

Discussion

The main conclusion from the results presented above is that there is no petrographic linkage between the conglomerate clasts in the Hornelen Basin and the Western Gneiss Region. The Hornelen Group must have been emplaced against the Western Gneiss Region at some time after Hornelen Group deposition (the exact timing of this remains poorly constrained). The metamorphic grade of the clast sourceland implies that it never reached the extreme burial depth attained by the Western Gneiss Region, and that substantial excision of crustal section must have taken place to bring the clasts and Western Gneiss Region together. The form of the uplift paths derived from the petrological studies shows that while some of this convergence took place *following* Hornelen Group deposition in the Middle Devonian (probably due to extensional faulting), much of it must have happened *prior* to deposition. This crustal extension and thinning is consistent with the presence of the thick mylonites which cap the Western Gneiss Region (Fig. 1), and with the metamorphic break between the Western Gneiss Region and the basement on Bremangerlandet.

Despite the great stratigraphic thickness of the Hornelen Group, the basin appears to have been sampling parts of the same footwall terrain throughout its recorded history. The clasts show no significant systematic increase in metamorphic grade; indeed, a superficial inspection of the clast distribution reveals an apparent decrease in grade up section. It seems, then, that the Hornelen Basin did not excavate a very large crustal thickness during its depositional phase, hence the sole fault must have been a very low-angle structure.

The late Caledonian evolution of the Hornelen Basin region can be summarized as follows:
(i) During Scandian thrusting the gneisses

now forming the Western Gneiss Region were subducted to depths of more than 65 km below a thick stack of thrust sheets, causing the eclogite-facies metamorphism (Cuthbert *et al.* 1983). The Bremangerlandet gneisses and schists lay at an intermediate level in this thrust stack (30–40 km?) and the schists were formed by metamorphism of a pelitic melange.

(ii) In response to this extreme crustal thickening, uplift took place in the Late Silurian and Early Devonian. Thinning of the orogenic prism by extensional faulting along a major shear zone resulted in a loss of crustal section and convergence of the Western Gneiss Region rocks (footwall) and the Bremangerlandet rocks (hangingwall).

(iii) By the Middle Devonian, the clast source-rocks had reached the surface and the Western Gneiss Region rocks lay at about 10 km depth. A new, low angle normal fault system broke the syn-orogenic surface; the hangingwall subsided, forming the Hornelen Basin, and the equivalent rocks in the footwall shed detritus into the basin as alluvial fans (Fig. 5a) to form the Hornelen Group.

(iv) This fault system (the Nordfjord–Sogn Detachment of Norton 1987) captured the Western Gneiss Region rocks and eventually brought them into juxtaposition with the Hornelen Group and the top of the hangingwall block. During this time uplift and erosion of the footwall block removed the clast source from its position next to the basin (Fig. 5b).

Concluding remarks

This contribution has been an attempt to demonstrate that provenance studies can be used not only to reconstruct the palaeogeology of a clast sourceland, but also to provide some quantitative information on the tectonothermal evolution of the source prior to its surface exposure. While it is clear that interpretations based on disembodied fragments cannot be treated with the same confidence as those based on in situ basement, careful selection of sampling sites should overcome some of these disadvantages. The benefits to be reaped are the insights into the early uplift histories of metamorphic belts. The results from the Hornelen Basin are encouraging as, when the clasts are compared with the exposed basement, they reveal a surprising amount of complexity in the uplift history of the area. The provident preservation of part of the clast source block makes this a useful example as it helps to confirm the validity of results from the clasts.

This work was carried out during tenure of a NERC Postdoctoral Fellowship at Glasgow University. I am grateful for help with analytical facilities to C. Farrow and G. Bruce (Glasgow), M. MacIntyre and J. Gray (SURRC, East Kilbride), P. Hill (Edinburgh, microprobe) and D. Rex (Leeds, K–Ar dating). M. Norton and W. Wilks (RHBNC), and R. Johnson and I. Bray (Glasgow) are thanked for collaboration in the field. I thank I. Bryhni (Oslo) for sending me his preliminary map of the Naustdal area. Constructive criticism by I. Sanders and J. Graham resulted in significant improvements in the manuscript.

References

BALLANCE, P. F. 1980. Models of sediment distribution in non-marine and shallow marine environments in oblique-slip fault zones. *Special Publication of the International Association of Sedimentologists,* **4**, 229–236.

BREKKE, H. & SOLBERG, P. O. 1987. The geology of Atløy, western Norway. *Norges Geologiske Undersøkelse Bulletin,* **410**, 73–94.

BRUECKNER, H. K. 1972. Interpretation of Rb–Sr ages from the Precambrian and Palaeozoic rocks of southern Norway. *American Journal of Science,* **272**, 334–358.

BRYHNI, I. 1966. Reconnaissance studies of gneisses, ultrabasites, eclogites and anorthosites in outer Nordfjord, Western Norway. *Norges Geologiske Undersøkelse Bulletin,* **241**, 1–68.

—— 1978. Flood deposits in the Hornelen Basin, west Norway (Old Red Sandstone). *Norsk Geologisk Tidsskrift,* **58**, 273–300.

—— & GRIMSTAD, E. 1970. Supracrustal and infracrustal rocks in the Gneiss Region of the Caledonides west of Breimsvatn. *Norges Geologiske Undersøkelse Bulletin,* **266**, 105–150.

—— & LYSE, K. 1985. The Kalvåg Melange, Norwegian Caledonides. *In:* GEE, D. G. & STURT, B. A. (eds) *The Caledonide Orogen—Scandinavia and Related Areas.* John Wiley, Chichester, 417–427.

——, BOCKELIE, J. FR. & NYSTUEN, J. P. 1981. The Southern Norwegian Caledonides Oslo–Sognefjord–Ålesund. *Excursions in the Scandinavian Caledonides.* UCS excursion No. A1. *Uppsala Caledonide Symposium 1981.*

——, BOLLINGBERG, H. & GRAFF, P.-R. 1969. Eclogites in quartzo-feldspathic gneisses of Nordfjord, West Norway. *Norsk Geologisk Tidsskrift,* **49**, 193–225.

CARSWELL, D. A., KROGH, E. J. & GRIFFIN, W. L. 1985. Norwegian orthopyroxene eclogites: calculated equilibrium temperatures and petrogenetic implications. *In:* GEE, D. G. & STURT, B. A. (eds) *The Caledonide Orogen—Scandinavia and Related Areas.* John Wiley, Chichester, 823–841.

CLIFF, R. A. 1985. Isotopic dating in metamorphic belts. *Journal of the Geological Society, London,* **142**, 97–110.

COTKIN, S. J., VALLEY, J. W. & ESSENE, E. J. 1988.

Petrology of a margarite-bearing meta-anorthosite from Seljeneset, Nordfjord, western Norway: Implications for the *P–T* history of the Western Gneiss Region during Caledonian uplift. *Lithos*, **21**, 117–128.

CUTHBERT, S. J. & CARSWELL, D. A. 1990. Formation and exhumation of medium-temperature eclogites in the Scandinavian Caledonides. *In*: CARSWELL, D. A. (ed.) *Eclogite Facies Rocks*. Blackie, Glasgow, 180–204.

——, HARVEY, M. A. & CARSWELL, D. A. 1983. A tectonic model for the metamorphic evolution of the Basal Gneiss Complex, western south Norway. *Journal of Metamorphic Geology*, **1**, 63–90.

DAVIS, G. H. & CONEY, P. J. 1979. Geological development of metamorphic core complexes. *Geology*, **7**, 120–124.

DODSON, M. H. 1979. Theory of cooling ages. *In*: JÄGER, E. & HUNZIKER, J. C. (eds) *Lectures in Isotope Geology*. Springer, Berlin, 194–202.

ENGLAND, P. C. & RICHARDSON, S. W. 1977. The influence of erosion upon the mineral facies of rocks from different metamorphic environments. *Journal of the Geological Society, London*, **134**, 201–213.

EVENSEN, S. 1978. *Petrology of conglomerates and sandstones in Hornelen Basin (Devonian), western Norway*. Cand. Real. Thesis, University of Bergen.

GANGULY, J. & SAXENA, S. K. 1984. Mixing properties of aluminosilicate garnets: constraints from natural and experimental data, and applications to geothermo-barometry. *American Mineralogist*, **69**, 88–97.

GEE, D. G., KUMPULAINEN, R., ROBERTS, D., STEPHENS, M. B., THON, A. & ZACHRISSON, E. 1985. Scandinavian Caledonides tectonostratigraphic map. *In*: GEE, D. G. & STURT, B. A. (eds) *The Caledonide Orogen—Scandinavia and Related Areas*. John Wiley, Chichester, map 1.

GLOPPEN, T. G. & STEEL, R. J. 1981. The deposits, internal structure and geometry in six alluvial fan–fan delta bodies (Devonian—Norway)—a study in the significance of bedding sequence in conglomerates. *Society of Economic Paleontologists and Mineralogists Special Publication*, **31**, 49–69.

GREEN, T. H. & HELLMAN, P. L. 1982. Fe–Mg partitioning between coexisting garnet and phengite at high pressure, and comments on a garnet–phengite geothermometer. *Lithos*, **15**, 253–266.

GRIFFIN, W. L. & BRUECKNER, H. K. 1980. Caledonide Sm–Nd ages and a crustal origin for Norwegian eclogites. *Nature*, **285**, 319–320.

—— & —— 1985. REE, Rb–Sr and Sm–Nd studies of Norwegian eclogites. *Chemical Geology (Isotope Geoscience Section)*, **52**, 249–271.

—— & MØRK, M. B. E. 1981. Eclogites and Basal Gneisses in West Norway. *Excursions in the Scandinavian Caledonides No. B1. Uppsala Caledonide Symposium, 1981*.

——, AUSTRHEIM, H., BRASTAD, K., BRYHNI, I., KRILL, A. G., KROGH, E. J., MØRK, M. B. E., QVALE, H. & TØRUDBAKKEN, B. 1985. High-pressure metamorphism in the Scandinavian Caledonides. *In*: GEE, D. G. & STURT, B. A. (eds) *The Caledonide*

Orogen—Scandinavia and Related Areas. John Wiley, Chichester, 783–801.

HARLAND, W. B., COX, A. V., LLEWELLYN, P. G., PICTON, C. A. G., SMITH, A. G. & WALTERS, R. 1982. *A Geological Time Scale*. Cambridge University Press, Cambridge.

HARRISON, T. M. 1981. Diffusion of argon in hornblende. *Contributions to Mineralogy and Petrology*, **78**, 324–331.

HAUGHTON, P. D. W. 1988. A cryptic Caledonian flysch terrane in Scotland. *Journal of the Geological Society, London*, **145**, 685–703.

——, ROGERS, G. & HALLIDAY, A. N. 1990. Provenance of Lower Old Red Sandstone conglomerates, S. E. Kincardineshire: evidence for timing of Caledonian terrane accretion in central Scotland. *Journal of the Geological Society, London*, **147**, 105–120.

HODGES, K. V. & ROYDEN, L. 1984. Geological thermobarometry of retrograded metamorphic rocks: An indication of the uplift trajectory of a portion of the northern Scandinavian Caledonides. *Journal of Geophysical Research*, **89B**, 7077–7090.

HOEG, O. A. 1936. Norges fossile flora. *Naturen*, **7–21**, 47–64.

HOSSACK, J. R. 1984. The geometry of listric growth faults in the Devonian basins of Sunnfjord, W. Norway. *Journal of the Geological Society, London*, **141**, 629–637.

INDARES, A. & MARTIGNOLE, J. 1985. Biotite–garnet geothermometry in the granulite facies: the influence of Ti and Al in biotite. *American Mineralogist*, **70**, 272–278.

JÄGER, E. 1979. Introduction to Geochronology. *In*: JÄGER, E. & HUNZIKER, J. C. (eds) *Lectures in Isotope Geology*. Springer, Berlin, 1–12.

KILDAL, E. S. 1970. *Geologisk Kart over Norge, 1 : 250,000 Måløy*. Norges Geologiske Undersøkelse.

KULLERUD, L., TØRUDBAKKEN, B. O. & ILEBERK, S. 1986. A compilation of radiometric age determinations from the Western Gneiss Region, South Norway. *Norges Geologiske Undersøkelse Bulletin*, **406**, 17–42.

LARSEN, V. & STEEL, R. J. 1978. The sedimentary history of a debris-flow dominated Devonian alluvial fan—a study of textural inversion. *Sedimentology*, **25**, 37–59.

——, SPINNANGR, Å., STEEL, R., AASHEIM, S., GLOPPEN, T. G. & MAEHLE, S. 1987. *A Field Guide to Hornelen Basin—A Deep, Late-Orogenic Basin (Devonian) in Western Norway*. Statoil/University of Bergen.

LUX, D. R. 1985. K/Ar ages from the Basal Gneiss Region, Statlandet area, Western Norway. *Norsk Geologisk Tidsskrift*, **65**, 277–286.

MAEHLE, S. 1975. *Devonian conglomerate-sandstone facies relationships and their palaeogeographic importance along the margin of the Hornelen Basin, between Storevann and Grøndalen, Sunnfjord, Norway*. Cand. Real. Thesis, University of Bergen.

McCLAY, K. R., NORTON, M. G., CONEY, P. & DAVIS, G. H. 1986. Collapse of the Caledonide orogen and the Old Red Sandstone. *Nature*, **323**, 147–149.

MILLER, M. G. & JOHN, B. E. 1988. Detached strata in a Tertiary low-angle normal fault terrane, southeastern California: A sedimentary record of unroofing, breaching, and continued slip. *Geology*, **16**, 645–648.

MØRK, M. B. E. & MEARNS, E. W. 1986. Sm–Nd systematics of a gabbro-eclogite transition. *Lithos*, **19**, 255–267.

NORTON, M. G. 1986. Late Caledonian extension in western Norway: A response to extreme crustal thickening. *Tectonics*, **5**, 195–204.

—— 1987. The Nordfjord—Sogn Detachment, W. Norway. *Norsk Geologisk Tidsskrift*, **67**, 93–106.

——, Wilks, W. & Cuthbert, S. J. 1990. *Discussion on Palaeomagnetism, magnetic fabrics and the structural style of the Hornelen Old Red Sandstone, western Norway. Journal of the Geological Society, London*, **147**, 411–412.

PETTIJOHN, F. J. 1975. *Sedimentary Rocks*. Harper & Row, New York, third edition.

POWELL, R. & HOLLAND, T. J. B. 1988. An internally consistent dataset with uncertainties and correlations: 3. Applications to geobarometry, worked examples and a computer program. *Journal of Metamorphic Geology*, **6**, 173–204.

SERANNE, M. 1988. *Tectonique des bassins Devoniens de Norvege: Mise en evidence de bassins sedimentaires en extension formes par amincissement d'une croute orogenique epaissie*. Diplome de Doctorat thesis, Universite des Sciences et Techniques du Languedoc.

—— & SEGURET, M. 1987. The Devonian basins of western Norway: tectonics and kinematics of an extending crust. *In*: COWARD, M. P., DEWEY, J. F.

& HANCOCK, P. L. (eds) *Continental Extensional Tectonics. Geological Society, London, Special Publication*. **28**, 537–548.

SMITH, D. C. 1984. Coesite in clinopyroxene in the Caledonides and its implications for geodynamics. *Nature*, **310**, 641–644.

SPINNANGR, A. 1975. *Some sedimentary and stratigraphic studies of the Devonian strata, across the western part of the Hornelen Basin, western Norway*. Cand. Real. Thesis, University of Bergen.

STEEL, R. J. 1976. Devonian basins of western Norway—sedimentary response to tectonism and to varying tectonic context. *Tectonophysics*, **36**, 207–224.

—— & GLOPPEN, T. G. 1980. Late Caledonian (Devonian) basin formation, western Norway: signs of strike-slip tectonics during infilling. *Special Publication of the International Association of Sedimentologists*, **4**, 79–103.

——, MAEHLE, S., NILSEN, H., ROE, S. L. & SPINNANGR, A. 1977. Coarsening-upward cycles in the alluvium of Hornelen Basin (Devonian) Norway: Sedimentary response to tectonic events. *Geological Society of America Bulletin*, **88**, 1124–1134.

——, SIEDLECKA, A. & ROBERTS, D. 1985. The Old Red Sandstone basins of Norway and their deformation: a review. *In*: GEE, D. G. & STURT, B. A. (eds) *The Caledonide Orogen—Scandinavia and Related Areas*. John Wiley, Chichester, 293–315.

STEIGER, R. H. & JÄGER, E. 1977. Subcommission on Geochronology: convention on the use of decay constants in geo- and cosmochronology. *Earth and Planetary Science Letters*, **36**, 359–362.

Index

361

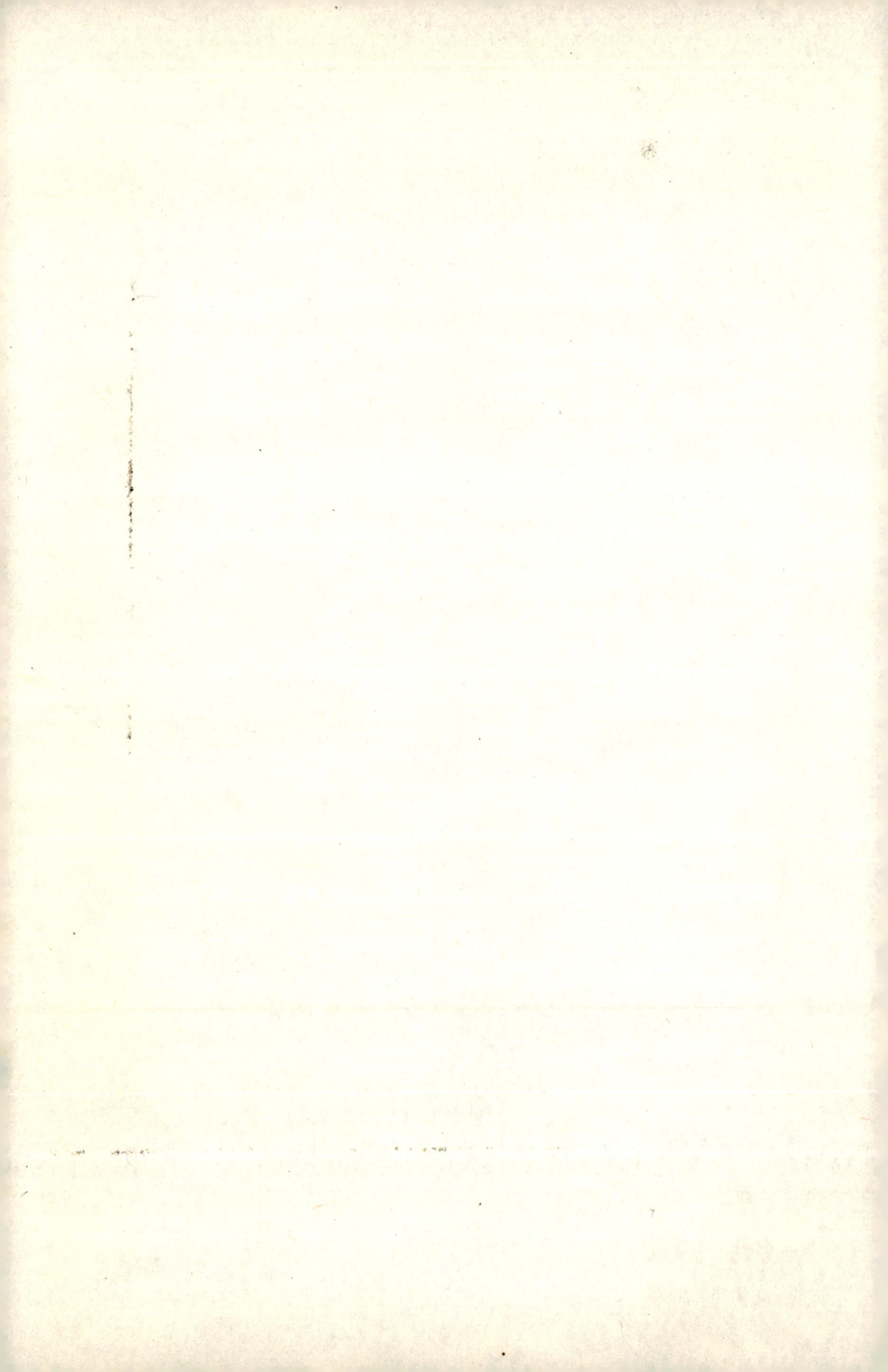